AF869149

LAURA KERSEY LIBRARY OF
ENGINEERING & TECHNOLOGY
UNIVERSITY OF LOUISVILLE

IMPACTS OF NUCLEAR RELEASES INTO THE AQUATIC ENVIRONMENT

Laura Kersey
Library

The following States are Members of the International Atomic Energy Agency:

AFGHANISTAN
ALBANIA
ALGERIA
ARGENTINA
AUSTRALIA
AUSTRIA
BANGLADESH
BELGIUM
BOLIVIA
BRAZIL
BULGARIA
BURMA
BYELORUSSIAN SOVIET SOCIALIST REPUBLIC
CAMBODIA
CANADA
CHILE
COLOMBIA
COSTA RICA
CUBA
CYPRUS
CZECHOSLOVAKIA
DEMOCRATIC PEOPLE'S REPUBLIC OF KOREA
DENMARK
DOMINICAN REPUBLIC
ECUADOR
EGYPT
EL SALVADOR
ETHIOPIA
FINLAND
FRANCE
GABON
GERMAN DEMOCRATIC REPUBLIC
GERMANY, FEDERAL REPUBLIC OF
GHANA
GREECE
GUATEMALA
HAITI
HOLY SEE
HUNGARY
ICELAND
INDIA
INDONESIA
IRAN
IRAQ
IRELAND
ISRAEL
ITALY
IVORY COAST
JAMAICA
JAPAN
JORDAN
KENYA
KOREA, REPUBLIC OF
KUWAIT
LEBANON
LIBERIA
LIBYAN ARAB REPUBLIC
LIECHTENSTEIN
LUXEMBOURG
MADAGASCAR
MALAYSIA
MALI
MAURITIUS
MEXICO
MONACO
MONGOLIA
MOROCCO
NETHERLANDS
NEW ZEALAND
NIGER
NIGERIA
NORWAY
PAKISTAN
PANAMA
PARAGUAY
PERU
PHILIPPINES
POLAND
PORTUGAL
REPUBLIC OF SOUTH VIET-NAM
ROMANIA
SAUDI ARABIA
SENEGAL
SIERRA LEONE
SINGAPORE
SOUTH AFRICA
SPAIN
SRI LANKA
SUDAN
SWEDEN
SWITZERLAND
SYRIAN ARAB REPUBLIC
THAILAND
TUNISIA
TURKEY
UGANDA
UKRAINIAN SOVIET SOCIALIST REPUBLIC
UNION OF SOVIET SOCIALIST REPUBLICS
UNITED KINGDOM OF GREAT BRITAIN AND NORTHERN IRELAND
UNITED REPUBLIC OF CAMEROON
UNITED STATES OF AMERICA
URUGUAY
VENEZUELA
YUGOSLAVIA
ZAIRE
ZAMBIA

The Agency's Statute was approved on 23 October 1956 by the Conference on the Statute of the IAEA held at United Nations Headquarters, New York; it entered into force on 29 July 1957. The Headquarters of the Agency are situated in Vienna. Its principal objective is "to accelerate and enlarge the contribution of atomic energy to peace, health and prosperity throughout the world".

Printed by the IAEA in Austria
December 1975

PROCEEDINGS SERIES

IMPACTS OF NUCLEAR RELEASES INTO THE AQUATIC ENVIRONMENT

PROCEEDINGS OF AN INTERNATIONAL SYMPOSIUM
ON RADIOLOGICAL IMPACTS OF RELEASES
FROM NUCLEAR FACILITIES INTO AQUATIC ENVIRONMENTS
HELD BY THE
INTERNATIONAL ATOMIC ENERGY AGENCY
AT OTANIEMI, FINLAND, 30 JUNE - 4 JULY 1975

INTERNATIONAL ATOMIC ENERGY AGENCY
VIENNA, 1975

Speed
School

QH
543.5
.A1I57
1975
c.2

IMPACTS OF NUCLEAR RELEASES INTO THE AQUATIC ENVIRONMENT
IAEA, VIENNA, 1975
STI/PUB/406
ISBN 92-0-020375-2

FOREWORD

The rapid increase in technological development has brought extensive benefit to society but at the same time has caused certain environmental and socio-economic problems. There is a need to assess the impacts of the new technologies, especially those represented by the nuclear industry, since a rapid increase in the number of nuclear power plants in the world is foreseen.

It is essential to ensure that the total releases of radioactivity from an increased number of nuclear plants will be limited and controlled to such a low level that there will be – as has been the case up to the present – no adverse effect on man and his environment. Several decades of experience are now available in the study of the environmental impact of radioactive materials. To review this experience and to survey recent work on the safety of releases from nuclear facilities into fresh water, estuaries and sea water, the International Atomic Energy Agency convened an international symposium from 30 June to 4 July 1975. The Symposium on the Radiological Impacts of Releases from Nuclear Facilities into Aquatic Environments was held at Otaniemi near Helsinki at the invitation of the Finnish Government.

A related IAEA Symposium, on the Interaction of Radioactive Contaminants with Constituents in the Marine Environment, held in Seattle, from 10 to 14 July 1972, covered much of what was then known concerning the fundamental aspects of marine radioactivity studies from the physical, chemical and biological viewpoints. The present symposium, in addition to surveying the progress made over the last three years in research on the behaviour of both radionuclides and their corresponding stable isotopes in the aquatic systems, was aimed particularly at the assessment of the radiological impacts and discussion of the present and future problems of safe control.

On the basis of a broad review of past experience, several speakers stressed that no harmful effects to man from nuclear facilities have so far been observed. Even effects from fall-out, studies of which indicate a larger dose to man than from nuclear power, have been shown to be minimal.

The Agency gratefully acknowledges the financial support of the United Nations Environment Programme and the co-operation of the Government of Finland in the preparation, running and successful outcome of the meeting.

29134 99A

EDITORIAL NOTE

The papers and discussions have been edited by the editorial staff of the International Atomic Energy Agency to the extent considered necessary for the reader's assistance. The views expressed and the general style adopted remain, however, the responsibility of the named authors or participants. In addition, the views are not necessarily those of the governments of the nominating Member States or of the nominating organizations.

Where papers have been incorporated into these Proceedings without resetting by the Agency, this has been done with the knowledge of the authors and their government authorities, and their cooperation is gratefully acknowledged. The Proceedings have been printed by composition typing and photo-offset lithography. Within the limitations imposed by this method, every effort has been made to maintain a high editorial standard, in particular to achieve, wherever practicable, consistency of units and symbols and conformity to the standards recommended by competent international bodies.

The use in these Proceedings of particular designations of countries or territories does not imply any judgement by the publisher, the IAEA, as to the legal status of such countries or territories, of their authorities and institutions or of the delimitation of their boundaries.

The mention of specific companies or of their products or brand names does not imply any endorsement or recommendation on the part of the IAEA.

Authors are themselves responsible for obtaining the necessary permission to reproduce copyright material from other sources.

CONTENTS

METHOD OF MONITORING AND EVALUATION OF EFFECTS ON AND DOSES TO MAN AND AQUATIC BIOTA (Sessions VIII and IX)

FACTORS DETERMINING THE ENVIRONMENTAL CAPACITY FOR RECEIVING RELEASES FROM NUCLEAR FACILITIES IN THE AQUATIC ENVIRONMENT (Session X)

RADIOLOGICAL IMPACTS OF RELEASES FROM NUCLEAR FACILITIES INTO AQUATIC ENVIRONMENTS

(Session I)

Chairman

J.K. MIETTINEN (Finland)

Invited paper

THE RADIOLOGICAL CONSEQUENCES OF RELEASES FROM NUCLEAR FACILITIES TO THE AQUATIC ENVIRONMENT

A. PRESTON
Ministry of Agriculture, Fisheries and Food,
Fisheries Laboratory,
Lowestoft, Suffolk,
United Kingdom

Abstract

THE RADIOLOGICAL CONSEQUENCES OF RELEASES FROM NUCLEAR FACILITIES TO THE AQUATIC ENVIRONMENT.

The release of radioactive materials to the environment is an inescapable consequence of the utilization of nuclear energy. The objective therefore is to decide on what basis and against what criteria regulatory action should be taken to protect the environment against the impact of radioactive substances. In properly regulated situations releases of such material will be minor in character and their radiological implications will rest largely in the field of public health. There are now some three decades of experience in respect of the environmental impact of radioactive materials, and certain major conclusions can be drawn. This paper reviews this experience in broad terms, and draws conclusions relevant to the regulatory problem. Future problems, especially in the context of an expanding use of nuclear power, are considered and priority research needs and opportunities indicated.

INTRODUCTION

The release of radioactive materials to the environment is an inevitable consequence of the use of nuclear energy. The problem therefore is to so limit and control such releases as to reduce adverse effects on man and his environment to acceptable proportions.

The radiological consequences stemming from the release of radioactive materials to the aquatic environment are biological in character and relate either to human health or to damage to aquatic resources. Artificial radioactivity has been introduced to the aquatic environment for some three decades, and a very substantial body of data relating to its behaviour and effects has been built up [1-6]. The primary attention has quite properly been directed to the protection of human health, and all experience gained to date in relation to the resource problem suggests that the control of the human health problem will provide a satisfactory basis for the control of the resource problem.

It is not the intention in this paper to review in detail this large body of information on artificial radionuclides in the aquatic environment, or the factors which determine their behavioural characteristics and lead to their observed spatial and temporal distributions. The body of published material has established an adequate picture in this respect and provides a suitable basis, when taken together with recommendations on acceptable human radiation exposure, on which to regulate their introduction to the environment. This paper will therefore only review certain important areas of the subject in order to outline the basis on which regulated introduction of radioactive materials can be accomplished, and indicate what assumptions are made and where outstanding research needs now lie.

In order to review the situation in relation to radiological effects it is first necessary to briefly consider certain basic radiological protection premises and the systems for implementing requirements in relation to dose limits. Some of this will already be familiar, but it is as well to have it thrust to the forefront of our minds at the beginning of a symposium such as this so that what follows may fit more easily into perspective.

ACCEPTABLE DOSE

The biological effects of ionizing radiation are probably better understood, certainly so in a human health context, than the effects of almost any other environmental contaminant. Data in relation to the biological effects of ionizing radiation are regularly reviewed at an international level by UNSCEAR[7], and acceptable levels of exposure are recommended on the basis of these evaluated data by ICRP[8]. In the context of public health the relevant recommendations of ICRP are currently those of its report of 1966[8], and guidance on their application to environmental situations is given in that report and in the ICRP report "Principles of Environmental Monitoring Related to the Handling of Radioactive Materials" of 1965[9,10].

Acceptable standards of radiation exposure are predicated on the basis of a linear relationship, without threshold, between absorbed radiation dose and effect, but in the absence of any direct evidence of radiation damage at the low doses and low dose rates relevant to the controlled environmental situations compatible with the recommended dose limits. Evidence on radiation damage relates to situations of high accumulated dose and high dose rate, and the relationship between dose and effect in this type of regime is then extrapolated back to zero effect at zero dose. Whilst there are a number of arguments relating to the validity of this assumption when comparing the effects of high and low LET radiation at low doses, and adjustments in relation to the RBE of these different LET radiations are required, the body of informed professional scientific opinion considering these aspects[7,11] takes the view that the linearity assumption is probably over-conservative, i.e. that the magnitude of effects will be lower than those estimated on the basis of the linear dose response model without threshold[8,12].

Notwithstanding this, there are from time to time attempts to attach a significance to the linear dose response relationship hypothesis out of proportion to its role and significance in radiation protection philosophy and, without due cognizance of its limitations, to calculate upper risk limits to carcinogenesis etc. based on it. These estimates are then used to support a policy for more restrictive dose limitations. This is either to misunderstand or to deliberately misuse the hypothesis.

There are, however, two immediate consequences of the hypothesis which are important considerations when applying dose limit recommendations to environmental situations. One is that, since any dose carries with it some finite, even if very low, risk of effect, doses should not only be below the appropriate dose limits but should be as far below as is practicable or readily achievable, bearing in mind cost/risk and cost/benefit considerations. The second is that, although the overriding requirement is to see that no individual receives a dose in excess of the recommended dose limit, the collective dose (man·rems) to the whole of the exposed population should in principle be assessed in order to estimate total radiation detriment, since all doses carry at least an implied risk of effect. In practice, and on the basis of experience to date[13], individual doses appear likely to pose the major constraint. This problem of the relative importance of individual versus collective dose is, or may be, of particular importance in the context of the long-lived radioisotopes whose distribution may become effectively world-wide either by virtue of their input pathway (e.g. krypton-85 to the atmosphere) or their very long half-life (e.g. plutonium-239).

However, even in these instances a great deal of care needs to be exercised when considering extremely low per capita doses delivered at very low dose rates and then summing them over large population groups. There is a lower limit beyond which such exercises are academic when one bears in mind that the risk estimates per unit of dose being used were derived at high doses and high dose rates - sometimes 3 to 4 orders of magnitude higher!

These concepts of collective dose and total detriment and their application to environmental situations are the subject of an ICRP report published recently[10], and the continuing subject of deliberations within an IAEA expert panel dealing with the problems of "environmental capacity", its definition, derivation and application.

APPLICATION OF ACCEPTABLE DOSE CONCEPT TO ENVIRONMENTAL SITUATIONS

Acceptable rates of radiation exposure of both individuals and populations are in addition to exposure from natural background and the medical uses of ionizing radiation. Exposures may result from sources external or internal to the body, and the ICRP dose limits (Table I) are annual limits of radiation dose intended to include both internal and external contributions.

In the case of internal radiation, dose rates are related to permissible body (or organ) burdens for individual radionuclides sufficient to generate the dose rates, and these in turn are related to the continuous daily intake rate required to establish and maintain the burdens. These daily intakes may also be related to permissible concentrations in air, drinking water and food: these are the more usual criteria in environmental situations, provided of course that respiration, drinking and eating rates are known. These media concentrations are termed "derived working limits", and for the purpose of defining air and water concentrations figures have been developed by ICRP on the basis of a so-called standard man whose metabolic parameters are based on the average for a typical occupational individual. In the case of food consumption separate investigations are required to establish rates for particular situations, and the same requirement, of course, obtains for external exposure regimes where the pattern of potential exposure time requires elucidation in order to apply the dose limits.

TABLE I. ICRP DOSE LIMITS FOR INDIVIDUAL MEMBERS OF THE PUBLIC

Organ or tissue	Dose limit, rems/year
Gonads, red bone-marrow (and in the case of uniform irradiation, the whole body)	0.5
Skin, bone, thyroid	3[a]
Hands and forearms; feet and ankles	7.5
Other single organs	1.5

[a] 1.5 rems/year to the thyroid of children up to 16 years of age.

The investigation of food intakes required for a realistic application of dose limits is itself a problem of some magnitude. The data derived from dietary surveys usually reveal that average per capita daily consumption rates are log normally distributed, and the problem then arises as to which point in the distribution the daily radionuclide intake data should be applied. ICRP[9] gives only general guidance on this point, suggesting that the average behaviour of a critical group of consumers should be used and that the group should be so selected as to ensure that there is reasonable uniformity of the characteristics determining dose. If the sample of the population is large, then standard statistical techniques such as those of extreme value theory can be applied to identify the critical group and quantify its habits[14]. If however the sample is too small to establish the distribution reliably, more arbitrary techniques are called for. Detailed discussion of these problems is available in relation to United Kingdom situations, and in particular that at the Windscale fuel reprocessing plant[15, 16]. In the event of a small but representative statistical sample, the habits of the most extreme individual in the sample may be taken. If the exposed population is extremely small, the most exceptional individual may actually be identified and his or her habits quantified.

Data of the type required to establish derived working limits are generated for each release situation from specially conducted habits surveys designed to identify all possible pathways by which radiation exposure of man may occur. The situation revealed by these surveys will often appear complex at first sight, but when considering the data in the context of the environment and the proposed discharge it may be satisfactorily resolved by application of critical path techniques[3, 4, 6, 12].

CRITICAL PATHWAY ANALYSIS

Although there are a number of techniques relevant to the application of controls to environmental quality problems in the context of radioactivity[17, 18], the

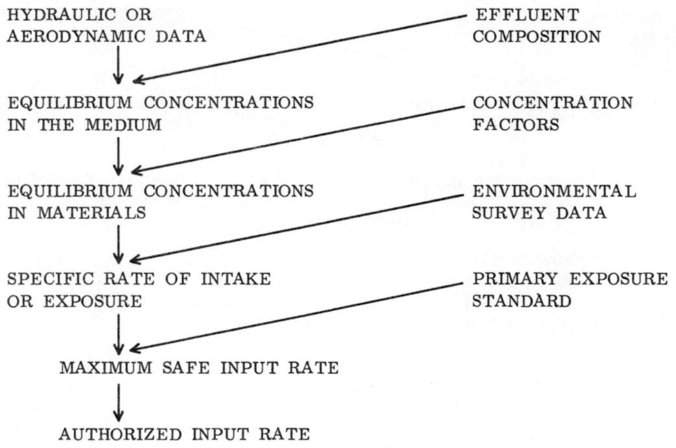

FIG. 1. Diagram of provisional waste disposal assessment.

most widely used and carefully rehearsed is that of critical path analysis. A full description of the application of this technique would be out of place here, and in any case it should be known to the majority of those present; only a brief account follows.

In the application of this method, each individual disposal situation is assessed in quantitative terms in order to arrive at a direct relationship between the rate of introduction of radioactive materials and the resultant radiation dose to man, and thus to establish against the recommended standards of the appropriate ICRP dose limits the limiting environmental capacity of the receiving environment.

The assessment requires detailed quantitative consideration of several processes (Fig. 1), the first of which is concerned with an estimate of concentrations in the receiving water mass for each radionuclide relative to a unit rate of discharge of the specified effluent composition. This requires information on the hydraulic characteristics of the receiving water mass and allowance for partition between water and suspended matter. Using appropriate concentration factors, these equilibrium concentrations in the water may be translated into equivalent values in materials revealed as playing a critical or limiting role in potential human radiation exposure as determined by the habits surveys - the so-called critical materials. This enables an estimate of an individual's rate of intake of radioactivity or rate of exposure to radiation to be made. ICRP data are then used to determine maximum rate of intake of radionuclides consistent with appropriate ICRP dose limits. Finally, by comparison with the fraction of ICRP dose limit generated from a unit rate of discharge, maximum rates of discharge can be derived.

OPERATIONAL ASSESSMENTS AND MONITORING OPERATIONS

Once a discharge is in routine operation, and assuming levels of radioactivity are measurable in the environment and especially in potentially critical materials, the situation can be assessed on a much more realistic basis (Fig. 2). Actual

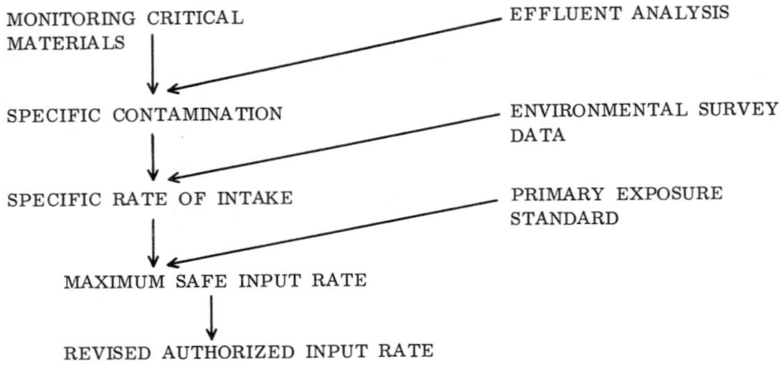

FIG. 2. Diagram of post-operational waste disposal assessment.

TABLE II. DERIVED WORKING LIMITS IN CRITICAL MATERIALS RELATING TO SELECTED DISCHARGES IN THE UNITED KINGDOM

Site	Critical material	Critical nuclide(s)	Rate of consumption (g/day) and exposure (h/year)	Derived working limit
Windscale	Porphyra	^{106}Ru	65 g/day[a] (≡ 130 g laverbread)	340 pCi/g
	Silt	^{106}Ru, $^{95}Zr/^{95}Nb$, ^{137}Cs	300 h/year	1700 μR/h
	Fish flesh	^{137}Cs	300 g/day[a]	150 pCi/g
		^{134}Cs		65 pCi/g
Bradwell	Oyster flesh	$^{110}Ag^{m}$	75 g/day	880 pCi/g
		^{65}Zn		2900 pCi/g
Hinkley Point	Fish and shrimp flesh	^{137}Cs	368 g/day	120 pCi/g
		^{134}Cs		55 pCi/g
	Silt	^{137}Cs, ^{134}Cs	720 h/year	690 μR/h
Trawsfynydd	Fish flesh	^{137}Cs	100 g/day	440 pCi/g
		^{134}Cs		200 pCi/g

[a] Critical groups.

measurements, as distinct from estimates, of effluent composition can be used and hydraulic data and concentration factors can be dispensed with. Actual discharge rates of specific radionuclides can be related directly to concentrations in specific environmental materials, and the relationship between discharge rate and dose estimates verified accordingly.

The establishment of potentially critical pathways at an early stage in the history of a discharge permits the most sensible deployment of resources on monitoring. Monitoring operations in the natural environment, in contrast to the monitoring of occupational exposure in the working environment, seldom permit the direct measurement of radiation exposure, and resort must be made to estimating exposure indirectly using measurements of environmental contamination in conjunction with data from habits surveys and the critical path analysis. This is done by comparing measured levels with the working standards, DWLs, whose derivation was described earlier. The DWL is a useful and convenient index against which to judge the significance of a measurement or series of measurements. It is normally calculated for individual radionuclides or, in the case of external exposure, from the appropriate dose limit. The DWL is only valid against the consumption rate or exposure time used in the calculation. The value used is usually a maximum exposure value or ingestion rate for the exposed individual, or the average of the critical group where one can be identified. It is therefore in effect a maximum permissible exposure level and should never be achieved since, in pursuit of the as low as readily achievable policy, all authorized rates of discharge will be at some level lower than that needed to satisfy the calculated environmental capacity. When a single nuclide is not responsible for almost all the exposure, then the fraction of the DWL for each nuclide present at radiologically significant concentrations must be calculated and the total degree of exposure assessed by summation of the fractions. Examples of DWLs are in Table II.

It is essential to recognize that critical path analysis is a continuing operation in the context of every individual discharge situation, and regular reassessment of environmental capacity and critical pathways against changing effluent composition, and re-evaluation of the habits of the critical population are essential requirements. Every significant discharge situation also presents a continuing opportunity for research into the environmental behaviour of introduced radionuclides, and often also presents opportunities for research relevant to wider fields of endeavour in oceanography and marine biology[19].

The remarks so far made on monitoring in the context of critical path analysis might of course be construed as relating only to the local situation caused by an individual discharge of radioactive materials. A little reflection will of course show that this is not so and that, as already stated, a requirement of a critical path analysis is to make realistic assessments of collective dose. To the extent that this is necessary and sensible, contamination due to individual discharges or the combined effects of several individual discharges will sometimes need to be assessed over significant geographical areas, as with tritium discharged to a large international river, or as with caesium radionuclides in United Kingdom coastal waters.

However, there has been much attention given in recent years to monitoring of pollutants on a global scale, particularly in the marine environment, and artificial radioactivity has received its share of attention.

TABLE III. ESTIMATES OF PUBLIC RADIATION EXPOSURE FROM LIQUID RADIOACTIVE WASTE DISPOSALS IN THE UNITED KINGDOM, 1972-73

Site	Pathway	Maximum exposure[a] of an individual[b] (% of ICRP recommended dose limit)	Approximate total population gonad dose[a], man·rem not exceeding:
BRITISH NUCLEAR FUELS LTD			
Windscale	*Porphyra*/laverbread	2 (to critical group)	10^0
	External dose	7	10^0
	Fish	3 (to critical group)	3×10^3
Springfields	External dose	< 1	10^{-1}
Chapelcross	External dose	≪ 1	10^{-2}
	Shellfish	≪ 1	10^{-1}
UNITED KINGDOM ATOMIC ENERGY AUTHORITY			
Winfrith	Shellfish	< 1	10^{-1}
Harwell[a]	Drinking water	< 1	10^2
Dounreay	External dose (foreshore)	< 1	10^0
	Beta dose (fishermen)	4	10^{-2}
	Shellfish	≪ 1	10^0
CENTRAL ELECTRICITY GENERATING BOARD			
Berkeley/Oldbury	External dose	< 0.3	10^{-1}
	Fish/shellfish	< 0.1	10^{-1}
Bradwell	Oyster	0.1	10^{-1}
Dungeness	External dose	≪ 0.1	10^{-1}
	Fish	≪ 0.1	10^{-1}
Hinkley Point	External dose	0.1	10^{-1}
	Fish/shellfish	0.1	10^0
Sizewell	External dose	≪ 0.1	10^{-1}
	Fish/shellfish	≪ 0.1	10^{-1}
Trawsfynydd	Lake fish	6	10^0
Wylfa	External dose	≪ 0.1	10^{-1}
	Fish/shellfish	≪ 0.1	10^{-1}
SOUTH OF SCOTLAND ELECTRICITY BOARD			
Hunterston	External dose	< 0.1	10^{-1}
	Fish/shellfish	< 0.1	10^{-1}
MINISTRY OF DEFENCE (NAVY DEPARTMENT)			
Chatham	External dose	< 0.1	10^{-1}
Devonport	External dose	≪ 0.1	10^{-1}
Faslane	External dose	< 0.1	10^{-1}
Rosyth	External dose	< 0.1	10^{-1}

[a] Assessed for discharges from the site named except Harwell where the entry includes an estimate for other discharges to the River Thames.

[b] Where critical group is quoted the exposure refers to the median value for that group.

It is important to realize that apart from nuclear weapon testing, which in any case has already been monitored on a global scale as "fallout", any significant introduction of radioactive materials has been carefully assessed prior to release, and often followed up with post-release monitoring. The greatest responsibility in terms of monitoring will therefore always lie with national authorities who, through discharge of their own responsibilities, will have to take care of their own situations. Any requirement for monitoring the environment on a global scale or establishing baselines can therefore already be met, in terms of the overriding criteria of public health, either from existing fallout data[20] or current monitoring programmes. All that is required is to review and properly integrate results, a task already under way within UNSCEAR in terms of its significance for genetic dose. The most sensible arrangements that could at present be mounted would relate to the establishment of an international register of marine radioactive waste disposals, a task already under active consideration by IAEA and one called for in so far as dumping is concerned under the terms of the London Convention. Such a register would permit an examination of disposal rates against the known order of magnitude of the limiting capacity of the relevant sections of the marine environment, and would indicate whether or not international surveillance of environmental monitoring operations was required. A two-stage exercise of this kind would be a much more rational approach to global monitoring and one much more conservative in its deployment of resources.

THE RADIOLOGICAL IMPACT OF RADIOACTIVE WASTE RELEASES TO THE AQUATIC ENVIRONMENT

It is difficult to obtain adequate information from published sources which would permit a thorough review of radiological impact resulting from the introduction of artificial radioactive materials to aquatic environments, and the publication of much more material in this respect would be welcome. However, estimates in respect of the aquatic situation resulting from the discharge of liquid radioactive wastes are presented in the UNSCEAR report of 1972[21] and suggest that, in so far as the data made available permit, radiation levels are very low both in terms of the fraction of the ICRP dose limit and the population dose commitment, the primary source of dose being waste discharge from fuel reprocessing rather than individual power reactor operation.

The most complete information available is that from United Kingdom sources[22] and recent data are summarized in Table III. These show that dose commitment in relation to nuclear power stations is negligible and that the significant dose is that arising from fuel reprocessing activities. Dose rates to members of the critical groups in the laverbread, fish and external dose pathways in connection with the Windscale discharge are shown in Table IV. Set against United Kingdom standards for population dose commitment the only significant contribution has come from the caesium radionuclides in the fish pathway, and the most recent estimates of this indicate the present level as 1500 man·rems per year which might reach an equilibrium level of 6000 man·rems if current rates of discharge are maintained. In collective dose terms, taking due account of the geographical extent of the spread of caesium-137 and the related low-level contamination of fish stocks exploited for human consumption, it is possible to derive estimates on the basis of published statistics relating to western European commercial fish landings from contaminated areas which would place the current

TABLE IV. PUBLIC RADIATION EXPOSURE FROM LIQUID RADIOACTIVE WASTE DISPOSALS TO THE SEA AT BNFL, WINDSCALE, 1970-74

Pathway and critical organ	Range of maximum exposure of an individual(a), rems/year	% of ICRP recommended dose limit
Porphyra/laverbread (GI tract)	0.003-0.50	0.2-33
Silt (whole body)	0.035-0.060	7-12
Fish (whole body)	0.005-0.055	1-11

(a) Laverbread and fish pathways based on the average individual of the critical group.

collective dose commitment for those countries for whom data are available at 3400 man·rems per year, rising to an equilibrium dose commitment of 14000 man·rems per year if current discharge rates persist.

These figures need careful interpretation, and it should be remembered that the United Kingdom policy for control of radioactive waste disposal stipulates that the population dose commitment to its own population must not exceed 1 rem per person per 30 years, or approximately 2.10^6 man·rems per year to the whole population. The current and foreseeable levels of exposure are therefore only of the order of 10^{-1} per cent of this upper limit. In the wider European context national objectives for limitation of population dose commitment are not generally known, and provisional guidance can only be sought by reference to ICRP[23] where it is suggested, for planning purposes, and ignoring the contribution from natural background and the medical uses of ionizing radiation, that not more than 5 rems per person per 30 years should be allowed. It appears reasonable therefore at a first approximation to assume that the United Kingdom allocation of 20 per cent of this illustrative apportionment by ICRP might be considered a sensible allotment to waste disposal from nuclear power programmes, in which case the Western European commitment is also only of the order of 10^{-1} per cent of this level.

The United Kingdom population dose commitment via the aquatic environment thus amounts to about 10^3 man·rems per year or, with the present nuclear generation of 4000 MW(e) per year, 0.38 man·rem per MW(e) per year. Present estimates indicate an installed nuclear capacity of the order of 10^4 MW(e) by 1990, and since in the United Kingdom the main route of public radiation exposure from the nuclear power programme has been via the aquatic, and especially the marine, environment, one half of the acceptable population dose commitment (0.5 rem per person per 30 years) might be allotted to this route for such a programme. An annual population dose commitment limit of 1.10^6 man·rem per year is thus implied, against a possible commitment, based on estimates of exposure at equilibrium with present rates of discharge, of 6000 man·rems per year or 0.6 per cent of the assumed limit from the aquatic sector. It is however intended to reduce current rates of discharge of the critical radionuclide, caesium-137, and population dose commitment is not likely to remain above about 0.1 per cent in the longer term.

It is clear therefore that individual dose commitment would be the limiting factor via this sector of the environment, and provided that this is kept within the ICRP dose limits, the population dose commitment will be well within acceptable limits.

It is clear also that reductions in dose commitment at either the individual or population level might be achieved by further expenditure on treatment of waste streams or by the judicious choice of further sites for fuel reprocessing in order to distribute the environmental load more evenly. The cost of extra treatment or of new site construction would need to be weighed carefully against the values of the benefits to be derived, or the costs to society of the relative risks. Furthermore, the risks attendant upon the transport of fissile material and radioactive waste products engendered by undue proliferation of fuel reprocessing sites would need to be assessed. These topics are all likely to require discussion internationally on at least a regional basis before satisfactory resolution of the problems involved can be achieved. Present trends within international conventions for the regulation of marine pollution problems[24, 25] indicate that radioactivity will come in for its share of attention, and they may perhaps be regarded as initial moves in this direction.

DAMAGE TO MARINE RESOURCES

As previously stated, the limiting consideration in controlling radioactive waste disposal is the radiation dose commitment to man, and it is implicit that, provided the human radiation exposure problem is catered for, the consequential dose to aquatic organisms will carry with it a negligible risk of biological damage.

The evidence for this statement has been extensively reviewed in recent years[26, 27, 28] and has been the subject of investigation by an IAEA expert panel due to report in the near future. The evidence available has been carefully evaluated and supports the view that fish are the most radiosensitive component of aquatic ecosystems, and that damage to resources, if it occurred, would most likely stem from the direct effects of radiation on fish rather than effects produced in organisms at lower levels in the food webs. Moreover, the elements most at risk in the life cycle of fish are the radiosensitive maturing and mature gametes and developing embryos.

Using geometrically idealized models and somewhat incomplete data on radionuclide tissue distributions and concentrations, it has been possible to derive reasonable estimates of dose for phytoplankton, zooplankton, molluscs, crustacea and fish[28] (Table V). Within the limitations imposed by the data, the results show that in a global context weapon-test fallout is the major source of the increased exposure of aquatic organisms arising from artificial radionuclides, with a dose rate contribution of the same order as that from the natural background.

TABLE V. SUMMARY OF DOSE RATES (μrad·h^{-1}) TO MARINE ORGANISMS FROM ENVIRONMENTAL RADIOACTIVITY

Source	Phytoplankton	Zooplankton	Molluscs[(a)]	Crustacea[(a)]	Fish[(a)]
	At 20 m depth remote from sea bed		At 20 m depth on the sea bed		
Natural background	2.8-8.2	3.3-16	9.5-32	10-38	4.8-21
Fallout	0.26-25	1.4-150	0.10-8.0	0.36-0.46	0.14-1.8
Waste disposal at Windscale	200-2100	530-6900	52-3400	43-3400	37-3300

(a) These values do not include the β-radiation component from the underlying sediment. For natural activity and waste disposal radionuclides the β-radiation dose rates at the sediment-seawater boundary are in the ranges 1.6-21 and 210-5400 μrad·h^{-1} respectively.

Higher dose rates are occasionally produced over a relatively limited area in the immediate locality of waste disposal operations, but even here, in terms of the collective dose to a population of organisms, the contribution is negligible.

In the case of developing fish embryos it has been the possibility of significant dose rates from the activity accumulated either on or within the egg that has generated concern. Even in this situation, judged to be one of major concern on the basis of some early experimental results, investigation of radionuclide accumulation by the eggs of the plaice shows that the radiation dose experienced by eggs in a major waste disposal area is much lower than those that have been found necessary to produce significant effects in laboratory experiments - the dose to the developing embryo is below that due to natural background [27].

A useful perspective in considering the biological effects of ionizing radiation on aquatic organisms can be developed by comparison of these effects with losses due to natural mortality or fishing. Certainly in the context of ecosystems and their stability it is effects at a population level rather than an individual level which require attention. Such effects would include effects on fertility, fecundity, growth rate, life shortening, and damage to genetic material. At the dose rates presently estimated, no deleterious effects would be expected at the population level, and certainly none would be detected under natural conditions. From a consideration of stock recruitment mechanisms in exploited fish populations it is evident that any effects caused by low level radiation exposure would be compensated for by the density dependent responses of the population. In terms of genetic damage it should be remembered that a majority of aquatic animals have a very high natural mortality and a compensatingly high fecundity, and these factors would make a substantial contribution to the early elimination of deleterious mutations which might otherwise affect population quality and size.

DEEP SEA DISPOSAL

The deep sea disposal of packaged radioactive waste has been practised in various contexts for a period now approaching 30 years. Recognition of this method of disposal has progressed from the 1958 UN Conference on the Law of the Sea where, under Article 25 of the "Convention on the High Seas", all States were called upon to make provision for measures to prevent pollution of the seas from the dumping of radioactive wastes, and where an associated resolution called upon the IAEA to provide guidance to member states in such matters, to the formal incorporation of control requirements in the London Convention[24], which again named the IAEA as the competent international body in the field. The Agency has since held a number of expert group meetings and has defined, as called upon to do so by Annex 1 of the Convention, high-level radioactive waste unsuitable for dumping at sea and has provided the guidance called for under Annex 2 for the dumping of other radioactive materials. The Agency is currently reviewing its recommendations, as again it is called upon to do so under the terms of the Convention, and is in particular evaluating the oceanographic model used to quantitatively assess the dispersion and dilution of radioactive materials introduced to the deep sea. Other aspects of the critical path analysis based upon the model are also under review. The model undoubtedly oversimplifies the situation obtaining in the deep sea in respect of those factors which will govern the initial dispersion and subsequent behaviour of released radioactive materials. The model is also known to relate to rather specific conditions in the European basin of the north-east Atlantic whose applicability to other deep ocean areas is questionable. The main point to be made, however, is that our lack of detailed knowledge of deep ocean mixing processes and of the chemistry and biology of these deep ocean areas does not permit the construction of much better models at this time. Clearly more research is necessary in this area to provide a sounder scientific basis for the assessment of the significance of deep sea disposal practices, but equally clearly and by any standards, present disposal operations and those on any foreseeable scale in the near future are of entirely negligible radiological significance.

There is another aspect to the requirement for a better understanding of the deep sea and the processes at play there governing the dispersion and behaviour of radioactivity, and that is its assessment as a possible option in the management, and perhaps the ultimate disposal, of high level radioactive waste. This is not advocacy for the pursuit of a policy for deep ocean disposal or seabed burial of high level wastes, but a statement of fact, that if the best decision is to be reached on the handling of these materials then all possible options should be thoroughly evaluated. There is no prima facie scientific case for excluding the deep sea or the deep sea floor from such an evaluation and the conduct of the necessary research is therefore important.

THE ACCIDENT SITUATION

There is a very substantial literature on nuclear accidents assessment, their magnitude in terms of release of radioactive materials and the probability associated with events of different scales. The field has recently been thoroughly reviewed in terms of LWRs[29] by the Rasmussen Committee. Whilst this is again a very controversial area in the context of the future development of nuclear power, the probable results of accidents do not appear to warrant the prominence given to the possibility of large-scale nuclear events.

TABLE VI. DOSES TO CRITICAL ORGANS OF PERSONS CONSUMING 1 kg OF THE FLESH OF FISH CAUGHT AT VARIOUS DISTANCES FROM THE POINT OF RELEASE (UP TO 2 DAYS AFTER RELEASE)

Critical organ	Distance from point of release, km	Estimated radiation dose based on 20% of radioactivity being released, rem	Permissible emergency dose, rem[30]
Thyroid	1	5	25
	5	0.1	
	10	0.004	
Lower large intestine	1	0.6	15
	5	0.01	
	10	0.0004	

The major concern has been with accidents to land-based installations and gaseous releases to atmosphere, with comparatively little attention being paid to other modes of release or to other sectors of the environment[2, 30, 31, 32, 33]. It is, however, clear that in an estuarial or coastal water environment release of gaseous or particulate radioactivity to atmosphere with subsequent deposition to the water surface will not result in a situation of any great radiological significance compared with that in the locality of the initiating event on land. Furthermore, the timescale associated with the emerging radiological pattern will be longer than that associated with atmospheric or terrestrial pathways, and the time available to assess the significance of the situation and take remedial action will be proportionately greater.

The most detailed examination of the consequences of a major nuclear accident in shallow coastal waters is that related to a marine propulsion reactor accident[2, 32]. The study shows that under the conditions postulated, the most likely risks resulting from such a type of accident would come from the consumption of fish flesh. Compared with levels of emergency radiation exposure recommended, for example, for use in the United Kingdom[34], the doses incurred from a single meal of fish flesh are not sufficient to cause alarm (Table VI).

The study also predicted that the most likely results from a similar accident in a shallow silty estuary would be those associated with the consumption of molluscan shellfish or the use of beaches. It might be necessary to close beaches in the vicinity for a short period but, as with fish, the consumption of molluscan shellfish in the period before countermeasures could be imposed would not result in serious exposures when judged against the same emergency dose criteria.

THE PLUTONIUM PROBLEM

A recent seminar held by the United Kingdom's National Radiological Protection Board gave figures relating to the production of plutonium by the early magnox reactors of the United Kingdom power programme. These reactors are among the more efficient producers of plutonium and have to date produced some 30 tonnes of which some 100 kg or 0.3 per cent have been discharged to sea. It is widely believed that if a fast breeder reactor programme were developed, quantities of plutonium entering the marine environment would be much greater per MW(e) produced. However, it can be shown that if the programme were to develop with substantial use of fast reactors the amount of plutonium produced might increase to 150 tonnes by the end of this century whereas, on the other hand, the same installed capacity provided by the more conventional reactor types would produce between 500 and 1000 tonnes of plutonium. Because, however, of the need to recycle fuel, some 500 tonnes of plutonium, even from a fast reactor programme, would have been subject to fuel reprocessing, and thus in either case approximately 1500 kg, or 10^5 Ci of plutonium might have been discharged to sea by the end of the century.

On the basis of present United Kingdom experience with such waste disposal[35] in the Irish Sea, approximately 95 per cent of the plutonium would become incorporated into the bed sediments. The strong association of plutonium with sediments is well documented in other contexts both in respect of reactor-produced material and weapon-produced material[36, 37, 38, 39, 40, 41] and is predictable in terms of known plutonium chemistry. In the Irish Sea the fraction of plutonium remaining in sea water and passing a 0.22 μm membrane filter exhibits a seawater distribution analogous to that of caesium[35, 42] and appears to be the major reservoir from which uptake to biological systems occurs.

In data to be reported subsequently to this meeting[35] it will be shown that the radiological significance of the 10^4 Ci of plutonium so far discharged to the Irish Sea has been negligible, and that there is no evidence of long-term build up in biological materials. There remains, of course, the long-term picture of a steadily increasing quantity of plutonium on the sea bed of the north Irish Sea and the future significance of this in radiological terms. It has been suggested by Bowen and his co-workers, on the basis of sediment concentration data from Buzzards Bay[43], that plutonium may be remobilized from sediment and into biological materials, and thus in principle enter food chains leading to man. There is no evidence that similar processes are taking place in the Irish Sea sediments and, if they are, then the fraction of plutonium involved is extremely small and quite insignificant compared with the plutonium entering food chains directly as a consequence of the discharge[35, 36].

The recently published data on US surveys of the Pacific weapons proving grounds at Bikini and Eniwetok[40, 41] confirm the sediment orientated behaviour of plutonium but also indicate a substantial degree of redistribution from sediments to the water leading to a net flux of plutonium from the Bikini lagoon to the North Equatorial Current of approximately 6 Ci per year, a figure somewhat less than the average transport out of the Irish Sea in recent years of 40 Ci per annum[35]. No figures however are given for the total plutonium inventory of either atoll lagoon. However, a further important point emerges from these recent coral atoll studies and that is that, although these must be the most plutonium-polluted waters in the world, the plutonium activity in biological materials is only a small fraction of the natural polonium activity, which is thus responsible for a larger dose commitment to marine organisms and consumers of marine foodstuffs than the bomb-derived plutonium.

Whilst no contemporary radiological problems are presented by current rates of discharge of transuranic nuclides, it is clear that further work on the biogeochemistry of plutonium and other transuranics in the aquatic, and especially the marine, environment is required. The problem of plutonium behaviour in sediments and of possible remobilization and its significance for food chain contamination, need to be addressed further, and this is being done in the United Kingdom and also, it is believed, in the United States. Meanwhile careful monitoring of the environment in association with all forms of radioactive waste discharges should continue, and the opportunities presented for research by such situations should be thoroughly exploited.

CONCLUSIONS

1. The standards of radiation exposure recommended by ICRP provide a suitable basis for regulating the introduction of radioactive materials to aquatic environments.

2. In applying these dose limits due attention should be given to the admonition to keep doses as low as readily achievable.

3. In the application of the collective dose concept, logical interpretation of the linear dose response hypothesis should not be allowed to go beyond the bounds of realism in estimating collective dose and associated detriment.

4. The application of critical path analysis techniques to environmental impact assessments provides a suitable basis for applying and interpreting dose limits in environmental terms.

5. The most useful deployment of resources on monitoring will arise through careful conduct of critical path analysis and will be local in character with emphasis first on input monitoring and second, where appropriate, on environmental monitoring. It will occasionally be necessary to monitor on a wider geographical scale, but global monitoring is not called for at this time and is unlikely to provide any better picture than will arise from the summation of national and regional efforts.

6. The available data carefully evaluated are consistent with the view that the protection of public health provides adequate protection for the marine ecosystem. More data are required on the distribution of natural and artificial radionuclides in aquatic organisms in order to enable better estimates of radiation exposure to be made, but it appears pointless at this time to look directly for radiation effects. It is more sensible to support further research effort on the factors determining the stability of ecosystems or the mechanisms responsible for determining the size of year-classes in fish populations that support commercial fisheries. The provision of such information would provide a much sounder basis for the interpretation of possible effects for a whole range of pollutants.

7. There is a need for a concerted research effort in respect of deep sea processes which will influence the behaviour of introduced radioactive materials. This will be of especial value in assessing the deep sea as an option for the storage or ultimate disposal of high-level radioactive waste.

8. The problems posed by the introduction of plutonium and other long-lived transuranic elements to aquatic environments are no different in principle to those posed by, and solved for, other radionuclides. In so far as they are different at all, this difference lies in the time scale of their radioactive half-lives, and the problems which this extreme persistence pose for the environment. However, these elements are no more persistent than lead, and provided that their introduction to the environment is so controlled as to ensure their incorporation into the geochemical cycle without undue local perturbations, all should be well. There are however some problems requiring attention, particularly the significance of potential remobilization from sediments. However, the major areas of uncertainty probably lie in such areas as ocean mixing processes and sedimentation phenomena in shallow seas and deep oceans.

REFERENCES

[1] INTERNATIONAL ATOMIC ENERGY AGENCY, Radioactive Waste Disposal into the Sea, Saf. Ser. int. atom. Energy Ag. No. 5 (1961).

[2] PRESTON, A., JEFFERIES, D. F., "Aquatic aspects in chronic and acute contamination situations", Environmental Contamination by Radioactive Materials (Proc. Symp. Vienna, 1969), IAEA, Vienna (1969) 183.

[3] SLANSKY, C. M. (Ed.), "Principles for limiting the introduction of radioactive waste into the sea", Atom. Energy Rev. 9 (1971) 853.

[4] NATIONAL ACADEMY OF SCIENCES, Radioactivity in the Marine Environment, NAS, Washington D.C. (1971).

[5] PRESTON, A., FUKAI, R., VOLCHOK, H. L., YAMAGATA, N., "Radioactivity", Report of Panel 7 in Report of the Seminar on Methods of Detection, Measurement and Monitoring of Pollutants in the Marine Environment, Rome, 1970, Fish. Rep. FAO, No. 99, Suppl. 1 (1971) 87.

[6] PRESTON, A., Contamination of the seas and oceans by artificial radioactivity, Underwater J. 4 (1972) 49.

[7] UNSCEAR, Report of the United Nations Scientific Committee on the Effects of Atomic Radiation, United Nations Supplement No. 13 (A/7613), UN, New York (1969).

[8] ICRP, Recommendations of the International Commission on Radiological Protection (Adopted September 17, 1965), ICRP Publication 9, Pergamon Press, Oxford (1966).

[9] ICRP, Principles of Environmental Monitoring related to the Handling of Radioactive Materials, ICRP Publication 7, Pergamon Press, Oxford (1965).

[10] ICRP, Implications of Commission Recommendations that Doses be kept as Low as Readily Achievable, ICRP Publication 22, Pergamon Press, Oxford (1973).

[11] NATIONAL ACADEMY OF SCIENCES-NATIONAL RESEARCH COUNCIL, The Biological Effects of Ionizing Radiation, NAS/NAC, Washington D.C. (1956).

[12] NATIONAL COUNCIL ON RADIOLOGICAL PROTECTION AND MEASUREMENTS, Review of the Current State of Radiation Protection Philosophy, NRCP Report No. 43, Washington D.C. (1975).

[13] PRESTON, A., MITCHELL, N. T., "Evaluation of public radiation exposure from the controlled marine disposal of radioactive waste (with special reference to the United Kingdom)", Radioactive Contamination of the Marine Environment (Proc. Symp. Seattle, 1972), IAEA, Vienna (1973) 575.

[14] BEACH, S. A., The identification of an homogeneous critical group using statistical extreme value: application to laverbread consumers and the Windscale liquid effluent discharges, Rep. natn. radiol. Prot. Bd, NRPB R-19, HMSO, London (1974).

[15] PRESTON, A., JEFFERIES, D. F., The ICRP critical group concept in relation to the Windscale sea discharges, Hlth Phys. 16 (1969) 33.

[16] PRESTON, A., MITCHELL, N. T., JEFFERIES, D. F., "Experience gained in applying the ICRP critical group concept to the assessment of public radiation exposure in control of liquid radioactive waste disposal", Population Dose Evaluation and Standards for Man and his Environment (Proc. Symp. Porto Roz, 1974), IAEA, Vienna (1974) 131.

[17] FOSTER, R. F., OPHEL, I. L., PRESTON, A., "Evaluation of human radiation exposure", Ch. 10, Radioactivity in the Marine Environment, National Academy of Sciences, Washington D. C. (1971).

[18] PRESTON, A., "Aquatic monitoring programmes", Environmental Contamination by Radioactive Materials (Proc. Symp. Vienna, 1969), IAEA, Vienna (1969) 309.

[19] PRESTON, A., WOODHEAD, D. S., MITCHELL, N. T., PENTREATH, R. J., The impact of artificial radioactivity on the oceans and on oceanography, Proc. R. Soc. Edinb. (B) 72 (1971/72) 411.

[20] VOLCHOK, H. L. et al., "Oceanic distributions of radionuclides from nuclear explosions", Ch. 3, Radioactivity in the Marine Environment, National Academy of Sciences, Washington D. C. (1971).

[21] UNSCEAR, Ionizing Radiation: Levels and Effects, Report of the United Nations Scientific Committee on the Effects of Atomic Radiation to the General Assembly, with annexes, 1: Levels, UN, New York (1972).

[22] MITCHELL, N. T., Radioactivity in Surface and Coastal Waters of the British Isles 1971, Tech. Rep. Fish. radiobiol. Lab., Lowestoft 9 (1973).

[23] ICRP, Recommendations of the International Commission on Radiological Protection, ICRP Publication 6, Pergamon Press, Oxford (1964).

[24] GREAT BRITAIN, PARLIAMENT, Final Act of the Inter-governmental Conference on the Convention on the Dumping of Wastes at Sea (London, 1972), Cmnd 5169, HMSO, London (1972).

[25] GREAT BRITAIN, PARLIAMENT, Convention for the Prevention of Marine Pollution from Land-Based Sources (Paris, 1974), Cmnd 5803, HMSO, London (1975).

[26] TEMPLETON, W. L., NAKATANI, R. E., HELD, E. E., "Radiation effects", Ch. 9, Radioactivity in the Marine Environment, National Academy of Sciences, Washington D. C. (1971).

[27] WOODHEAD, D. S., The assessment of the radiation dose to developing fish embryos due to the accumulation of radioactivity by the egg, Radiat. Res. 43 (1970) 582.

[28] WOODHEAD, D. S., "Levels of radioactivity in the marine environment and the dose commitment to marine organisms", Radioactive Contamination of the Marine Environment (Proc. Symp. Seattle, 1972), IAEA, Vienna (1973).

[29] RIPPON, S., The Rasmussen study on reactor safety, Nucl. Engng, Lond. 19 (1974) 1001.

[30] JEFFERIES, D. F., "Exposure to radiation from gamma-emitting fission-product radionuclides in estuarine sediments from the north-east Irish Sea", Ch. 26, Environmental Surveillance in the Vicinity of Nuclear Facilities (REINIG, W. C., Ed.), Charles C. Thomas, Springfield, Illinois (1970) 205.

[31] PARKER, F. L., SCHMIDT, G. D., COTTRELL, W. B., MANN, L. A., Dispersion of radiocontaminants in an estuary, Hlth Phys. 6 (1961) 66.

[32] PRESTON, A., JEFFERIES, D. F., "An evaluation of the public health aspects of a marine reactor accident in coastal waters", Carriage of Dangerous Goods by Sea and Inland Waterways (Proc. Conf. Jacksonville, Florida, USA, October 1975). To be published.

[33] MORGAN, F., "The effects on marine foodstuffs of a large release of radioactivity into the sea", Nuclear Detonations and Marine Radioactivity (Proc. Symp. Kjeller, Norway, SMALL, S. H., Ed.), Norwegian Defence Research Establishment (1964) 153.

[34] GREAT BRITAIN, PARLIAMENT, The Hazards to Man of Nuclear and Allied Radiations. A Second Report to the Medical Research Council, Cmnd 1225, HMSO, London (1960).

[35] HETHERINGTON, J. A., JEFFERIES, D. F., LOVETT, M. B., "Some investigations into the behaviour of plutonium in the marine environment", Paper IAEA-SM-198/29, these Proceedings.

[36] HETHERINGTON, J. A., "The behaviour of plutonium nuclides in the Irish Sea", Eighth International Conf. on Environmental Toxicity (Proc. Conf. Rochester, N.Y., 1975). To be published.

[37] AARKROG, A., Radioecological investigations of plutonium in an arctic marine environment, Hlth Phys. 20 (1971) 31.

[38] NOSHKIN, V. E., BOWEN, V. T., "Concentrations and distributions of long-lived fallout radionuclides in open ocean sediments", Radioactive Contamination of the Marine Environment (Proc. Symp. Seattle, 1972), IAEA, Vienna (1973) 671.

[39] BOWEN, V. T., Transuranic elements and nuclear wastes, Oceanus 18 (1974) 42.

[40] NEVISSI, A., SCHELL, W. R., Distribution of plutonium and americium in Bikini Atoll Lagoon, Hlth Phys. 28 (1975) 539.

[41] NEVISSI, A., SCHELL, W. R., ^{210}Po and ^{239}Pu, ^{240}Pu in biological and water samples from the Bikini and Eniwetok atolls, Nature, Lond. 255 (1975) 321.

[42] JEFFERIES, D. F., PRESTON, A., STEELE, A. K., Distribution of caesium-137 in British coastal waters, Mar. Pollut. Bull. 4 (1973) 118.

[43] INTERNATIONAL ATOMIC ENERGY AGENCY, Report of Advisory Group on the Development of a Generalized Oceanographic Model Related to the Dumping of Radioactive Wastes at Sea, IAEA, Vienna. To be published.

DISCUSSION

S. BONOTTO: What kind of radionuclides were incorporated into the marine alga Porphyra?

A. PRESTON: The data on dose from laverbread given in my paper relate to exposure of the lower large intestine and are therefore concerned with the high-energy beta emitters: ^{106}Ru-^{106}Rh and ^{144}Ce-^{144}Po.

There are, of course, other fission nuclides present such as ^{95}Zr/^{95}Nb, ^{90}Sr, ^{137}Cs, ^{99}Tc and transuranics. These are at levels of little significance for gastro-intestinal tract dose, but some of them are important in the context of bone dose. However, the dose to bone is not as limiting as the dose to the gastro-intestinal tract. Data on all nuclide concentrations may be found in the published reports of the Fisheries Radiobiological Laboratory at Lowestoft, United Kingdom.

Yu.V. SIVINTSEV: I believe you said that current observations and calculations show that these releases can be permitted on the assumption that the most vulnerable factor in the ecosystem is man. If some other factor is found to be more vulnerable how will your conclusions be affected?

A. PRESTON: My assertion is that all the available data, carefully evaluated, are consistent with the view that the protection of human health will provide an adequate degree of protection for other biological systems in aquatic environments.

It is clear, however, that a continuous and careful review of the situation is required. Critical path analyses, suitably broadened to include data relevant to the resource problem, will provide a basis for this kind of review. The data I presented show the results of such a review at this time and support the conclusions drawn. If a resource problem were involved, its significance would have to be evaluated and regulatory action, consistent with the scale of the problem and the relative costs of risks and benefits, would have to be taken. However, I believe that, on the basis of the data to hand, such a situation is very unlikely to arise.

Yu.V. SIVINTSEV: Why do you think that fast reactors will involve large releases of plutonium into the aquatic environment? Why should one not assume that some means of removing plutonium from waste releases will be developed for these reactors?

A. PRESTON: There appears to be a misunderstanding here. My paper in fact supports the view that in a United Kingdom context the use of fast reactors will not generate any more Pu for sea disposal than conventional reactor systems.

Of course, further treatment of effluent streams might effect a substantial reduction of Pu inputs to the environment, but the adoption of such schemes must take into account relative costs in regard to environmental risk and effluent treatment. Present indications are that expensive treatment is not justified at this time.

V.V. PISAREV: You correctly pointed out the long duration of the process of contamination of the bottom sediments. Are any studies being performed on geological, geochemical or other natural processes causing irreversible retention of radioactive wastes in the bottom sediments? Perhaps the processes enhancing reversible movement of radionuclides (^{239}Pu) back to the aquatic environment predominate.

A. PRESTON: Our present research activities indeed include a programme devoted to an investigation of the processes leading to removal of radionuclides to sediments in suspension and on the sea-bed. This programme has been reported on and will be the subject of further comments from my colleague Mr. Hetherington in Paper IAEA-SM-198/29 of these Proceedings.

However, I would also draw your attention to certain points in my paper which may not have come over clearly in the presentation. In particular, the evidence for remobilization relates to Eniwetok and Bikini atolls, where in fact we may be concerned with resuspension of finely divided coral material on which Pu is absorbed, rather than true remobilization, while the evidence from Buzzards Bay is highly dependent on information on cores from a single point sampled over time, and there are alternative explanations for the phenomena observed.

Neither the extensive Lake Michigan studies nor those on the north Irish Sea reveal any evidence of remobilization. The balance of the evidence to date therefore favours retention of sediments, and not later release to the water.

Invited paper

RADIOLOGICAL IMPACTS OF RELEASES FROM NUCLEAR FACILITIES INTO AQUATIC ENVIRONMENTS – USA VIEWS

C.L. OSTERBERG
Division of Biomedical and Environmental Research,
Energy Research and Development Administration,
Washington, D.C.,
United States of America

Abstract

RADIOLOGICAL IMPACTS OF RELEASES FROM NUCLEAR FACILITIES INTO AQUATIC ENVIRONMENTS – USA VIEWS.

We may be entering an age which will see a rapid proliferation of nuclear power plants. Since we learn from past experience, the author looks back at two major events that led to large-scale releases of radionuclides to the environment. While these two events did not lead to major exposure doses to man, they would have been precluded in peacetime by the many environmental protection laws now in existence in the United States. Although during the peak, the Hanford Laboratories released about 1000 curies a day of neutron-induced radionuclides to the Columbia River and the Pacific Ocean, no harmful effects have been observed. Studies of worldwide fallout show a larger dose to man than from nuclear power, but food chains in the ocean seem to filter out much of the radioactivity from fission products so that they only appear minimally in the diet of man. These past events appear encouraging for the future of nuclear power.

INTRODUCTION

We are entering a new period in which continuing oil shortages may be driving us – ready or not – into a large-scale nuclear power programme. Many look at this possibility of logarithmic growth with fear and trepidation. Fortunately, behind us are countless man-years of professional knowledge of artificial radioactivity; most of it has been published. Man learns from his experience and he has learned much from experience with radioactivity. As with other things, the bark of radioactivity is worse than its bite.

I was with an organization – the US Atomic Energy Commission – that presented an unready world with a large stockpile of new sources of radioactivity – the atomic bombs – for national defence. The same organization, with the expenditure of billions of dollars, harnessed the energy of the atom to do man's work, through the creation of an abundant supply of electricity. The group I represented, the old Division of Biology and Medicine, invested well over a billion dollars over the better part of three decades to ensure that nuclear energy could be delivered safely, with minimum harm to man and other living things in the environment. Other countries have also contributed much to our store of knowledge.

In that thirty-year period since the first nuclear explosion at Alamagordo, New Mexico, we witnessed events that will probably never happen again in these days of the Environmental Protection Agency and the National Environmental Policy Act and the Air Quality Act and the Water Quality Act. Imagine the difficulty of writing an environmental impact statement — required of every major Federal action in my country — for a series of nuclear bomb tests in the atmosphere at a Pacific Island.

Or think of starting from the beginning to write a statement for the construction of a major nuclear facility that would release 1000 curies of radioactivity a day into a river containing the largest salmon population in any of the lower 48 states? We have the information to write one now, although public opposition would be fierce and could prevent us from actually completing the project. In the mid-1940s when the first Pacific Tests were held and in the early 1940s when the Hanford Reactor was built, we couldn't have written an environmental impact statement. But we did the next best thing. We depended on the wisdom of a small group of capable scientists, we began the process of learning, we started a research programme, and, environmentally speaking, we were fortunate. These actions were necessitated by the Second World War. The physics was there waiting to be used by either side. We admit to some things in weapons development and testing which introduced far more radioactivity into the waters of the world than anything we're doing today. We have learned from these past experiences, and we'll continue to learn. The results should help to ease our apprehensions of the future.

Man's view of earth from space has reminded us that we're all in this together. The world is slowly but constantly changing everywhere, as it has done without and with mankind. At the US Government level we now find the Atomic Energy Commission gone; it was abolished on 19 January 1975; now we're the Energy Research and Development Administration (ERDA). ERDA is involved with conventional sources of energy such as coal, and with advanced ideas such as harnessing the temperature differences of the ocean to generate electricity, and with what we refer to as the peaceful atom.

The fact remains that, as we look at all our energy options, nuclear power cannot be discounted. We can look forward to continuing to live with some artificial radioactivity indefinitely. The findings and the lessons of the past three decades should continue to be studied.

I'll discuss in some detail two past events: The Columbia River story and fallout from the weapons tests in the atmosphere. I'll conclude with some data on nuclear power in the USA and its releases to the environment.

THE COLUMBIA RIVER

The Columbia River has the largest supply of cool water in the Western United States. Large amounts of water for cooling and large amounts of electric power available nearby forced selection of the site during World War II.

In the early 1940s, the USA began building nuclear plants to produce weapons-grade plutonium on the Columbia River at Hanford. When these activities were at their peak, some 20 years later, eight reactors were using Columbia River water in single pass cooling, releasing approximately 1000 curies a day of radioactivity to the River. By the time the reactor effluent reached the river mouth 350 miles downstream, ^{51}Cr and ^{65}Zn were the only two artificial gamma emitters from other than fallout that could be measured directly in situ, that is, by placing a 3 in. × 3 in. detector in the water and counting for up to 400 minutes.

With radiochemical techniques, by which the ^{51}Cr was removed for counting from 600-litre samples of sea water, the river could be followed far to sea. Measurements of the radioactively labelled river water gave much new information about the seasonal movement of the Columbia River plume at sea, and the availability of the radionuclides to the biota.

Even though levels of radioactivity in the river's plume, after dilution at sea, were very low, techniques were developed for their measurement. Surface water was pumped into 600-litre tanks for chemical removal of the radionuclides of interest, principally ^{51}Cr. Because ^{51}Cr in sea water exists mostly as Cr VI, which is not removed by our precipitation techniques, it had to be reduced to Cr III, using Sb III as a reducing agent. Ferric hydroxide precipitation then very efficiently removed the ^{51}Cr. The precipitate was allowed to settle, then drained and eventually reduced by decanting and evaporating to approximately 8 litres. This concentrated sample was then analysed by gamma-ray spectrometry. By this technique, ^{51}Cr from Hanford could be detected at sea over 700 km from the mouth of the Columbia River.

Because the plume of the River was fresher and less dense than the ambient sea water, it remained near the surface. However, biological processes moved some of the radioactivity downward. When a series of sea cucumber samples from the bottom at various water depths (50, 100, 125, 150, 175, 200, 400, 600, 800, and 2086 metres) was analysed, ^{65}Zn from the Columbia River could be found in all the samples from 400 metres or less in depth. This is a different picture from what was observed with fission products from fallout, as will be seen later.

We have not seen any adverse effects of the radioactivity on marine communities. Frankly, we wouldn't expect to see any unless the effects were disastrous. We don't know the marine system well enough to detect changes unless they are major. Apparently, no catastrophies have occurred.

A better place to look for changes would be in the salmon populations, because they have been studied intensely and fluctuations in the populations are more easily discovered. For one thing, the commercial and sport fisherman will let us know if salmon populations dwindle.

The Columbia River has always been a major salmon fishery, and its well-being was of great concern to the Manhattan District, Corps of Engineers (predecessor to the AEC), when they built Hanford Laboratories. One of their first acts was to contract with the University of Washington School of Fisheries to determine the nature of the impact of radioactivity on the salmon. Dr. Lauren Donaldson, a world-renowned salmon expert, directed this work. To facilitate his studies, he started an experimental population of salmon whose home was a small pond outside his laboratory. Each spawning season, after three or four years at sea, the fish would return to the University of Washington campus.

Spawners were stripped of eggs. These were fertilized and, after hatching, the young migrated to the Pacific Ocean to mature. In the laboratory some eggs were irradiated lightly with a gamma source, some rather heavily, and some, the controls, were not irradiated. Flaws in development and early mortality were noted but the real interest was in how the laboratory-raised irradiated fish would perform in the real competitive world – the Pacific Ocean. After the first fish were sent to sea, there would be a long wait (3-4 years) during which Dr. Donaldson must have wondered if he would see his experimental salmon again.

But that's all ancient history. As you probably know, Dr. Donaldson did succeed in establishing a race of fish that homed on his laboratory pond. They were tagged for ready identification and the amount of radiation they had received was recorded. Their success in coping with the hazards of life at sea and returning as adult salmon was studied.

Results show that some of the higher doses of radiation administered in the laboratory did cause abnormalities. Let me quote: "Although abnormalities in young fish were increased by irradiation (at 0.5 R/day from the time of fertilization to feeding stage[1]), the number of adults returning was not affected. On the contrary, the irradiated stock returned in greater numbers and produced a greater total of viable eggs than the control stock" [1].

While work was going on in the University of Washington, other scientists from the Hanford group were studying the salmon in the river. Most of the early research — designed to measure effects of a large nuclear facility on aquatic organisms — was negative. No effects could be seen. In fact, an early conclusion of the Hanford group was that the only harmful effects on young salmon raised in reactor effluent were due to chemical toxicity from the sodium dichromate used to keep the reactor cooling system free from corrosion. So far so good for radioactivity, if not for a more common chemical.

The lower salinity waters along the Oregon-Washington coasts, which result from the Columbia River outflow, are an excellent habitat for oysters, clams, and crabs. While radioactivity from Hanford was always measurable in these organisms, and they were in fact monitored by the health departments of both states, I know of no case where sea food from the area was not acceptable as human food.

So, after many years of study of what must have been one of the more radioactive rivers in the world, all of our research data on environmental effects of radiation are negative. That doesn't mean the salmon populations of the Columbia River are perfectly healthy. Other things have happened to the river: dams have flooded their spawning grounds and interrupted their spawning runs, for example. But it seems certain, since the reactors have been shut down now for four or five years, that the reactors have had minimal effect. The late Dr. Paulik, a University of Washington fisheries statistician who analysed many years of Hanford data on fish nests on the Columbia River, reported: "These data provide no evidence of a reactor effect; if such an effect does exist, it is masked by other events occurring during the period of observation." He further states that "dam construction during the past 23 years was probably the critical factor controlling the numbers of fall chinook spawning in the Hanford areas" [2].

I conclude that man's activities on the Columbia River have adversely affected the biota, particularly the salmon, but that the nuclear reactors were probably not at fault. And I ask, if we see no adverse effect from the production reactors releasing 1000 curies a day, can we expect much of an effect from the modern-day commercial reactor — a power plant that releases at most a few hundred curies per year? I am encouraged by what we've learned on the Columbia River.

---

[1] 0.5 R/day is a large exposure. Maximum doses in the Columbia River were much less.

FALLOUT

Our second lesson was from fallout. The Columbia River nuclides are practically dead, and the fallout is dying. But since long-lived fission products made up much of the fallout while short-lived neutron-induced radionuclides were in Hanford's effluents, fallout is taking longer to disappear. Numbers that I will show later do indicate that fallout is slowly fading away, however.

Not all the lessons are in on fallout, but it does seem obvious that fallout's major threat is to terrestrial systems rather than aquatic. Certainly the ocean is least vulnerable because of its depth and volume, and because of the chemical competition and isotope dilution, which cause ^{90}Sr and ^{137}Cs to be less of an environmental problem in the sea than in fresh water and much less than on dry land.

We can lay to rest one story that swept through the United States. For some time there was a conviction that radioactivity, like DDT, is magnified as it passes through food chains and ends up on man's table in rather large doses. It is true that ^{32}P in the Columbia River is concentrated by rather large factors by aquatic biota, but that river is relatively free from stable phosphorus — it receives very little sewage or industrial waste. The concentration factors for ^{32}P would tend to be much less in other, more polluted and more typical US streams. But even ^{32}P is found in equal abundance in phytoplankton and is not magnified at higher trophic levels. According to the recent summary by Lowman, Rice and Richards [3], it is the plankton, both phytoplankton and zooplankton, that have the maximum concentrations of trace elements.

During the days when fallout levels were elevated from nuclear testing by the USA and other nuclear nations, I made some observations that convinced me that certain fission products (mostly ^{95}Zr-^{95}Nb and $^{141-144}Ce$) reached their peak in the second trophic level organisms.

Samples of many organisms were taken in a mid-water trawl from near the surface of the ocean, and sorted by groups. When various groups were analysed for gamma emitters and those were compared with a filter through which sea water had been pumped, a certain pattern evolved. If we call the filter representative of trophic level I, since it presumably contained mostly particulates and diatoms (larger plankton had been removed with a prefilter), we see that ^{51}Cr (probably on the diatoms) and several particulate fission products appear on the filter.

Trophic level II, the grazers of the sea, was represented by *Euphausia pacifica*. An analysis of the gamma spectrum of this small shrimp-like organism revealed little ^{51}Cr but a large peak due to ^{95}Zn-^{95}Nb, a fission product. Apparently, *Euphausia pacifica* does not concentrate ^{51}Cr, which is mostly in solution in sea water, but does concentrate ^{95}Zn-^{95}Nb, which is in particulate form.

Trophic level III, a group of small predators that feeds on copepods, euphausiids, etc., was represented by two groups, *Lampanyctus leucopsarus* (a small lantern fish) and *Pasiphaea pacifica* (a shrimp-like organism). The spectra of these two predators showed almost no radioactivity from fission products. The conclusion is that many radionuclides (in this case ^{51}Cr, ^{95}Zr-^{95}Nb, $^{141-144}Ce$, and ^{103}Ru) are cleansed from the system as they are passed through marine food chains.[2]

[2] This work is discussed in detail in OSTERBERG, C., PEARCY, W.G., CURL, H., Jr., Radioactivity and its relationship to oceanic food chains, J. Mar. Res. 22(1) (1964) 2-12.

TABLE I. CAESIUM-137 ACTIVITY IN LAKE MICHIGAN (in curies) [5]

	1970	1975
Weapons testing fallout cumulative deposition	5937	5786
Activity present in water	< 1200 - 2000	< 900 - 1800
Probable annual reactor input		10

TABLE II. COMPARISON OF DISCHARGES OF THE PRINCIPAL RADIONUCLIDES FROM INDIAN POINT I AND FALLOUT FROM WEAPONS TESTS (in curies) [5]

	Sr-90	Mn-54	Cs-137	H-3 tritium
Annual discharge from Indian Point I (1971)	0.01	10.2	22.5	725.0
Fallout from weapons (1963)				
On Hudson River watershed (35 000 km^2)	825.0	1236.0	1320.0	205 000.0
On Hudson River surface	3.7	5.5	5.9	920.0

TABLE III. ACTIVITIES OF INTERMEDIATE AND LONG-LIVED RADIONUCLIDES RELEASED IN LIQUID REACTOR WASTE DURING 1971 [5]

Nuclide	H-3	Cs-137	Cs-134	Mn-54	Co-60	Co-58	Sr-90
Activity (Ci)	725	22.5	16.4	10.2	4.7	2.1	<0.01

TABLE IV. POTENTIAL SOURCES OF RADIONUCLIDES IN THE AQUATIC ENVIRONMENT [5]

Source	Major nuclides
Mining	Ra-226
Milling	Ra-226
UF_6 production	U
Enrichment	U
Fuel fabrication	U
Reactors	Cs-137, Cs-134, Co-58, Co-60, Mn-54, Fe-55, H-3, I-131
Reprocessing plants	Sr-90, H-3, Cs-137, Cs-134, Co-60, Transuranics
Waste storage	Sr-90, Cs-137, Transuranics

Where does the radioactivity go to? Apparently it passes through the digestive tract of trophic level II grazers and falls to the ocean floor in faecal pellets. Evidence for this is seen in a comparison of spectra from two groups of sea cucumbers — one group from the bottom at 200 metres, the other from 2800 metres. Comparison shows almost equal amounts of particulate fission products. Since the half-life of ^{95}Zr is 65 days, and since we know that small fallout particles could be expected to take 1 to 100 years to sink to 2800 metres, we conclude that they are arriving on the ocean bottom in faecal pellets. According to experiments, faecal pellets should sink 2800 metres rapidly enough to account for this phenomenon.

The conclusions are summarized by Bowen et al. [4], who state: "Lower trophic levels in the sea are likely to have greater concentrations of radioactivity than higher trophic levels." The verdict seems to be unanimous on that score. There is no concentration-magnification mechanism operating in the ocean. In the same article Bowen et al. point out that our harvesting of fishes of the higher trophic levels has almost reached a maximum. Because overfishing will soon occur, it is suggested that to increase our marine food supply we must move down the food chain and eat the small grazers — euphausiids, copepods, salps, etc. When and if we do, radioactive fallout becomes more of a problem.

When moving from fallout radioactivity to that released from commercial power reactors, we have to learn to think small. Table I shows the amount of fallout radiocaesium (^{137}Cs) in Lake Michigan compared with that expected from the approximately 10 reactors along the shore. Here we see that fallout caesium-137 is decreasing, and it will be a very long time before the input from reactors will be comparable.

Table II shows that a similar difference between fallout and reactor input occurs on the Hudson River in New York.

When we look at the releases from an actual example, the BWR at Dresden, Illinois, it is seen that the annual rate of liquid releases is less than 4 curies/year, in the following order in mCi/second (not dose): $^{3}H > ^{60}Co > ^{58}Co > ^{89}Sr > ^{137}Cs > ^{140}Ba > ^{134}Cs > ^{131}I > ^{90}Sr$. Of these, the principal radionuclides relating to dose to man are caesium, both ^{137}Cs and ^{134}Cs.

The reactor at Indian Point I on the Hudson River in 1971, its worst year, released the amounts seen in Table III, with radiocaesium again providing the major dose. Wrenn estimates that the major pathway of the radiocaesium to man is through the consumption of Hudson River fish. A person eating 30 g of such fish a day for a year would receive an annual whole body dose of 0.19 mrem, 98% of which would be due to radiocaesium. Lesser doses are 0.06 mrem from 200 hours' exposure to Verplanck Beach soil, 0.01 mrem from 200 hours of swimming at the same beach, with radiocaesium as the principal contributor.

Of course other parts of the fuel cycle contribute radioactivity to the environment. Table IV shows the spectrum of radionuclides expected from the fuel cycle. Here we see different radionuclides from those typical of reactors entering the picture — mostly ^{226}Ra, U, and the transuranics. These, particularly ^{226}Ra, are expected to provide the major dose to the population. Table V shows that irrigated crops and water are expected to be the major pathways to man, with bone as the critical organ.[3]

[3] There is at least some feeling in the USA that carbon-14 will be another of the larger contributors to dose to man.

TABLE V. EXPECTED ANNUAL PER CAPITA DOSES TO POPULATIONS FROM NUCLEAR POWER (ca. 1980) [5]

Activity	Nuclide	Path	Critical organ	Annual dose range (mrem)
U mining and milling				
Colorado watershed	Ra-226	Water irrigated crops	Bone	~10
Nuclear electric generating stations				
Estuarine (Hudson River)	Cs-137, 134	Fish	Whole body	~ 0.1
Great Lakes	Cs-137, 134	Fish	Whole body	< 0.1
Midwest	Cs-137, 134	Fish	Whole body	~ 1

TABLE VI. DOSE COMMITMENT ACCRUED IN 1968 IN THE VICINITY OF NUCLEAR FUEL SERVICES [5]

			Dose commitment[a] (mrem)	
Radionuclide	Critical organ	Media	Typical individual	Maximum individual
Sr-90[b]	Bone	Fish	16	215[c]
Cs-137	Whole body	Fish	0.2	8.6
Co-60	G.I. tract	Fish	0.05	0.5
Ru-Rh-106	G.I. tract	Fish	0.9	16.4
H-3	Whole body	Fish	0.01	0.3
		Drinking water	0.17	1.0

a Dose per 50 years or per lifetime.

b For Sr-90 ~ 7% of the dose commitment to bone is delivered during the first year after intake. Fish: 1.1 mrem (typical) and 15 mrem (maximum).
For other radionuclides: > 90% of the dose delivered in the first year after intake.

c Represents whole fish. The majority of Sr-90 concentrates in bone which is usually considered inedible. Therefore, the dose commitment may represent a high estimate.

Some of you by now may have noticed that, so far, nothing has been said about releases from nuclear fuel processing. At present, no commercial fuel processing plants are in operation in the USA. There are some 1968 data from the vicinity of the Nuclear Fuel Services plant near West Valley, New York State. The doses (Table VI) given are for 50 years or per lifetime, and this should be taken into consideration.

Thus we see that levels of radioactivity that we deal with today are small compared with what we have had over the past thirty years.

Integrating the artificial radioactivity in the biosphere and determining the rates of release for a few notable years of the past three decades might well have led the scientist to predict high risks and dire events for all living things. A couple of hundred years ago de Montesquieu pointed out the human attraction toward vain hopes and groundless fears. One does not have to look far nor wait long today to learn of something which will make things better, or more often, make them everlastingly worse. Radioactivity produced by the works of man has been a focal point of much official and public attention. Some see it as a by-product of work that is unquestionably for the good of mankind. Other people view radioactivity with apprehension an and as a threat which can only become more threatening.

In the history of the world, thirty years is a brief period. But I will emphasize that the last thirty years are the period during which we have learned so much of what is known about radioactivity and where it goes and what it does in the biosphere. As of this time, and taking into account the possibilities of changes which may confront us tomorrow, I look to the future of nuclear energy with confidence, encouragement, and hope.

REFERENCES

[1] Radioactivity in the Marine Environment, National Academy of Sciences, Washington D.C. Chapter 9 (1971) p.226.

[2] Fall Chinook Salmon Spawning in the Columbia River near Hanford, 1947 - 1969. D.G. Watson, BNWL-1515-UCA8 (Oct. 1970) Richland, Washington, p.35.

[3] Radioactivity in the Marine Environment, op.cit., pp.168-169.

[4] Radioactivity in the Marine Environment, op.cit., p.218.

[5] WRENN, McD.E., JINKS, S.M., "Dosimetric implications of releases to the aquatic environment from the nuclear power industry in the decade 1972 - 1982", Rochester Inter.Conf. on Environmental Toxicity, 1975 (in press).

DISCUSSION

J.A. HETHERINGTON: You mentioned the potential increased importance of ^{14}C as a waste product in the nuclear fuel cycle. Can you say at what stage in the cycle you think it will be important and whether it will appear in terrestrial or marine food chains?

C.L. OSTERBERG: Carbon-14 is made in the reactor by neutron activation of nitrogen impurities in the fuel and in the air. About 10 Ci/GW·a is average but, of course, there are differences between reactors. About one third of the carbon-14 is released at the reactor site, the balance being released during fuel processing. Emission controls at the fuel processing plant will reduce this release, leaving the reactor as the primary source to the environment. With the exercise of care to avoid nitrogenous impurities in the reactor, releases can be reduced here too. I doubt whether terrestrial or marine food chain pathways will prove to be important sources of carbon-14 to man.

M.N. KHALANSKI: You said that the primary level of the food chains is generally the most contaminated. In that case, the marine food chain would act as an ecological filter of radioactivity transferred to man. There may, however, be exceptions to this — for example zinc, according to the slides you presented. Could you say what elements might be exceptions to this rule?

C.L. OSTERBERG: The only radionuclide I know of that appears to be concentrated as it passes up the food chain is caesium-137, and the magnification factor is fairly small. Examination of the higher trophic levels for radionuclides shows the presence, above all, of cations, such as ^{54}Mn, ^{65}Zn, and ^{60}Co. Except for ^{137}Cs, there is no apparent evidence that the values are higher than at lower trophic levels.

J.C. ANCELLIN: How should we regard the data on phosphorus-32 published some 15 years ago in relation to the Columbia River site and often mentioned as an example of the increasing accumulation of a radionuclide with rising trophic levels?

C.L. OSTERBERG: The data you apparently refer to are controversial and, in my opinion, have not been borne out in subsequent analyses. Instead, it seems there is as much ^{32}P in lower trophic levels as in the higher forms of predators, or maybe even more. Admittedly, concentrations in the Columbia River were high, for the reasons I gave; the river is relatively pure, and free from stable phosphorus from sewage and farm run-off.

A.R. BERG: I agree about the scientific interest of studying the Columbia River after it has received discharges amounting to thousands of curies per year. However, it is not reasonable to conclude from the results that there is no need to worry about the possible proliferation of nuclear facilities. The results obtained on the Columbia site cannot be extrapolated since they were mostly concerned with ^{51}Cr and ^{65}Zn. The first of these radionuclides is not intensively accumulated by living organisms. The second, ^{65}Zn, is not magnified along the food chain since it undergoes a process of homeostatic regulation by fish; moreover, its physical half-life is short. We cannot predict what will happen when plutonium is discharged. This element might actually be concentrated along the food chain owing to its non-metabolic role, and its half-life is of a geological order of magnitude.

C.L. OSTERBERG: The Hanford reactors created essentially the same spectrum of fission products as do commercial power plants. The increase in neutron-induced radionuclides was due to direct single-pass cooling of the reactors with river water, a technique not used in commercial power plants. Although fission products were there, they were overwhelmed by similar radionuclides present in fall-out. I know of no radionuclide present in the effluent of a commercial power plant that is not also present in the Hanford effluent.

Your concern for plutonium is shared by many, and it is true that my paper did not refer to it. However, it is being studied with intense care, and I remain optimistic.

A. PRESTON: In this same connection I should like to point out to Mr. Berg that although Mr. Osterberg and I could necessarily mention only a few nuclides to illustrate the points we made, the same sort of information is available for a whole range of other radionuclides in widely differing environments; it is true to say that we know enough about the behaviour of these materials and their distribution in the environment to be able to use this information in conjunction with ICRP recommendations to carry out modelling in terms which are consistent with proper regulation. What Mr. Osterberg and I have said about caesium, zinc and chromium is in general also true in relation to other nuclides. There are also other reasons for excluding the biomagnification concept in relation to marine food chains. Considering the rate of accumulation of radionuclides from water and from food and the biological half-lives of nuclides accumulated either from water or from food, and taking a simple trophic pyramid, the animal that is acting as predator is physically incapable of achieving a rate of activity input into its system that could lead to biomagnification. Now much of this argument is too detailed to go into here, but a great deal of relevant information has been published, and I would refer Mr. Berg to the literature, in which these concepts have been modelled in mathematical terms.

V.V. PISAREV: How quickly does the self-purification process take place in the river after the reduction in the discharge of long-lived products?

C.L. OSTERBERG: Most of the ^{51}Cr and ^{65}Zn are now gone from the river. Probably ^{60}Co and ^{137}Cs are the dominant gamma-emitters remaining. These are found mostly in the fine muds on the bottom.

BEHAVIOUR OF RADIOACTIVE CONTAMINANTS AND THEIR STABLE NATURAL ISOTOPES IN AQUATIC SYSTEMS

(Sessions II and III)

Chairmen

C.L. OSTERBERG (USA)

M. SAIKI (Japan)

ETUDE DE L'INFLUENCE DU CADMIUM STABLE SUR LE TRANSFERT DU ZINC-65 EN ECOSYSTEME IRRIGUE PAR SUBMERSION (RIZIERE IRRIGUEE)

C. MYTTENAERE, Margaret MERLINI, R. BITTEL*,
P. DABIN, J.M. MOUSNY, G. POZZI
Commission des Communautés européennes,
Euratom, Centre commun de recherches,
Ispra (Varèse), Italie

Abstract–Résumé

STUDY OF THE INFLUENCE OF STABLE CADMIUM ON THE TRANSFER OF ZINC-65 IN AN ECOSYSTEM IRRIGATED BY SUBMERSION (IRRIGATED RICE FIELD).

Irrigation water contains varying amounts of stable cadmium from industrial disposal. The presence of this element is capable of modifying the transfer of and affecting the mechanisms of absorption of zinc-65, a radionuclide which is discharged into water by nuclear power stations. This type of interaction between a nuclear contaminant and a conventional contaminant was studied in a rice-field irrigated by submersion. This ecosystem was reproduced under controlled 'mini-rice-field' conditions: the water was enriched in stable zinc (1 ppm) and in stable cadmium (2.5×10^{-3}; 5×10^{-3}; 50×10^{-3} ppm) so as to reproduce the actual conditions; the stable zinc was traced by means of zinc-65. Fish (20 *Carassius auratus* L.) were introduced into each compartment of the ecosystem. Samples of irrigation water and surface water were taken during cultivation, and at the end of cultivation the content of stable zinc, radioactive zinc and stable cadmium was determined in each component of the ecosystem. The results show the important influence of a conventional contaminant on the transfer of a radionuclide and justify the use of the term 'associate contaminant' to describe it.

ETUDE DE L'INFLUENCE DU CADMIUM STABLE SUR LE TRANSFERT DU ZINC-65 EN ECOSYSTEME IRRIGUE PAR SUBMERSION (RIZIERE IRRIGUEE).

Les eaux d'irrigation contiennent des quantités variables de cadmium stable provenant des rejets industriels. La présence de cet élément est susceptible de modifier le transfert et d'influencer les mécanismes d'absorption du zinc-65, radionucléide rejeté dans les eaux par les centrales nucléaires. Une telle interaction entre un polluant nucléaire et un polluant conventionnel a été étudiée en rizière irriguée par submersion. A cette fin, cet écosystème fut reproduit en conditions contrôlées («mini-rizières»); l'eau fut enrichie en zinc stable (1 ppm) et en cadmium stable ($2,5 \cdot 10^{-3}$; $5 \cdot 10^{-3}$ et $50 \cdot 10^{-3}$ ppm) de façon à reproduire les conditions réelles; le zinc stable fut tracé à l'aide de zinc-65. Des poissons (20 *Carassius auratus* L.) furent introduits dans chacun des compartiments de l'écosystème. Des prélèvements d'eau d'irrigation et de surface furent effectués en cours de culture et, en fin de culture, les teneurs en zinc stable et radioactif et en cadmium stable de chaque composant de l'écosystème furent déterminées. L'examen des résultats permet de souligner l'importance de la présence d'un polluant conventionnel sur le transfert d'un radionucléide et justifie dans le cas du premier l'appellation de «polluant associé».

INTRODUCTION

Parmi les radionucléides présents dans les effluents liquides des installations nucléaires, le zinc-65 paraît actuellement l'un des plus importants, en raison des activités rejetées, de ses caractéristiques nucléaires, de son état physico-chimique assez complexe qui lui permet d'échapper en partie aux mécanismes artificiels et naturels d'épuration, enfin, de la place de l'élément zinc parmi les oligoéléments essentiels pour les êtres vivants.

---

* CEA, Département de protection, Association Euratom/CEA, Fontenay-aux-Roses, France

Dans la liste des substances chimiques plus ou moins défavorables à la vie, le cadmium a une place particulière en raison de sa toxicité, de la multiplicité des sources potentielles de pollution, de sa très grande mobilité dans l'environnement et de sa parenté chimique avec des oligoéléments indispensables, le zinc en particulier [1-5]. Dans la nature, le cadmium accompagne d'ailleurs le zinc stable et les seules manifestations de toxicité ayant conduit à de très graves préjudices à la santé des populations ont été les conséquences de rejets, dans les eaux de surface, d'eaux d'exhaure de mines de zinc [4, 5]. Il paraît donc intéressant de considérer le cadmium comme un polluant chimique associé au zinc stable et au zinc-65.

Parmi les écosystèmes concernés par d'éventuelles pollutions en cadmium stable et en zinc-65 figurent les milieux aquatiques et les milieux irrigués. Les écosystèmes irrigués par submersion, et en particulier les rizières, méritent une attention spéciale, en raison des quantités énormes d'eau utilisées; de plus, l'expérience a montré que la submersion créait des conditions favorables à l'absorption des éléments au niveau des racines superficielles et de la base des tiges [6].

Il est assez fréquent que les rizières submergées soient utilisées pour l'élevage de certains poissons [7]. C'est ce qui est encore réalisé en Italie du nord, où les rizières servent parfois à l'élevage de carpes, et ce qui connaît un développement considérable dans certaines régions tropicales. Les rizières constituent donc des écosystèmes dulçaquicoles assez complets pour lesquels se posent des problèmes de bilan des éléments nutritifs ou polluants entre l'eau, les sols, les végétaux et des animaux aquatiques.

L'objet du présent mémoire est donc d'étudier le prélèvement du zinc-65 par le riz et le poisson, le paramètre principal étant la teneur des eaux d'irrigation en cadmium et en zinc stable.

1. METHODOLOGIE

En vue d'imiter, dans la mesure du possible, les conditions naturelles réalisées dans la région rizicole de la plaine du Pô en Italie, on a eu recours à un modèle de rizière constitué par un ensemble de bacs remplis d'un sol prélevé près de Vercelli (Italie) et sur lequel on a semé du riz, variété Arborio, à raison de six poquets par bac. On a réalisé, un mois après le semis, une irrigation par submersion, à raison de 150 litres par jour durant quinze semaines, en utilisant de l'eau du lac Majeur, l'eau circulant du bac le plus élevé A au bac le plus bas F (tableau I). On a enrichi l'eau en zinc stable de façon à obtenir une concentration initiale (bac supérieur) égale soit à 5 mg/l (expériences de 1973), soit à 1 mg/l (expériences de 1974). Comme traceur du zinc stable on a utilisé ^{65}Zn (1,6 μCi en 1973 et 5 μCi en 1974 par mg de zinc stable). On a, d'autre part, enrichi l'eau en cadmium stable (marqué au ^{109}Cd en 1973), de manière à obtenir dans l'eau d'irrigation initiale soit $2,5 \cdot 10^{-3}$ (essai I), soit $5 \cdot 10^{-3}$ (essai II), soit $50 \cdot 10^{-3}$ mg/l (essai III), ces concentrations étant choisies en fonction des limites recommandées pour les eaux d'irrigation, selon qu'il s'agit d'irrigations intenses ou continues ($5 \cdot 10^{-3}$ ppm Cd) ou d'irrigations discontinues ($50 \cdot 10^{-3}$ ppm Cd) [8].

Au cours des essais réalisés en 1974, on a introduit des cyprinidés dans chacun des bacs (<u>Carassius auratus</u>), à raison de cinq par bac, et l'expérimentation a été poursuivie durant cinquante-six jours.

TABLEAU I. CARACTERISTIQUES CHIMIQUES DE L'EAU DU LAC MAJEUR

Conductivité (μmho)	100 - 140
pH	7,2 - 7,8
Ca (ppm)	19 - 23
Co (ppm)	10^{-4}
K (ppm)	1,7
Mg (ppm)	3 - 3,8
Sr (ppm)	0,17 - 0,26
Na (ppm)	2
Fe total (ppm)	0,02
Zn (ppm)	0,04
Mn (ppm)	0,02
Cu (ppm)	0,02

Les mesures de radioactivité ont été réalisées par spectrométrie gamma, les éléments stables ont été déterminés par absorption atomique, les destructions de matière organique sur les échantillons biologiques ont été effectuées par voie humide (acide nitrique-perhydrol).

2. RESULTATS EXPERIMENTAUX

2.1. Eau circulant sur les bacs

Comme on pouvait le prévoir, les teneurs de l'eau circulant dans les bacs diminuent progressivement de l'amont vers l'aval, qu'il s'agisse du zinc stable, du cadmium stable ou du zinc-65. D'autre part, l'élimination du zinc est d'autant plus intense en valeur relative que la teneur de l'eau en cadmium est plus élevée (tableau II). Il en est de même pour le zinc-65 (tableau III).

2.2. Sols

Comme on pouvait également le prévoir, les couches les plus superficielles (0 à 2 cm) des sols des différents bacs fixent intensément le zinc et le cadmium, ces éléments s'infiltrant peu dans les couches plus profondes (2 à 4 cm).

2.3. Riz

2.3.1. Remarques générales

En ce qui concerne les expérimentations réalisées en 1973, on a constaté pour le zinc-65 un transfert modéré du zinc-65 absorbé, de l'eau

TABLEAU II. TENEURS EN ZINC STABLE DE L'EAU DES DIFFERENTS BACS (origine, A à F) (ppm)

Essai	Origine	A	B	C	D	E	F
I	1,06	1,11	0,81	0,64	0,52	0,36	0,27
II	1,02	0,91	0,67	0,44	0,28	0,20	0,18
III	0,92	0,89	0,52	0,34	0,25	0,18	0,18

TABLEAU III. TENEURS EN ZINC-65 DE L'EAU DES DIFFERENTS BACS (origine, A à F) (unités arbitraires)

Essai	Origine	A	B	C	D	E	F
I	100	105	80	60	48	30	20
II	100	94	60	40	24	16	16
III	100	90	56	33	21	14	13

à la plante, qui augmente assez régulièrement quand on passe du bac amont au bac aval. On remarque, d'autre part, que la migration du cadmium absorbé vers le grain, en valeur relative, est à peu près constante de l'amont vers l'aval. Elle est plus forte que celle du zinc-65 pour les bacs amont et lui est sensiblement égale pour les bacs aval.

En 1974, on a suivi le transfert du seul zinc-65. On a constaté qu'une fraction relativement importante de la quantité totale absorbée par la plante migre dans le grain, cette migration étant en valeur relative d'autant plus intense qu'on progresse vers l'aval. Ainsi, pour l'essai I, l'albumen des graines contient 3,6%, 8,2%, ou 11,8% du zinc-65 absorbé suivant qu'on se trouve dans le bac le plus élevé (A), un bac intermédiaire (C) ou le bac inférieur (E). Des constatations analogues peuvent être faites dans les cas des essais II et III. Pour des bacs placés en position identique, on ne constate aucune différence vraiment significative pour les teneurs en ^{65}Zn des graines entre les essais I et II (en moyenne pour l'essai II (à $5 \cdot 10^{-3}$ ppm Cd), la migration moyenne pour le zinc-65 est légèrement plus grande). Par contre, pour l'essai III, avec $50 \cdot 10^{-3}$ ppm de cadmium, on constate une migration relative plus faible.

Il convient maintenant de dégager l'incidence de la variabilité des teneurs en cadmium et en zinc de l'eau d'irrigation sur celle des résultats obtenus.

2.3.2. Incidence de la teneur en cadmium de l'eau

Si on schématise les résultats en calculant les facteurs F de concentration eau d'irrigation → plantes, on constate (tableau IV, expériences de 1974) que,

TABLEAU IV. FACTEURS DE CONCENTRATION ENTRE L'EAU ET L'ALBUMEN DU GRAIN DE RIZ POUR ^{65}Zn (1974)

Essai \ Bac	A	B	C	D	E
I	20	25	21	22	30
II	23	23	28	23	28
III	23	14	15	13	12

pour ^{65}Zn, ceux-ci sont compris entre 20 et 30 pour les traitements I et II ($2,5 \cdot 10^{-3}$ et $5 \cdot 10^{-3}$ ppm Cd) et qu'ils sont compris entre 12 et 23 pour le troisième traitement ($50 \cdot 10^{-3}$ ppm Cd). Le cadmium n'abaisse donc le facteur de concentration du zinc-65 qu'à partir d'un niveau relativement élevé, en tout cas supérieur à $5 \cdot 10^{-3}$ ppm.

2.3.3. Incidence de la teneur en zinc stable de l'eau

La confrontation des résultats obtenus, d'une part en 1973 avec une eau à 5 ppm Zn stable (activité spécifique du zinc: 1,6 mCi/mg), d'autre part en 1974 avec une eau à 1 ppm Zn stable, $5 \cdot 10^{-3}$ ppm Cd stable (activité spécifique du zinc: 0,5 mCi/μg Zn stable), montre l'incidence de la concentration de l'eau en zinc stable: pour une concentration initiale de l'eau en zinc stable égale à 1 ppm (1974), le transfert du zinc-65 de l'eau au caryopse est de deux à quatre fois, en valeur relative, plus intense que pour 5 ppm de zinc, ce qui indique que, dans les présentes expérimentations, un équilibre isotopique semble être réalisé, au moins partiellement, ce qui n'est pas toujours le cas.

2.3.4. Facteurs de concentration

Il est commode d'évaluer les transferts eau d'irrigation → organismes aquatiques en termes de facteur de concentration F (tableau IV)

$$F = \frac{\text{concentration ou activité du tissu (en } \mu g/g \text{ ou pCi/g)}}{\text{concentration ou activité de l'eau (en } \mu g/g \text{ ou pCi/g)}}$$

2.4. Poissons (Carassius auratus)

2.4.1. Dynamique de l'échange du zinc

En présence de faibles teneurs en cadmium de l'eau (essai II à $5 \cdot 10^{-3}$ ppm), le facteur de concentration du zinc-65 est pratiquement indépendant de la teneur en zinc de l'eau (essai à 1 ppm, soit en réalité: 0,67 à 0,20 ppm Zn stable), avec une moyenne de F = 75 (tableau V).

Il est intéressant de comparer cette valeur avec celles obtenues dans des essais antérieurs dans un intervalle physiologique de concentration en zinc (0,01 à 0,05 ppm) où existe une relation inverse entre le facteur de

TABLEAU V. RESULTATS RELATIFS AUX POISSONS

Essai	Bac	Zn/eau (ppm)	F_{Zn-65}	Zn/poisson (ppm)	Cd/eau (10^{-3} ppm)	F_{Cd}
II	B	0,67	83	72	3	136
	C	0,44	95	74	2	265
	D	0,28	52	65	1,5	193
	E	0,20	72	81	1	490
III	B	0,52	67	73	26	73
	C	0,34	68	81	17	79
	D	0,25	76	69	12	115
	E	0,18	67	82	9	80

TABLEAU VI. FACTEURS DE CONCENTRATION DU ZINC STABLE ENTRE L'EAU ET LE POISSON A DIFFERENTES CONCENTRATIONS DE ZINC DANS L'EAU

Concentrations en Zn de l'eau (ppm)		F
Physiologiques	0,01	6 000
	0,05	1200
Expérimentales	0,20	375
	0,67	112

concentration du zinc et la teneur dans l'eau (tableau VI). En terme de zinc stable, l'échange du zinc est constant dans l'intervalle physiologique, alors qu'il est proportionnel à la concentration de l'élément dans l'eau entre 0,20 et 0,67 ppm. Il en résulte des variations concomitantes de la période biologique chez Carassius (218 jours jusqu'à moins de 50 jours).

2.4.2. Transfert du cadmium

Pour $5 \cdot 10^{-3}$ ppm de cadmium introduit dans l'eau, la concentration en cadmium du poisson est de l'ordre de 0,3 à 0,5 ppm et se rapproche de celles observées par ailleurs pour Lepomis gibbosus en l'absence de cadmium introduit, l'adsorption du cadmium par les sols ayant abaissé les concentrations à des valeurs très faibles. Dans les essais à $50 \cdot 10^{-3}$ ppm de cadmium dans l'eau, les teneurs en cadmium du poisson passent à des valeurs plus élevées (0,7 à 1,9 ppm), en relation directe avec la concentration dans l'eau. En l'absence de mesures directes de la concentration en cadmium de l'eau, on pourrait tenter d'estimer cette concentration sur la base des teneurs décroissantes en zinc, observées dans les deux séries (essais II et III), de l'amont vers l'aval, par suite de l'adsorption par le sol. Les valeurs du facteur de concentration en cadmium (tableau V) qui en découlent sont du même ordre de grandeur que celles obtenues dans un autre essai pour Alburnus alborella, c'est-à-dire de l'ordre de quelques centaines aux faibles concentrations du cadmium dans l'eau (10^{-3} ppm) pour descendre au-dessous de 100 aux concentrations supérieures à $10 \cdot 10^{-3}$ ppm. Il convient de noter, pour le cadmium comme pour le zinc, une relation inverse du facteur de concentration avec la teneur de l'eau en éléments aux faibles concentrations (bac B: Cd 3 μg/l, F = 136; bac E: Cd 1 μg/l, F = 490). Par contre, aux teneurs de l'eau en cadmium supérieures (9 à 26 μg/l), ce facteur de concentration reste à peu près constant (compris entre 73 et 115). Etant donné que l'état d'équilibre entre le poisson et l'eau pour le cadmium est certainement atteint en 56 jours, ces valeurs de F représenteraient des valeurs d'équilibre, nettement inférieures à celles obtenues pour le zinc.

2.5. Interférence cadmium-zinc

On dispose des essais II et III poursuivis en 1974 ($5 \cdot 10^{-3}$ et $50 \cdot 10^{-3}$ ppm Cd introduit). Dans ces intervalles de concentrations en cadmium, il n'existe pas d'interférence du cadmium sur le zinc, alors que les essais antérieurs réalisés dans l'intervalle physiologique du zinc indiquent une très nette interférence de cadmium sur le zinc. Il faut en conclure que les fortes teneurs de zinc utilisées ici empêchent la manifestation des phénomènes de compétition entre le cadmium et le zinc, facilement décelables au contraire dans l'intervalle physiologique.

3. DISCUSSION ET CONCLUSION

On pourrait s'attendre à une interférence quantitativement importante du cadmium stable vis-à-vis du zinc-65. Les résultats précédents montrent en fait que, dans les conditions expérimentales, le cadmium a une action assez complexe, mais moins importante que prévue sur le comportement du zinc dans l'écosystème artificiel qui a été réalisé. Les résultats les plus importants à souligner sont les suivants:

- l'élimination du zinc à partir de l'eau, d'autant plus intense que la teneur de l'eau est plus élevée
- l'incidence relativement faible de la teneur en cadmium des eaux sur le transfert du zinc-65 de l'eau à l'albumen du grain de riz
- l'incidence importante de la teneur en zinc stable de l'eau sur l'absorption du zinc-65 par le riz et sur sa mobilité dans la plante (transfert vers le grain).

En ce qui concerne le poisson, la teneur en zinc stable de l'eau conditionne également l'accumulation du zinc par l'organisme, tout au moins dans l'intervalle physiologique (0,01 à 0,05 ppm Zn dans l'eau). Au contraire, dans l'intervalle expérimental utilisé (0,20 - 0,67 ppm), le transfert du zinc-65 dépend peu de la teneur en zinc stable de l'eau (tableau V).

En ce qui concerne les débits de dose délivrée à l'homme du fait de l'ingestion de riz provenant de rizières contaminées, un calcul classique montre que, pour le zinc-65, si on admet un facteur de transfert eau → grain de l'ordre de 20, lorsque l'activité spécifique de l'eau est de 1 pCi/cm^3 et pour une consommation annuelle de 1 kg de riz, le débit de dose est de l'ordre de 120 μrem/an. Rappelons ici qu'une évaluation identique conduit pour le césium-137 à 1 mrem/an. En fait, pour des populations rurales des zones rizicoles, où on peut consommer jusqu'à 20 kg de riz par an et par individu, le débit de dose pour ^{65}Zn serait de 3 mrem/an. L'ingestion de 1 kg de riz par an implique, si on admet un facteur de concentration pour le cadmium de l'ordre de 100, l'ingestion de 100 · x (μg Cd), x étant la concentration du cadmium (μg/l) dans l'eau d'irrigation. Si on admet provisoirement que la quantité de cadmium ingérée en moyenne par an et par individu, de l'ordre de 20 mg, représente la dose maximale, le niveau limite dérivé pour une consommation de 1 kg de riz serait de 200 μg/l; pour une consommation de 20 kg, elle serait de 10 μg/l, c'est-à-dire de l'ordre de la limite actuellement admise pour les eaux servant à une irrigation continue (5 μg/l).

Les mêmes considérations peuvent être faites dans les cas de consommation de carpes «élevées» en rizières. Si on admet que les facteurs de concentration trouvés expérimentalement ici ($F_{Zn} \simeq 70$, $F_{Cd} \simeq 100$) sont valables in situ, on obtient, pour une consommation individuelle annuelle de 20 kg de poisson, un débit de dose dû au ^{65}Zn égal à 10 mrem/an pour 1 pCi/cm^3 d'eau, et une concentration limite pour le cadmium égale à 2 μg/l.

En conclusion, cette étude, réalisée sur un écosystème assez complet (eau, sol, végétaux, poissons), montre l'intérêt qu'il y a à envisager simultanément le cas d'éléments aussi voisins que le zinc et le cadmium. Ce type de recherches devra être poursuivi par l'étude des microlocalisations des deux éléments, en fonction de paramètres écologiques et toxicologiques, en particulier de la présence, dans les eaux, de composés capables de modifier les échanges transmembranaires.

REFERENCES

[1] BROWNING, E., Toxicity of Industrial Metals, Butterworth, London (1967).
[2] LEE, D.H.K., Metallic Contaminants and Human Health, Academic Press, New York (1972).
[3] UNDERWOOD, E.J., Trace Elements in Human and Animal Nutrition, Academic Press, New York (1971).
[4] FRIBERG, L., PISCATOR, M., NORDBERG, G., Cadmium in the Environment, Chemical Rubber, Cleveland (1971).
[5] FULKERSON, W., GOELLER, H.E., Cadmium, The Dissipated Element, USAEC Rep. ORNL-NSF-EP 21 (1973).
[6] MYTTENAERE, C., BOURDEAU, Ph., «Quelques aspects du problème de la radiocontamination des cultures irriguées», La radioprotection du milieu devant le développement des utilisations pacifiques de l'énergie nucléaire, Société française de radioprotection, Paris (1968) 289-305.
[7] HICKLING, O.F., Fish Culture, Faber and Faber, London (1971).
[8] UNITED STATES DEPARTMENT OF INTERIOR, Federal Water Pollution Central Administration, Report of the Committee on Water Quality Criteria, Washington (1968).

DISCUSSION

S. BONOTTO: You showed that if the zinc concentration in the water increases, the concentration factor for fish decreases. Do the higher concentrations employed have a toxic effect on fish?

R. BITTEL: In the 'biological' domain the transfer factors are inversely proportional to the concentrations or activities in the water, owing to a regulatory phenomenon. At the experimental concentrations, although no toxicity is observed, the regulatory mechanism is probably not operative; these are sub-toxic concentrations.

A.R. BERG: The physiological range of Zn in which we normally work in our laboratory is between 20 and 50 ppb. In this range we observe homeostatic regulation of Zn in fish, and the exchange of stable Zn is independent of its concentration in the water. In this range, the Zn concentration in the fish is constant.

We have also detected a sub-lethal effect on fish for Zn concentrations starting at 500 ppb, i.e. in the range used for the work you report in your paper. In this case, however, we find no homeostatic Zn regulation in the fish, the Zn exchange is proportional to its concentration in the environment, and the Zn concentration in the fish is slightly higher than normal.

ETUDE DE LA CONTAMINATION DE CONSTITUANTS DU MILIEU MARIN PAR DES FORMES SOLUBLES ET INSOLUBLES DE RADIONUCLEIDES

A. FRAIZIER, J.C. ANCELLIN
CEA, Laboratoire de radioécologie marine,
Cherbourg, France

Abstract–Résumé

STUDY OF THE CONTAMINATION OF COMPONENTS OF THE MARINE ENVIRONMENT BY SOLUBLE AND INSOLUBLE FORMS OF RADIONUCLIDES.

The experimental contamination of various physical components and organisms of the marine environment was carried out using radionuclides such as ^{137}Cs, ^{51}Cr, ^{60}Co, ^{106}Ru and ^{59}Fe. The relationships between the physico-chemical states of the radionuclides, the variations in the environmental conditions, and the properties of the experimental samples were clarified. Marine organisms were more readily contaminated by the insoluble forms of ^{106}Ru and ^{59}Fe than by the soluble forms. It appears that the physiology of the marine organisms can have a bearing on the degree and evolution of the contamination whatever the physico-chemical state of the radionuclides may be, but in certain circumstances the contamination level is independant of the variations in environmental conditions and the related variations in the physiology of the organism.

ETUDE DE LA CONTAMINATION DE CONSTITUANTS DU MILIEU MARIN PAR DES FORMES SOLUBLES ET INSOLUBLES DE RADIONUCLEIDES.

La contamination expérimentale de divers constituants physiques et biologiques du milieu marin a été réalisée au moyen de radionucléides tels que ^{137}Cs, ^{51}Cr, ^{60}Co, ^{106}Ru, ^{59}Fe, et a permis de préciser les relations entre les caractéristiques physico-chimiques du contaminant, les variations de conditions de milieu et les propriétés des différents supports contaminés. Il se confirme en particulier que les formes insolubles de ^{106}Ru et de ^{59}Fe ont, pour les organismes, un pouvoir contaminant supérieur à celui de leurs formes solubles. Par ailleurs le degré et l'évolution de la contamination peuvent être en rapport avec la physiologie des organismes quelle que soit la forme physico-chimique des radionucléides. Dans d'autres circonstances le niveau de contamination demeure indépendant des variations des conditions de milieu et des variations concomitantes de la physiologie des organismes.

INTRODUCTION

Les données concernant le pouvoir d'accumulation des espèces marines vis-à-vis des éléments présents naturellement à l'état de traces dans le milieu ne sont pas forcément transposables au cas des éléments surajoutés, provenant des activités humaines. En effet, alors que les composants naturels du milieu se présentent en état d'équilibre physico-chimique avec celui-ci, il n'en est pas nécessairement de même pour les éléments surajoutés, entrant dans la catégorie des polluants ; suivant les cas des cycles géochimiques différents seraient à prendre en considération [1] [2] [3] [4].

Il doit donc être tenu tout particulièrement compte de cette différence de comportement en ce qui concerne la fixation des radioéléments par les espèces marines : en effet, si la mise à l'équilibre des diverses formes physico-chimiques est lente, l'accumulation, la distribution et le devenir des radioéléments dans la biosphère marine peuvent ne pas être identiques à ceux de leurs isotopes stables. Ceci implique que les recherches concernant les

TABLEAU I. ETUDE DES FORMES PHYSICO-CHIMIQUES DU CESIUM 137 EN MILIEU EXPERIMENTAL.

Formes physico-chimiques / Délais d'observation	Anioniques ou neutres	Indéterminées	Cationiques	Insolubles
1 heure	0	0	100 %	0
2 mois	0	0	100 %	0

TABLEAU II. ETUDE DES FORMES PHYSICO-CHIMIQUES DU COBALT 60 EN MILIEU EXPERIMENTAL

Formes physico-chimiques / Délais d'observation	Anioniques ou neutres	Indéterminées	Cationiques	Insolubles
1 heure	2 %	13 %	82 %	3 %
2 mois	33 %	5 %	48 %	14 %

modalités de contamination soient précédées d'études relatives à la chimie des radioéléments en milieu marin, et que soit précisé le rôle des formes physico-chimiques de ces radioéléments.

1. ETUDE DES FORMES PHYSICO-CHIMIQUES DE RADIOELEMENTS EN MILIEU MARIN

Pour l'analyse de formes physico-chimiques de radioéléments en milieu marin expérimental nous avons employé la méthode de chromatographie sur résine échangeuse d'ions décrite par GUEGUENIAT [5].

Cette méthode permet la séparation et l'étude :
- des formes solubles cationiques
- des formes solubles anioniques ou neutres et de complexes organiques
- des formes insolubles colloïdales et particulaires.

Nous rappellerons dans ce qui suit à titre d'exemple quelques-unes des données obtenues par cette méthode et relatives à différents radionucléides : ^{137}Cs, ^{60}Co, ^{51}Cr, ^{106}Ru, ^{59}Fe.

1.1. Cas du césium 137, du cobalt 60 et du chrome 51

Dans les tableaux I, II et III, sont présentés les résultats d'analyses (composition en pourcentages des différentes formes) relatifs au ^{137}Cs, au ^{60}Co et au ^{51}Cr.

1.2. Cas du ruthénium 106

Une analyse de ^{106}Ru en milieu acide HNO_3 6 N, sous forme de nitrato complexes peu stables et facilement hydrolysables :

$$RuNO(NO_3)_x\ (OH)_y\ (H_2O)_z \quad \text{avec } x + y + z = 5$$

a permis de mettre en évidence les formes suivantes et leur proportion :

- formes anioniques di nitrato (une charge positive) : 45 %
- formes cationiques mono nitrato (deux charges positives) : 10 %
- formes polymérisées : 3 %.

1.3. Cas du fer 59

En ce qui concerne le ^{59}Fe introduit en milieu expérimental les analyses ont montré que les formes insolubles représentaient initialement un pourcentage important (80 à 90 %) ; ce pourcentage se réduisait dans le temps en faveur de formes anioniques ou neutres (fig. 1).

TABLEAU III. ETUDE DES FORMES PHYSICO-CHIMIQUES DU CHROME 51 EN MILIEU DE CULTURE DE L'ALGUE DUNALIELLA B.

Délais d'observation	Formes physico-chimiques	Milieu de culture	Milieu sans algues
1 heure	Solubles	96,6 %	94,2 %
	Insolubles	3,4 %	5,8 %
24 heures	Solubles	97,4 %	93,6 %
	Insolubles	2,6 %	6,4 %

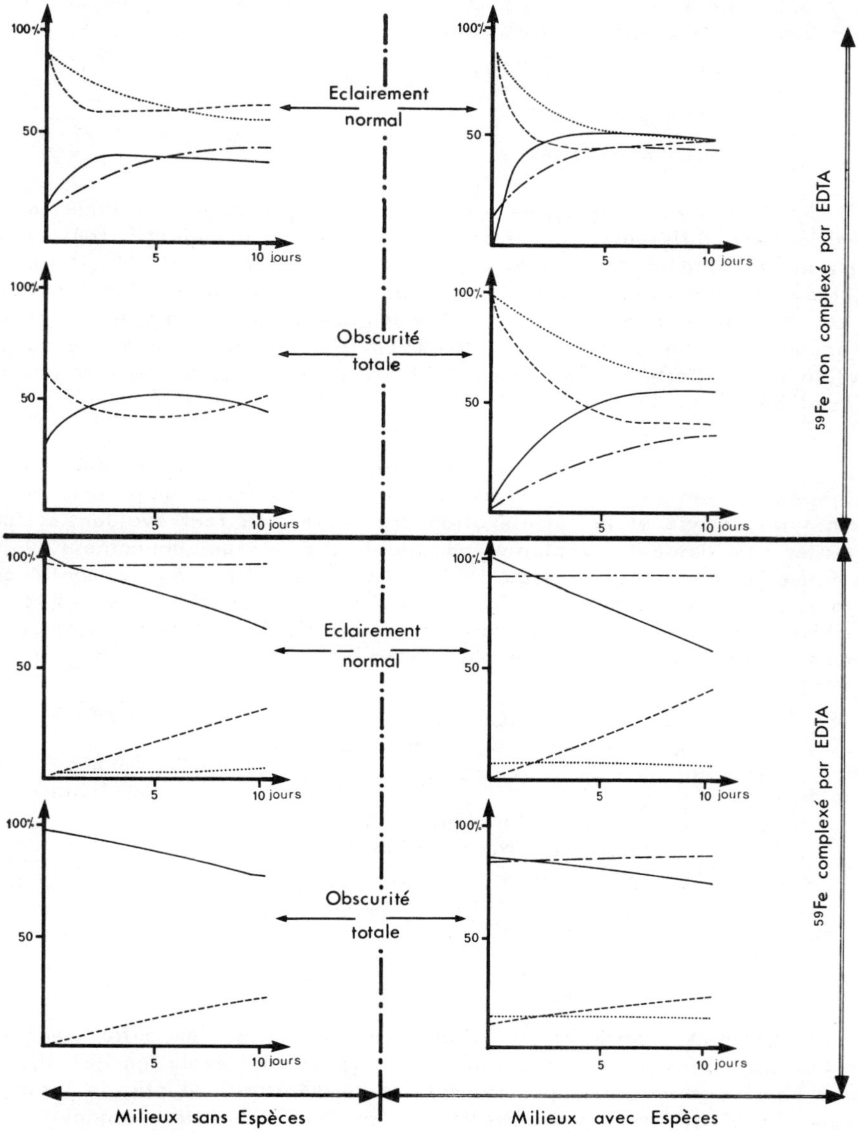

FIG. 1. Evolution dans le temps des pourcentages des formes insolubles et solubles du fer-59 présentes dans l'eau des aquariums pour différentes conditions expérimentales (présence ou non d'espèces, température 15°C ou 5°C, éclairement normal ou obscurité totale, contaminations par des formes insolubles ou solubles).
Formes insolubles ---- (à 15°C) .... (à 5°C).
Formes solubles ——— (à 15°C) — - - (à 5°C).

Ces observations correspondent bien aux constatations de ROBERTSON [3] qui indique que les éléments présents à l'état de traces dans le milieu marin peuvent se classer en trois catégories :

- formes en solution vraie
- formes colloïdales
- formes particulaires.

On notera qu'un même élément peut se présenter simultanément dans le milieu sous différentes formes dont les pourcentages respectifs sont susceptibles de varier dans de grandes proportions dans le temps si l'équilibre physico-chimique avec le milieu n'est pas réalisé : ainsi le ^{59}Fe introduit en eau de mer se transforme très rapidement en composés insolubles, phénomène en accord avec les constatations d'HARVEY [6] ; après une phase de précipitation se manifeste une tendance à la mise en équilibre des formes solubles et insolubles (fig. 1).

Deux processus principaux gouvernent les modifications des formes physico-chimiques des éléments présents à l'état de traces dans le milieu marin : l'hydrolyse et la complexation organique ; ils font évoluer les formes présentes vers des états insolubles ou solubles. En ce qui concerne l'évolution des formes physico-chimiques du ^{106}Ru, GUEGUENIAT [5] a montré que l'hydrolyse de ce radioélément fait apparaître à un certain stade des réactions de polymérisation suivies de la formation de colloïdes ; l'auteur propose le schéma suivant pour illustrer ce phénomène :

Ru sol. $\rightleftarrows$ A $\rightleftarrows$ B $\rightleftarrows$ C $\rightleftarrows$ Monomères $\longrightarrow$ Polymères

(Formes solubles) — (Produits d'hydrolyse) — (Formes insolubles)

Formes dites dans un état intermédiaire de solubilité
A - fortement solubles
B - moyennement solubles
C - peu solubles

Certains paramètres du milieu physique peuvent en outre jouer un rôle au point de vue de l'importance relative et de l'évolution des différentes formes physico-chimiques : par exemple la température, le pH, la force ionique du milieu et la concentration de celui-ci en matière organique complexante [7]. Nous avons constaté pour notre part (fig. 1) que l'évolution des formes physico-chimiques du fer 59 étaient liées à la température et dans une moindre mesure à l'éclairement. L'augmentation de température (passage de 5°C à 15°C) et la présence d'un éclairement normal (en opposition à l'obscurité totale) favorisent vraisemblablement les réactions de polymérisation, la coagulation des colloïdes et par suite la précipitation du fer. En ce qui concerne des composés organiques solubles du fer 59 (obtenus par chélation par l'EDTA), ces facteurs physiques doivent intervenir dans les destructions des complexes.

2. INFLUENCE DES FORMES PHYSICO-CHIMIQUES DE RADIOELEMENTS AU POINT DE VUE DE LA CONTAMINATION D'ESPECES MARINES

Différents procédés - séparation sur résines échangeuses d'ions [8] ou complexation organique [9] - ont été utilisés pour préparer des solutions contaminantes comportant une forte proportion soit de formes solubles, soit de formes insolubles, et au moyen desquelles on a pu suivre la cinétique et le taux de contamination d'espèces marines en fonction de la forme physico-chimique du contaminant.

Des résultats relatifs au ^{106}Ru et au ^{59}Fe sont présentés dans les tableaux IV et V et figures 2 et 3.

Ces résultats montrent que les formes polymérisées sont nettement plus contaminantes que les formes solubles.

En ce qui concerne la cinétique d'accumulation, on constate que celle-ci est plus rapide pour des formes polymérisées du ruthénium que pour des formes solubles, pour autant que ce phénomène puisse être mis en évidence en raison de la formation de composés insolubles au cours du temps dans le contaminant se présentant initialement sous forme soluble [8].

En ce qui concerne le ^{59}Fe, là encore, le taux de fixation du radio-élément est plus élevé lorsque celui-ci se présente sous forme insoluble (Tableau IV).

Les cinétiques d'accumulation montrent des différences marquées suivant qu'il s'agit de formes solubles ou insolubles de ce radioélément (fig. 3). On note pour les trois supports étudiés qu'une fixation initiale très rapide s'opère pour les formes insolubles, l'accumulation des formes solubles étant pour sa part plus progressive. En ce qui concerne l'accumulation des formes insolubles par la coquille l'état d'équilibre est rapidement atteint, alors que pour la chair de Moule et la Blennie le niveau d'accumulation du radioélément après avoir atteint un maximum montre une décroissance plus ou moins importante.

Les résultats d'une étude de la répartition de la contamination par le ruthénium 106 et le fer 59 dans les organes de différentes espèces sont présentés dans les tableaux VI, VII et VIII.

Au point de vue de la contamination des organes par le fer 59, aussi bien pour les formes solubles que pour les formes insolubles, on observe une radioactivité plus élevée au niveau d'organes qui sont le siège d'activité respiratoire ou de transit alimentaire. Par ailleurs la contamination par les formes insolubles du fer 59 est relativement favorisée au niveau d'organes comme le manteau chez *Mytilus edulis* et la peau chez *Blennius pholis*, ce qui suggère l'existence d'un mode passif de fixation. Pour le ruthénium il semble que les formes solubles puissent être accumulées au niveau d'organes internes alors que la contamination par des formes polymérisées insolubles intéresserait seulement les organes en contact avec le milieu extérieur.

TABLEAU IV. INDICES DE CONCENTRATION DE DIFFERENTES ESPECES MARINES POUR DES FORMES SOLUBLES ET POLYMERISEES DU RUTHENIUM 106.

Espèces	Indices de concentration (Ic)[a]		Pouvoir contaminant relatif (Pc)[b]
	Formes solubles	Formes polymérisées	
Corallines	89	428	4,8
Chondrus	40	166	4,1
Crevettes	0,63	4,8	7,6
Moules (chair)	3,9	35	8,9
Moules (coquille)	4,7	9,2	1,9
Plies (*in toto*)	0,04	0,16	4

[a] $$Ic = \frac{\text{Activité spécifique moyenne de l'espèce}}{\text{Activité spécifique moyenne de l'eau}}$$ (pendant la durée de l'expérimentation)

[b] $$Pc = \frac{\text{Ic (pour des formes insolubles)}}{\text{Ic (pour des formes solubles)}}$$

TABLEAU V. INDICES DE CONCENTRATION DE DEUX ESPECES MARINES POUR DES FORMES SOLUBLES ET INSOLUBLES DU FER 59.

Espèces	Indices de concentration		Pouvoir contaminant relatif
	Formes solubles	Formes insolubles	
Moules (chair)	17,2	109	6,3
Moules (coquille)	8,2	32	3,9
Blennies (*in toto*)	1,8	5,9	3,2

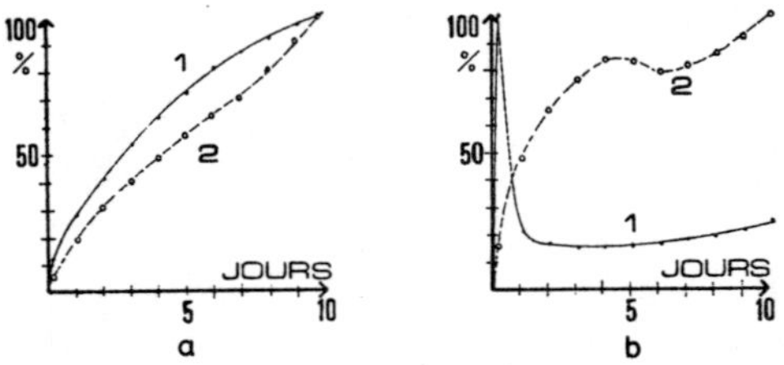

FIG.2. Evolution de la contamination de la coquille (a) et de la chair (b) de la moule (Mytilus edulis) par des formes polymérisées insolubles (courbes 1) ou solubles (courbes 2) du ruthénium-106.

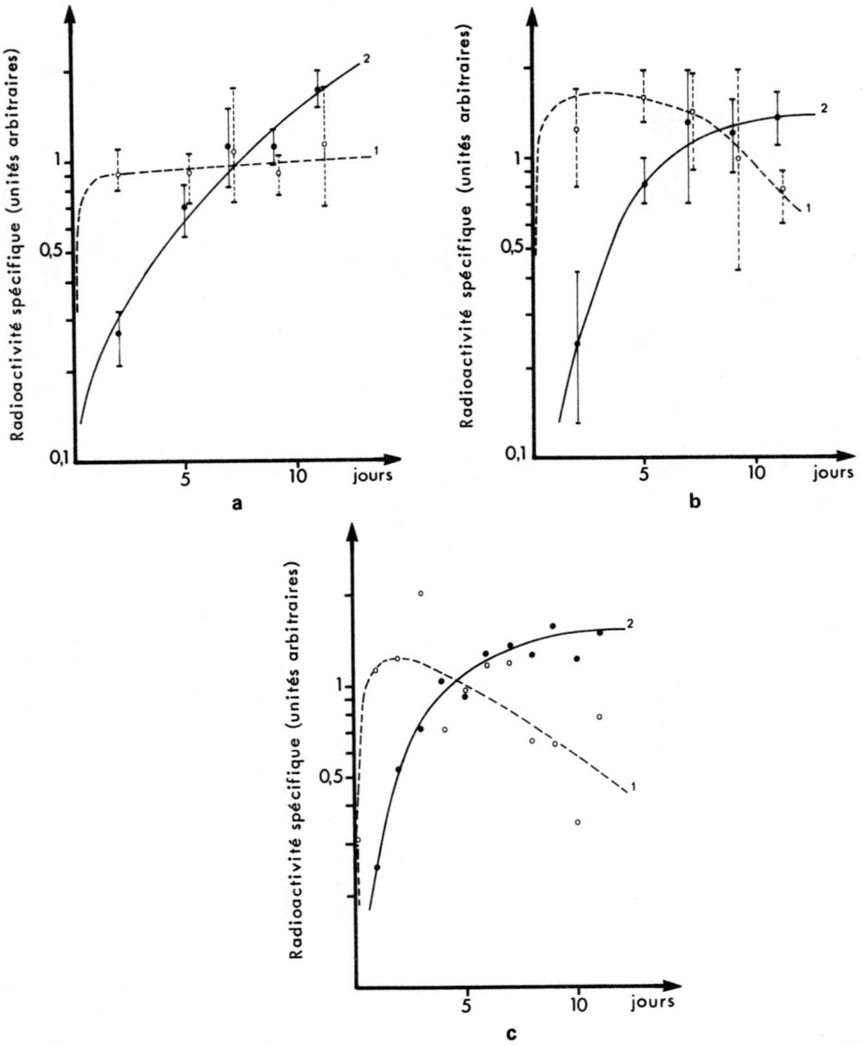

FIG.3. Evolution de la contamination de la coquille (a) et de la chair (b) de la moule (Mytilus edulis) et du poisson (Blennius pholis) (c) par des formes insolubles (courbes 1) ou solubles (courbes 2) du fer-59.

TABLEAU VI. INDICES DE CONCENTRATION DE DIFFERENTS ORGANES DE LA PLIE (PLEURONECTES PLATESSA) POUR DES FORMES SOLUBLES ET POLYMERISEES DU RUTHENIUM 106.

Organes et tissus	Indices de concentration	
	Formes solubles	Formes polymérisées
Tube digestif	2,5	4,2
Rate	1,4	NS*
Cœur	0,8	NS*
Branchies	0,3	1,3
Foie	0,2	NS*
Chair	0,1	NS*
Peau	0,05	0,1
Arête	0,03	NS*
Rein	0,02	NS*

*NS: non significatif

TABLEAU VII. INDICES DE CONCENTRATION DE DIFFERENTS ORGANES DE LA MOULE (MYTILUS EDULIS) POUR DES FORMES SOLUBLES ET INSOLUBLES DU FER 59.

Organes et tissus	Indices de concentration	
	Formes solubles	Formes insolubles
Masse viscérale	80	620
Branchies	14,6	131
Pied	12,2	69
Coquille	11,1	53
Palpes labiaux	6,6	61
Muscle adducteur	2,8	38
Manteau	2,1	37

TABLEAU VIII. INDICES DE CONCENTRATION DE DIFFERENTS ORGANES DE LA BLENNIE (BLENNIUS PHOLIS) POUR DES FORMES SOLUBLES ET INSOLUBLES DU FER 59

Organes et tissus	Indices de concentration	
	Formes solubles	Formes insolubles
Branchies	6,9	15
Tube digestif	6,9	14
Rate	2,8	1,9
Foie et vésicule biliaire	0,6	1
Peau	0,6	3,2
Chair	NS*	NS*
Arête	NS*	NS*

* NS: non significatif

3. REMARQUES GENERALES ET CONCLUSIONS

La question se pose de savoir dans quelle mesure la fixation du contaminant liée comme nous venons de le voir, à la forme physico-chimique de celui-ci, peut également être influencée par des paramètres propres aux supports biologiques et au milieu ambiant. Diverses expériences ont montré que pour le fer 59 une élévation de température détermine un accroissement du niveau de contamination, en particulier pour les formes solubles ; d'autre part une élévation de température accélère très sensiblement le phénomène de fixation des formes insolubles ; ces phénomènes seraient en relation avec une activité physiologique plus importante due elle même à l'élévation de température [9].

Il est à noter par ailleurs qu'on observe pour les seules formes insolubles et dans certains cas, un phénomène de rejet succédant à une phase d'accumulation maximale. Ce rejet pourrait être attribué à un processus d'auto-épuration de l'organisme (élimination de particules par les pseudo-féces par exemple pour la Moule) ou à un processus de solubilisation des formes insolubles par les sécrétions organiques favorisant le recyclage des radioéléments.

Ces quelques exemples montrent que, s'il est important de tenir compte de la forme physico-chimique du contaminant, l'étude des modalités de la contamination doit s'accompagner d'investigations concernant divers autres paramètres, propres à l'espèce et au milieu. En complément des recherches sur les cinétiques de fixation, les niveaux de concentration atteints et l'évaluation du pouvoir contaminant relatif des différentes formes physico-chimiques les unes par rapport aux autres, on s'efforcera de prendre en considération la nature des échanges, actifs ou passifs, s'établissant entre les espèces et le contaminant. Cette dernière étude pouvant elle-même être complétée par l'observation de la distribution dans les organes en tenant compte des voies d'entrée du contaminant et de l'élimination biologique.

REFERENCES

[1] POLIKARPOV, G.G., The role of aquatic organisms in migration of radioactive substances. In : "Radioecology of aquatic organisms", North Holland Publishing Company, (1966) 175-179.

[2] FUKAI, R., LANG HUYNH-NGOC, Studies on the chemical behaviour of radionuclides in sea water. Radioactivity in the sea, I.A.E.A., Vienna, Publ. n° 22 (1968) 26.

[3] ROBERTSON, D.E., FORSTER, W.O., RIECK, H.G. and LANGFORD, J.C., A study of the trace element and radionuclide behaviour in a Northeast Pacific Ecosystem, 350 miles off Newport, Oregon. In : BNWL-715, part 2 (1968) 92-108.

[4] ANCELLIN, J., AVARGUES, M., BOVARD, P., GUEGUENIAT, P., VILQUIN, A., Aspects biologiques et physico-chimiques de la contamin tion d'espèces et de sédiments marins. In "Radioactive contamination of the marine environment" Proceedings of a Symposium, Seattle, 10-14 July 1972, I.A.E.A., Vienna (1973) 225-243.

[5] GUEGUENIAT, P., Etude sur la physico-chimie du ruthénium dans l'eau de mer, Rapport CEA-R-4125 (1971) 43.

[6] HARVEY, H.W., Note on colloidal ferric hydroxyde in sea water, J. mar. Biol. Ass. U.K. 22 (1937) 221-225.

[7] MARCHAND, M., Contribution à l'étude de l'influence de la matière organique dissoute sur la physico-chimie d'éléments à l'état de traces dans l'eau de mer, Thèse de Doctorat, Université de Paris (1972) 132.

[8] FRAIZIER, A., Relation entre la forme physico-chimique du ruthénium 106 et les phénomènes de contamination observés expérimentalement chez diverses espèces marines. In : "Comparative Studies of Food and Environmental Contamination" Otaniemi 27-31 August 1973, I.A.E.A. Vienne (1974) 135-148.

[9] FRAIZIER, A., ANCELLIN, J., Influence de la température sur la contamination d'espèces marines par le fer 59. Actes d'un colloque sur les effets combinés des rejets radioactifs, chimiques et thermiques de l'industrie nucléaire sur l'environnement, A.I.E.A-OCDE (AEN) Stockholm 2-6 juin 1975 (sous presse).

DISCUSSION

J.A. HETHERINGTON: By what means did you differentiate between the soluble and insoluble forms of the nuclides you have studied?

A. FRAIZIER: We used ion-exchange chromatography. After elution, first of anionic forms, then of cationic forms, the insoluble forms are retained on the resin by an absorption process. In this connection, I would refer you to the work of Guegueniat described in Ref. [5] of the paper.

S. FOWLER: Have you studied how the different physico-chemical forms are handled, for example excreted, once they have been metabolically incorporated by the animals?

A. FRAIZIER: No, we have not studied this.

A.R. BERG: Do you not think that the distribution of the physico-chemical forms of the elements may undergo a shift during filtration in the ion-exchange column? I should also like to ask you whether solubles and insolubles are distinguished by the size of the molecules or by particle association.

A. FRAIZIER: In using ion-exchange chromatography to separate different physico-chemical forms of radionuclides present in contaminating seawater, the resin must be preconditioned to the seawater so that the results cannot be distorted by changes in physico-chemical form — the elution fluid is also seawater. Only the eluted forms are regarded as soluble.

BEHAVIOUR OF CAESIUM-137 IN THE MARINE ENVIRONMENT

E. GILAT
Department of Fisheries,
Ministry of Agriculture,
Haifa

Y. LAICHTER, N.H. SHAFRIR
Department of Nuclear Engineering,
Technion – Israel Institute of Technology,
Haifa,
Israel

Abstract

BEHAVIOUR OF CAESIUM-137 IN THE MARINE ENVIRONMENT

The radiological impacts of radioactive release from nuclear facilities into aquatic environments can be studied by following the behaviour of fallout radionuclides in the sea, provided that they exist in the same physico-chemical state. Caesium-137 entering the sea from fallout, although found to be at extremely low levels in the marine environment, is still detectable and can be considered representative for the behaviour of ^{137}Cs appearing in cooling waters and effluents of nuclear installations and processing plants located near the sea. The relatively high radiotoxicity, long physical half-life and high fission yield of ^{137}Cs justify further investigation of its behaviour in aquatic environments. In this investigation the sampling of sea water, sediments and organisms in the Mediterranean Sea was undertaken. Caesium-137 determinations were carried out using an extremely low level Ge(Li) gamma spectrometry system specifically designed for studies of environmental marine radioactivity. The activity of ^{137}Cs on the surface of the sea was found to be 0.49 ± 0.17 pCi/l. Caesium-137 concentrations at depth profiles down to 1000 m below the surface show variations by a factor of ten (0.65 - 0.07), maximum concentrations being found at a depth of 100 - 300 m. These concentration peaks may be explained by good mixing in the upper layers and by the existence of a thermocline or pycnocline which serve as mixing barriers at depths down to approximately 300 m, at least during part of the year. Caesium-137 concentrations in a number of marine biota representing a marine food chain composed of benthic algae, invertebrates and fishes, were determined. With reference to the average ^{137}Cs content in the upper water layers of the sea, concentration factors for these organisms were calculated. In all the three taxonomic groups mentioned, values within the range 10 - 300 were found. Biological concentration factors for any particular radionuclide are independent of its content in the sea water within a wide range of concentrations; thus the results obtained can be considered valid also for monitoring levels existing near nuclear installations. For sediment samples a concentration factor for ^{137}Cs of approximately 200 was found. If one adopts, for the most common edible link of the food chain as an upper limit for the ^{137}Cs concentration factor, according to this study, a value of 150, the maximum allowable concentration for this radionuclide in sea water is estimated to be approximately 5×10^{-2} μCi/kg. This result suggests a considerable reduction of the recommended MPC value in sea water.

1. INTRODUCTION

The release of radioactive pollutants from nuclear installations and nuclear-propelled vessels causes the accumulation of certain radionuclides in marine food chains. Among the fission products ^{137}Cs can be considered of great interest and importance because of its relatively high radiotoxicity. It is readily absorbed from the gastrointestinal tract and circulates freely through the body, so that the radiation doses received by most organs will be comparable. Therefore it is a potential cause of genetic as well as somatic injury.

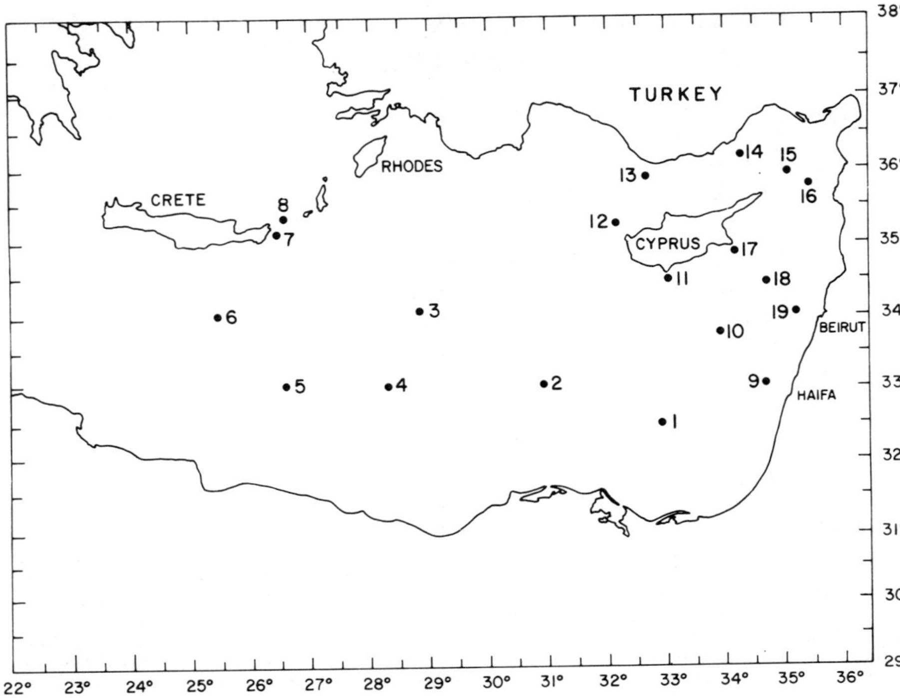

FIG.1. Sampling area for surface water. ● Water sampling stations.

The radiological impacts of radioactive release from nuclear facilities into aquatic environments can be studied by following the behaviour of fallout radionuclides in the sea, provided that they exist in the same physical and chemical state. During the last three decades approximately 34 MCi of this radionuclide have been introduced into the atmosphere by nuclear detonations, the larger fraction of it entering into the sea.

Since caesium exists in sea water mainly in the ionic state, independent of its origin, ^{137}Cs from fallout can be considered as representative for the behaviour of this radionuclide in cooling waters and effluents from nuclear facilities.

The aim of this study was to follow the pathway and concentration of ^{137}Cs in the marine environment by measuring its concentration in sea water and biota. A compilation of results of ^{137}Cs concentration measurements in sea water has recently been published [1]. However, there seem to exist relatively few measurements of biological concentration factors in the natural marine environment. Most of the results published refer to laboratory experiments performed under conditions which cannot be truly representative of natural environments [2]. Caesium-137 originating from fallout was measured by means of low-level Ge(Li) gamma spectrometry, and concentration factors for a variety of marine organisms representing different trophic levels are calculated.

2. METHODS AND EXPERIMENTAL

Sampling

In the course of this investigation a planned sampling of sea water and organisms in a grid of stations in the eastern basin of the Mediterranean Sea was undertaken. Water sampling at various depth levels in relation to thermocline, upwelling and other mixing processes on the continental shelf was carried out on board the R/V Shikmona[1].

A PVC water sampler collecting a volume of 30 litres was used. Samples were taken from the surface (Fig.1) as well as for depth profiles down to 1 000 m. Biota samples composed of benthic organisms, mainly invertebrates, as well as a few sediment samples, were collected by means of a Peterson Bottom Grab or dredge. Fish were caught by commercial equipment.

The station grid for benthic samples was determined mainly according to variability of sea bottom structure and depth. The species in which the ^{137}Cs content was determined belong to the following taxonomic groups: Anthozoa, Crustacea Decapoda, Cephalopoda, Echinodermata and Pisces. Of the Anthozoa, the species Pennatula rubra, a coral-like sedentary animal, is a filter feeder and indicator of contamination due to suspended particles of organic and inorganic character.

Crustacea Decapoda include the shrimps, e.g. Parapenaeus longirostris, which appear in commercial trawl catches in large quantities. This is a species of Atlantic Mediterranean distribution and forms an important item in the human diet of the Mediterranean countries. The crab Charybdis longicollis (Decapoda) of Indo-Pacific origin is an important member of the benthic communities in the eastern Mediterranean and one of the organisms in the food chain which has both pelagic and demersal stages [3]. The adults of this species have relatively little vertical movement and represent the benthic fauna.

The Mollusca are represented in the above-mentioned taxonomic groups by squids, Sepia officinalis, which are common in the trawl catches in the Mediterranean. These organisms are migratory animals moving also vertically and thus play a role in the transport of radionuclides between the surface and sea bottom. The Echinodermata include sea urchins, Brissopsis lyrifera, which burrow relatively deep into the sediment and are detritus feeders. They indicate the uptake of radionuclides from sediments.

The fish in which ^{137}Cs was determined include, among others, sardines Sardinella aurita, which feed on zooplankton fishes migrating in the open sea. Other species such as Mugil cephalus are inshore fish and feed in their adult stage on detritus extracting organic matter from the sea sediment.

^{137}Cs determinations

The ^{137}Cs determinations were carried out using a low-level Ge(Li) gamma spectrometry system designed for studies of environmental radioactivity in general and for marine environment studies in particular. The system comprises a 42-cm^3 Ge(Li) 5-sided rectangular main detector operated in anticoincidence with a 40 cm × 40 cm diameter plastic scintillator Compton suppression annulus. A 10-cm-thick lead enclosure serves

[1] The Israel Oceanographic and Limnological Research Company.

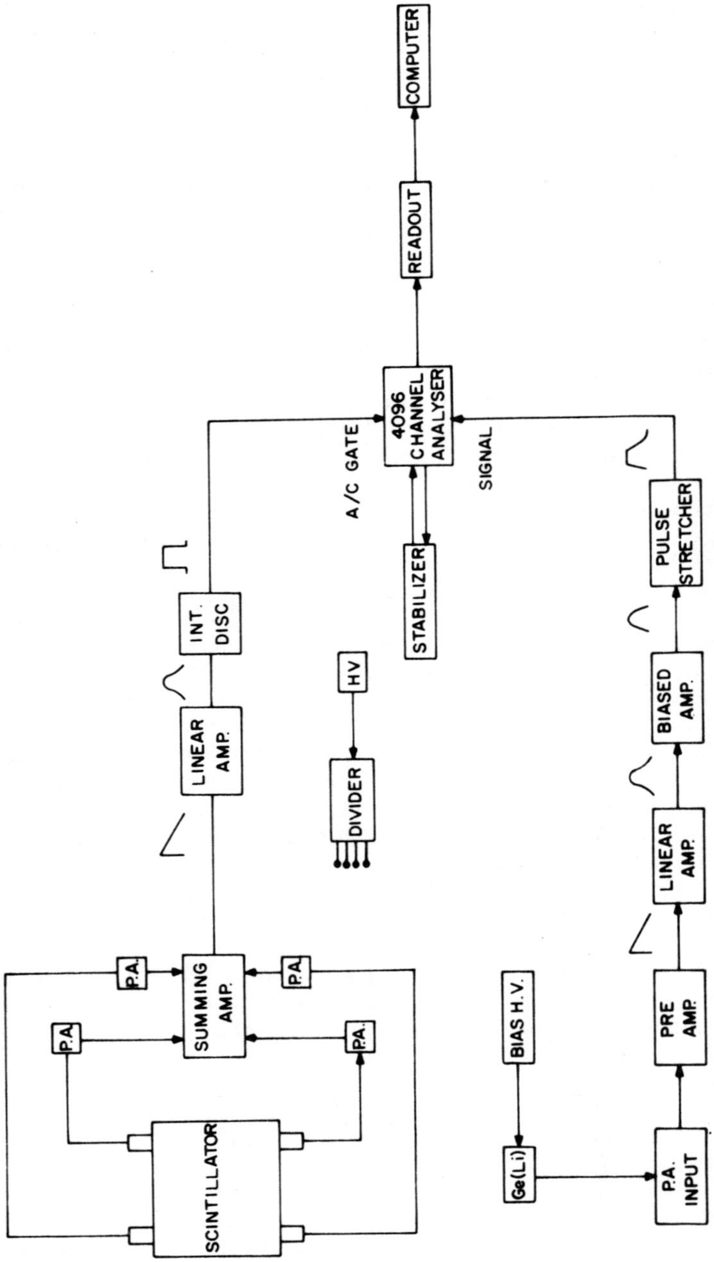

FIG.2. Block diagram of the low-level gamma spectrometry system.

TABLE I. COMPARISON OF ^{137}Cs DETERMINATIONS WITH EXPECTED VALUES AS OBTAINED FROM THE RESULTS OF THE IAEA INTERCALIBRATION EXERCISE [8-10]

Sample Type	Sample Number	Activity per Unit Weight	Results Measured Value	Results IAEA Expected Value
Sea Water	SW-I-1	pCi/kg	13.7 ± 0.4	14 ± 2
Sea Water	SW-I-2	pCi/kg	195 ± 0.7	195 ± 14
Sea Weed	AG-I-1	pCi/g	71.7 ± 0.3	74.9 ± 0.9
Sediment	SDB-1	pCi/g	357.5 ± 3.6	356 ± 11

as the basic cosmic radiation shielding. The detector provides an efficiency of 5.6% at 1.33 MeV relative to NaI(Tl). The resolution at the same energy is 3.1 keV fwhm with a peak-to-Compton-edge ratio of 18.5 : 1. The surface area of the crystal is 9.4 cm^2 and its length 45 mm. The n-layer thickness as revealed by copper staining is 0.4 mm. The drift depth is 12 mm minimum.

The detector is housed in a vacuum chamber mounted horizontally on a 31-litre liquid nitrogen dewar. To accommodate passage through the lead shielding and into the mantle, the cap was made 12 in. (30.5 cm) long. Its diameter is 2.5 in. (6.35 cm). The configuration of the dewar was dictated by experimental conditions. The Ge(Li) detector itself is located on the end of an OFHC copper cold finger at 13 mm from the 0.020-in.-thick type 1100 aluminium endcap. A lateral well accommodates the Ge(Li) endcap as well as over 200 cm^3 of samples. Four 5-m photomultipliers view the phosphor: two above and two below, shifted 90°, housed in anti-magnetic mu-metal. The phosphor is vertically mounted on a stainless-steel table with roller bearings to facilitate sample insertion. Thus no additional openings are required, and a good coverage of scattering angles is preserved. It is coated with white reflector paint and housed in stainless steel. There is no metal between the phosphor and the vacuum chamber.

A schematic diagram of the electronics used for the low-level anti-coincidence mode is shown in Fig.2. A detailed description and discussion of the electronics of an essentially similar gamma spectrometry system and its performance has been published elsewhere by Lewis and Shafrir [4] and by Shafrir et al. [5].

Sea-water samples were prepared for gamma spectrometry in the following way: the samples were acidified to pH 1.5 by adding concentrated hydrochloric acid immediately after collection on board the research vessel. In this way radionuclide adsorption at the container walls can be considered as insignificant — even for long periods of storage [6]. Sea water samples were evaporated using silicon-clad Pyrex vessels at a temperature of 100°C, which is slightly lower than the initial boiling point of sea water. During the whole evaporation process the liquid sample was magnetically stirred. In a separate series of experiments the loss of radionuclides due to evaporation was determined to be almost independent of the evaporation temperature

TABLE II. ^{137}Cs CONCENTRATION IN SEA WATER SAMPLES COLLECTED AT THE SURFACE IN THE EASTERN MEDITERRANEAN

Sampling Date	Station[a]	Sonic Depth (fathoms)	Activity per Water Weight (pCi/kg)
May 1967	1	625	0.42 ± 0.13
"	2	1200	0.50 ± 0.17
"	3	1400	0.46 ± 0.15
"	4	1540	0.57 ± 0.17
"	5	1200	0.44 ± 0.14
"	6	1550	0.45 ± 0.19
"	7	560	0.52 ± 0.18
"	8	150	0.27 ± 0.17
June 1967	9	300 - 400	0.57 ± 0.17
"	10	1000	0.55 ± 0.18
"	11	300 - 400	0.45 ± 0.14
"	12	1000	0.43 ± 0.13
"	13	700	0.47 ± 0.17
"	14	300 - 400	0.60 ± 0.19
"	15	500	0.44 ± 0.13
"	16	800	0.41 ± 0.17
"	17	300	0.55 ± 0.18
"	18	1000	0.56 ± 0.21
"	19	600	0.62 ± 0.17

[a] Station positions are shown in Fig. 1.

as long as boiling is eliminated. The loss was found to be within the experimental error of the measurements and could thus be neglected. The residue, while still wet, was transferred to a silicon-clad Pyrex tray and dried by infra-red top heating until a constant weight of the salt mixture was obtained which was about 44 g per litre sea water. The dry substance was ground in a vibrational stainless-steel ball mill and then transferred to a high-pressure hydraulic press applying a pressure of 1.5 tons/cm^2. This procedure was performed in a cylindrical container 60 mm in diameter and 35 mm in length. The sample was enclosed in nylon protecting it from contamination and moisture absorption. A specific weight of 1.25 g/cm^3 was thus obtained. Finally the compressed salt mixture was sealed in a plastic container of the same dimensions. The available space for counting samples was about 200 cm^3 and could thus accommodate two plastic containers in series.

TABLE III. ^{137}Cs CONCENTRATIONS IN SEA WATER DEPTH PROFILES

Sample Number	Sampling Date	Geographic Position	Sonic Depth (fathoms)	Sampling Depth (m)	Volume (ℓ)	Activity per Water Weight (pCi/kg)
60	June 1968	33°10'N;33°50'E	1750	0	90	0.47 ± 0.20
61	"	"	"	100	"	0.65 ± 0.19
62	"	"	"	200	"	0.26 ± 0.12
63	"	"	"	300	"	0.25 ± 0.13
64	"	"	"	500	"	0.15 ± 0.13
65	March 1969	33°45'N;34°07'E	1200	0	80	0.30 ± 0.20
66	"	"	"	100	"	0.59 ± 0.17
67	"	"	"	200	"	0.57 ± 0.17
68	"	"	"	300	"	0.36 ± 0.13
69	"	"	"	500	"	0.22 ± 0.12
70	"	"	"	1000	"	0.07 ± 0.12
71	November 1969	34°10'N;33°28'E	1000	0	40	0.15 ± 0.15
72	"	"	"	100	"	0.30 ± 0.15
73	"	"	"	200	"	0.20 ± 0.15
74	"	"	"	300	"	0.40 ± 0.16
75	"	"	"	500	"	0.33 ± 0.20
76	"	"	"	1000	"	0.20 ± 0.20

Biological and sediment solid samples were prepared by drying at 110°C until constant weight. The dry powdered substance was ground and compressed as described above. Specific gravities of up to 1.5 g/cm^3 for biological samples and up to 2.2 g/cm^3 for terrestrial solid samples were obtained.

For the determination of the radionuclide content in sea-water samples, solid, large-volume calibration standards were prepared. The process involves contamination of sea-water samples by known amounts of radionuclides and after evaporation to dryness, solid calibration standards of excellent uniformity are obtained. These standards were also used for the determination of biological and sediment samples — even of undetermined elementary matrix — after performing proper correction for differences in density and mass attenuation. This general approach to the calibration of Ge(Li) gamma spectrometry systems for measurement of environmental and biological samples has been published previously in detail [7].

The ^{137}Cs activity determinations were performed by using the full energy peak of the 661.6 keV photon. The minimum detectable level of the system, by applying the sample preparation procedure as described, is 0.8 pCi ± 20% ^{137}Cs per litre sea water for a counting time of 5 000 min.

TABLE IV. CONCENTRATION AND CONCENTRATION FACTORS OF ^{137}Cs IN ALGAE COLLECTED AT THE NORTHERN SECTION OF THE ISRAELI COAST

Sample Number	Dry Weight (g)	Activity per Dry Weight (pCi/g)	Concentration Factor
117	200.0	0.026 ± 0.005	53
118	185.4	0.068 ± 0.011	138

TABLE V. CONCENTRATION AND CONCENTRATION FACTORS OF ^{137}Cs IN MARINE INVERTEBRATES

Sample Number	S P E C I E S	Dry Weight (g)	Activity per Dry Weight (pCi/g)	Concentration Factor
	ANTHOZOA			
105	Pennatula rubra	97.2	0.035 ± 0.008	71
	CRUSTACEA DECAPODA			
93	Parapenaeus longirostris	50.0	0.050 ± 0.010	102
90	Metapenaeus monoceros	127.0	0.028 ± 0.007	57
92	Penaeus japonicus	133.0	0.017 ± 0.008	35
86	Penaeus japonicus	208.0	0.028 ± 0.009	57
102	Charybdis longicollis	90.5	0.028 ± 0.006	57
	MOLLUSCA			
91	Sepia offcinalis (shell)	245.5	0.032 ± 0.007	65
91	Sepia officinalis	245.5	0.023 ± 0.006	47
	ECHINODERMATA			
103	Astropecten aranciacus	43.4	0.004 ± 0.001	8
88	Brissopsis lyrifera	333.0	0.128 ± 0.009	261
96	Brissopsis lyrifera	153.0	0.166 ± 0.015	339

TABLE VI. CONCENTRATION AND CONCENTRATION FACTORS OF ^{137}Cs IN MARINE FISHES

Sample Number	SPECIES	Dry Weight (g)	Activity per Dry Weight (pCi/g)	Concentration Factor
	PISCES			
107	Sardinella aurita	113.9	0.065 ± 0.009	133
110	Sardinella aurita	204.6	0.065 ± 0.008	133
115	Sardinella aurita	198.2	0.030 ± 0.006	61
116	Sardinella aurita	198.3	0.025 ± 0.011	51
109	Sardina pilchardus	127.0	0.054 ± 0.007	110
112	Scomber japonicus	90.0	0.050 ± 0.009	102
108	Mugil cephalus & M. capito	238.0	0.052 ± 0.008	106
106	Siganus luridus	88.4	0.020 ± 0.008	41
114	Boops boops	91.7	0.102 ± 0.009	208
113	Sphyraena sp.	100.2	0.090 ± 0.010	184
111	Upeneus moluccensis[a]	121.0	0.018 ± 0.007	37

[a]This sample originates from the Red Sea

TABLE VII. ^{137}Cs CONCENTRATIONS IN SEDIMENT FROM THE ISRAELI COAST

Sample Number	Sample Depth (m)	Weight (g)	Activity per Dry Weight (pCi/g)
1577	1	227.0	0.120 ± 0.002
SLM-724	1	121.0	0.134 ± 0.001

In Table I results of ^{137}Cs determinations in sea water, biota and sediment samples, performed by two of the authors of this paper[2] using the same methods and equipment as mentioned above, are shown. These measurements were performed as part of an IAEA Intercalibration Program and are compared with the expected values as obtained from the published results of the Program [8-10]. Comparison of the measured and expected results clearly demonstrates the reliability of the sample preparation procedures and the performance and calibration procedure of the gamma spectrometry system.

[2] N.H. Shafrir and Y. Laichter, IAEA Intercalibration Program, 1971-1973.

3. RESULTS

Results of ^{137}Cs determinations in sea water, biota and sediment samples are given in Tables II - VII. Surface water concentrations of ^{137}Cs in the eastern Mediterranean are given in Table II. The sampling stations cover the area between the Israeli coast and the Island of Crete (Fig.1).

The results have a mean value of 0.49 ± 0.17 pCi/kg. No other artificial gamma-emitting radionuclides were detected, i.e. their concentration was below the minimum detectable level of the gamma spectrometry system. Results of ^{137}Cs concentrations for three water depth profiles down to 1 000 m are given in Table III.

Concentration and concentration factors of ^{137}Cs in algae, invertebrates and fish are given in Tables IV, V and VI, respectively. The estimation of the concentration factors is based on the ^{137}Cs content in the dry weight of biota and on the mean value of the ^{137}Cs concentration in the upper layer of the sea as taken from Table II. Caesium-137 concentrations in sediment samples in Israeli coastal waters are given in Table VII.

4. DISCUSSION

Caesium-137 originating from nuclear device detonations and from nuclear reactor installations and fuel reprocessing plants has been measured in pelagic and demersal marine ecosystems. The availability of this radionuclide to marine biota depends largely on its chemical and physical form in the sea water. It seems safe to assume that ^{137}Cs originating from fallout as well as from effluents of nuclear facilities is largely present in sea water in soluble form with a single valency, although some fraction may exist in particulate and to a smaller extent in colloidal form [11, 12].

To find the concentration and pathway of this radionuclide in the marine environment we have investigated the behaviour of ^{137}Cs originating from fallout.

The results of the ^{137}Cs concentrations obtained for surface waters and for a number of profiles down to a depth of 1 000 m show variations by a factor of ten (0.65 - 0.07), maximum concentrations being found at a depth of 100 - 300 m (Table III). These concentration peaks may be explained by good mixing in the upper layers and by the existence of a thermocline or pycnocline which serve as mixing barriers at depths down to approximately 300 m, at least during part of the year [13, 14].

The presence of trace quantities of ^{137}Cs in deep ocean water has been reported by several authors [1, 11, 13]. Soluble and colloidal states of radionuclides tend to follow the mixing processes of the water masses and will thus normally not penetrate through the thermocline. This explains the results of some authors who used ion exchange or any other techniques that, in the course of sample preparation and determination of ^{137}Cs, are selective for the soluble states only; they did not find any ^{137}Cs below a depth of about 300 m [11, 13].

The penetration of ^{137}Cs into deeper layers below the thermocline may supply some evidence for the existence of a particulate fraction of ^{137}Cs in sea water. Another contribution to the deep water occurrence of ^{137}Cs could arise from association processes of ionic and colloidal forms with particles due to physico-chemical and biological activity which affect the penetration towards the deeper layers. The accumulation of radionuclides close to the sea bottom by benthic biota takes place mainly through plankton activity and the sinking of organic and inorganic particulate matter.

The main ecosystems in the marine environment are pelagic, benthic and near-shore. These major realms are a function of water depth. This study deals mainly with the benthic invertebrate communities and with some demersal fish existing between the depths of about 10 m and 200 m. Such depths are characteristic for the continental shelf in major areas of the Mediterranean Sea.

The benthic invertebrates form an intermediate link in the marine food chain. They are dependent on the availability of food from the phytoplankton and zooplankton as well as on the organic matter in sediment and water. The stability in distribution of benthic communities on the continental shelf and their negligible seasonal fluctuations in subtropical areas make them indicators for the concentration of radionuclides in certain areas. It has been found in field and laboratory experiments that marine phytoplankton and zooplankton accumulate ^{137}Cs to a small extent only [2, 15]. This indicates that the plankton composed of primary and secondary producers has little effect on the passage of this radionuclide to higher levels of the food chain.

In the pelagic realm we dealt with a number of edible fish which are an important item in the worldwide human diet. The uptake of ^{137}Cs by marine plants and animals is known to be a rapid process. It has been shown by laboratory experiments that equilibrium conditions in sea water are reached within a few days or weeks, depending on the specific activity of ^{137}Cs and the characteristics of the taxonomic group of the organism [16,17].

The benthic invertebrates represent certain taxonomic groups: corals (Anthozoa), prawns and crabs (Crustacea Decapoda), squids (Mollusca), sea stars and sea urchins (Echinodermata).

The results obtained in this investigation for ^{137}Cs concentration factors (CF) measured in the natural marine environment show a similar range of values for benthic invertebrates and for fish (Tables V and VI).

Since some of the benthic invertebrates may belong to a lower trophic level than the fish, the results indicate that the concentration of ^{137}Cs varies only little in these two trophic levels. Similar results have been obtained in laboratory experiments in sea water for phytoplankton, zooplankton and fish constituting three trophic levels [2,15]. However, a more detailed analysis of the results obtained for some organisms belonging to two successive trophic levels, prawns and fish, permits us to draw certain conclusions. The carnivorous fish show somewhat higher CF as compared with the prawns and crabs, which are representatives of a lower trophic level. In this study the species of fish, *Scomber japonicus*, *Sphyraena sp.* and *Boops boops*, have CF values in the range of 100 to 200. The prawns, which during a certain period of their life history are their prey, *Penaeus japonicus* and *Metapenaeus monoceros* have CF values around 50. From this we may conclude that in this particular pathway there is an increase in the CF from one trophic level (prawns) to the next (fish) by a factor of 2 to 5.

The fish *Mugil cephalus*, feeding on the detritus in sediment originating mainly from phytoplankton, is representative of a short food chain consisting of the primary producers and a vertebrate. There is no significant difference in CF in this fish as compared with a predator species. The highest CF values, lying between about 200 and 300, were obtained for the sea urchin

Brissopsis lyrifera, an important member of the benthic communities in our waters. This organism feeds on the organic matter extracted from the sediment. However, the probability of ^{137}Cs reaching the last link of the food chain through this organism is limited since it is rare in the diet of demersal fish. The common pelagic species of fish, *Sardinella aurita*, feeding on the prawns, show a CF twice as large as their prey. These fish form a considerable part of the total catches in our waters and are a relatively common sea product.

The benthic invertebrates accumulate ^{137}Cs in their tissues, as shown in laboratory experiments, resulting in a CF of about 50 [15]. This supports the results obtained in this study for shrimp and crabs which are edible to predatory fish.

Results for ^{137}Cs concentration factors obtained in freshwater laboratory and field studies show increased concentration in higher trophic levels [18, 19]. The CF values obtained for sea weeds collected in the coastal area were 53 and 138, which are in the same range as those for invertebrates and fish (Table IV).

The ^{137}Cs concentrations in sediment samples of a particle size range 32 to 64 are in our waters similar to those obtained in the North Sea and the western Mediterranean (Table VII) [20]. In terms of concentration factors these results correspond to the upper limit values obtained for benthic invertebrates and fish.

According to the specific activity approach, the maximum permissible concentration of a certain radionuclide in an edible marine product is limited by the allowable specific activity of this element in man [21]. If one adopts, for the most common edible link of the food chain as an upper limit for the ^{137}Cs concentration factor, according to this study, a value of 150, the maximum permissible concentration (MPC) for this radionuclide in sea water can be estimated. Using for the maximum permissible specific activity of ^{137}Cs in man, the value of 3.1×10^{6} μCi/g [21], we obtain a value of 0.5×10^{-1} μCi/kg for the MPC of ^{137}Cs in sea water. This result suggests a considerable reduction of the recommended ^{137}Cs MPC value in sea water, which according to Ref. [21] is 1.5×10^{-1} μCi/kg. However, to establish a reliable value for the release of ^{137}Cs into the sea, further studies in the entire marine environment and especially of a large variety of edible marine organisms, covering also other geographical areas, have to be performed.

REFERENCES

[1] VOLCHOK, H.L., et al., in Radioactivity in the Marine Environment, National Academy of Sciences (1971) 42.

[2] BOWEN, V.T., et al., in Radioactivity in the Marine Environment, National Academy of Sciences (1971) 200.

[3] GILAT, E., Bull Inst. océanogr. Monaco 62 (1964) 1290.

[4] LEWIS, S.R., SHAFRIR, N.H., Nucl. Instr. Meth. 93 (1971) 317.

[5] SHAFRIR, N.H., LAICHTER, Y., GILAT, E., in Rapid Methods for Measuring Radioactivity in the Environment, (Proc. Conf. Neuherberg, 1971) IAEA, Vienna (1971) 291.

[6] ROBERTSON, D.E., Anal. Chim. Acta 42 (1968) 533.

[7] LAICHTER, Y., NOTEA, A., SHAFRIR, N.H., Nucl. Instr. Meth. 113 (1973) 61.

[8] FUKAI, R., BALLESTRA, S., MURRAY, C.N., in Radioactive Contamination of the Marine Environment, (Proc. Symp. Seattle, 1972) IAEA, Vienna (1973) 3.

[9] FUKAI, R., BALLSTRA, S., "Intercalibration of methods for radionuclide measurements on a seaweed sample", in Symp. on the Determination of Radionuclides in Environmental and Biological Materials, London, April, 1973.

[10] FUKAI, R., et al., in Environmental Surveillance Around Nuclear Installations (Proc. Symp. Warsaw, 1973) 1, IAEA, Vienna (1974) 313.

[11] SCHUERT, E.A., Radioactivity in the Marine Environment, National Academy of Sciences (1971) 69.

[12] GREEDALE, A.E., BALLOU, N.E., U.S. Naval Radiological Defense Lab. San Francisco, Rep. USNRDL 436 (1954).

[13] PRITCHARD, D.W., et al., Radioactivity in the Marine Environment, National Academy of Sciences (1971) 90.

[14] BROECKER, W.S., J. Geophys. Res. 71 24 (1966) 5827.

[15] CHIPMAN, W.A., Radioactivity and Human Diet (RUSSELL, R.S., Ed.), Pergamon Press (1966) 421.

[16] POLIKARPOV, G.G., Radioecology of Aquatic Organisms, North-Holland Publishing Company (1966) 65.

[17] HARRISON, F.L., in Radioactive Contamination of the Marine Environment (Proc. Symp. Seattle, 1972) IAEA, Vienna (1973) 453.

[18] OSTERBERG, C., PEARCY, W.G., CURL, H., Jr., J. Mar. Res. 22 (1964) 2.

[19] PENDLETON, R.C., HANSON, W.C., in Int. Conf. Peaceful Uses of Atomic Energy (Proc. 2nd. Conf. in Geneva, 1958) 18, UN, New York (1958) 419.

[20] DUURSMA, E.K., GROSS, M.G., in Radioactivity in the Marine Environment, National Academy of Sciences (1971) 147.

[21] NATIONAL ACADEMY OF SCIENCES, Disposal of Low-Level Radioactive Waste into Coastal Waters, Publication 985 (1962).

DISCUSSION

A. PRESTON: Are the values reported for ^{137}Cs concentrations based on single samples, as the tables seem to imply? If so, then are you not drawing very detailed conclusions from a small amount of data? I would suggest that your data merely indicate, in conformity with the established data on ^{137}Cs concentration factors in marine organisms, that a CF of 30 - 150 (say of the order of 10^2) may be expected in fish.

N.H. SHAFRIR: The ^{137}Cs concentration factors given are not based on single-sample measurements but on a small number of samples of the same kind. The accuracy of the results can be considered as a good deal better than just an order of magnitude. What I wanted to say is that the results have a trend that suggests a reduction in the recommended ^{137}Cs MPC value for sea water. However, I agree that in order to establish a reliable value for releases of ^{137}Cs into the sea, further measurements in the natural marine environment are needed, and for a large variety of edible marine organisms.

A. PRESTON: Were the echinoderms particulate or filter feeders? If they were, then the high CF may be accounted for by the particulate ^{137}Cs which you postulate is reaching deep water rapidly.

E. GILAT (answer supplied after the meeting): The echinoderms composed of the species Brissopsis lyrifera feed on the detritus in the sediment. They ingest large quantities of sediment, extracting the organic material included in the sand and silt of the sea bottom.

J.C. ANCELLIN: Is there any direct correlation between the CFs given in your paper and those already published for marine species? The point is that your results are for dry weight, while the published results for marine species are nearly always for wet weight.

N.H. SHAFRIR: We tried to extrapolate published results to something like the dry weight, and this gave us a CF of about twice the published values.

However, there have been very few measurements of CFs in natural marine environments, so not much data is available.

A. PRESTON: In connection with Mr. Shafrir's reply, perhaps I could point out that there is in fact a great deal of published in-situ environmental data on concentration factors.

STUDIES ON RARE EARTH ELEMENTS IN SEAWATER AND UPTAKE BY MARINE ORGANISMS

H. SUZUKI, T. KOYANAGI, M. SAIKI
National Institute of Radiological Sciences,
Chiba-shi, Japan

Abstract

STUDIES ON RARE EARTH ELEMENTS IN SEAWATER AND UPTAKE BY MARINE ORGANISMS.

The contents of rare earth elements in marine environmental samples were determined by neutron activation analysis to examine the existing state in coastal seawater and the concentration by marine organisms of the elements. Seawater was filtered through a Millipore filter GS (pore size 0.22 μm), before the analysis. Some of the seawater was treated with HCl solution before filtration and some after filtration. Certain marine organisms were also analysed for determination of rare earth elements. These were: flounder (*Paralichthys olivaceus*); yellowtails (*Seriola quinqueradiata*); immature anchovy (*Engraulis japonica*); clams (*Meretrix lusoria*); green algae (*Ulva pertusa*); brown algae (*Hizikia fusiforme, Sargassum fulvellum, Undaria pinnatifida*). In the seawater without HCl treatment before filtration, considerable amounts of the elements existed in residue on the filter, whereas in the seawater treated with HCl before filtration, the greater part remained in the dissolved state. Concentration factors calculated from the contents of stable elements, therefore, are affected remarkably by the existing state of the elements in seawater. If only the dissolved state is assumed available for marine organisms, values one order higher are attained compared with the case where total amounts of the elements were used for the calculation. However, the contribution of the insoluble state seems to be not negligible with some organisms. The higher concentration factors for immature anchovy and clams observed in this study were considered to be caused by surface adsorption of elements in particulate form and also ingested sediment with high element concentration.

1. INTRODUCTION

Power reactors and a spent fuel reprocessing plant are under construction in coastal areas of Japan, in addition to the reactors that have been built, also mainly on seashore sites, over the past two decades. The environmental considerations of the installed reactors were discussed in a previous paper [1]. Although the discharge of low-level radioactive liquid wastes from these facilities into the coastal seas has been considered unavoidable to some extent owing to the geographical and topographical factors in Japan, radioactive contamination of the marine environment has aroused public concern since the sea is an important source of livelihood to the Japanese. Seafoods contribute more than 50% of the total intake of animal protein and most of the salt produced in Japan comes from seawater. For these reasons the estimate of radiation doses to man resulting from the radioactivity introduced into the sea in the form of wastes is of great importance.

The radioactive effluent from fuel reprocessing plants contains the radionuclides of rare earth elements represented by $^{141+144}Ce$, as the constituent of fission products with other important nuclides. However, data on their marine radioecological behaviour are not numerous enough for estimating the contamination level of marine organisms by radioactive rare earth elements. Also, information on the concentration and distribution of stable rare earth elements in samples of the marine environment is also very meagre, although a knowledge of the ratio of the concentration in the organisms to that in the seawater is an essential parameter for estimating the radiation dose to man due to the uptake of contaminated seafoods.

Balashov and Khitrov [2], using the rapid determination method of precipitation of the elements with methyl violet, reported that total rare earth element contents of surface waters collected from the Indian Ocean were 0.1-4.7 μg/l. For the concentration of cerium in seawater, although the very high value of 0.4 μg/l was previously reported by Wattenberg [3], far lower values of 10^{-3}-10^{-2} μg/l were given in later data reported by Goldberg et al. [4], Høgdahl et al. [5], Shigematsu et al. [6], Nagatsuka et al. [7] and Sastry et al. [8]. Practical data on the concentration of stable rare earth elements in marine organisms are almost non-existent, apart from those of Kameda [9].

On the other hand, it has been frequently pointed out that the biological concentration of elements in marine organisms not only depends on the difference between one element and another, but is also affected by the chemical and physical forms of the elements in seawater. In particular, since the action of rare earth elements in the ocean is quite complicated, several investigations have been carried out on the physico-chemical aspects to clarify the states and concentrations that actually occur. Balashov and Khitrov [2] determined the concentration of total rare earth elements in the waters of the Indian Ocean and found that (a) the concentration increased gradually with depth, (b) surface waters of the near-shore zone were enriched with the rare earth elements and (c) not more than 50% of the rare earth element content in a unit volume of the water was associated with suspended matter (particles > 0.7 μm). However, Shigematsu et al. [6] determined the concentration of stable cerium in seawater collected at near-shore points and bays around Japan, finding that up to 70% of the total amount of cerium was removed by membrane filter.

By radioisotope tracer experiments, Carpenter and Grant [10] reported that radioactive cerium chloride added to coastal waters became associated with particulate materials with apparent diameters of 0.02 to 0.1 μm.

According to the results of studies with electrodialysis by Nishiwaki, Honda et al. [11], positively charged colloidal forms of ^{144}Ce seem to predominate initially but decrease with time, while the electro-neutral species increase with aging in seawater. After aging for 3 days, the negatively charged species also increase with time. This may be due to the formation of electro-neutral and then negatively charged species by hydrolysis and complex formation, e.g. sulphato complex formation with aging [12]. Most of the electrically charged forms of ^{144}Ce in seawater showed as non-dialytic, presumably because of the formation of colloids or particulates.

Hirano et al. [13] also found that radioactive cerium chloride introduced into seawater existed mainly in dissolved form and 20-25% of the cerium turned into the particulate state with a diameter of more than 0.01 μm after the aging period of 40 days. They pointed out that the radioactive cerium introduced in ionic form or easily dissociable form was hydrolysed slowly at a pH greater than 6 and formed complexes with hydroxide, chloride or other anions present in seawater, or existed in polymerized forms. From the residence time of natural cerium in the sea, and the form of precipitation for natural cerium, reported as about 80 years and CeO_2 respectively [4],

Hirano et al. inferred that the equilibrium between natural cerium and radioactive cerium chloride introduce into seawater might be achieved after a fairly long time.

The present investigation was carried out to determine the quantity of cerium, europium, terbium, ytterbium and lutetium in near-shore seawater and marine organisms, and the distribution and biological concentration of these elements will be discussed.

2. EXPERIMENTS

2.1. Sampling and ashing

A total of 40 litres of seawater was collected from the surface 5 km off the seashore of Ibaraki Prefecture, on the east coast of Mid-Japan about 10 km from the planned discharge point of radioactive liquid wastes from the fuel reprocessing plant, and transported to the laboratory immediately. Each 20 litres of the seawater was filtered through a Millipore filter of GS-type (pore size 0.22 μm) before or after adjustment of the pH value of the seawater to less than 1.0 with HCl solution prepared by saturating the redistilled water with HCl gas. The filter and residue were ashed together in a quartz crucible and used for the analysis along with the seawater.

The following marine organisms were also collected at the same area as the seawater was sampled: flounder (*Paralichthys olivaceus*); yellowtails (*Seriola quinqueradiata*); immature anchovy (*Engraulis japonica*); clams (*Meretrix lusoria*); green algae (*Ulva pertusa*); brown algae (*Sargassum fulvellum*, *Undaria pinnatifida* and *Hizikia fusiforme*). Except for the muscle samples taken from flounder and yellowtails, the samples were washed with redistilled water to remove adhering substances or sand, and moisture was blotted with filter paper. After being dried in an oven at 110°C, the samples were ashed by an electric furnace at 450°C.

2.2. Pre-irradiation procedure

To 10 litres of the filtered seawater, 40 mg of iron (ferric ion) was added as a collector for the elements. A known amount of carrier-free ^{88}Y, which is expected to show the same behaviour as the rare earth elements, was also added as a tracer to determine the overall chemical yield of the elements through pre- and post-irradiation procedures. The elements were collected from the seawater by means of co-precipitation into ferric hydroxide precipitate, the pH being adjusted with HCl gas or NH_3 gas as shown in Fig.1. Iron of 99.999% purity and gases of HCl or NH_3 were used for the analysis to avoid the possible contamination of rare earth elements by the reagents. The contribution of the reagents to the concentration the levels obtained was also checked by the blank test performed with redistilled water and the reagents and ascertained as small enough to be neglected.

An anion-exchange separation step was applied successively to remove uranium which might be co-precipitated with ferric hydroxide and might produce the fission product ^{141}Ce, identical to the neutron capture product of stable cerium by thermal neutron irradiation. To the eluate from the anion-exchange resin, Dowex 1 × 8, iron carrier and NH_3 gas were added

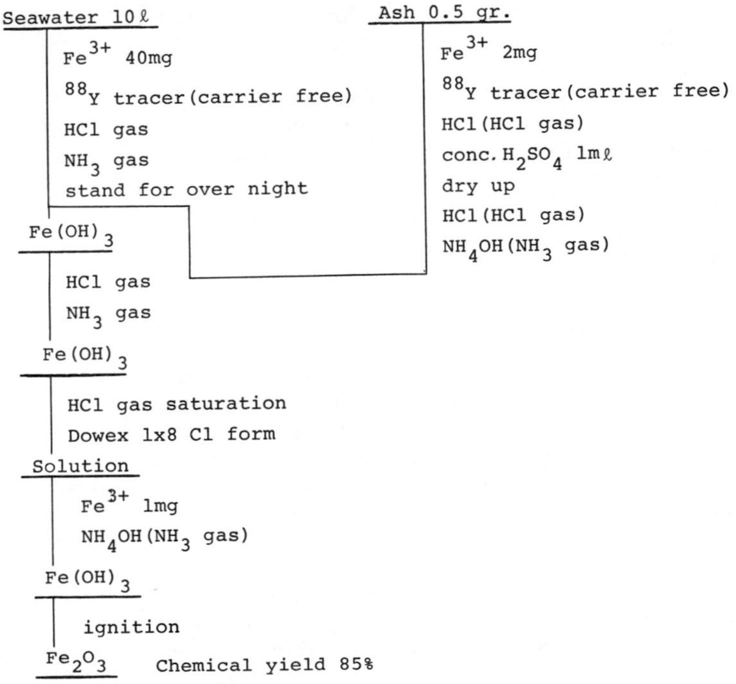

FIG. 1. Scheme for pre-irradiation procedure.

to reprecipitate the rare earth elements with ferric hydroxide. The final precipitate filtered with No. 5A filter paper was dried and ignited in a quartz crucible after being wrapped in high-purity aluminium foil. The ashed sample was wrapped in another foil before irradiation. For the samples other than the seawater, the co-precipitation method was also used for collecting the rare earth elements after the sample had been treated by HCl solution. The chemical yield of the elements in the pre-irradiation procedure was more than 80% for all samples.

The prepared samples were irradiated in the JRR-2 reactor of the Japan Atomic Energy Research Institute for 290 hours at a neutron flux of $2 \times 10^{13}\ n \cdot cm^{-2} \cdot s^{-1}$, with the reference standards of the elements.

2.3. Post-irradiation procedure

After cooling for about 2 weeks, the irradiated samples were analysed for separation of the activated rare earth elements by the method shown in Fig. 2. The sample taken out from the aluminium foil was transferred into a beaker containing 10 mg of lanthanum carrier and hold-back carriers of Na^+, Ba^{2+}, Co^{3+} and Zn^{2+}. It was then dissolved by thermal decomposition with HCl and HNO_3 solutions. The rare earth elements in the solution were purified by means of reprecipitation in the order of hydroxide, fluoride,

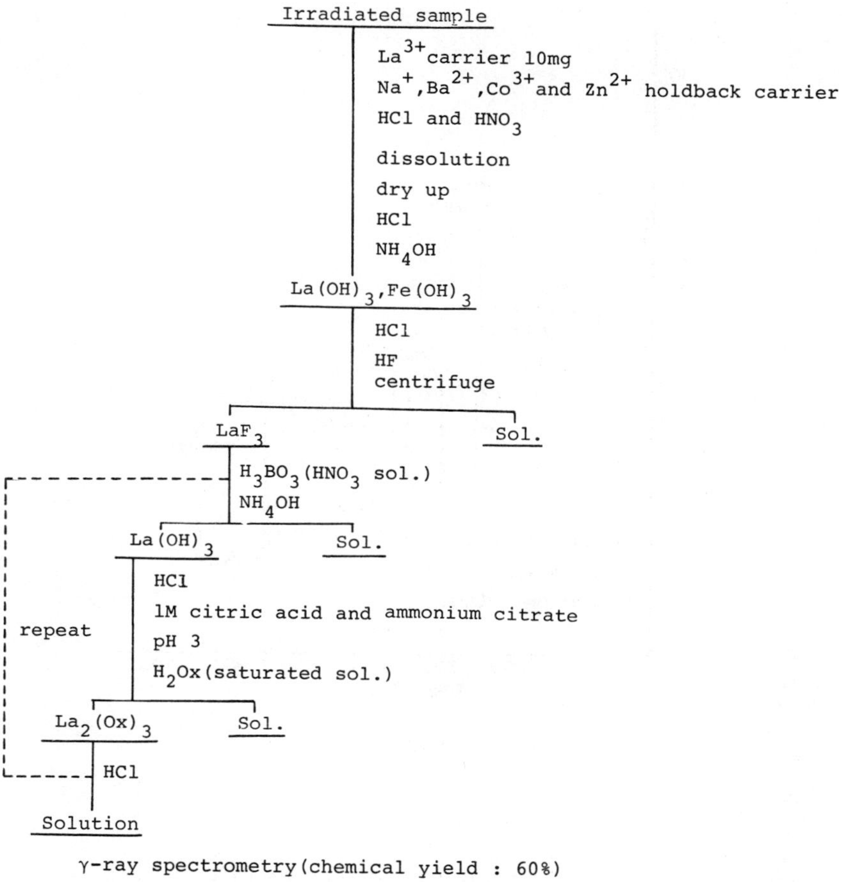

FIG.2. Scheme for post-irradiation procedure.

hydroxide and oxalate together with the lanthanum carrier, and the last two re-precipitation steps were repeated for greater purification. The final precipitate of oxalate was dissolved in a definite volume of HCl solution for measuring radioactivity.

2.4. Radioactivity measurements

Radioactivity of the sample solution was measured by using a 1024-channel gamma-ray pulse height analyser equipped with a Ge(Li) detector. The typical gamma-ray spectra obtained for the seawater and the green algae samples are shown in Figs 3 and 4, respectively. For the radioactive rare earth elements, ^{141}Ce, ^{152}Eu, ^{160}Tb, ^{169}Yb and ^{177}Lu, each photopeak at 145 keV, 344 keV, 298 keV, 177 keV and 208 keV was used for the calibration

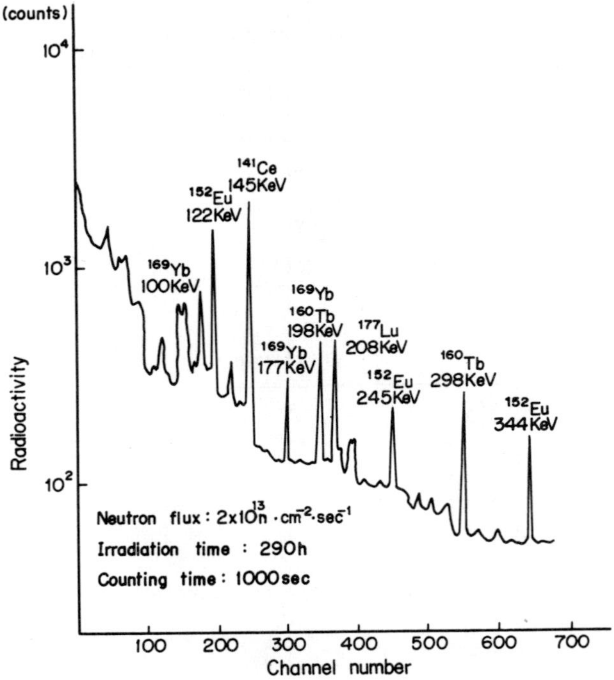

FIG. 3. Gamma-ray spectrum of the seawater sample.

of radioactivities. The net area of each photopeak was obtained by subtracting the area of base line from the total area under the peak, and the concentration of the elements in the sample was calculated by comparing the net areas between the sample and the reference standards.

Overall chemical yield estimated by the radioactivity of the ^{88}Y tracer was about 60% for the seawater and more than 70% for the other samples, respectively.

3. RESULTS AND DISCUSSION

Table 1 shows the concentrations of the rare earth elements in seawater filtered before and after HCl treatment. The concentrations of the rare earth elements in seawater acidified with HCl solution after filtration were slightly lower than those in seawater acidified before filtration. This may be due to the loss caused by the partial adsorption of those elements in seawater without HCl treatment on the surface of polyethylene bottles during transportation and of filtering apparatus. Therefore the concentrations of cerium, europium, ytterbium and lutetium in seawater, acidified before filtration, 0.076, 0.0024, 0.0072 and 0.0008 μg/l, respectively, were considered to represent the

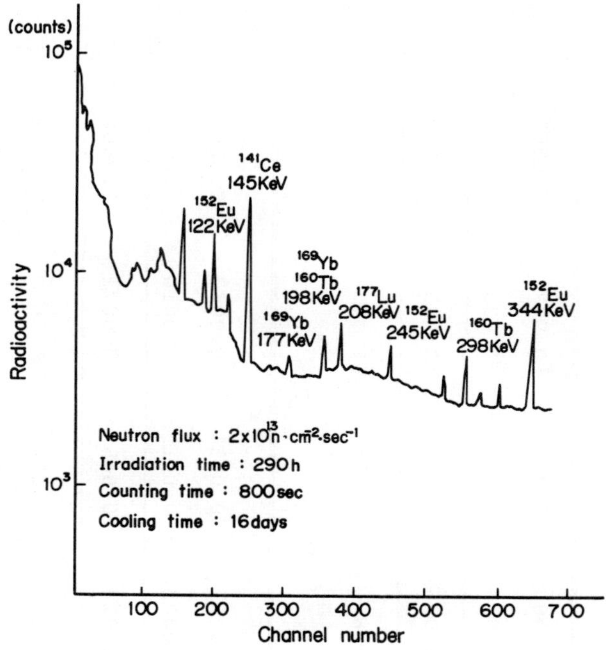

FIG.4. Gamma-ray spectrum of green algae (Ulva pertusa).

total amounts of these elements in seawater. The amounts are considerably higher than those in the surface seawater of the Pacific Ocean near California investigated by Goldberg et al. [4], but rather close to those in the near-shore surface seawater at Arasaki in the Kanagawa Prefecture in Japan investigated by Nagatsuka et al. [7] and also those in the coastal surface seawater of the Indian Ocean investigated by Sastry et al. [8].

Nagatsuka found different concentrations of rare earth elements between near-shore seawater with chlorinity of 18.54 ‰ and that with chlorinity of 19.25 ‰. The chlorinity of the near-shore surface seawater used for our investigation, 18.73 ‰, was closer to the former, and our results are closer to those obtained with the former than the latter. This suggests that the distribution pattern of the rare earth elements in seawater might be depend upon the chlorinity.

From the geochemical point of view, chondrite normalized patterns were compared among the concentrations of the rare earth elements in the near-shore surface seawater, in the surface seawater of open sea analysed by Goldberg [4], and in the shale analysed by Minami [14]. As shown in Fig. 5, the distribution pattern of the rare earth elements in the near-shore surface seawater is different from that of the open sea, but very close to that of the

Table 1. Concentration of rare earth elements (Ce, Eu, Tb, Yb and Lu) in seawater (Cl : 18.73)

Sample		Concentration of rare earth elements ($\mu g/\ell$)				
		Ce	Eu	Tb	Yb	Lu
Acidified with HCl, after filtration	Filtrate (F_a)	0.007	0.0003	0.0008	0.0021	0.0003
	Residue	0.051	0.0013	0.0034	0.0035	0.0004
	Total (T_a)	0.058	0.0016	0.0042	0.0056	0.0007
	[Ratio (F_a/T_a x 100)	12	18	19	37	42]
Acidified with HCl, before filtration	Filtrate (F_b)	0.060	0.0021	0.0036	0.0062	0.0007
	Residue	0.016	0.0003	0.0019	0.0010	0.0001
	Total (T_b)	0.076	0.0024	0.0055	0.0072	0.0008
	[Ratio (F_b/T_b x 100	78	87	65	86	87]
	Ratio (F_a/T_b x 100)	9	12	14	29	37

Ratio : percent of solute in total concentration

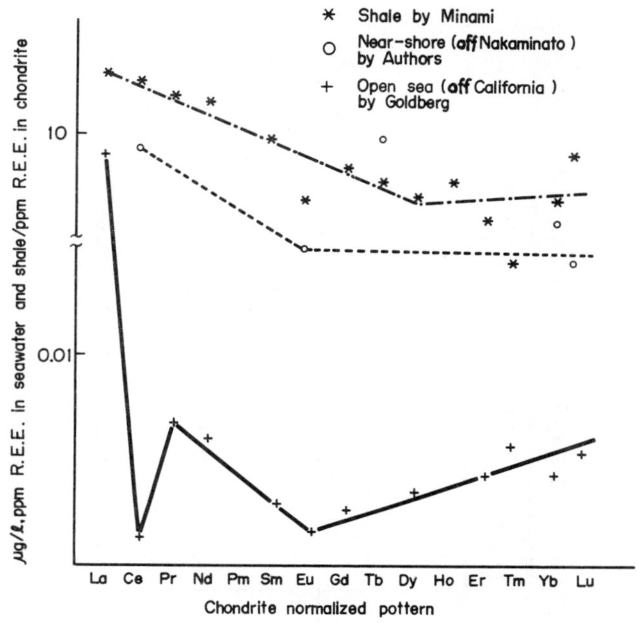

FIG. 5. Ratio of abundance of each rare earth element in seawater and shale to the abundance of the same element in chondrite [15].

shale. This suggests that the composition of the rare earth elements in near-shore surface seawater might be affected by the contribution of terrestrial materials, represented by the shale.

As shown previously, with the seawater without HCl treatment before filtration, considerable amounts of the rare earth elements existed in residue on the Millipore filter but the greater part of the elements remained in the dissolved state in the seawater treated with HCl solution before filtration. It was assumed that rare earth elements in near-shore seawater would be present in the undissolved state, as suggested by Shigematsu et al. [6] and Balashov and Khitrov [2].

Table 2 shows the contents of the rare earth elements in marine organisms. Algae showed high levels of the elements with the exception of one of the brown algae (Undaria pinnatifida). Higher contents were observed in immature anchovies and the soft parts of clams than in fish muscles, which suggests that surface adsorption of the rare earth elements and ingested sediment play a role when there is a high concentration of the elements in an organism. Of these elements, the level of cerium in marine organisms was highest.

Tables 3 and 4 show the concentration factors for the rare earth elements calculated, respectively, from the total amount and from the amount in the dissolved state. It is clear that the concentration factors are greatly affected by the existing state of the elements in the seawater. If only the

Table 2. Contents of rare earth elements (Ce,Eu,Tb,Yb and Lu) in marine organisms

Sample	Organ	Element (μg/kg wet)				
		Ce	Eu	Tb	Yb	Lu
Flounder (*Paralichthys olivaceus*)	Muscle	1.38	0.04	0.03	0.28	0.01
Yellowtails (*Seriola quinqueradiata*)	Muscle	2.38	0.14	0.07	0.39	0.01
Immature anchovy (*Engraulis japonica*)	Whole body	7.34	0.22	0.83	1.50	0.06
Clams (*Meretrix lusoria*)	Soft part	39.3	2.06	2.54	5.43	0.23
Green algae (*Ulva pertusa*)	Whole body	39.1	0.77	0.97	1.67	0.24
Brown algae (*Sargassum fulvellum*)	Whole body	22.3	0.54	1.78	0.97	0.28
Brown algae (*Undaria pinnatifida*)	Whole body	3.70	0.09	0.30	0.61	0.09
Brown algae (*Hizikia fusiforme*)	Whole body	19.2	0.24	0.62	0.68	0.22

Table 3. Concentration factor calculated from total amount of rare earth elements in seawater.

Sample		Concentration factor				
Kind of marine organism	Organ	Ce	Eu	Tb	Yb	Lu
Flounder (_Paralichthys olivaceus_)	Muscle	18	17	5	39	13
Yellowtails (_Seriola quinqueradiata_)	Muscle	31	58	13	54	13
Immature anchovy (_Engraulis japonica_)	Whole body	97	92	150	210	75
Clams (_Meretrix lusoria_)	Soft part	520	860	460	750	300
Green algae (_Ulva pertusa_)	Whole body	520	320	180	230	300
Brown algae (_Sargassum fulvellum_)	Whole body	290	230	320	140	350
Brown algae (_Undaria Pinnatifida_)	Whole body	49	38	54	85	110
Brown algae (_Hizikia fusiforme_)	Whole body	250	100	110	94	280

Table 4. Concentration factor calculated from soluble forms of rare earth elements in seawater.

Sample		Concentration factor				
Kind of marine organisms	Organ	Ce	Eu	Tb	Yb	Lu
Flounder (*Paralichthys olivaceus*)	Muscle	200	130	38	130	33
Yellowtails (*Seriola quinqueradiata*)	Muscle	340	470	88	190	33
Immature anchovy (*Engraulis japonica*)	Whole body	1050	730	1040	710	200
Clams (*Meretrix lusoria*)	Soft part	5610	6870	3180	2590	770
Green algae (*Ulva pertusa*)	Whole body	5590	2570	1210	800	800
Brown algae (*Sargassum fulvellum*)	Whole body	3190	1800	2230	460	930
Brown algae (*Undaria pinnatifida*)	Whole body	530	300	380	290	300
Brown algae (*Hizikia fusiforme*)	Whole body	2740	800	780	320	730

dissolved state is assumed available for marine organisms, values about one order of magnitude higher are obtained, especially for cerium, europium and terbium.

Taking these results along with the results of Hirano et al. [13], the authors consider that the higher values of concentration factor, based on the concentration in the dissolved form, should be used when studying the bio-accumulation of rare earth elements at points quite near to the discharge point of radioactive liquid wastes.

REFERENCES

[1] SAIKI, M., KOYANAGI, T., OHMOMO, Y., UCHIYAMA, M., "Environmental considerations and public acceptance aspects of nuclear plant site selection in coastal areas in Japan", Environmental Aspects of Nuclear Power Stations (Proc. Symp. New York, 1970), IAEA, Vienna (1971) 821-846.

[2] BALASHOV, Yu.A., KHITROV, L.M., Distribution of the rare earths in the waters of the Indian Ocean, Geochemistry 9 (1961) 877-890.

[3] WATTENBERG, H., Chemistry of seawater. Recent investigations on the dissolved gases, Z. Anorg. Allg. Chem. 251 (1943) 71-85.

[4] GOLDBERG, E.D., KOIDE, M., SCHMITT, R.A., SMITH, R.H., Rare-earth distributions in the marine environment, Geophys. Res. 68, 14 (1963) 4209-4217.

[5] HØGDAHL, O.T., MELSOM, S., BOWEN, V.T., Neutron activation analysis of lanthanide elements in seawater, Adv. Chem. Ser. 73 (1968) 308-325.

[6] SHIGEMATSU, T., NISHIKAWA, Y., HIRAKI, K., GODA, S., TSUJIMOTO, Y., Fluorometric determination of cerium in seawater, Japan Analyst 20 (1971) 575-581.

[7] NAGATSUKA, S., SUZUKI, H., NAKAJIMA, K., Activation analysis of lanthanide elements in natural water, Radioisotopes (Tokyo) 20, 7 (1971) 305-309.

[8] SASTRY, V.N., KRISHNAMOORTHY, T.M., DOSHI, G.R., SARMA, T.P., Manganese dioxide as sorber in the determination of cerium, ruthenium, zirconium and cesium fission nuclides from seawater by neutron activation analysis, Indian J. Chem. 12, 2 (1974) 200-202.

[9] KAMEDA, K., Study on abundance of rare earth elements in marine organisms, J. Radiat. Res. 3 (1962) 82-103.

[10] CARPENTER, J.H., GRANT, V.E., Concentration and state of cerium in coastal waters, J. Mar. Res. 25, 3 (1967) 228-38.

[11] NISHIWAKI, Y., HONDA, Y., KIMURA, Y., MORISHITA, H., KOGA, T., MIYAGUCHI, Y., KAWAI, H., "Behaviour and distribution of radioactive substances in coastal and estuarine waters", IAEA Proceedings of a Symposium Radioactive Contamination of the Marine Environment (Proc. Symp. Seattle, 1972), IAEA, Vienna (1973) 177-193.

[12] KEČKEŠ, S., PUČAR, Z., MARAZOVIĆ, Lj., Radioecological Concentration Processes (ÅBERG, B., HUNGATE, F.P., Eds.) Pergamon Press, N.Y. (1966) 999.

[13] HIRANO, S., KOYANAGI, T., SAIKI, M., "The physico-chemical behaviour of radioactive cerium in seawater", Radioactive Contamination of the Marine Environment (Proc. Symp. Seattle, 1972), IAEA, Vienna (1973) 47-55.

[14] MINAMI, E., Gehalte an seltenen Erden in europäischen und japanischen Tonschiefern, Nachr. Ges. Wiss. Göttingen, Math.-Phys.Kl. 4, 1 (1935) 155-170.

[15] SCHMITT, R.A., SMITH, R.H., LASCH, J.E., MOSEN, A.W., OLEHY, D.A., VASILEVSKIS, J., Abundances of the fourteen rare earth elements, scandium and yttrium in meteoritic and terrestial matter, Geochim. Cosmochim. Acta 27 (1963) 577-622.

DISCUSSION

K.C. PILLAI: Have you any information on the concentrations of rare earths in coastal sediments?

M. SAIKI: Our present investigations indicate that a considerable amount of natural Ce and other rare earth elements (REE) exists in an undissolved state in seawater so that the concentrations of natural REE in the suspended matter and in the sediment may be high. In radioisotope tracer experiments, it was shown that several thousand times more ^{144}Ce was concentrated in sedimental mud than in seawater. The same tendency was observed in radioactivity survey data on fall-out due to nuclear explosion tests. My colleague, Mr. Nagaya, has studied this problem, and probably he could add something.

Y. NAGAYA: We have some experience in determining ^{144}Ce concentrations in Japanese coastal sediments. In general, the ^{144}Ce contents of the sediments show wide horizontal variation, even in small areas. The variation is assumed to be due to the different physical and chemical properties of the sediments in the area and not to differences in the states of ^{144}Ce in the seawater. From field studies, it is estimated that the amount of ^{144}Ce in particulate form in Kashimanada seawater is 80%.

S. FOWLER: I was surprised to note that in your cerium radiotracer experiments you were able to achieve biological concentration factors that were similar to those for stable element concentration. This was not the case in some of the work we did using the same isotope. For how long were your experiments carried out in order to achieve this isotopic equilibrium?

M. SAIKI: In our oral presentation we reported the interesting finding that for immature anchovy the same Ce concentration factors were obtained by two different methods. In experiments carried out for a week using a radioisotope tracer technique to study ^{144}Ce concentration by living immature anchovies, factors of about 800 were obtained. The values were far higher than the C.F. of 97 given in Table 3 of our paper, but were almost the same as those given in Table 4 (1050) calculated from the amount of dissolved Ce. Sometimes, however, it takes a long time to establish isotopic equilibrium between fish organs and seawater. According to our RI tracer experiment with ^{85}Sr, the concentration of ^{85}Sr in fish bone does not reach equilibrium with the seawater concentration within an experimental time of 100 days; the experimental period required may depend on the kind of RI and its chemical state, and on various other factors. In our paper we only point out that the concentration factors are strongly affected by the state of the rare earth elements in seawater.

Perhaps I can add some further information on the uptake of radionuclides through the digestive tracts of fish. Sometimes this pathway has been ignored in RI tracer experiments in the aquarium. Nakahara, Koyanagi and Saiki (in "Radioecological Studies on Marine Environment", NIRS-R-3, National Institute of Radiological Science, Chiba-shi, Japan (1974) 45-46) reported the absorption rate of fish with RI administered via the digestive tract. They gave high absorption rates of 20-80% for ^{137}Cs and ^{65}Zn, medium values for ^{131}I, ^{60}Co and ^{54}Mn, and low values for ^{85}Sr and ^{144}Ce; only 2% of the ^{144}Ce administered was absorbed by adult fish.

Finally, perhaps I can refer to our work on concentration factors for immature anchovy. Since the entire fish is consumed in Japan, the immature anchovy is regarded as the critical food for bone-seeking radionuclides. As you will realize, immature anchovy represents one particular stage in the fish's development. Owing to its poor swimming ability immature anchovy tends to move with contaminated water or effluents from the fuel-reprocessing plant. The actual contamination period for this organism was thus relatively restricted. A detailed method of estimating internal doses to man from immature anchovy has been previously reported by Saiki et al. (Ref. [1] of our present paper).

M.N. KHALANSKI: Did you not observe a rapid decrease in the activity of the seawater as soon as the cerium chloride was introduced, owing to adsorption on the wall and surface of your experimental tank?

M. SAIKI: We observed a rapid decrease in the concentration of soluble ^{144}Ce in the seawater in the aquarium, soon after the RI was introduced. However, our RI tracer experiments on living fish were started when the concentration of soluble ^{144}Ce in the aquarium water had almost reached the equilibrium condition.

J.A. HETHERINGTON: In your measurements on the distribution of cerium between solution and suspended matter, for how long were your samples stored after collection before you analysed them? We have measured the partition of ^{144}Ce between solution and suspended material in seawater and we find, like you, that approximately 70% of the total ^{144}Ce is associated with material retained by a 0.22 μm filter. We also found, however, that the fraction increased with the time between collection and filtration; after one week's storage, for example, an amount of the order of 90% was retained on a 0.22 μm filter.

M. SAIKI: We are glad to hear of your experimental results showing the same tendency as those reported in our paper. We also recognize the importance of the time elapsing between collection of seawater and analysis of the sample. We intend to reduce the storage period after collection, but it is at present about two hours.

A TRACE ELEMENT STUDY IN THE THAMES ESTUARY

C.D. BEVAN,* S.A. HARBISON, +
L.A. NELSON, J.R.A. LAKEY
Department of Nuclear Science and Technology,
Royal Naval College,
Greenwich, London,
United Kingdom

Abstract

A TRACE ELEMENT STUDY IN THE THAMES ESTUARY.

The paper reports a study of the trace element concentrations in part of the Thames Estuary, a partially enclosed area of water. The concentration of the trace elements in the 'dissolved fraction' and the 'non-dissolved fraction' of the water and in the sediment is reported and relationships between these elements are sought. This information may be important in predicting transport mechanisms for trace elements and gives an indication of their fate. Neutron activation analysis was selected for this study but preconcentration was required for elements in the dissolved fraction. Chelex-100 resin was used to selectively remove cobalt, zinc, chromium, mercury and scandium from the water matrix. All elements except mercury were reproducible with a coefficient of variation less than 20%. Ultraviolet absorbance (to give an indication of organic content) and salinity were recorded for samples taken at points spanning the Estuary in Sea Reach. All other fractions were subject to instrumental neutron activation analysis and all results were correlated for every pair of elements in each sample component. Six groups of elements were identified with correlations which indicated relationships within each group. The most important group was common to the muddy sediment and the non-dissolved fraction in the water. There were no strong correlations in the dissolved fraction but cobalt, zinc and chromium were linked with UV absorbance and salinity. Iron, cobalt, chromium and hafnium did not align with the main grouping and possible mechanisms for these differences are suggested.

1. INTRODUCTION

The natural concentration of trace and minor elements in sea water is required when limits are set to control the discharges of both radioactive and non-radioactive material to the sea. This is because the relationship between the element and identifiable components of the marine system helps to identify critical pathways through which the discharge might have an impact on man. The bottom sediment is a component in which many elements are deposited and it also may play an important part in the transport of the trace elements. It is also possible that the trace element content of a particular component of the system may help to identify the component or its source. A trace element study should therefore encompass as many matrices as possible.

Estuaries are often at the focal point of man's activities and are a major source of trace element input to the sea. Some of this may be associated with pollution but this has been reduced in the Thames Estuary due to strong controls applied over the last 10 years. The discharge of domestic and industrial effluent continues, however, and dumping of sewage sludge containing heavy metals occurs in the outer estuary [1]. Some of

* Now with Hoechst Pharmaceuticals Research Laboratories of Hoechst UK Ltd.
+ Now with the Nuclear Installations Inspectorate, London.

this may be returned to the estuary by tidal action and this process together with the direct discharge to the estuary (producing components with a residence time of about 2 months), makes a study of this estuary worth while.

It is thought that the suspended particulate component of sea water may be responsible for part of the trace element input to the sea [2], where it eventually makes up the bottom sediment.

The transport of heavy elements is not fully understood and many reports have been based on limited numbers of measurements. In this study multi-element activation analysis of components of the marine system is used together with correlation of the trace elements to indicate associations with other trace elements as well as with iron, organic content (represented by ultraviolet absorbance) and salinity.

2. SAMPLE COLLECTION

Sea water, suspended particulate and bottom sediment samples were obtained from stations in the Sea Reach Area of the Thames Estuary (see Fig. 1) over a period from June to December 1972. Surface water samples were collected with a polythene bucket and were filtered through pre-washed Millipore ® membrane filters of average pore size 0.22 μm [3]. Salinity and ultraviolet absorbance measurements were made on 0.3 litres of the sample and the remaining 2.5 litres were rapidly frozen and stored frozen. The filter membranes with the filtered sediment particulates were bagged and stored. Sediment samples were collected using a 0.1 m^2 grab [4] and a Craib corer [5]. The corer collected samples approximately 0.15 metres long with minimum disturbance of the surface layer. One sub-sample was removed at random from each grab sample and cores were sectioned into 0.025-m-long sub-samples. All sediments were oven-dried at 90°C.

The sediment samples were found to be similar in appearance to those reported in a previous survey of the Thames Estuary [6].

Samples taken at stations 1 and 2 were sandy clay mixed with shells and small stones. In the middle of the estuary at station 3 the sediment was a black mud with some pieces of stone and apparently depleted in oxygen. Samples from the northern half of the line at stations 4 and 5 consisted of shells mixed with fine sand and about 1% silt.

Since the stones and shells were not likely to play a part in trace element transport and could distort the relationships between the mineralogical components of the sediment and the trace elements, all samples of sediment were sieved through 500 μm mesh before analysis. A 5-g sample from each sediment sub-sample was ground in agate and 0.15 g taken for activation analysis.

3. ANALYTICAL TECHNIQUES

3.1. Preconcentration of elements from sea water

The trace elements mercury, cobalt, zinc, chromium and scandium were selectively removed from each sample of sea water by using chelating

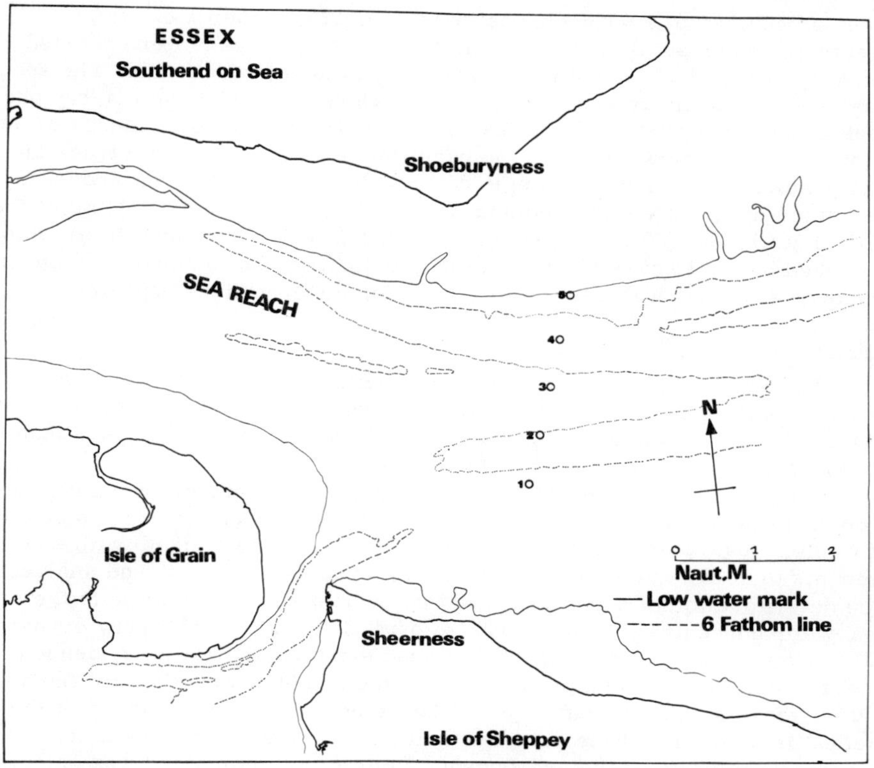

FIG. 1. Thames Estuary sampling area.

ion exchange materials [7, 8]. The medium employed was a purified form of Chelex-100 resin,[1] which has a high affinity for many of the transition elements and a low affinity for alkali and alkaline earth metals and the halogens. Each frozen sea water sample of 1.5 to 2.5 litres was allowed to thaw and the water was pumped through a small column at flow rates of 1 ml/min, after which the resin was drained and lyophilized. The resin was transferred to a silica phial which was heat-sealed, canned and irradiated together with multi-element standards, blanks and iron-wire neutron flux monitors. The irradiation time was 4 days at an average neutron flux of 6×10^{12} $n \cdot cm^{-2} \cdot s^{-1}$.

3.2. Activation analysis

After a decay period of several weeks the flux monitors were counted and the resin transferred to polyethylene capsules for counting. A high-resolution gamma-ray spectrometer using a lithium drifted germanium

[1] Biorad Laboratories Ltd.

detector coupled to a 4096-channel pulse-height analyser was used for spectrum analysis. Samples and multi-element standards were counted for a live time of 10^4 s and dead time losses did not exceed 5%. The spectrum was recorded over the energy range 60 to 1800 keV. The gamma-ray photo peak areas were evaluated and the backgrounds subtracted by means of the SAMPO computer codes [9]. The quantitative analysis of the 5 trace elements was performed by using the single comparator method [10] and iron was chosen as the most suitable comparator. The gamma-ray spectrum of the activated Chelex-100 concentrate is dominated by gamma lines from isotopes produced by the fission of uranium-235, indicating that natural uranium is removed from sea water by this resin under the conditions reported.

3.3. Precision and accuracy

The reproducibility of this sea water analysis method was tested by at least four repeated measurements on a single large sample from Sea Reach. The coefficient of variation for zinc, cobalt and chromium was less than 8%, rising to 27% for scandium and 50% for mercury.

To check the accuracy of the procedure, trace elements were stripped from a large volume of filtered Thames water by passage through a column of Chelex-100. Samples ($2\frac{1}{2}$ litres) of this water were 'spiked' with known amounts of cobalt, zinc, scandium, mercury and chromium, and analysed. The determination of added trace metals at levels approximating to their sea water concentrations was satisfactory. In all cases recovery exceeded 83%. Gamma-ray spectra of these spiked samples showed no evidence of interference from fission products since the uranium was stripped from the water before spiking. Thus it would be expected that the accuracy of the method is better for these samples than for the real Thames samples.

The sensitivity of the preconcentration/activation method was calculated by the method of Currie [11]. For this purpose tests were performed on a sample with salinity of 33% and the sample volume was set at 2 litres with an irradiation time of 96 hours followed by 37 days decay. The limits are specified in Table I on the basis of a 10% standard deviation of the background count.

TABLE I. LIMITS FOR QUANTITATIVE DETERMINATION OF TRACE ELEMENTS IN SEA WATER

Element	Isotope produced	γ-ray employed for determination (energy, keV)	Limit of quantitative determination (expressed as μg of element)
Mercury	Hg-203	279.1	0.025
Chromium	Cr-51	320.0	0.128
Scandium	Sc-46	889.6	0.0013
Scandium	Sc-46	1120.0	0.0010
Zinc	Zn-65	1115.4	0.52
Cobalt	Co-60	1173.1	0.011
Cobalt	Co-60	1332.4	0.012

3.4. Salinity and ultra-violet method

Salinity is usually derived from the chlorinity by Unesco methods [12]. Chlorinity was measured by the Mohr-Knudsen titrimetric method [13]. Estuarine waters are composed of varying mixtures of sea water and river water which is originally derived from land drainage. Whilst salinity methods are useful in determining the fraction of sea water, a parameter related to the quantity of land drainage was required, so measurements were made of the ultraviolet absorption of the sea water in 100-mm cuvettes at wavelengths between 225 and 350 nm. Photo-oxidation of the organic matter in the water by exposure to ultra-violet radiation at 254 nm produced the expected reduction in absorbance between these wavelengths as the oxidation proceeded.

4. SUMMARY OF TRACE ELEMENT MEASUREMENTS

The range and mean value of all measurements is given in Table II, the dissolved component being expressed as $\mu g/l$ and the remainder in ppm dry weight except iron which is in $\% Fe_2O_3$.

The concentrations reported in the dissolved component are in the main similar to other results reported on the UK coast, the exception being mercury which is high compared to other measurements in the same area [14]. In Ref. [14] the sampling and analysis techniques were markedly different from the present study and higher precision was claimed. Measurements on chromium and scandium are sparse, the former being subject to uncertainty about the oxidation state of the element in the sea [15]. Scandium is predominantly found in particulate and low levels have been found generally in the dissolved component.

Ten elements were identified in the suspended particulate fraction including iron. This was not readily detected in solution since it cannot exist as water-soluble ionic iron at concentrations greater than 0.5 pg/l in sea water [16] and is mainly present as colloidal and particulate hydrous iron oxide. It is this hydrous iron oxide which is said to exhibit a scavenging effect far out of proportion to its total concentration [17] and it is significant that iron is at a higher concentration in the suspended particulate component than in the sediments.

Cerium, chromium and thorium are on the same level in the suspended particulate fraction as in muddy deposits, whereas cobalt, caesium, europium and scandium are enriched in the suspended particulate fractions. Hafnium is exceptional in being at a higher concentration in the sediments. In general, trace element concentrations are lower in the sandy deposits than in the muddy deposits altough the differences are insignificant for iron and cobalt.

5. DISCUSSION OF CORRELATION ANALYSIS

The correlation analysis was applied to each pair of elements in turn for the measurements made in all sea water samples (dissolved fraction and suspended particulate) and sediments from three sampling locations. The method used an IBM Subroutine Multiple Linear Regression (IBM 1130 Scientific Subroutine Package Program Number 1130-CM-02X).

TABLE II. IRON AND TRACE ELEMENTS IN THE WATER, SUSPENDED PARTICULATES AND SEDIMENTS OF SEA REACH

MATRIX	Sea water (dissolved component)		Surface water (suspended particulate)		Bottom mud <75 μm 49 - 56%		Bottom mud <75 μm 20 - 57%		Bottom sand <75 μm 1 - 2%	
	(values quoted as μg/l)		(values quoted as ppm dry weight, except Fe, which is %Fe_2O_3)							
Positions sampled	1,2,3,4,5		1,2,3,4,5		2		3		4	
Element	Range	Mean	Range	Mean	Range	Mean	Range	Mean	Range	Mean
Ce	-	-	38 - 80	68	18 - 178	63	25 - 162	72	13 - 54	30
Co	0.022 - 0.133	0.059	10 - 26	19	4.4 - 8.3	6.4	3.3 - 18	6.5	2.4 - 15	6.2
Cr	0.02 - 0.48	0.106	.50 - 103	71	16 - 162	76	41 - 178	83	14 - 82	31
Cs	-	-	2.3 - 4.7	4.2	0.95 - 4.5	2.6	1.1 - 7.6	2.8	0.78 - 1.9	1.2
Eu	-	-	1.0 - 1.6	1.3	0.39 - 1.7	0.84	0.49 - 2.1	0.91	0.29 - 0.89	0.49
Fe	-	-	2.4 - 6.6	5.5	1.4 - 3.1	2.3	1.1 - 5.6	2.3	0.77 - 5.1	2.1
Hf	-	-	1.7 - 5.2	3.7	2.8 - 49	20	7.4 - 60	23	1.1 - 37	8.8
Hg	0.02 - 0.61	0.15	-	-	-	-	-	-	-	-
Sc	0.0003 - 0.0026	0.0012	7.0 - 16	13	1.8 - 8.6	4.9	2.1 - 8.2	5.0	1.3 - 3.1	2.2
Th	-	-	4.5 - 9.7	8.2	0.60 - 25	7.8	2.7 - 25	9.4	1.3 - 7.0	3.0
Zn	5.0 - 27	11.6	5.3 - 33	13	-	-	-	-	-	-

TABLE III. CORRELATION COEFFICIENTS BETWEEN PAIRS OF DISSOLVED TRACE ELEMENTS AND BETWEEN DISSOLVED TRACE ELEMENTS AND OTHER PARAMETERS

	Co	Cr	Hg	Sc	Zn	Salinity S‰	u.v.	Fe particulate in water
Co	1.00							
Cr	0.27	1.00						
Hg	0.04	0.38	1.00					
Sc	0.35	0.02	0.11	1.00				
Zn	0.44	0.47	0.12	0.00	1.00			
Salinity S‰	-0.39	-0.34	-0.11	0.18	-0.31	1.00		
u.v.	0.33	0.15	-0.13	-0.08	0.55	-0.68	1.00	
Fe particulate in water	0.01	-0.29	-0.06	-0.02	0.24	0.24	0.13	1.00

5.1. Sea water samples

The correlations for the suspended particulate fraction are given in Table IV and for the dissolved fraction in Table III together with the ultraviolet absorbance and salinity. There is no strong correlation between cobalt, chromium, mercury, scandium and zinc in the dissolved fraction although there is a moderate correlation between ultraviolet absorbance and cobalt, zinc and, to a lesser extent, chromium, so these are identified as a group (A). The iron content of the dissolved fraction must be below 0.5 pg/l but it could be measured in the suspended particulate fraction, where strong correlation was found between iron and cerium, cobalt, caesium, europium, hafnium, scandium and thorium. All these elements correlate with each other and to a lesser extent with chromium and so are identified as a group (B).

Strong negative correlation is found between u.v. absorbance of the sea water and its salinity which is most probably created by the mixing of river and sea water in the estuary. This in turn suggests that the source of cobalt, zinc and chromium may be associated with the organic content of the river water. A slight negative correlation between cobalt, zinc and chromium with salinity is observed.

The strong negative correlation of zinc (which is not found in the suspended particulate group B) with iron and a lesser negative correlation with other elements in the suspended particulate group is interesting and provides a third group (C).

In general the suspended particulate group (B) appears to represent elements associated in the same phase of the sea water. The different performance of chromium in this group is notable since it is not depleted in the suspended particulate fraction compared to sediments. Under alkaline oxidizing conditions the soluble anionic chromate complexes are formed of the type $(CrO_4)^{2-}$ and so chromium may be removed into solution in the well aerated surface water.

TABLE IV. INTER-ELEMENT CORRELATIONS OF Fe AND TRACE ELEMENTS IN THE SUSPENDED PARTICULATES AND BOTTOM SEDIMENTS OF SEA REACH

		Ce	Co	Cr	Cs	Eu	Fe	Hf	Sc	Th	Zn
Surface Water	Ce	1.00									
(Suspended particulates)	Co	0.80	1.00								
	Cr	0.57	0.39	1.00							
Positions 1, 2, 3, 4, 5	Cs	0.70	0.80	0.53	1.00						
	Eu	0.77	0.83	0.51	0.80	1.00					
	Fe	0.72	0.79	0.46	0.89	0.79	1.00				
	Hf	0.69	0.65	0.46	0.56	0.66	0.69	1.00			
	Sc	0.80	0.85	0.58	0.87	0.87	0.90	0.71	1.00		
	Th	0.82	0.76	0.69	0.82	0.82	0.81	0.70	0.89	1.00	
	Zn	-0.13	0.06	-0.25	-0.36	-0.13	-0.52	-0.30	-0.27	-0.34	1.00
Bottom Sediment (muddy)	Ce	1.00									
Position 2	Co	0.59	1.00								
<75 μm 49 - 56%	Cr.	0.91	0.54	1.00							
	Cs	0.66	0.70	0.69	1.00						
	Eu	0.91	0.61	0.86	0.77	1.00					
	Fe	0.59	0.85	0.52	0.70	0.66	1.00				
	Hf	0.91	0.40	0.93	0.50	0.86	0.42	1.00			
	Sc	0.85	0.69	0.86	0.91	0.88	0.70	0.73	1.00		
	Th	0.99	0.59	0.91	0.70	0.92	0.58	0.94	0.87	1.00	
	Zn	-	-	-	-	-	-	-	-	-	
Bottom Sediment (muddy)	Ce	1.00									
Position 3	Co	0.54	1.00								
<75 μm 20-57%	Cr	0.86	0.79	1.00							
	Cs	0.58	0.95	0.78	1.00						
	Eu	0.78	0.77	0.90	0.74	1.00					
	Fe	0.62	0.98	0.85	0.94	0.83	1.00				
	Hf	0.85	0.36	0.74	0.32	0.70	0.46	1.00			
	Sc	0.76	0.92	0.92	0.93	0.86	0.95	0.56	1.00		
	Th	0.97	0.49	0.84	0.53	0.77	0.57	0.87	0.73	1.00	
	Zn	-	-	-	-	-	-	-	-	-	

TABLE IV. (Cont.)

		Ce	Co	Cr	Cs	Eu	Fe	Hf	Sc	Th	Zn
Bottom Sediment (sandy)	Ce	1.00									
Position 4	Co	-0.38	1.00								
<75 μm 1-2%	Cr	0.58	-0.38	1.00							
	Cs	0.20	0.27	0.28	1.00						
	Eu	0.78	-0.43	0.41	0.09	1.00					
	Fe	-0.38	0.99	-0.41	0.23	-0.41	1.00				
	Hf	0.81	-0.52	0.63	-0.17	0.84	-0.51	1.00			
	Sc	0.79	-0.21	0.70	0.49	0.77	-0.23	0.78	1.00		
	Th	0.93	-0.47	0.60	0.17	0.83	-0.47	0.90	0.86	1.00	
	Zn	-	-	-	-	-	-	-	-	-	

TABLE V. SUMMARY OF CORRELATION ANALYSIS

Correlation[1]	SEA WATER		SEDIMENT	
	Dissolved	Suspended particulate	Mud	Sand
Elements omitted[2]	Ce, Cs, Eu, Fe, Hf, Th	Hg	Hg, Zn	Hg, Zn
Strong +ve (above 0.5)	None	[Ce, Co, (Cr), Cs, Eu, Fe, Hf, Sc, Th] Group B	[Ce, Co, Cr, Cs, Eu, Fe(Hf), Sc, Th] Group D	[Ce—Cr—(Eu), —Hf, Sc, Th] Group E [Co, Fe] Group F
Moderate +ve (0.25 - 0.49)	[Co, Zn, (Cr) u.v.] Group A	Cr with Co, Fe, Hf	Hf, with Co, Fe	Cs with Cr, Sc, Co
Moderate -ve -0.25 to -0.49	Salinity S‰ with Co, Zn and Cr	Zn with Th, Hf, Cs, Sc		
Strong -ve above -0.5	u.v., Salinity S‰	[Zn, Fe] Group C		Hf with Co and Fe

[1] Boxes denote groups in which all elements correlate with all others in the group. Minor exceptions which correlate with the majority of other members are bracketted. The letter of the group is indicated beside each box within a broken line to permit cross-reference to the text.

[2] Analysis applied to Ce, Co, Cr, Cs, Eu, Fe, Hf, Hg, Sc, Th, Zn except where omitted (at head of column) for analytical reasons.

5.2. Sediment

The correlation analysis given in Table IV was performed only on the bottom sediment from sampling stations 2, 3 and 4 selected because 2 and 3 were described as muddy and 4 was predominantly sandy. Measurements were not possible on zinc and mercury owing to the limitation of the activation analysis technique but all the elements analysed (and to a lesser extent hafnium) correlated strongly with each other in the muddy samples and are therefore assigned to a fourth group (D). The only weakness in correlation is exhibited for hafnium with iron and with cobalt. It is suggested that hafnium may be associated with zirconium because it has similar chemical behaviour (e.g. zircon).

A markedly different correlation is found for the sandy deposit in which cerium, chromium, hafnium, scandium and thorium form a strongly correlated group (E) with which europium is associated.

Cobalt and iron form a sixth separate group (F), which has a strong negative correlation with hafnium. These differences are proposed as a useful method of identification of the sediment types and may have some geochemical origin. The strong resemblance between the mud group (D) and the suspended particulate group of sea water (B) suggests a common geochemical nature for these two groups.

In a preliminary correlation analysis, Nelson [18] suggests that caesium, europium and scandium, together with cobalt and iron, are strongly associated with clay. The muddy sediments (2 and 3) contain between 5% and 20% clay whilst the sandy sediment has less than 1% clay. This is consistent with the correlations although the separation into two groups (E and F) for the sandy sediments have not been studied.

6. CONCLUSION

6.1. Correlation results

The correlation study is summarized in Table V, in which six distinct groups of elements are identified. These groupings provide some evidence as to the phase in which the elements are associated and the implications of these groups form the conclusion of this report.

Group (A): cobalt, zinc and chromium in solution

The only significant correlation in solution was found in this group which correlates positively with ultraviolet absorbance and negatively with salinity. Gibbs [19] shows that most of the transport of iron in fresh water involves metallic coatings or crystalline particles, so this correlation may be due to mixing.

Group (B): cerium, cobalt, chromium, caesium, europium, iron, hafnium, scandium and thorium in particulate

These elements are possibly associated with clay. There is no correlation between chromium, cobalt and scandium and their soluble conterparts. Gibbs [19] suggests that a large proportion of chromium is transported in the crystalline particles but cobalt may be co-precipitated with iron in the metallic coatings.

Group (C): zinc and iron in particulate

These elements have a strong negative correlation which contrasts with the correlation of Groups (A) and (B), but no mechanism can be postulated.

Group (D): cerium, cobalt, chromium, caesium, europium, iron, hafnium, scandium, thorium in muddy deposits

These elements are identical to group (B) but hafnium is less strongly related possibly owing to its expected association with zirconium. Five of the elements in groups (B) and (D) are correlated positively with the clay content of the sediment.

Group (E): cerium, chromium, europium, hafnium, scandium, thorium in sandy deposits

These elements contain most of the group (D) elements which do not correlate with the clay content and must be in a separate phase.

Group (F): cobalt and iron in sandy deposits

These elements correlate with clay and some elements in Group (D), the remainder of group (D) being found in group (E). This separation of group (D) is a significant phenomenon and should direct attention to geochemical studies of the sediment in relation to their trace element content.

6.2. Analytical methods

The activation analysis method chosen for this study is considered to be satisfactory for the elements reported although it seems to be least reliable for mercury. The use of u.v. absorbance as an indicator of organic content is a quick method but further work is needed to confirm its value.

The conclusions on correlation results indicate the importance of the geochemical analysis of the components other than those soluble in sea water.

Particle size analysis has been used in this paper to specify the sediment type but more specific identification of at least the clay minerals is needed. In particular the chemical properties of chromium and hafnium must be elucidated.

ACKNOWLEDGEMENTS

Thanks are due to Professor J. Edwards, Head of the Department of Nuclear Science and Technology, and to the external supervisors, Mr. A. Preston and Dr. J.E. Portmann, for their encouragement and support; and to Mrs. M.M. Byford and Dr. M.D. Rogers for their valuable contributions. The authors gratefully acknowledge the financial support of the Science Research Council through a research grant and two research studentships.

© Controller, Her Majesty's Stationery Office, London 1975.

REFERENCES

[1] SHELTON, R.G.J., Mar. Poll. Bull. 2 (1971) 24.
[2] RILEY, J.P., CHESTER, R., Introduction to Marine Chemistry, Ch. 13, Academic Press, London (1971).
[3] INTERNATIONAL ATOMIC ENERGY AGENCY, Reference Methods for Marine Radioactivity Studies, Technical Reports Series No. 118, IAEA, Vienna (1971).

[4] VAN VEEN, J.H., Dock and Harbour Authority 27 (1946) 167, 198.
[5] CRAIB, J.S., J. Cons. Perm. Int. Explor. Mer 30 (1965) 34.
[6] WATER POLLUTION RESEARCH LABORATORY, "Effects of polluting discharges on the Thames Estuary", Ch.11, Bottom Deposits, Her Majesty's Stationery Office, London (1964).
[7] RILEY, J.P., TAYLOR, D., Anal. Chim. Acta. 40 (1968) 479.
[8] GOYA, H.A., LAI, M.G., Adsorption of Trace Elements from Sea Waters by Chelex-100, USNRDL-TR-67-129 (1967).
[9] ROUTTI, J.T., SAMPØ, A FORTRAN IV Program for Computer Analysis of Gamma Spectra from Ge(Li) Detectors and for other Spectra with Peaks, UCRL-19452, UC-34 PHYSICS. T10-4500, Lawrence Radiation Laboratory, University of California, Berkeley (1969).
[10] GIRARDI, F., GUZZI, G., PAULY, J., Anal. Chem. 37 (1965) 1085.
[11] CURRIE, L.A., Anal. Chem. 40 (1968) 587.
[12] UNESCO, Report of the Joint Panel on the Equation of State of Sea Water, Ref.NS/9/114B, Paris (1962).
[13] KNUDSEN, M., FORCH, C., SØRENSEN, S.P.L., Wiss. Meeresuntersuch, N.F. 6 (1902) 123.
[14] SMITH, J.D., NICHOLSON, R.A., MOORE, P.J., Nature (London) 232 (1971) 393.
[15] FUKAI, R., Nature (London) 213 (1967) 901.
[16] COOPER, L.H.N., J. Mar. Biol. Assoc. U.K. 22 (1948) 221.
[17] GIBBS, R.J., Science 156 (1967) 1734.
[18] NELSON, L.A., A Study of Minor Elements in the Sediments of the Thames Estuary, Ph.D. Thesis, Royal Naval College, Greenwich, London, U.K. (1975).
[19] GIBBS, R.J., Mechanisms of Trace Element Transport in Rivers, Science 180 (1973) 71.

DISCUSSION

E.I. HAMILTON: Your data for levels of Hf in bottom muds are most interesting; they seem to be very high compared with those expected for the Thames estuary. Using a natural geochemical abundance ratio of ~50 for the Zr-Hf pair, your data indicate a possible zirconium content of ~0.1% for the muds. As most of the hafnium and zirconium in natural systems is present in the heavy mineral zircon, which will tend to concentrate in sand deposits, your data suggest that there is no noticeable enrichment of Hf in sandy facies. It would be interesting to identify the physical and chemical form of hafnium in muds and, in particular, to consider a 'non-natural' origin for the hafnium. What were your experimental blank values for Hf?

J.R.A. LAKEY: I agree with your deductions based on the chemical relationship between hafnium and zirconium; it was this which led us to suggest the location of hafnium in the heavy mineral zircon. The high concentrations in the muddy, compared with sandy, sediments are therefore difficult to understand. There might be some clue to this in the differences in the correlation of hafnium in the two types of deposit. We are not aware of any artificial input of hafnium to the Thames.

The nuclear reaction $^{180}Hf(n,\gamma)^{181}Hf$ was employed, and the 482-keV gamma line from ^{181}Hf was used in the determination. The limits for quantitative determination of hafnium were 0.11 μg, 0.15 μg and 0.037 μg in muddy sediment, sandy sediment and suspended particulate, respectively. These are equivalent to concentrations below 0.5 ppm. We tested our methods on United States Geological Survey standard rock mineral samples, and our estimates of hafnium content were well inside the range of other reported measurements.

J.A. HETHERINGTON: May I take up one point you made in relation to the Fe/Mn scavenging process, which is considered to be an important mechanism for the removal of a number of trace elements from seawater. If I understood correctly, you said that your conclusion — that the concentrations of Fe and Mn were higher in suspended matter than in bottom sediment — was compatible with this mechanism. I do not understand why you believe so; it would appear to me that if Fe and Mn are reaching the bed in association with settling suspended matter, then the concentrations should be the same. If they are not, this could perhaps be regarded as evidence that Fe and Mn are being remobilized again some time after they have become consolidated with the sediments.

J.R.A. LAKEY: We found iron at a higher concentration in the suspended particulate than in the sediment (Table II), and you will see that the other elements we measured were also found at different concentrations. Some of these differences may be associated with the difference in iron concentration, and in this connection I referred to the work of Gibbs, reported in Ref.[19] of my paper. He stresses the importance of the metallic coating of the particle in trace-element distributions in fresh water and this suggests that we need to understand the genesis of the particles making up the suspended particulate before proposing a mechanism for removal of trace elements from the sea water. Our samples of sea water were taken at the surface of the sea and there is opportunity for further changes to occur before this material reaches the sea-bed. Therefore I see no reason to expect identical concentrations and I do not think that these results are strong evidence for remobilization.

D.N. EDGINGTON: I should like to stress the importance of differences in particle size in regard to their effect on trace-element content. I feel that if this factor is taken into account there will be considerable differences in the trace-element correlations.

J.R.A. LAKEY: I agree that particle size is important, and I am sure that trace-element correlations of fractions of the sediment within a limited particle-size range would differ from the bulk sample. The differences between trace-element concentrations and correlations in sandy and muddy sediments confirm that particle-size effects are important. Further evidence on this effect is given in Ref.[18] of the paper.

Yu.V. SIVINTSEV: Since your work was performed using neutron-activation analysis, how did you separate the thorium effects from the background of uranium effects? As far as I know, in this type of analysis both U and Th are detected from the Ba/^{140}La gamma-line. How did you separate these effects?

J.R.A. LAKEY: In our work the thorium was measured by identification of the 312-keV gamma line of protactinium-233, formed by a (n, γ, β) reaction after thermal-neutron activation of thorium-232. This line was successfully resolved in the suspended particulate and the sediments, where it suffers minor interference from ytterbium-169 and chromium-51; however, there is no interference from the fission products of uranium at this gamma energy. The gamma spectrum measurements reported in the paper involved the use of a 50 cm^3 Ge(Li) detector with a resolution of 2.4 keV for the 1332.4 keV gamma ray of cobalt-60 at full width half maximum, and a peak-to-Compton ratio of 24.5 to 1.

ENVIRONMENTAL BEHAVIOUR OF THE BINUCLEAR OXYGEN-BRIDGED NITRATONITROSYLRUTHENIUM COMPLEX

Y. NISHIWAKI*
Atomic Energy Research Institute, Kinki University

Y. HONDA, Y. KIMURA
Department of Nuclear Reactor Engineering,
Faculty of Science and Technology, Kinki University,
Higashi-Osaka City, Osaka

T. ISHIYAMA, T. MATSUMURA, T. MAMURO
Radiation Center of Osaka Prefecture,
Sakai City, Osaka,
Japan

Abstract

ENVIRONMENTAL BEHAVIOUR OF THE BINUCLEAR OXYGEN-BRIDGED NITRATONITROSYL-RUTHENIUM COMPLEX.

The physico-chemical behaviour of prepared binuclear oxygen-bridged nitratonitrosylruthenium, $[RuO_2(NO)_2(NO_3)]_2O$, was investigated by filter paper electrophoresis, paper chromatography, and infrared absorption spectra of a decomposition product in fresh water. The binuclear (Ru-O-Ru) nitrate complex in aquatic systems was electrophoretically fractionated into at least four 'anionic' species, one 'neutral' species and one 'cationic' species, with aging, and about 75% of the total Ru was present in the anionic fractions. It was found that about 90% of the total Ru in the solution was taken up in the fraction with R_f less than 0.2, corresponding to the anionic mononitro-, mononitrato- and probably the polynuclear species from paper chromatography. The proportions of R_f 0-0.2 and R_f 0.2-1.0 to total Ru were scarcely altered by increasing the aging time more than one day. On the basis of infrared spectra of decomposition product by aging the (Ru-O-Ru) nitrate, it was confirmed that two oxide anions, nitrosyl, nitro, nitrato groups and water molecule are all coordinated to Ru metal, assuming $[RuO_2(NO)(NO_2)(NO_3)(H_2O)]^-$ as one of the possible chemical forms of the anionic species. RuNO-nitrato, RuNO-nitro and (Ru-O-Ru) nitrate complexes were entirely adsorbed by the addition of 500 - 700 ppm of activated carbon produced from coconut shell by batch equilibrium test. In spite of the high concentration of natural ion exchangers, the percentages of adsorption of these complexes were not observed to be very high.

1. INTRODUCTION

The study of the behaviour of radioactive contaminants and their stable natural isotopes in aquatic systems is important for the assessment of the radiological impacts of releases from nuclear facilities. Ruthenium is one of the predominant fission products found in reactor fuel-reprocessing liquid wastes. The process solutions obtained by dissolving irradiated nuclear reactor fuel in nitric acid are known to contain nitrosylruthenium (RuNO) compounds, namely, the nitrato-, nitro-, (Ru-O-Ru) nitrate complexes and uncomplexed RuNO [1]. Because of the complexity of the chemical state of ruthenium and its importance from the radioecological point of view, the physico-chemical forms of radioruthenium in aquatic environments and their accumulation by aquatic biota have been studied extensively [2-31]. Story and Gloyna [4] and Bhagat and Gloyna [9] reported the

* Present address: International Atomic Energy Agency, Vienna.

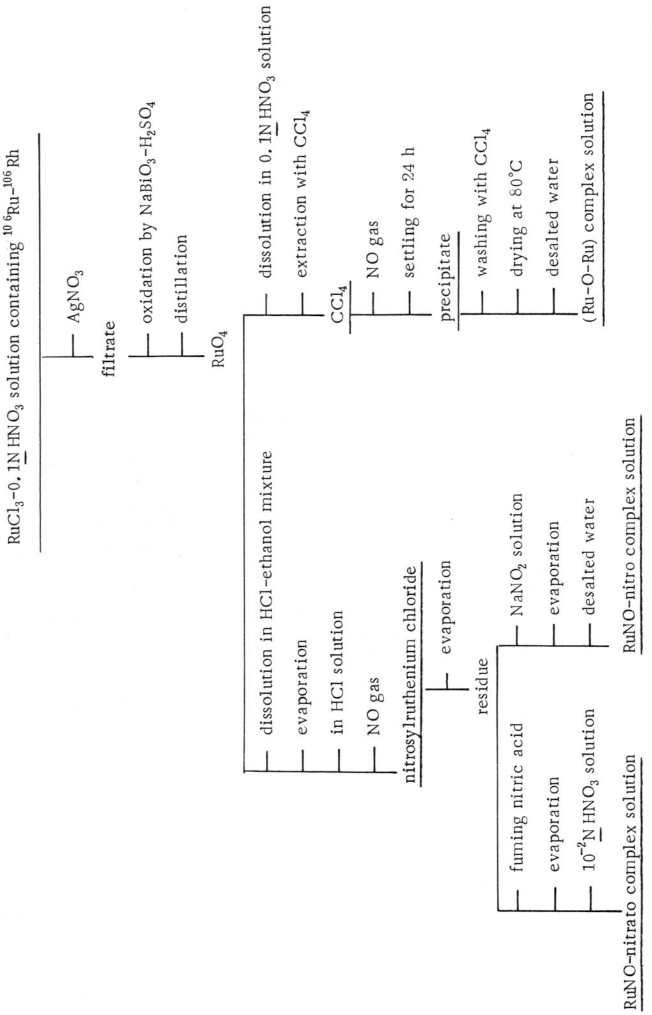

FIG. 1. Preparation methods of nitrosylruthenium complexes.

environmental behaviour of RuNO-nitrato and RuNO-nitro, RuNO-nitrato, respectively, from studies using paper chromatography, absorption spectroscopy and uptake by ion-exchange resins, clay, soil and organic debris. Marazović et al. [6,8], Ishikawa et al. [17, 21, 22], and Honda et al. [18] reported the physico-chemical behaviour of Ru-nitro, RuNO-nitrato, RuNO-chloro and Ru-chloro complexes in sea water by paper electrophoresis and electrodialysis studies. Peperstraete et al. [23] and Vos et al. [28] reported that the electrophoretic separation pattern of RuNO-nitrato, RuNO-nitro, RuNO-chloro, RuNO-hydroxo and Ru-chloro complexs in fresh and sea water was changed by aging, but dilution had no marked influence on this pattern. In spite of a number of studies on radioruthenium in aquatic systems, little work has been done on the environmental behaviour of the (Ru-O-Ru)nitrate complexes.

Ishiyama [32] reported that the (Ru-O-Ru) nitrate complex was entirely adsorbed to diethylaminoethyl cellulose (anion exchanger) in neutral media containing sodium nitrate. In the present study the physico-chemical behaviour of prepared binuclear oxygen-bridged nitratonitrosylruthenium complex, (Ru-O-Ru) nitrate, whose chemical formula was given as $[RuO_2(NO)_2(NO_3)]_2O$ by Martin et al. [33], was investigated in natural fresh-water environments by high-voltage paper electrophoresis, paper chromatography and infrared spectroscopy together with the adsorption behaviour to activated carbon and several natural ion exchangers.

2. MATERIALS AND EXPERIMENTAL METHODS

2.1. Radioactive materials

The radioisotope used was obtained from The Radiochemical Centre, Amersham, as a form of carrier-free ruthenium-106 chloro-complex in 8-10N hydrochloric acid. The stock solutions of RuNO-nitrato, RuNO-nitro and (Ru-O-Ru) nitrate complexes were prepared in accordance with the method described in Ref. [34] except for the use of sodium bismuthate(V) as an oxidizing agent. The preparation methods are shown in Fig.1. The concentrations of radioactivity and carrier of the three stock solutions were adjusted to 1 μCi/ml and 10^{-5} mol/l, 1 μCi/ml and 10^{-5} mol/l, and 3 μCi/ml and 3.5×10^{-3} mol/l, respectively.

2.2. One-dimensional electrophoresis of the stock solutions

To confirm the proportion of filterable form to total Ru, a filtration test of the three kinds of ^{106}RuNO-complexes in fresh water was carried out by using Millipore filters equivalent to a 0.45 μm size, before the paper electrophoresis experiments. A commercially available apparatus for high-voltage paper electrophoresis made by Toyo Filter Co. Ltd was used. The experimental conditions for the electrophoresis were as follows: basic electrolyte, 0.1N $NaClO_4$; filter paper strip (Toyo No.50) 40 × 2 cm; voltage, 1150 V; voltage gradient, 33.8 V/cm; current density, 3.4-5.2 mA/cm; duration of electrophoresis, 20 min. Three filter paper strips wetted with 0.1N $NaClO_4$ were placed horizontally between the electrode vessels and were run parallel in the apparatus. To prevent thermal effects, the inside of electrophoretic apparatus was maintained

TABLE I. LIMNOLOGICAL CHARACTERISTICS OF THE FRESH WATER

Tested item	Mean value
Temperature	11 ± 1°C
pH	7.8
Alkalinity	30.3 ppm
Na	2.6 ppm
K	0.2 ppm
Ca	25 ppm
Cl^-	9.6 ppm
Total dissolved solids	114.7 ppm
Residue after ignition (inorganic)	79.9 ppm
Dissolved organic matter (loss on ashing)	34.8 ppm
Total suspended matter	1.8 ppm
Electrical conductivity	241 μmho/cm

at 11 ± 1°C. Radioactive (Ru-O-Ru) nitrate complex stock solution used in this experiment was prepared by dissolving original solution in filtered fresh water using Millipore filter (0.45 μm). The limnological characteristics of the water sampled at a river in Samegai Trout Hatchery, Shiga Prefecture, are given in Table I. Aging of the solution was carried out by aeration at a flow rate of about 100 ml/min and a temperature of 15 ± 2°C. Each time about 30 μl of the sample was put on the starting point at the middle of the filter paper strip. After completion of the electrophoresis, the filter paper strips were dried with an infrared lamp and exposed to an ordinary X-ray film. The electrophoretic mobilities were determined autoradiographically. The electrophoretic pattern of (Ru-O-Ru) nitrate complex was obtained from the result of the β-counting for ^{106}Ru of each cut fraction.

2.3. Paper chromatography of the stock solutions

Of the stock solution under test, about 10 μl was applied to the bottom section of a Toyo-Roshi No.5A paper strip, 1 × 18 cm, corresponding closely to Whatman No.41. The paper strip was immediately placed in a closed 1000-ml beaker containing 1 cm depth of MIPK (methyl isopropyl ketone) equilibrated with 3N HNO_3 as solvent [35]. After drying, the paper strip was cut into the ten sections and each one was β-counted for ^{106}Ru. A histogram representing the distribution of ruthenium along the paper strip was constructed from these results. The effect of aging on the fate of (Ru-O-Ru) species with time was also determined after adjusting the pH to 7.8.

2.4. Preparation of a decomposition product formed by aging (Ru-O-Ru) nitrate complex in fresh water, and its infrared spectra

The (Ru-O-Ru) nitrate complex of 0.75 g thus prepared was dissolved in a small quantity of distilled water. The saturated solution of the complex was allowed to age at room temperature for several days by aeration. To this aqueous solution, 0.5 g of sodium nitrate was added. The solution was then cooled in a refrigerator, and the crystals formed were recrystallized with distilled water and dried in vacuo. The crystal thus prepared is too hygroscopic in air for elemental analysis to be carried out.

Infrared spectra were obtained by the KBr disc method on an infrared spectrophotometer Model IR-S of the Japan Spectroscopic Co. To confirm the presence of coordinated water the product was deuterated in heavy water on standing at room temperature for three days. The deuterated product was recrystallized in the heavy water.

2.5. Activated carbon and natural ion exchangers

Commercially available activated carbon produced from coconut shell was ground to a standard mesh size lower than 60. The powdered activated carbon thus prepared was used for a batch equilibrium test. Natural ion exchangers used in this experiment were kaolin, sericite, vermiculite and humic acid, respectively. They were ground to a standard mesh size lower than 100.

2.6. Uptake experiments by adsorbers

To measure uptake of the above Ru-complexes by activated carbon, the solution of radioactive Ru-complexes was prepared. The activated carbon was placed in a 50-ml flask and brought in contact with the radioactive solution. The pH values were adjusted with 1N hydrochloric acid and 1N sodium hydroxide. The material was shaken for about one hour and the suspension was filtered through filter paper. An amount of 2 ml of filtrate was dried, and the activity was measured by a GM counter (mica window 1.7 mg/cm^2). Uptakes by natural ion exchangers were also measured by the method just described.

3. RESULTS AND DISCUSSION

3.1. Electrophoresis of (Ru-O-Ru) nitrate complex

Figure 2 shows the apparent electrophoretic mobilities of (Ru-O-Ru) species in fresh water separated by one-dimensional electrophoresis, and Fig.3 shows the influence of aging time on the proportion of 'anionic', 'neutral' and 'cationic' species of the three RuNO-complexes. The cationic fraction, which has mobility towards the cathode, diminished, whereas both the nonionic and anionic fractions increased after one hour's aging in fresh water. The cationic fraction further decreased, and a significant decrease of the nonionic fraction started, but the anionic fraction continued

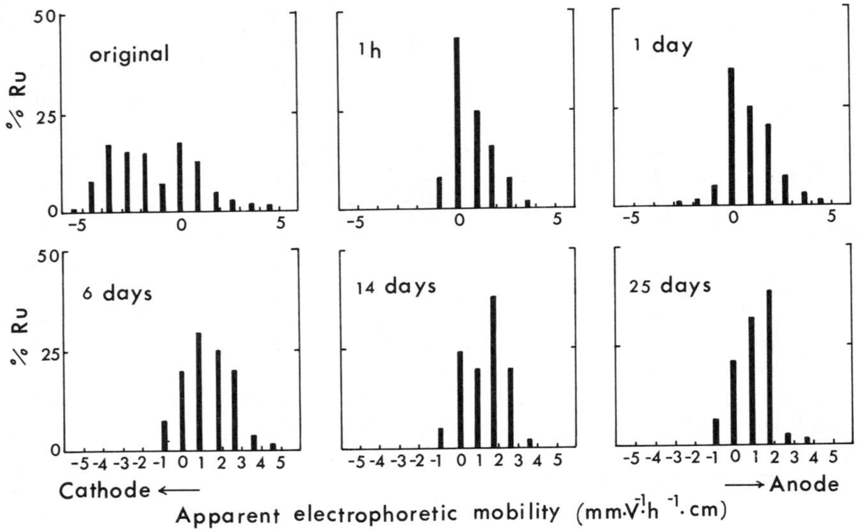

FIG. 2. Electrophoretic patterns of (Ru-O-Ru) complex in fresh water at different aging time.

to increase after one day's aging, and reached 75% of total Ru. These electrophoretic data indicate that the binuclear oxygen-bridged nitratonitrosylruthenium, (Ru-O-Ru) nitrate complex, slowly decomposed with the increase of aging time. In a comparison of the proportions of 'anionic', 'neutral' and 'cationic' species among the three RuNO-complexes, the nitrato complex had a dominant neutral species as reported by Vos et al. [28], while the nitro- and (Ru-O-Ru) nitrate complexes showed dominant anionic species. The fact that the apparent distance in electrophoretic patterns of the separated anionic fractions was relatively short could indicate that the anionic fractions might have a low negative charge or polymeric form. The presence of a dominant anionic species is in agreement with the adsorption behaviour to an anion exchanger [32]. Furthermore, the filtration results for the complexes in fresh water (Fig.4) indicate that most of the nitrosylruthenium complexes in the solutions are in non-filterable form, in agreement with results obtained for the fallout ^{106}Ru present in the Columbia River by Robertson et al. [24].

3.2. Paper chromatographic determination of (Ru-O-Ru) nitrate complex species

Before determining the ionic and adsorption behaviour of (Ru-O-Ru) nitrate complex in a fresh water environment, it was necessary to separate and identify the species of decomposition products of the complex. Figure 5 illustrates how (Ru-O-Ru) nitrate complex may be separated by paper-partition chromatography, together with those of RuNO-nitrato and RuNO-nitro complexes. It was found that about 90% of total Ru in the fresh water environment was taken up in the fraction with flow rate R_f less than 0.2,

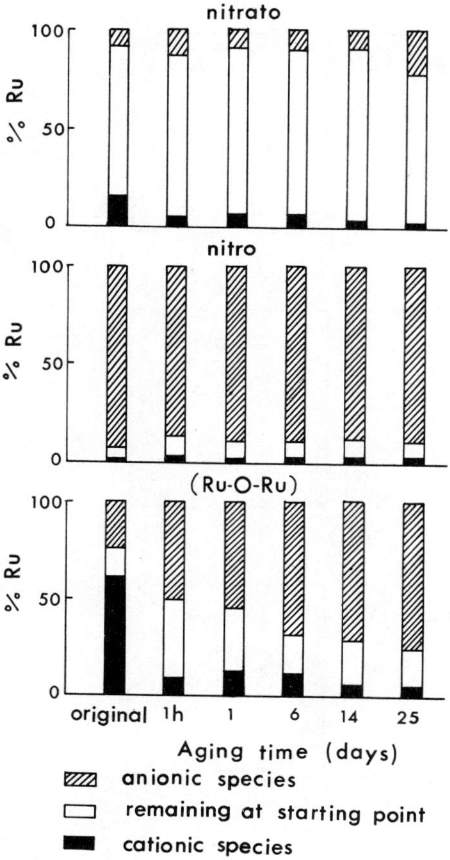

FIG.3. Effect of aging time on the distribution of Ru complexes.

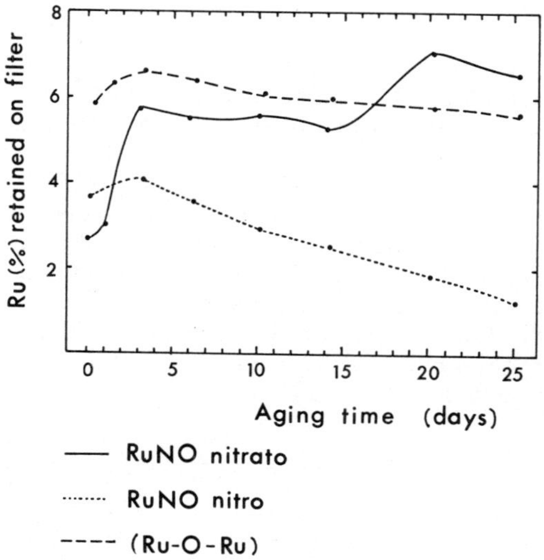

FIG.4. Filtration of Ru complexes in the fresh water using Millipore HA filter.

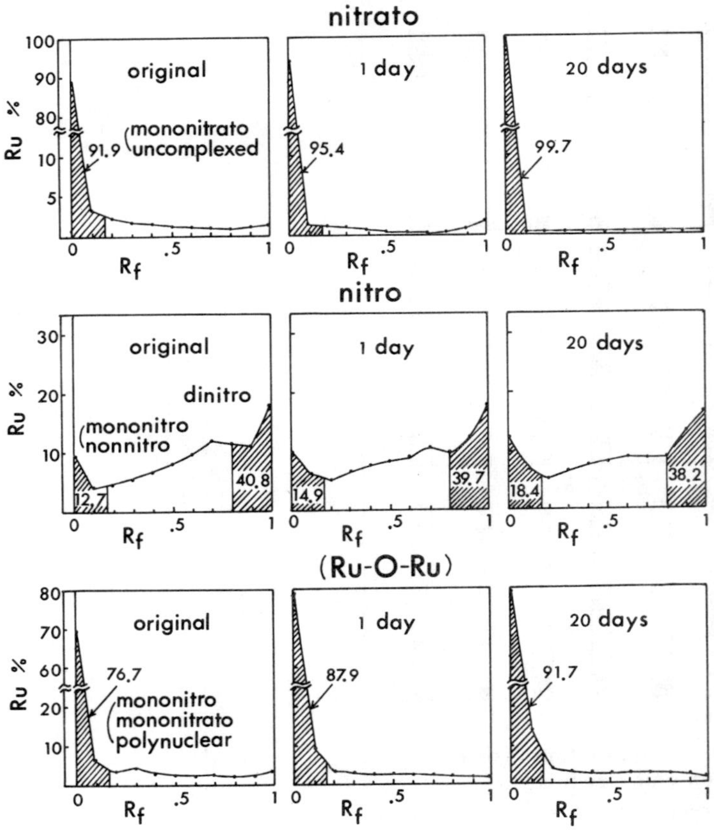

FIG. 5. Chromatographic separation of Ru complexes in MIPK (methyl isopropyl ketone).

corresponding to the anionic mononitro-, mononitratonitrosyl and uncomplexed species. This fact is presumably due to suppression of the negative charge of the filter paper by nitric acid contained in the developer (MIPK).

On the other hand, dinitro-, dinitratonitrosylruthenium or uncharged species (R_f 0.2-1.0) were within 10%. The proportion of each fraction to total Ru was scarcely altered by increasing the aging time more than one day. The paper chromatographic pattern of the (Ru-O-Ru) nitrate complex resembles that of RuNO-nitrato complex except for the proportion of the fraction with R_f less than 0.2. The paper chromatographic behaviour of the RuNO-nitrato and RuNO-nitro complexes used in this experiment is in good agreement with the work reported by Story and Gloyna [4] and Bhagat and Gloyna [9].

3.3. Infrared studies

To verify the results obtained by paper chromatography, infrared studies were undertaken. These were essentially valuable in interpreting

TABLE II. INFRARED DATA OF ORIGINAL (Ru-O-Ru) COMPLEX AND ITS DECOMPOSITION PRODUCT (cm^{-1})

Assignments	Original complex $[RuO_2(NO)_2NO_3]_2O$	Decomposition product with aging for several days
$\nu(H_2O)$		3400 vs
$\nu(D_2O)$		2535 s
$\nu(NO^+)$	1976 vs	1860 vs
$\delta(H_2O)$		1635 s
$\nu(NO_3^-)$	1543 s	1510 m
	1276 s	1290 sh
$\delta(D_2O)$		1260 s
$\nu(NO_2^-)$		1425 sh
		1395 s
		1335 w
ν(Ru-O)	970 s, br	845 s
$\delta(NO_2^-)$		835 s

s = strong, vs = very strong, m = medium, w = weak, br = broad, sh = shoulder

the chemical changes that take place in the (Ru-O-Ru) nitrate complex, $[RuO_2(NO)_2(NO_3)]_2O$, under different aqueous environments. The main infrared absorption bands of the decomposition product in aqueous solution are shown in Table II, together with those of the original (Ru-O-Ru) nitrate complex. The absorption bands at 3400 and 1635 cm^{-1} were assigned to the H_2O stretching mode and bending mode attributed to coordinated water, respectively. These bands have not been observed in the original (Ru-O-Ru) nitrate complex. The two bands were both shifted to 2535 and 1260 cm^{-1}, respectively, by deuteration. The absorption band at 1860 cm^{-1} was assigned to the NO^+ stretching vibration of the nitrosyl group. This band shifted from 1976 to 1860 cm^{-1} after aging of the original (Ru-O-Ru) nitrate complex aqueous solution for several days. The bands at 1510 and 1290 cm^{-1} were both assigned to the NO_3^- stretching of the nitrato group. The absorption bands at 1425, 1395, 1335 and 835 cm^{-1} were assigned to the NO_2^- stretching and bending mode, respectively. These bands of the nitro group have not been observed in the original (Ru-O-Ru) nitrate complex [33]. This fact indicates that one of the NO groups of the original complex was altered to an NO_2 group by air oxidation. The band at 845 cm^{-1} was assigned to the O-Ru-O stretching vibration, since the peak was not shifted by deuteration. Similar O-M-O bands have been observed at 830 and 790 cm^{-1} for $Cs_2[OsO_2(CN)_4]$ and $K_2[OsO_2(OH)_4]$, respectively [36].

The fact that the bands attributed to the O-Ru-O, NO^+, NO_2^-, NO_3^- and H_2O were observed suggests that one of the possible chemical formulae of the anionic species obtained by the decomposition of (Ru-O-Ru) nitrate complex may be given as $[RuO_2(NO)(NO_2)(NO_3)(H_2O)]^-$ anion. This assumption is not contrary to the electrophoresis and paper chromatography results.

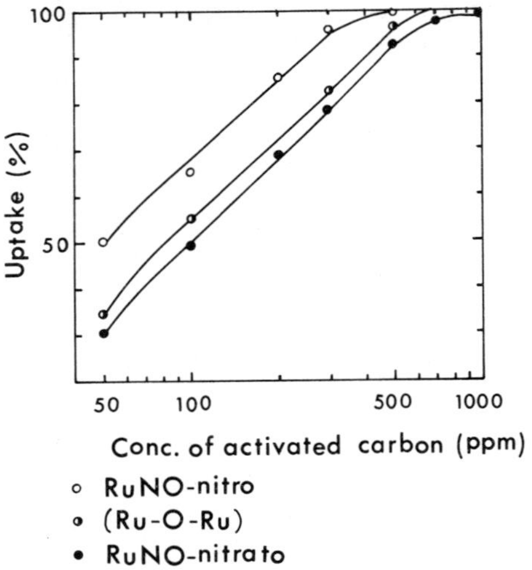

FIG. 6. Effect of concentration of activated carbon on uptake of Ru complexes.

TABLE III. UPTAKES BY NATURAL ION EXCHANGERS

Natural ion exchangers	Place of origin	Concentration of ion exchanger (ppm)	Uptake (%)		
			nitrato	nitro	(Ru-O-Ru)
Kaolin	Nagano-pref.	10 000	50	9	30
Sericite	Niigata-pref.	10 000	40	16	30
Vermiculite	Fukushima-pref.	10 000	40	10	2
Humic acid	Hokkaido	10 000	40	8	3

3.4. Uptake by activated carbon and natural ion exchangers

The effect of activated carbon concentration on uptake at pH 7.0 is shown in Fig.6. Uptakes increased with concentration of activated carbon and reached final steady values. RuNO-nitrato, RuNO-nitro and (Ru-O-Ru) nitrate complexes were entirely adsorbed by the addition of activated carbon of 700, 500 and 600 ppm, respectively.

The percentages of adsorption of RuNO-nitrato, RuNO-nitro and (Ru-O-Ru) nitrate complexes were not observed to be very high (kaolin, sericite, vermiculite and humic acid), as shown in Table III. The adsorption behaviour of RuNO-nitrato and RuNO-nitro complexes to these natural ion exchangers is similar to that of cation exchange resins and Ca-bentonite as reported by Story and Gloyna [4].

REFERENCES

[1] BROWN, P.G.M., FLETCHER, J.M., HARDY, C.J., KENNEDY, J., SCARGILL, D., WAIN, A.G., WOODHEAD, J.L., in 2nd Int. Conf. Peaceful Uses Atom. Energy (Proc. Conf. Geneva, 1958) 17, UN, New York (1958) 118.

[2] JONES, R.F., The accumulation of nitrosyl ruthenium by fine particles and marine organisms, Limnol. Oceanogr. 5 (1960) 312.

[3] AUERBACH, D.J., OLSON, J.S., "Biological and environmental behavior of ruthenium and rhodium", Radioecology (SCHULTZ, V., KLEMENT, A.W., Eds), Reinhold, New York (1963) 509.

[4] STORY, A.H., GLOYNA, E.F., Environmental Behavior of Nitrosylruthenium, USAEC Report ORO-620, (1963).

[5] KEČKEŠ, S., PUČAR, Z., MARAZOVIĆ, Lj., "The influence of the physico-chemical form of ^{106}Ru on its uptake by mussels from sea water", Radioecological Concentration Processes (ABERG, B., HUNGATE, F.P., Eds, Proc. Int. Symp. Radioecological Concentration Processes, Stockholm, 1966) Pergamon, New York (1966) 993.

[6] MARAZOVIĆ, Lj., PUČAR, Z., Electrodialysis of ^{106}Ru, ^{56,57,58}Co and ^{65}Zn in sea water through ion-exchange membranes, Croat. Chem. Acta 38 (1966) 183.

[7] KEČKEŠ, S., PUČAR, Z., MARAZOVIĆ, Lj., Accumulation of electrodialytically separated physico-chemical forms of ^{106}Ru by mussels, Int. J. Oceanol. Limnol. 1 (1967) 246.

[8] MARAZOVIĆ, Lj., PUČAR, Z., Two-dimensional electrochromatography of ^{106}Ru and some other radiomicroconstituents in sea water, J. Chromatogr. 27 (1967) 450.

[9] BHAGAT, K.K., GLOYNA, E.F., Nitrosylruthenium-nitro complexes in an aqueous environment, J. Wat. Pollut. Control Fed. 39 (1967) 334.

[10] WILSON, P.D., "Some aspects of the chemistry of ruthenium in sea water, PG Report 819(W), UKAEA (1968).

[11] KIMURA, K., ICHIKAWA, R., Uptake of ruthenium-106 by short-necked clam. Bull. Jap. Soc. Scient. Fish. 36 (1970) 653.

[12] VAN DER BORGHT, O., VAN PUYMBROECK, S., Initial uptake, distribution and loss of soluble ^{106}Ru in marine and fresh water organisms in laboratory conditions, Health Phys. 19 (1970) 801.

[13] GUEGUENIAT, P., Etude sur la physico-chimie du ruthénium dans l'eau de mer, CEA-R-4125, Centre de la Hague (1970).

[14] ROBERTSON, D.E., Influence of the physico-chemical forms of radionuclides and stable trace elements in sea water in relation to uptake by the marine biosphere, Marine Radioecology (Proc. 2nd ENEA Seminar, Hamburg, 1971), OECD, Paris (1971) 21.

[15] HONDA, Y., KOGA, T., MORISHIMA, H., SATO, Y., KITAWAKI, S., NISHIWAKI, Y., Uptake and loss of radioactive ruthenium by fresh water organisms (in Japanese), Radioisotopes (Tokyo) 20 (1971) 376.

[16] BEQUE, H., VAN PUYMBROECK, S., JAUMIER, J., BITTEL, R., VAN DER BORGHT, O., Initial biological availability of the heavy metal ruthenium in fresh water, I. Influence of salts, animal species, and temperature under controlled conditions, Environ. Physiol. 1 (1971) 37.

[17] ISHIKAWA, M., PUČAR, Z., High-voltage electrophoresis of ^{106}Ru-sea water system, J. Radioanal. Chem. 11 (1972) 197.

[18] HONDA, Y., KIMURA, Y., TSURII, N., Electrodialysis of ^{60}Co, ^{106}Ru and ^{144}Ce in sea water through ion-exchange membranes: On the effects of aging of ^{60}Co, ^{106}Ru and ^{144}Ce in sea water, Radioisotopes (Tokyo) 21 (1972) 269.

[19] KOGA, T., KISHIMOTO, K., HONDA, Y., Chemical behaviour of nitrosylruthenium in fresh water environment (in Japanese), Annu. Rep. Kinki Univ. Atomic Energy Res. Inst. 9 (1973) 11.

[20] KOGA, T., TSUKADA, S., HONDA, Y., Studies on the behaviour and distribution of radioactive ruthenium in the fresh water environment: The role of food chains (in Japanese), Annu. Rep. Kinki Univ. Atomic Energy Res. Inst. 10 (1973) 5.

[21] ISHIKAWA, M., PUČAR, Z., Continuous electrophoretic separation of ^{106}Ru species in sea water and uptake of the separated fractions by sea algae, J. Radioanal. Chem. 14 (1973) 53.

[22] ISHIKAWA, M., SUMIYA, M., SAIKI, M., "Chemical behaviour of ^{106}Ru in sea water and uptake by marine organisms", Radioactive Contamination of the Marine Environment (Proc. Symp. Seattle, 1972) IAEA, Vienna (1973) 359.

[23] PEPERSTRAETE, H., VOS, J., VAN PUYMBROECK, S., VAN DER BORGHT, O., "Physico-chemical characteristics of five ruthenium salts during fresh water to marine transition", Radioactive Contamination of the Marine Environment (Proc. Symp. Seattle, 1972) IAEA, Vienna (1973) 219.

[24] ROBERTSON, D.E., SILKER, W.B., LANGFORD, J.C., PETERSON, M.R., PERKINS, R.W., "Transport and depletion of radionuclides in the Columbia River", Radioactive Contamination of the Marine Environment (Proc. Symp. Seattle, 1972) IAEA, Vienna (1973) 141.

[25] ANCELLIN, J., AVARAGUES, M., BOVARD, P., GUEGUENIAT, P., VILQUIN, A., "Aspects biologiques et physico-chimiques de la contamination radioactive d'espèces et de sédiments marins", Radioactive Contamination of the Marine Environment (Proc. Symp. Seattle, 1972) IAEA, Vienna, (1973) 225.

[26] PENTREATH, R.J., "The role of food and water in the accumulation of radionuclides by marine teleost and elasmobranch fish", Radioactive Contamination of the Marine Environment (Proc. Symp. Seattle, 1972), IAEA, Vienna (1973) 421.

[27] WOODHEAD, D.S., "Levels of radioactivity in the marine environment and the dose commitment to marine organisms", Radioactive Contamination of the Marine Environment (Proc. Symp. Seattle, 1972), IAEA, Vienna, (1973) 499.

[28] VOS, J., VAN PUYMBROECK, S., VAN DER BORGHT, O., PEPERSTRAETE, H., D'HONT, M., Anionic electrophoretic pattern of five ruthenium salts in fresh and sea water, effects of aging and dilution, Anal. Chim. Acta 67 (1973) 480.

[29] VAN DER BORGHT, O., VAN PUYMBROECK, S., COLARD, J., BITTEL, R., "Long-term laboratory experiments on the fixation of various ruthenium compounds by fresh water organisms", Environmental Surveillance Around Nuclear Installations, (Proc. Symp. Warsaw, 1973) 2, IAEA, Vienna, (1974) 31.

[30] KAMADA, H., YUKAWA, M., SAIKI, M., "Studies on the removal of strontium-90, ruthenium-106, caesium-137 and cerium-144 on land and in fresh water", Environmental Surveillance Around Nuclear Installations, (Proc. Symp. Warsaw, 1973) 2, IAEA, Vienna, (1974) 137.

[31] ISHIKAWA, M., SAIKI, M., "Chemical behaviour of radioactive ruthenium in sea water and biological concentration" (in Japanese), Radioecological Studies on Marine Environment, NIRS-R-3, National Inst. Radiol. Sci., Chiba, Japan, (1974) 26.

[32] ISHIYAMA, T., Decontamination of radioactive ruthenium by cellulose derivative — bentonite flocculation (II) (in Japanese), Radioisotopes (Tokyo) 17 (1968) 581.

[33] MARTIN, F.S., FLETCHER, J.M., BROWN, P.G.M., GATEHOUSE, B.M., The nitratonitrosylruthenium complex, $RuN_3O_{7.5}$, J. Chem. Soc. 1959 (1959) 76.

[34] FLETCHER, J.M., JENKINS, I.L., LEVER, F.M., MARTIN, F.S., POWELL, A.R., TODD, R., Nitrato and nitro complexes of nitrosylruthenium, J. Inorg. Nucl. Chem. 1 (1955) 378.

[35] WAIN, A.G., BROWN, P.G.M., FLETCHER, J.M., The separation and estimation of nitro and non-nitro(nitrato) complexes of nitrosylruthenium in nitric acid solution by paper chromatography, J. Inorg. Nucl. Chem. 12 (1960) 346.

[36] GRIFFITH, W.P., Nitro-complexes of osmium, J. Chem. Soc. 1962 (1962) 3248.

DISCUSSION

G. STOTT: In my opinion it is unlikely that low-level liquid waste from future nuclear fuel reprocessing plants will be discharged into freshwater systems such as rivers. It is more likely that such discharges will be made into the sea. If seawater had been used in the experiments, would the results, such as aging, have been different from those found using freshwater?

Y. NISHIWAKI: To obtain a basic pattern of the physico-chemical behaviour of the Ru complex in an aquatic environment, the experiment was first conducted in freshwater. Depending on the quality and quantity of the chemical, physical and biological substances contained in seawater, i.e. types and content of organic matter, amino acids, chelating agents, etc., it may be that quite a different pattern would be observed. However, preliminary comparative experiments have been conducted with paper electrophoresis, and the pattern of physico-chemical behaviour of the Ru complex in normal surface seawater was observed to be more or less similar to that of freshwater. The pattern may also be affected by various chemical reagents released simultaneously by the nuclear fuel reprocessing plant into the aquatic environment.

R. BITTEL: Since the kinetics of physico-chemical transformations in ruthenium are sometimes slow, I should like to ask what was the time-scale of the experiments.

Y. NISHIWAKI: The experiments were continued up to the aging time of 25 years.

In view of the questions raised regarding the physico-chemical state and aging of radionuclides in the marine environment, I should like to take this opportunity of mentioning briefly a study on electrodialysis of radionuclides in seawater through ion-exchange membranes which was conducted some years ago at the Atomic Energy Research Institute of Kinki University, Osaka, Japan.

The electrodialysis cell, which is made of glass, was divided by ion-exchange membranes into three compartments: the cathode, middle, and anode compartments. The ion-exchange membranes were Selemion CMV and AMV types, produced by the Asahi Glass Co. Ltd. The surface area of the platinum electrode was about 15.9 cm^2.

The volume of the middle compartment was 500 ml and that of each electrode compartment including the circulatory vessel was 2000 ml. The walls of the electrode compartments have a double-layer structure for cooling by circulating tap-water. The anode and cathode compartments each contained 2000 ml of filtered seawater, while 500 ml of filtered seawater contaminated with each radionuclide was introduced into the middle compartment. To reduce adsorption on the ion-exchange membranes and electrodes, the solution in the middle compartment was stirred by rotating a stirrer at about 400 rev/min and the solution in each electrode compartment was cirulated by means of two suitably adjusted peristaltic pumps (Type QP-2, Iwaki Ltd.) at a flow rate of about 55 ml/min. A direct current supply of 30 V was applied to the electrodes.

During electrodialysis, hydrochloric acid or sodium hydroxide solutions were added to the electrode compartments to keep the pH of the solutions constant (pH $\sim$ 8).

The relationship between voltage and current density during electrodialysis followed a straight line, from which it could be seen that the limiting current density was not exceeded. To estimate the radioactivity adsorption on the ion-exchange membranes, after electrodialysis, autoradiograms of the ion-exchange membranes were made with X-ray films, and the radioactivities were determined using a low-background counter.

The chloro-complex of ^{106}Ru and the chlorides of ^{144}Ce and ^{60}Co were introduced into natural, surface seawater filtered with a Millipore filter of pore size 0.45 μm, to give a radioactive concentration of about 10^{-2} μCi/ml. The total sodium contents, chlorinity and pH of the seawater were estimated to be 10 ± 1‰, 18 ± 1‰ and 8.0 ± 0.2 respectively. The seawater, to which each radionuclide was added, was kept in a volumetric flask at room temperature of about 10°C for aging. The fractions of electro-neutral and negatively or positively charged species were estimated from electrodialysis through ion-exchange membranes after aging of the radionuclides in the seawater for various periods.

The behaviour of ^{106}Ru in seawater depends on the form of the original compound; chloride, nitrate, or complex, etc. From the electrodialysis results, ^{106}Ru in the original chloro-complex form seems to be composed mostly of electro-neutral (65% at 1 h, 54% at 25 days) and negatively charged colloidal species (28% at 1 h, 20% at 25 days) in seawater. The fraction of cationic or anionic species was rather small (1-2%). Most of the electrically charged forms of ^{106}Ru seemed to be non-dialytic species which were adsorbed on the ion exchange membranes. However, the fraction of positively charged colloidal species remained relatively small (4-7%). During aging for 25 days, a tendency was observed for the electro-neutral and the negatively charged forms of ^{106}Ru to decrease slowly, while the vessel-wall-adsorbed fraction was observed to increase markedly (negligible at 1 h, 16% at 25 days).

Radioactive cerium is also known to be one of the most important radioactive contaminants found in various marine organisms. From the results of these experiments, it seems that positively charged colloidal forms of ^{144}Ce predominate initially (56% at 1 h) but decrease with time (15% at 25 days), while the electro-neutral species increase with aging in seawater (34% at 1 h, 52% at 1 day). After aging for 3 days, the negatively charged species also increase with time (3% at 3 days, 25% ar 25 days). This may be due to the formation of electro-neutral and then negatively charged species by hydrolysis and complex formation, e.g. sulphato-complex formation with aging. Most of the electrically charged forms of ^{144}Ce in seawater were non-dialytic, presumably because of the formation of colloids or particulates.

Cerium-144 also showed more or less adsorbable species in the electrodialytic experiments. From a study of autoradiograms of the radionuclides adsorbed on the ion-exchange membranes during electrodialysis after aging for 25 days, ^{106}Ru and ^{144}Ce in seawater appear to be composed more or less of colloidal or particulate species. A considerable fractions of ^{144}Ce was also observed to be adsorbed on the vessel wall (negligible at 1 h, 20% at 25 days).

Cobalt as a microconstituent in seawater appears to behave as a cationic species (27% at 1 h, 85% at 3 days, 64% at 25 days). The ^{60}Co in seawater shows a definite dialytic, cationic form at 3 days of aging, when the percentage of electro-neutral fraction becomes minimum (35% at 1 h, 3% at 3 days, 25% at 25 days). After 3 days, however, the percentage of ^{60}Co left in the middle compartment (electro-neutral fraction) increases, with a slow decrease in the dialytic, cationic form. Although further studies are required for identification of the chemical forms in the electro-neutral fraction, one possible mechanism for the formation of electro-neutral species may be the formation of various hydroxo-complexes by hydrolysis of cobalt in seawater with aging.

Adsorption on a glass surface, which may be regarded as an ion-exchange adsorbent, is an indication of the presence of radionuclides in ionic form. However, the possibility of adsorption of colloids carrying a charge opposite so that of the vessel surface and of electro-neutral forms, including colloidal particles, is not ruled out. It is thus not certain to which of the forms originally present in the body of the solution the wall-adsorbed radionuclides fraction belongs.

Although more or less reproducible tendencies were obtained in these experiments, it seems rather difficult to conclude that the results give a sufficiently accurate reproduction of the proportions that would actually exist in the sea under natural conditions. Further studies are necessary to elucidate these complicated phenomena, and to enable us to link the results of laboratory experiments with those of field studies under natural environmental conditions.

LE TRANSFERT DU ZINC-65 DE SEDIMENTS A DES LARVES DE CHIRONOMIDES ET A UN POISSON D'EAU DOUCE ET L'EFFET DU CADMIUM SUR CE TRANSFERT*

A. R. BERG, G.M. WEISS
Commission des Communautés européennes, Euratom, Centre commun de recherches, Ispra (Varèse), Italie

Abstract–Résumé

TRANSFER OF ZINC-65 FROM SEDIMENTS TO CHIRONOMID LARVAE AND TO A FRESHWATER FISH AND THE EFFECT OF CADMIUM ON TRANSFER.

This transfer was studied by equilibrating – in the presence of ^{109}Cd – a sediment rich in organic matter with water from Lake Maggiore in three different cadmium concentrations (0, 10 and 20 pp10^9). The chironomid larvae reared on this sediment were then used as 'natural' food for a small tropical fish (*Haplochromis burtoni*), kept in an aquarium at these same concentrations of cadmium in the water. The transfer of radiozinc and radiocadmium was traced over various links of the ecological chain (sediment – water – larvae – fish). The results indicate that cadmium strongly inhibits the exchange of zinc for the larvae. This inhibition must constitute a significant sublethal effect on a population of benthic organisms of considerable ecological importance. In the case of the fish, however, this inhibition is not apparent at the cadmium concentrations employed. Branchial absorption plays a predominant role, compared with intestinal absorption, for both radiozinc and radiocadmium. While the equilibration time for the accumulation of radiozinc by the fish is very long, the cadmium accumulation reaches equilibrium rapidly (5–7 days). On the other hand, the excretion of the accumulated cadmium takes place at a relatively slow rate.

LE TRANSFERT DU ZINC-65 DE SEDIMENTS A DES LARVES DE CHIRONOMIDES ET A UN POISSON D'EAU DOUCE ET L'EFFET DU CADMIUM SUR CE TRANSFERT

Ce transfert a été étudié en équilibrant, en présence de cadmium-109, un sédiment riche en matière organique avec une eau du lac Majeur à trois concentrations de cadmium (0, 10 et 20 ppb). Les larves de chironomides élevées sur ce sédiment ont ensuite servi de nourriture «naturelle» à un petit poisson tropical (*Haplochromis burtoni*), maintenu en aquarium à ces mêmes concentrations de cadmium dans l'eau. Le transfert du radiozinc et du radiocadmium a été suivi sur les différents maillons de la chaîne écologique (sédiment – eau – larve – poisson). Les résultats indiquent chez les larves une forte inhibition par le cadmium de l'échange du zinc. Cette inhibition doit constituer un effet sublétal significatif sur une population d'organismes benthiques d'importance écologique considérable. Chez le poisson, cette inhibition n'est par contre pas apparente aux concentrations choisies de cadmium. L'absorption branchiale joue un rôle prééminent par rapport à l'absorption intestinale, aussi bien en radiozinc qu'en radiocadmium. Alors que le temps d'équilibration de l'accumulation du radiozinc par le poisson est très long, l'accumulation du cadmium atteint rapidement un équilibre (5 à 7 jours). Par contre, l'excrétion du cadmium accumulé se poursuit à un taux relativement faible.

INTRODUCTION

De nombreux travaux effectués en radioécologie ont mis en relief les interactions entre les facteurs du milieu et les radionucléides dans le cheminement de ces derniers dans l'environnement. Il suffira de citer en exemple l'interférence des éléments Ca et K sur l'accumulation respective du radiostrontium et du radiocésium.

* Contribution n° 1224 du Programme de biologie, Direction générale XII de la Commission des Communautés européennes.

L'immission dans l'environnement de substances polluantes nombreuses et variées vient augmenter la fréquence de telles interactions. Suivant les cas, on pourra distinguer les interactions d'ordre physico-chimique, intervenant avant même qu'entrent en jeu les processus biologiques d'accumulation, et les interactions d'ordre physiologique où les substances polluantes interfèrent sur les mécanismes biologiques proprement dits. Parmi ces interactions d'ordre physiologique, l'interférence du Cd sur l'accumulation du radiozinc présente un intérêt particulier, tant en radioécologie [1, 2] qu'en toxicologie où elle met en lumière, pour des espèces variées, les processus compétitifs entre les deux éléments Cd et Zn, particulièrement au niveau des sites enzymatiques [3].

Par ailleurs, il apparaît intéressant d'aborder expérimentalement le problème du cheminement en écosystème aquatique des radionucléides et celui des interactions, dans le cadre d'une chaîne alimentaire qui se rapproche le plus possible de conditions naturelles, ceci afin de ne pas courir le risque de sous-estimer l'importance de la voie d'entrée intestinale par rapport à la voie d'entrée branchiale. Dans cette optique, il a semblé utile de considérer une chaîne alimentaire aquatique dont un maillon présente une capacité élevée d'accumulation vis-à-vis des radioncléides à étudier, comme c'est le cas des sédiments riches en matières organiques vis-à-vis du Cd [4]. L'importance écologique des larves de chironomides dans la productivité de la faune benthique et dans la production piscicole naturelle [5, 6] nous a amenés par ailleurs à choisir ces invertébrés comme aliment naturel des poissons utilisés au cours des essais décrits ci-dessous.

1. MATERIEL ET METHODES

1.1. Traceurs radioactifs et techniques de comptage

Le zinc-65 et le cadmium-109, de haute activité spécifique, ont été utilisés comme traceurs dans tous les maillons de la chaîne écologique adoptée (sédiment – eau – larve – poisson). Les comptages de radioactivité ont été effectués par spectrométrie gamma sur cristal NaI(Tl) à puits (efficacité cpm/dpm 0,0755 pour ^{65}Zn et 0,0292 pour ^{109}Cd), afin de discerner les radioactivités respectives des deux radionucléides, en appliquant les corrections habituelles pour tenir compte de la géométrie (dans l'intervalle 0,5 à 4 cm^3) et de la décroissance physique.

1.2. Préparation du sédiment marqué

Le sédiment organique a été prélevé sur la rive est du lac Majeur (baie d'Ispra, lieu-dit Girolo). Après élimination du matériel grossier et centrifugation prolongée, le pourcentage d'eau résiduelle a été déterminé (52,5%) ainsi que la teneur en matériel organique par différence entre poids des cendres à 600°C et poids sec (25,0% du poids sec), et le pH (7,10).

Etant donné le choix, pour les essais projetés, de deux concentrations expérimentales en Cd dans l'eau (10 et 20 ppb), il convenait de préparer des échantillons de sédiment en équilibre avec de telles concentrations, en tenant compte de la forte capacité d'adsorption de sédiments organiques vis-à-vis du Cd [4]. Au cours d'essais successifs, réalisés avec des solutions

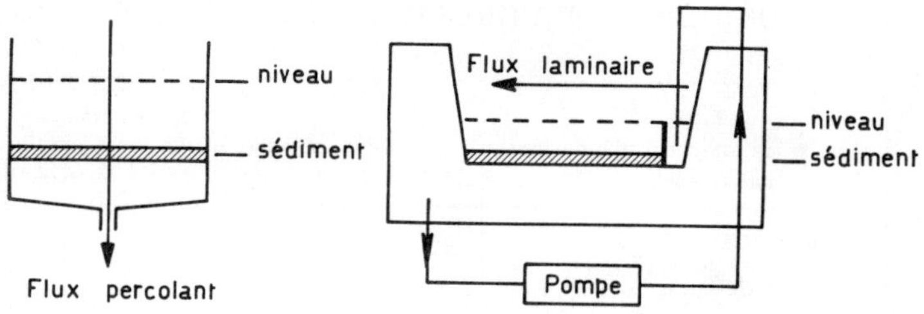

FIG. 1. Processus de flux percolant et de flux laminaire utilisés pour la préparation de sédiment marqué.

marquées en ^{109}Cd de radioactivité spécifique déterminée, une technique de mise en équilibre du sédiment expérimental a été mise au point, comportant deux étapes successives d'équilibration. Dans une première étape d'équilibration (grossière), 2,5 g (sec) de sédiment sont agités pendant 5 minutes dans 100 cm^3 d'eau du lac Majeur (filtrée à 1 μm) additionnés respectivement de 1250 et 2500 ppb de Cd. Dans ces conditions, la concentration en Cd stable de la solution surnageante (déterminée par le traceur radiocadmium) est voisine de 10 et 20 ppb respectivement. Dans une seconde étape d'équilibration (fine), ces mêmes échantillons sont ultérieurement soumis à un processus de percolation lente continue (à raison de 1000 cm^3 par jour) au moyen de solutions (eau lac Majeur filtrée à 1 μm) de 10 et 20 ppb Cd respectivement, de même radioactivité spécifique que celle des solutions utilisées dans la première étape (fig. 1). Un contrôle de la radioactivité de la solution percolante, avant et après passage sur le sédiment, permet de déterminer le moment où un équilibre satisfaisant est établi. Pour simplifier le mode d'équilibration de plus grandes quantités de sédiment (50 g nécessaires pour la production en masse de larves marquées), le processus de percolation utilisé dans la seconde étape d'équilibration fine a été remplacé par un processus de flux laminaire comportant la recirculation, par pompe péristaltique, d'un litre (renouvelé quotidiennement) de solution identique à la solution percolante, avec contrôle à l'aide du traceur radioactif des teneurs effectives en Cd stable (fig. 1).

Parallèlement aux échantillons additionnés de Cd stable, des échantillons sans addition de Cd ont été utilisés comme témoins dans tous les essais réalisés.

Il est à remarquer que le problème de l'équilibration, exposé ci-dessus pour le Cd, ne s'est pas posé en ce qui concerne le zinc, dont la concentration naturelle dans l'eau du lac Majeur (filtrée à 1 μm) a été maintenue, en présence d'une quantité arbitraire de traceur ^{65}Zn.

1.3. Equilibration et comptage des larves

Les larves de chironomides (du groupe Chironomus plumosus) proviennent d'anciens étangs de pisciculture situés à Ligny (Belgique) où elles font l'objet d'un élevage fournissant des appâts pour la pêche.

Les larves, après adaptation préalable au sédiment expérimental, sont introduites dans le sédiment en équilibre, soumis au système de percolation ou de flux laminaire, à un taux de renouvellement réduit de

TABLEAU I. CONDITIONS EXPERIMENTALES DES ESSAIS (POISSONS)

Traitement	EAU Eau du lac Majeur filtrée à 1 μm additionnée de:		NOURRITURE Larves obtenues en sédiment équilibré avec de l'eau du lac Majeur filtrée à 1 μm et additionnée de:	
1	20 ppb Cd		20 ppb Cd	+ 3 μCi/l ^{65}Zn + 4 μCi/l ^{109}Cd
2	10 ppb Cd		10 ppb Cd	
3	"0" ppb Cd		"0" ppb Cd	
4	20 ppb Cd	+ 3 μCi/l ^{65}Zn + 4 μCi/l ^{109}Cd	20 ppb Cd	
5	10 ppb Cd		10 ppb Cd	
6	"0" ppb Cd		"0" ppb Cd	

moitié (tous les deux jours au lieu de quotidien). Au cours de la phase d'équilibration des larves avec le sédiment marqué, ainsi qu'au cours de la phase d'excrétion dans le sédiment non marqué, la radioactivité des larves a été mesurée périodiquement (par groupes de 3 à 5 larves, dans un volume de 1 cm^3). Par contre, on n'a mesuré que le niveau final de radioactivité (après 4 à 5 jours) des larves utilisées pour l'alimentation des poissons.

1.4. Accumulation de radiozinc et radiocadmium par les poissons

Neuf individus de petite taille du cichlide africain, Haplochromis burtoni, d'un poids global compris entre 997 et 1007 mg, sont introduits par aquarium de 5 litres maintenu à une température de 23 ± 1°C. Afin de discerner les contributions respectives de l'accumulation branchiale (à partir de l'eau) et intestinale (à partir de la nourriture), deux séries de trois traitements (20, 10, "0" ppb Cd) ont été réalisées, la première avec les traceurs ajoutés à l'eau, la seconde avec les traceurs présents dans la nourriture (larves marquées comme indiqué ci-dessus). La technique d'équilibration du sédiment décrite plus haut a permis de rendre les activités spécifiques en Cd et en Zn très voisines dans les deux séries de traitements. En dehors du traçage, les autres conditions sont identiques dans les deux séries. Ces conditions expérimentales sont détaillées dans le tableau I.

Les valeurs des principaux paramètres chimiques de l'eau du lac Majeur utilisée dans les essais sont les suivantes: Ca 19 à 23; Mg 3,0 à 3,8; Na 2; K 1,5; Zn 0,01 à 0,03 mg/l.

Préalablement à l'essai, les poissons ont été adaptés pendant deux jours aux concentrations expérimentales de Cd. L'essai lui-même a comporté une phase d'accumulation en ^{65}Zn - ^{109}Cd de 14 jours, avec mesures périodiques de la radioactivité des poissons (par groupes de 2 ou 3 dans un volume de 2 cm^3) tous les 3 à 4 jours. L'eau des aquariums a été renouvelée avec la même fréquence et sa radioactivité contrôlée tout au long de l'essai, ainsi que sa concentration en Zn stable (par spectrophotométrie d'absorption atomique).

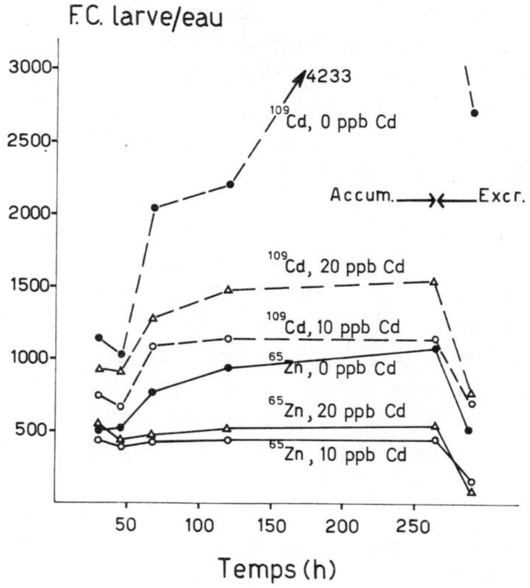

FIG.2. Courbes d'accumulation de ^{65}Zn et de ^{109}Cd par les larves de chironomides, en fonction du temps.

Pour tenir compte seulement de la fraction de radionucléide absorbée et effectivement fixée sur les tissus [7], la dernière mesure d'accumulation (après 14 jours) a été effectuée après 24 heures de rinçage dans l'eau expérimentale aux mêmes concentrations de Cd stable mais sans traceurs. Par contre, l'excrétion a été suivie durant 11 jours en l'absence de Cd ajouté.

Etant donné la petite taille des poissons expérimentaux, il n'a pas été possible d'administrer la nourriture naturelle sous forme de larves vivantes mais sous forme de poudre grossière, obtenue après dessiccation des larves à 40°C. La ration journalière de nourriture sèche (25 mg/j) a été calculée en excès (2,3 fois) par rapport au métabolisme de routine de poissons tropicaux [8], compte tenu de la taille et de la température. A cause du métabolisme élevé des individus de petite taille, cette ration représente une fraction considérable du poids frais du poisson (7,3% par jour).

2. ACCUMULATION DU RADIOZINC ET DU RADIOCADMIUM PAR LES LARVES DE CHIRONOMIDES

La figure 2 illustre l'allure des courbes d'accumulation des deux radioncléides dans les trois conditions expérimentales.

Afin de soustraire des mesures de radioactivité la fraction due au bol alimentaire (sédiment marqué), ingérée mais non absorbée par les larves, celles-ci sont soumises, après la phase d'accumulation de 10 jours, à un comptage ultérieur après 24 heures de maintien en sédiment non contaminé. Ce comptage représente donc la fraction du radionucléide effectivement absorbée dans les tissus de la larve.

Le paramètre d'accumulation utilisé ici est le facteur de concentration (FC) larve/eau en équilibre avec le sédiment.

On peut, de l'examen de la figure 2, déduire les intéressantes conclusions suivantes:

— Après une période d'accumulation irrégulière, en relation avec l'importance variable du bol alimentaire fortement radioactif, l'accumulation du ^{65}Zn atteint très rapidement un plateau en présence de Cd tandis que le témoin manifeste une accumulation constamment croissante.

— Le radiozinc effectivement absorbé est en relation inverse nette avec la concentration en Cd du milieu (FC = 87, 167 et 517 respectivement à 20, 10 et "0" ppb Cd). On note donc un effet inhibiteur très important du Cd stable du milieu sur l'échange de Zn par la larve, d'autant plus que la valeur obtenue pour le témoin (FC = 517) ne présente pas encore une condition d'équilibre. Cette inhibition exercée par le Cd sur un processus métabolique (échange du Zn stable) représente sans aucun doute un important effet à niveau sublétal sur une population d'organismes jouant dans le benthos un rôle écologique de premier plan.

— L'égalité des FC du radiocadmium effectivement absorbé à 20 et à 10 ppb Cd (783 et 709) indique une accumulation du Cd stable proportionnelle à la concentration en Cd du milieu. Par ailleurs, cette accumulation atteint très rapidement (en 3 à 4 jours) une situation d'équilibre. Au contraire, la concentration relative est beaucoup plus importante et en continuelle augmentation dans le temps, en présence de traces de Cd (FC = 2733).

— Les concentrations des larves en Cd effectivement absorbé, calculées à partir des niveaux d'accumulation du radiocadmium, se situent, après 11 jours de maintien dans le sédiment, à des valeurs très élevées (respectivement 15,7 et 7,1 ppm dans le poids frais des larves, à 20 et 10 ppb Cd dans le milieu).

— La capacité d'accumulation des larves est de 4 à 8 fois plus élevée vis-à-vis du Cd que vis-à-vis du Zn. Cette donnée doit être la résultante d'une composante physico-chimique (capacité d'adsorption du sédiment organique) et d'une composante physiologique des larves elles-mêmes. Or la capacité d'adsorption du sédiment organique (capacité exprimée par les facteurs de concentration «sédiment sec/eau», obtenus avec les traceurs radioactifs) est, aux concentrations expérimentales de Cd utilisées, assez voisine pour le Zn (environ 10 000) et pour le Cd (environ 8000). Il en découle que c'est la composante physiologique qui doit être mise en cause, faisant intervenir, dans l'occupation des sites métaboliques, une substitution partielle du Zn de la part du Cd. Ceci ne met que davantage en relief l'effet du Cd à niveau sublétal, décrit ci-dessus, sur une population d'organismes benthiques. Les chironomides pourraient donc constituer des bioindicateurs précieux dans les études de la pollution radioactive et conventionnelle par le radiozinc et le cadmium.

3. ACCUMULATION DU RADIOZINC ET DU RADIOCADMIUM PAR LES POISSONS

3.1. Evolution de l'accumulation dans le temps

Les comptages intermédiaires permettant de suivre cette évolution sont nécessairement soumis à la variabilité inhérente à l'absorption d'une fraction de radioactivité non fixée, destinée à être excrétée rapidement via les fèces [7]. Ils mettent cependant en évidence une croissance continue dans le temps de l'accumulation du radiozinc, aussi bien en accumulation directe qu'en accumulation intestinale, allure qui concorde avec le faible coefficient de renouvellement du zinc stable et donc sa longue demi-vie biologique [1, 2, 7, 9, 10]. L'accumulation du cadmium atteint au contraire un niveau maximum déjà après 5 jours en accumulation intestinale et après 7 à 9 jours en accumulation directe.

3.2. Facteurs de concentration après 14 jours d'essai

Ces facteurs de concentration ont été déterminés sur la base du rapport dpm/g poisson sur dpm/g eau «ambiante», qu'il s'agisse d'accumulation branchiale ou intestinale. Dans ce dernier cas en effet, l'eau «ambiante» représente le milieu en équilibre avec les sédiments et les larves ingérées par le poisson. Le facteur de concentration global est à considérer comme la somme d'un facteur de concentration direct (à partir de l'eau) et d'un facteur de concentration indirect (à partir de l'eau par les sédiments et la nourriture ingérée (tableau II)).

3.3. Discussion

3.3.1. Contribution de l'accumulation branchiale et intestinale

On observe pour les deux éléments Zn et Cd la contribution prééminente de l'accumulation directe par rapport à l'accumulation intestinale (88 à 93% pour Zn, 95 à 96% pour Cd).

En ce qui concerne le Zn, cette observation est en concordance avec des données obtenues dans notre laboratoire sur d'autres espèces (*Carassius auratus*, *Alburnus alborella*, *Lebistes reticulatus*). Il est à noter cependant que ces données avaient été obtenues avec une nourriture artificielle et restaient de ce fait sujettes à discussion.

Il est digne d'intérêt que la contribution de l'absorption intestinale de Zn reste limitée, même dans le cas d'une alimentation naturelle abondante, dont l'équilibration avec des sédiments à très forte capacité d'adsorption vis-à-vis du Zn pourrait laisser présager une contribution accrue dans l'échange du Zn. En fait, la régulation probable de la teneur en Zn des larves elles-mêmes rend leur composition en cet élément indépendante de la composition des sédiments.

Au contraire, en ce qui concerne un élément non métabolique comme le Cd, l'absence de régulation entraîne dans les larves une accumulation en Cd stable proportionnelle à la concentration dans l'eau et, dans ce cas, on pourrait escompter une importance accrue de l'absorption intestinale de poissons ingérant une telle nourriture riche en Cd. Les résultats obtenus indiquent au contraire la prééminence très nette de l'absorption branchiale de Cd. On peut en conclure que la perméabilité branchiale vis-à-vis du Cd est suffisamment élevée pour compenser l'apport important de l'élément par voie intestinale.

Notons encore que les données d'accumulation intestinale obtenues indiquent une rétention relative (retenu/distribué) de l'ordre de 4 à 6% pour le Cd (après 5 jours nécessaires pour atteindre l'équilibre) et de 4 à 5% pour le Zn (en 14 jours).

TABLEAU II. RESULTATS DES ESSAIS (POISSONS)
(FC = facteur de concentration poisson/eau)

Traitement	Type d'accumulation	ppb dans l'eau		cpm/g eau		cpm distribué/j		FC Zn	FC Cd
		Zn	Cd	^{65}Zn	^{109}Cd	^{65}Zn	^{109}Cd	(± err. st.)	(± err. st.)
1	Intest.	26,5	20,0	-	-	7 254	4 731	9,4 ± 0,4	3,9 ± 0,3
2	"	26,5	9,7	-	-	9 031	4 692	14,5 ± 0,6	4,2 ± 0,4
3	"	26,5	"0"	-	-	10 138	4 408	16,1 ± 0,7	5,4 ± 0,7
4	Branchiale	26,5	20,0	409	257	-	-	124,8 ± 4,1	90,0 ± 4,8
5	"	26,5	9,7	409	247	-	-	120,8 ± 8,3	103,3 ± 5,4
6	"	26,5	"0"	409	238	-	-	120,9 ± 7,1	114,5 ± 6,6
Intest. + Branchiale		26,5	20					134,2	93,9
" + "		26,5	9,7					135,3	107,5
" + "		26,5	"0"					137,0	119,9

3.3.2. Effet du Cd sur l'accumulation du radiozinc

Les résultats indiquent clairement l'absence d'inhibition de l'accumulation directe du radiozinc de la part du Cd ajouté à l'eau. Par contre, une relation négative est évidente entre la concentration en Cd de l'eau et l'accumulation intestinale de radiozinc. Il convient de noter cependant que cette relation ne reflète pour une bonne part que les différences de radiozinc ingéré, conséquence de l'effet inhibiteur du Cd sur l'accumulation des larves elles-mêmes. Dans ces conditions, il est vraisemblable qu'on ne peut incriminer, chez les poissons étudiés, un mécanisme d'inhibition directe du Cd sur l'échange du zinc, aussi bien en accumulation branchiale qu'en accumulation intestinale.

En ce qui concerne le facteur de concentration global obtenu en ^{65}Zn après 14 jours (~135), il convient d'insister sur le fait qu'il est loin de représenter une situation d'équilibre, étant donné la longue demi-vie biologique de l'élément. Sur la base des teneurs en Zn stable de l'espèce étudiée (74 ppm) et de l'eau (26,5 ppb), on pourrait en effet escompter un facteur de concentration à l'équilibre de l'ordre de 2800.

3.3.3. Accumulation du Cd stable en fonction de sa concentration dans l'eau

La diminution progressive des facteurs de concentration en ^{109}Cd en relation avec l'augmentation en Cd stable de l'eau indique que l'accumulation du Cd stable, tout en présentant une relation directe avec la concentration de l'élément dans le milieu, manifeste une certaine tendance à la saturation aux concentrations plus fortes.

Sur la base des facteurs de concentration globaux obtenus, les teneurs en Cd stable du poisson à l'équilibre sont respectivement de l'ordre de 1880 et 1040 ppb pour des concentrations en Cd dans l'eau de 20 et 10 ppb respectivement.

Les observations précédentes sont généralement en accord avec les données obtenues dans notre laboratoire sur deux autres espèces (Carassius auratus, Alburnus alborella) [11].

4. EXCRETION DU RADIOZINC ET DU RADIOCADMIUM PAR LES POISSONS

L'excrétion a été suivie pendant 12 jours en plaçant les poissons, après la phase d'accumulation des radionucléides, en eau du lac Majeur non additionnée de Cd. La figure 3 donne les valeurs du facteur de rétention relative de la radioactivité par rapport à la radioactivité effectivement accumulée par les poissons. Pour simplifier le graphique, les valeurs non significativement différentes sont représentées par leurs moyennes.

L'examen de la figure 3 permet de faire les observations suivantes:

— Le Zn est retenu très fortement, en conséquence de sa longue demi-vie biologique, et de manière non significativement différente pour les divers traitements.

— Bien que l'excrétion du Cd accumulé en présence de 10 ou 20 ppb Cd s'effectue à un taux plus rapide que celle du Zn, la fraction retenue après 12 jours est encore considérable (70 à 80%).

— Par contre le radiocadmium accumulé en présence de traces de Cd stable est moins fortement retenu par l'organisme (50% excrété après 9 jours).

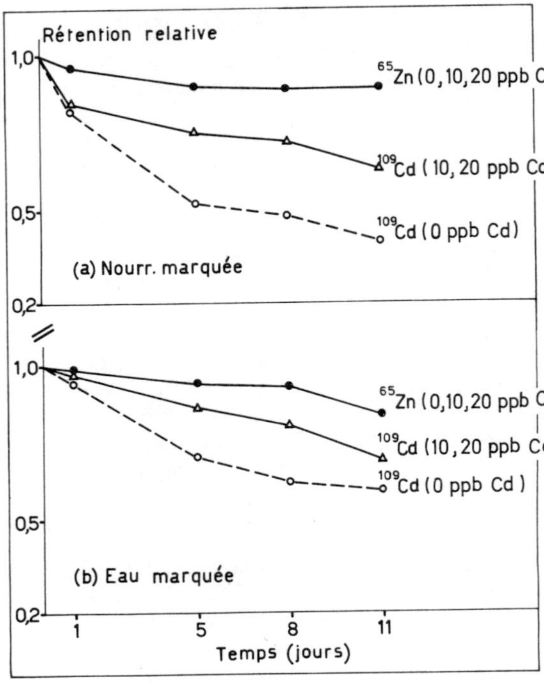

FIG. 3. Courbes d'excrétion de ^{65}Zn et de ^{109}Cd par le poisson, en fonction du temps.

5. DISCUSSION GENERALE ET CONCLUSIONS

5.1. L'évolution dans le temps de l'accumulation de ^{65}Zn indique, même chez les poissons de très petite taille (100 mg), un faible taux du métabolisme de l'élément stable. Il en découle que le facteur de concentration à l'équilibre du radionucléide ne peut être obtenu à partir d'essais à court terme, même si un taux modéré de croissance contribue à accélérer quelque peu l'accumulation comme dans les essais décrits ici.

5.2. L'accumulation branchiale du Zn contribue pour une part prépondérante à l'accumulation totale de l'élément par le poisson, même dans le cas où un maillon de départ de la chaîne alimentaire expérimentale (sédiment) présente une capacité d'adsorption considérable vis-à-vis de ce même élément.

Par ailleurs, alors que cette capacité d'adsorption du sédiment organique est particulièrement prononcée vis-à-vis du Cd et que, de ce fait, les larves en équilibre avec le sédiment présentent une teneur élevée en cet élément, l'absorption intestinale du Cd par le poisson reste une voie d'entrée négligeable par rapport à l'absorption branchiale.

Dans ces conditions, on pourrait suggérer, dans les études d'accumulation du Zn et du Cd par les poissons, d'éviter des dispositifs expérimentaux complexes comportant une chaîne alimentaire naturelle, qui restent toujours particulièrement laborieux à réaliser, des conclusions significatives pouvant être obtenues à partir de données d'accumulation branchiale et d'accumulation intestinale comportant une nourriture expérimentale artificielle. Il convient d'insister cependant sur la nécessité d'équilibrer, dans de tels essais, la radioactivité spécifique (rapport entre le radionucléide et l'élément stable) de l'eau et de la nourriture, afin de ne pas fausser systématiquement les contributions respectives des deux types d'accumulation.

5.3. L'interférence de la teneur en Cd sur l'accumulation du radiozinc, mise en évidence sur Lepomis gibbosus adulte à 40 ppb de Cd dans l'eau [2], n'a pu être détectée dans nos essais avec un jeune cichlide de petite taille, à une concentration de 20 ppb de Cd dans l'eau. Des recherches ultérieures sont donc nécessaires pour définir les facteurs agissant dans cette interférence (espèce, âge, seuil de concentration de Cd et teneurs en Zn). Il est probable qu'un mécanisme de détoxication, faisant intervenir la métallo-thionéine chez les poissons comme chez les mammifères [12], contribue à masquer jusqu'à un certain point l'interférence du Cd sur le métabolisme du Zn.

Par contre, nos essais indiquent une inhibition très nette, par des traces de Cd, de l'échange du Zn de la part de larves de chironomides. Cette inhibition est à interpréter comme un effet sublétal prononcé sur une population d'organismes benthiques qui pourraient de ce fait servir de bioindicateurs précieux. Dans ce maillon important de la chaîne écologique, la forte adsorption du Cd par le sédiment organique vient sans doute exalter l'inhibition observée. Il est significatif de constater que dans l'écosystème aquatique comme dans l'écosystème terrestre, le transfert du Cd vient acquérir une signification considérable en relation avec la teneur du substrat en matériel organique [13].

5.4. En opposition avec le métabolisme du zinc, l'accumulation du Cd semble représenter un processus rapide. Par contre, son excrétion se poursuit à un taux assez limité. Ces observations pourraient s'expliquer, d'une part, par l'induction rapide de métallo-thionéine liant spécifiquement le Cd en présence de traces de l'élément dans le milieu et, d'autre part, par le «turnover» lent de cette protéine spécifique.

REFERENCES

[1] MERLINI, M., et al., «The biological pathway of Zinc (^{65}Zn) in freshwater fish and its alteration by heavy metals», Proc. Third Nat. Symp. Radioecology 1, Oak Ridge, May 1971 (CONF-710501-P1) (1971) 285-306.

[2] MERLINI, M., et al., «The effects of sublethal amounts of cadmium and mercury on the metabolism of zinc-65 by freshwater fish», Int. Symp. Radioecology Applied to the Protection of Man and His Environment (Proc. Symp. Rome, Sept. 1971), EUR 4800 d-f-i-e, vol.2, CCE, Luxembourg (1972) 1327-37.

[3] FRIBERG, L., et al., Cadmium in the Environment, 2nd Ed. (1974) Cleveland – CRC Pr., 248 pp.

[4] GARDINER, J., The chemistry of cadmium in natural water, II. The adsorption of cadmium on river muds and naturally occurring solids, Water Res. 8 (1974) 157-64.

[5] GERKING, S.D., Production and food utilization in a population of bluegill sunfish, Ecol. Monogr. 32 (1962) 31-78.

[6] GERKING, S.D., et al., «Rate of sediment intake by midge larvae (Chironomus plumosus: Diptera) using a ^{134}Cs tracer», Proc. Fourth Nat. Symp. Radioecology (Corvallis, Oregon, May 1975) 29 pp (in press).

[7] BERG, A., BRAZZELLI, A., Etude expérimentale du transfert du ^{65}Zn à un poisson d'eau douce, avec référence particulière aux deux voies d'absorption et au métabolisme de l'élément stable, Radioprotection (1975) (sous presse).

[8] WINBERG, G.G., Rate of Metabolism and Food Requirements of Fishes, Fish. Res. Board Canada, Translation Series, No. 194 (1960) 202 pp.

[9] BAPTIST, J.P., et al., Retentions of ^{51}Cr, ^{59}Fe, ^{60}Co, ^{65}Zn, ^{85}Sr, ^{95}Nb, ^{141m}In and ^{131}I by Atlantic croaker (Micropogon undulatus), Health Phys. 18 2 (1970) 141-48.

[10] BRUNGS, W.A., The relative distribution of multiple radionuclides in a freshwater pond, Dissert. Abstr. 24 7 (1969) Abstract No. 3023.

[11] MYTTENAERE, C., et al., «Etude de l'influence du cadmium stable sur le transfert du zinc-65 en écosystème irrigué par submersion (rizière irriguée)», présents comptes rendus, IAEA-SM-198/6.

[12] MARAFANTE, E., et al., Detossicazione dei metalli pesanti nei pesci: isolamento della metallotioneina dal fegato di Carassius auratus, Boll. Soc. Ital. Biol. Sper. 48 (Suppl.) (1972) 109.

[13] DENAEYER-DE SMET, S., Premier aperçu de la distribution du cadmium dans divers écosystèmes terrestres non pollués et pollués, Oecol. Plant. 9 2 (1974) 169-82.

DISCUSSION

Yu.V. SIVINTSEV: How was the accumulation parameter obtained in your Table II? What specific activities were used in determining the concentration factor? In other words, what was in the numerator of the ratio and what was in the denominator?

A.R. BERG: The parameter used in Table II is the fish/water 'concentration factor'. The numerator of the ratio is the radioactivity per unit wet weight of fish, and the denominator is the radioactivity per unit weight of water. The same ratio is used in experiments with labelled water and with labelled food, since the larvae fed to the fish were themselves in equilibrium with sediments and with water. The total fish/water concentration factor is thus represented by the sum of two partial concentration factors — for water and for food.

Yu.V. SIVINTSEV: What is the relationship between the accumulation parameters given in Fig. 2 and those in Table II?

A.R. BERG: The concentration factor used in Fig. 2 for the larvae is identical to the one used for fish, with the radioactivity per unit weight of larva as the numerator of the ratio and the radioactivity per unit weight of water as the denominator.

S. BONOTTO: Do the microorganisms play any part in the transfer of radiozinc and radiocadmium (^{65}Zn and ^{109}Cd) from the sediment to the chironomid larvae?

A.R. BERG: Microorganisms could certainly play a part in the assimilation of the sediments ingested by the larvae and hence also in the transfer of radionuclides from sediments to larvae. We wanted to clarify this aspect by using sediments treated with antibiotics, but this work has not yet been performed.

Gwyneth D. HOWELLS: Did you measure the total body burden of stable Zn and Cd at equilibrium in the fish and chironomids, and did it vary with different treatments? Did the different levels of exposure concentration in the water affect these equilibrium levels or only the rates of uptake?

A.R. BERG: For chironomid larvae, the body burden of the stable elements Zn and Cd was calculated from the equilibrium radioactivity of the corresponding radionuclide. The results indicate a strong decrease — in the presence of 10 and 20 ppb Cd — both in the rate of uptake of stable Zn and in the body burden at equilibrium.

For fish the body burden of Zn was measured directly on the stable element (around 70 ppm, fresh weight). It is not influenced by the presence of Cd, and no inhibition in the rate of uptake of Zn in the presence of Cd could be detected.

О ПОВЕДЕНИИ ^{90}Sr И ^{137}Cs В ВОДАХ БАЛТИЙСКОГО МОРЯ

А.Г.ТРУСОВ
Государственный комитет по использованию атомной энергии СССР,
Москва
Л.М.ИВАНОВА, В.С.ЗОЛОТУХИНА,
Л.И.ГЕДЕОНОВ
Радиевый институт им.В.Г.Хлопина,
Ленинград,
Союз Советских Социалистических Республик

Доклад представлен Ю.В.Сивинцевым

Abstract–Аннотация

THE BEHAVIOUR OF ^{90}Sr AND ^{137}Cs IN THE WATERS OF THE BALTIC SEA.

On the basis of observed concentrations of ^{90}Sr and ^{137}Cs in the waters of the Baltic Sea and previously published data, the authors estimate the changes in the average values of these concentrations for several years. It is shown that variations in the concentration of ^{90}Sr in general follow changes in the rate of global fall-out of this radionuclide with a considerable delay. It is assumed that ^{90}Sr is very slowly removed from the water mass and that the concentration slowly decreases owing to the shallowness and the isolation of the Baltic Sea. Various other possible reasons are discussed. The distribution of ^{137}Cs in the waters of the Baltic Sea is quite different from that of ^{90}Sr, its content there being much smaller than that of the latter. This is explained by the sorption processes in fresh water and by the flow of caesium-depleted water from the rivers. Factual data are given on the content of ^{90}Sr and ^{137}Cs in the waters of the Gulf of Finland. The presence of these nuclides is due to global radioactive contamination. For the reasons given above the ^{137}Cs/^{90}Sr ratio is particularly low here.

О ПОВЕДЕНИИ ^{90}Sr И ^{137}Cs В ВОДАХ БАЛТИЙСКОГО МОРЯ.

На основе наблюдений за концентрацией ^{90}Sr и ^{137}Cs в водах Балтики с привлечением данных, опубликованных ранее, дана оценка изменений средних значений этой концентрации за несколько лет. Установлено, что изменения концентрации ^{90}Sr в общем следуют за изменениями темпа глобального выпадения этого радионуклида со значительным запаздыванием. Предполагается, что ^{90}Sr очень медленно удаляется из водной массы, и вследствие мелководности и замкнутости Балтийского моря концентрация ^{90}Sr убывает медленно. Рассмотрены и другие возможные причины. Распределение ^{137}Cs в водах Балтийского моря заметно отличается от распределения ^{90}Sr. Наблюдается много меньшее содержание его в воде, чем содержание ^{90}Sr. Причина – воздействие сорбционных процессов в пресной воде и сток обедненных цезием вод и рек. Приведены фактические данные о содержании ^{90}Sr и ^{137}Cs в воде Финского залива. Присутствие их обусловлено глобальным радиоактивным загрязнением. По указанным выше причинам отношение ^{137}Cs/^{90}Sr здесь особенно низко.

В настоящее время реализуется программа широкого развития атомной энергетики в прибалтийских странах. Целый ряд атомных электростанций (АЭС) строится непосредственно на побережье Балтийского моря. Другие размещаются на реках или озерах, имеющих сток в Балтику. По мере усовершенствования конструкций реакторов и всех мер технической безопасности убывает доля радиоактивных продуктов работы АЭС, сбрасываемых в водные системы. Однако рост общей мощности вводимых в эксплуатацию ядерных установок идет

ТАБЛИЦА I. СРЕДНЯЯ КОНЦЕНТРАЦИЯ ^{90}Sr и ^{137}Cs В ПОВЕРХНОСТНЫХ ВОДАХ ПО РАЙОНАМ БАЛТИЙСКОГО МОРЯ В 1972 И 1973 гг.

№№ п/п	Район	1972 г.			1973 г.		
		Концентрация (расп/мин · 100 л)		$^{137}Cs/^{90}Sr$	Концентрация (расп/мин · 100 л)		$^{137}Cs/^{90}Sr$
		^{90}Sr	^{137}Cs		^{90}Sr	^{137}Cs	
1	Северо-восточная часть собственно Балтийского моря, включая западную часть Финского залива	190 ± 30 (5)	110 ± 20 (5)	0,6	145 ± 20 (4)	107 ± 15 (4)	0,7
2	Восточная часть собственно Балтийского моря	180 ± 25 (8)	130 ± 20 (8)	0,7	130 ± 20 (4)	115 ± 15 (4)	0,9
3	Западная часть собственно Балтийского моря	150 ± 20 (14)	120 ± 25 (13)	0,8	140 ± 25 (5)	120 ± 20 (4)	0,9
4	Южная часть собственно Балтийского моря	120 ± 15 (5)	110 ± 20 (5)	0,9	120 ± 20 (5)	120 ± 20 (5)	1,0
5	Район, примыкающий к Датским проливам	110 ± 15 (3)	110 ± 20 (3)	1,0	120 ± 25 (4)	130 ± 20 (4)	1,05
6	Датские проливы	130*	-	-	170 ± 30	130 ± 25	0,7

* По данным работы [8]

еще быстрее. Поэтому не ослабевает, а усиливается внимание к поведению радиоактивных примесей в Балтийском море, а также реках и озерах его бассейна.

В масштабах всего Балтийского моря действие короткоживущих радиоактивных загрязнителей до сих пор не наблюдается. Нет данных и о полях или ореолах радиоактивной загрязненности вблизи сбросных каналов работающих ядерных установок. Поэтому для выработки представления о поведении некоторых радиоактивных примесей исследуются присутствующие в водах продукты глобального радиоактивного выпадения. Несмотря на очень высокую чувствительность применяемых нами методик определения радионуклидов в воде [1,2] и на возможность использования и других весьма чувствительных методов [3], реально в наши дни в водах Балтийского моря и его бассейна можно наблюдать только ^{90}Sr и ^{137}Cs, может быть ^{54}Mn и $^{239+240}Pu$.

В настоящей работе представляются и используются данные по первым двум из указанных радионуклидов.

Информация о поведении этих радионуклидов может дать представление также о судьбе и поведении ^{89}Sr, ^{134}Cs и других изотопов щелочноземельных и щелочных элементов, которые могут попасть в водную систему Балтики в составе сбросов и отходов.

Исследования радиоактивного загрязнения вод Балтики ведутся в целом ряде научных учреждений в разных странах [4-8]. Обзор результатов за 1959-1967 гг. был выполнен в известном обобщении Академии наук США [9]. В последние годы появились новые публикации, в числе которых назовем и наши работы [10,11]. В них сделаны попытки обобщения литературного материала, ставшего доступным к 1974 году. В настоящую работу включены и самые последние данные, полученные совсем недавно. В табл.I представлены результаты наших определений ^{90}Sr и ^{137}Cs в международных водах Балтийского моря в 1972 и 1973 гг. Усреднение величин концентраций дает следующие значения: ^{90}Sr в 1972 г. – (150±25) расп/мин, в 1973 г. – (130±20) расп/мин, ^{137}Cs в 1972 г. – (120±25) расп/мин.

Отношение средних значений концентрации ^{137}Cs/^{90}Sr оказалось в 1972 г. равным 0,8, а в 1973 г. – 0,9.

Сравнение средних для всей Балтики результатов наблюдений 1972 и 1973 гг. показывает, что они практически одинаковы. Концентрация ^{90}Sr несколько уменьшилась, но это уменьшение находится в пределах ошибки опыта.

Рассмотрение средних по отдельным регионам значений свидетельствует о некотором уменьшении в 1973 г. концентрации ^{90}Sr по сравнению с 1972 г. Для ^{137}Cs в пределах точности определений этого не наблюдается. Наряду с этим замечается, особенно по данным 1972 г., что содержание ^{90}Sr было наиболее высоким в северо-восточной части моря и в Финском заливе и самым низким – в районе проливов. Это объяснялось притоком вод из Северного моря с меньшей концентрацией в них радионуклидов и наибольшим удалением от проливов именно северо-восточной части моря и, следовательно, меньшей возможностью обмена вод этого района. А в мелководной и почти пресноводной северо-восточной области моря и в особенности в Финском заливе решающую роль в балансе веществ глобального радиоактивного выпадения играет сток рек, несущих ^{90}Sr с очень большой территории суши.

Результаты наблюдений 1972 и 1973 гг., рассмотренные выше, с привлечением данных, опубликованных ранее [7-11], позволяют продолжить оценку изменений со временем средних значений концентрации ^{90}Sr и ^{137}Cs, начатую в предшествующих работах. На рис.1 показано изменение средней концентрации ^{90}Sr в поверхностных водах Балтийского моря с 1959 г. по 1973 г. Для сопоставления здесь же представлены данные о выпадении и накоплении ^{90}Sr на суше на берегу Финского залива в Зеленогорске близ Ленинграда. Из рис.1 следует, что в отношении загрязнения ^{90}Sr воды Балтики представляют "инерционную" систему. Это понятно, если принять во внимание мелководность и замкнутость Балтийского моря. По-видимому, приносимый реками, а возможно и выпавший на поверхность моря, ^{90}Sr находится в растворе в ионной форме. При небольшой глубине вся вода перемешана. Уход ^{90}Sr из поверхностного слоя воды на большую глубину компенсирован за счет перемешивания. На дно ^{90}Sr осаждается крайне медленно. Известно, что в некоторых застойных мелководных районах отношение ^{137}Cs к ^{90}Sr в донных отложениях составляет 4 и более, в то время как в выпадениях это отношение равно в среднем 1,6 [12]. Следовательно, ^{90}Sr отлагается в донные осадки много медленнее, чем ^{137}Cs. Вследствие этого шансы удаления ^{90}Sr из водной массы Балтики крайне низки. В годы слабого стока рек, как это было в 1972 и 1973 гг., невелик и шанс

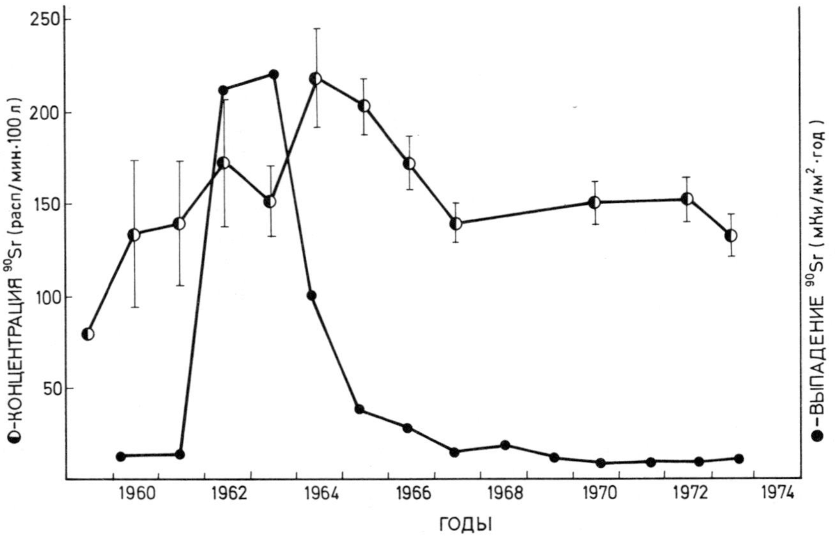

Рис.1. Изменение содержания ^{90}Sr в водах Балтийского моря и выпадения ^{90}Sr на поверхность с 1959 по 1973 гг.

ухода ^{90}Sr при водообмене с Северным морем. В некоторые годы вообще из Балтийского моря в Северное вода совсем не поступает. Испарение воды, само собой понятно, не удаляет ^{90}Sr. В результате, как мы и видим из рис.1, процесс удаления ^{90}Sr из вод Балтийского моря после усиленных поступлений 1964-65 гг. идет медленно. Значительное запоздание максимума концентрации ^{90}Sr по сравнению с максимумом плотности выпадений свидетельствует о большой роли стока с водами рек. Этот последний процесс запаздывает по сравнению с выпадением весьма сильно и определяет положение максимума концентрации. Не будь этого влияния, концентрация ^{90}Sr должна была бы следовать за изменениями плотности радиоактивного выпадения со значительно меньшим сдвигом фаз. Указанные причины накопления ^{90}Sr в водах Балтийского моря приводят к тому, что в этом море концентрация ^{90}Sr в 5-6 раз выше, чем в открытом океане на тех же широтах [10]. Приведенные в работе [11] сведения за 1969-1970 гг. показывают, что если принять концентрацию ^{90}Sr в открытом Атлантическом океане в поверхностных водах умеренных северных широт за 1, то для Средиземного моря она была 1,5; для Северного моря – 3,2; для Черного моря – 4 и Балтийского – 6.

Распределение ^{137}Cs в водах Балтийского моря заметно отличается от распределения ^{90}Sr. Диапазон значений концентрации ^{137}Cs, наблюдавшихся нами в 1972-1973 гг., несколько уже, чем для ^{90}Sr. Не замечено большой разницы в разных районах Балтики, как это отмечалось для ^{90}Sr в 1972 г. (табл.I). В сравнительно близких точках были найдены и минимальные и максимальные значения [11]. В поведении ^{137}Cs большое значение имеют сорбционные процессы и, следовательно, те условия,

ТАБЛИЦА II. ХАРАКТЕРИСТИКА НЕКОТОРЫХ ГРУПП ИЗ СОВОКУПНОСТИ РЕЗУЛЬТАТОВ ОПРЕДЕЛЕНИЯ ОТНОШЕНИЯ ^{137}Cs к ^{90}Sr В ВОДАХ МОРЕЙ И ОКЕАНОВ

Группа данных	Число определений	Интервал (размах) значений	Среднее арифметическое	Среднеквадратичное отклонение разового результата	Погрешность разового экспериментального результата (± 20%)
Все данные	318	0,4 - 4,4	1,7 ± 0,05	± 0,5	± 0,34
Атлантический океан (все данные)	172	0,6 - 4,4	1,8 ± 0,04	± 0,5	± 0,36
Поверхностные воды (1963-1970 гг.)	159	0,6 - 4,4	1,8 ± 0,04	± 0,5	± 0,36
Поверхностные воды (1963-1964 гг.)	78	0,7 - 3,7	1,7 ± 0,07	± 0,6	± 0,34
Поверхностные воды (1969-1970 гг.)	81	0,6 - 4,4	1,9 ± 0,07	± 0,6	± 0,38
Воды с глубин до 700 м (1969-1970 гг.)	13	0,7 - 2,6	1,6 ± 0,14	± 0,5	± 0,32
Тихий океан (все данные)	83	0,4 - 2,0	1,4 ± 0,07	± 0,4	± 0,28
Поверхностные воды (1966-1970 гг.)	53	0,8 - 2,0	1,5 ± 0,04	± 0,3	± 0,3
Воды с глубин до 700 м (1966-1967 гг.)	15	0,4 - 2,0	1,4 ± 0,12	± 0,5	± 0,28
Воды с больших глубин от 700 до 7000 м (1966-1970 гг.)	15	0,6 - 1,8	1,3 ± 0,1	± 0,4	± 0,26
Балтийское море (1970 г.)	63	0,4 - 1,4	0,9 ± 0,05	± 0,2	± 0,24

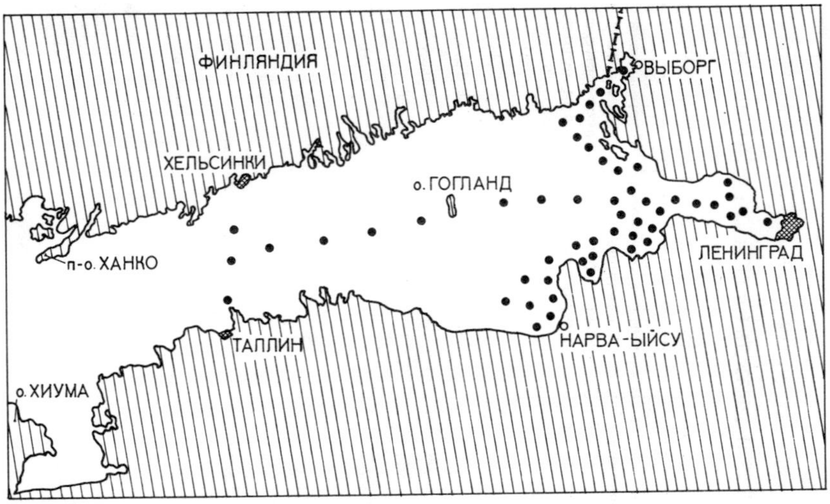

Рис.2. Карта-схема расположения точек отбора проб воды в Финском заливе Балтийского моря.

которые складываются в том или ином участке бассейна. Воды рек обеднены ^{137}Cs, и значит, чем в большей степени водосодержание моря определяется стоком рек, тем беднее его воды ^{137}Cs. Кроме того, процессы сорбции, идущие непосредственно в море, способствуют переходу ^{137}Cs в донные отложения.

Разница в поведении ^{90}Sr и ^{137}Cs в водах Балтийского моря обнаруживается и при анализе отношений концентрации $^{137}Cs/^{90}Sr$. В табл.II повторены цифры из нашей работы [13], но для Балтийского моря в ней теперь приводятся последние результаты. Из данных табл.II следует, что Балтийское море по значению величины рассматриваемого отношения резко отличается и от Атлантического и от Тихого океанов. Этот факт вполне соответствует тому, что сказано выше о медлительности удаления ^{90}Sr и дефиците ^{137}Cs в речных водах.

Из всех районов Балтийского моря для Советского Союза особый интерес представляет Финский залив, питаемый в основном речным стоком с территории северо-запада нашей страны и со значительной части территории Финляндии.

На протяжении последних лет Радиевым институтом им.В.Г.Хлопина ведется систематическое наблюдение за содержанием долгоживущих радионуклидов в водах Финского залива. Отбор проб производится в водах собственно Финского залива и в отдельных его частях: Невской губе, Копорской губе, Нарвском заливе, Выборгском заливе и Лужской губе. На рис.2 представлено расположение точек отбора проб воды. В табл.III сведены результаты определения ^{90}Sr и ^{137}Cs в водах Финского залива в 1971, 1972 и 1973 гг.

ТАБЛИЦА III. РЕЗУЛЬТАТЫ ОПРЕДЕЛЕНИЯ ^{90}Sr и ^{137}Cs В ПОВЕРХНОСТНЫХ ВОДАХ ФИНСКОГО ЗАЛИВА БАЛТИЙСКОГО МОРЯ В 1971, 1972 И 1973 гг.

Район отбора пробы	Дата отбора пробы			Концентрация (расп/мин · 100 л)						Соотношение $^{137}Cs/^{90}Sr$		
				^{90}Sr			^{137}Cs					
	1971 г.	1972 г.	1973 г.	1971 г.	1972 г.	1973 г.	1971 г.	1972 г.	1973 г.	1971 г.	1972 г.	1973 г.
1	2	3	4	5	6	7	8	9	10	11	12	13
Невская губа	15.07	15.09	5.09	200 ± 40	440 ± 60	240 ± 30	50 ± 10	20	40 ± 15	0,3	-	0,2
	15.07	15.09	5.09	160 ± 30	420 ± 60	220 ± 30	50 ± 10	20	40 ± 15	0,3	-	0,2
	15.07	15.09	5.09	220 ± 30	410 ± 60	250 ± 40	30 ± 10	20	20	0,1	-	-
	15.07	15.09	5.09	190 ± 30	410 ± 60	210 ± 30	50 ± 10	20	35 ± 15	0,3	-	0,2
	15.07	17.09	5.09	200 ± 30	350 ± 50	240 ± 30	30 ± 10	20	30 ± 10	0,2	-	0,1
	16.07	17.09		–	320 ± 10		30 ± 10	20		-	-	
	16.07	17.09		–	380 ± 50		30 ± 10	30 ± 15		-	0,08	
	16.07	18.09		320 ± 50	550 ± 80		30 ± 10	70 ± 20		0,1	0,15	-
	16.07			–			30 ± 10			-		
	16.07			120 ± 20			50 ± 10			0,4		
	Среднее по району:			210 ± 20	410 ± 60	230 ± 30	40 ± 10			0,2	0,1	0,2
Копорская губа	30.07	12.10	29.07	340 ± 50	240 ± 40	160 ± 30	60 ± 10	130 ± 20	80 ± 20	0,2	0,6	0,5
	30.07	12.10	29.07	260 ± 40	280 ± 40	170 ± 30	80 ± 10	130 ± 20	80 ± 20	0,3	0,5	0,5
	30.07	12.10	29.07	220 ± 30	230 ± 30	170 ± 30	100 ± 20	170 ± 30	90 ± 20	0,5	0,7	0,5
	30.07	12.10	29.07	320 ± 50	100 ± 20	160 ± 30	100 ± 20	70 ± 20	110 ± 20	0,3	0,7	0,7
	30.07	12.10	29.07	340 ± 50	100 ± 20	180 ± 30	110 ± 20	70 ± 20	120 ± 20	0,3	0,7	0,7
	30.07	12.10	29.07	330 ± 50	120 ± 20	190 ± 30	120 ± 20	80 ± 20	100 ± 20	0,4	0,7	0,5
			29.07			160 ± 30			100 ± 20			0,6
			29.07			200 ± 30			70 ± 20			0,4
			29.07			180 ± 30			60 ± 20			0,3
			29.07			180 ± 30			70 ± 20			0,4
	Среднее по району:			300 ± 45	180 ± 30	180 ± 30	100 ± 20	110 ± 20	90 ± 20	0,3	0,7	0,5
Нарвский залив	14.08	3.10	24.07	630 ± 90	180 ± 30	190 ± 30	80 ± 10	230 ± 40	100 ± 20	0,1	1,3	0,5
	14.08	3.10	24.07	330 ± 50	290 ± 40	200 ± 30	90 ± 10	140 ± 20	80 ± 20	0,3	0,5	0,4
	14.08	3.10	24.07	270 ± 40	280 ± 40	200 ± 30	70 ± 10	170 ± 30	70 ± 20	0,3	0,6	0,4
	14.08	3.10	26.07	290 ± 40	140 ± 20	160 ± 30	100 ± 20	220 ± 30	110 ± 20	0,3	1,6	0,7
		11.10	26.07		130 ± 20	150 ± 30		120 ± 20	100 ± 20		0,9	0,7
	Среднее по району:			380 ± 60	200 ± 30	180 ± 30	90 ± 10	180 ± 30	90 ± 20	0,3	1,0	0,5

ТАБЛИЦА III.(ПРОДОЛЖЕНИЕ)

1	2	3	4	5	6	7	8	9	10	11	12	13
Выборгский залив	24.07	20.09	20.09	280 ± 40	260 ± 40	170 ± 40	40 ± 10	110 ± 20	80 ± 20	0,1	0,4	0,5
	24.07	20.09	20.09	380 ± 50	300 ± 40	180 ± 30	40 ± 10	100 ± 20	90 ± 20	0,1	0,4	0,5
	24.07	20.09	20.09	—	270 ± 40	180 ± 30	40 ± 10	130 ± 20	80 ± 20	-	0,5	0,4
	24.07	20.09	23.09	440 ± 60	130 ± 20	180 ± 30	30 ± 10	110 ± 20	80 ± 20	0,1	0,8	0,5
	24.07	20.09	23.09	280 ± 40	310 ± 40	170 ± 30	50 ± 10	130 ± 20	60 ± 20	0,2	0,4	0,4
	24.07	20.09	23.09	250 ± 40	250 ± 40	220 ± 30	70 ± 10	120 ± 20	80 ± 20	0,3	0,5	0,4
	24.07	20.09	23.09	220 ± 40	180 ± 30	190 ± 30	80 ± 10	100 ± 20	90 ± 20	0,4	0,6	0,5
	24.07	20.09	23.09	270 ± 40	200 ± 30	210 ± 30	60 ± 10	90 ± 20	60 ± 20	0,2	0,5	0,3
	31.07	20.09	23.09	310 ± 50	240 ± 40	190 ± 30	70 ± 10	70 ± 20	40 ± 15	0,2	0,3	0,2
	31.07	21.09	24.09	380 ± 60	210 ± 30	190 ± 30	80 ± 10	70 ± 20	40 ± 15	0,2	0,3	0,2
	31.07	21.09		290 ± 50	240 ± 40		70 ± 10	150 ± 30		0,2	0,6	
	31.07	21.09		320 ± 50	210 ± 30		80 ± 10	60 ± 20		0,3	0,3	
	31.07			310 ± 50			80 ± 10			0,3		
	31.07			210 ± 30			80 ± 10			0,4		
	31.07			320 ± 50			80 ± 10			0,3		
	31.07			400 ± 60			80 ± 10			0,2		
	Среднее по району:			310 ± 50	230 ± 40	190 ± 30	60 ± 10	100 ± 20	70 ± 20	0,2	0,5	0,4
Лужская губа	-	-	27.07	-	-	200 ± 30	-	-	90 ± 20	-	-	0,5
	-	-	27.07	-	-	160 ± 30	-	-	120 ± 20	-	-	0,7
	-	-	27.07	-	-	170 ± 30	-	-	100 ± 20	-	-	0,6
	Среднее по району:					180 ± 30			100 ± 20			0,6

1	2	3	4	5	6	7	8	9	10	11	12	13
Финский залив	30.07	18.09	8.09	230 ± 40	300 ± 50	180 ± 30	70 ± 10	90 ± 20	120 ± 20	0,3	0,3	0,6
(Центральная	30.07	18.09	8.09	250 ± 40	180 ± 30	180 ± 30	70 ± 10	80 ± 20	100 ± 20	0,3	0,4	0,6
часть)	30.07	18.09	8.09	240 ± 40	170 ± 30	190 ± 30	60 ± 10	80 ± 20	90 ± 20	0,3	0,5	0,5
	30.07	18.09	8.09	400 ± 60	180 ± 30	150 ± 30	70 ± 10	80 ± 20	100 ± 20	0,2	0,4	0,7
	30.07	18.09	8.09	230 ± 40	210 ± 30	120 ± 20	90 ± 10	90 ± 20	90 ± 20	0,4	0,4	0,8
	30.07	18.09	8.09	280 ± 40	260 ± 40	170 ± 30	70 ± 10	100 ± 20	90 ± 20	0,3	0,4	0,5
	30.07	18.09	8.09	490 ± 70	350 ± 50	170 ± 30	100 ± 20	100 ± 20	100 ± 20	0,2	0,3	0,6
	30.07	19.09	11.09	330 ± 50	190 ± 30	150 ± 30	80 ± 10	110 ± 20	100 ± 20	0,2	0,6	0,7
		19.09	11.09		140 ± 20	180 ± 30		80 ± 20	90 ± 20		0,5	0,5
		20.09	11.09		220 ± 30	160 ± 30		150 ± 30	80 ± 20		0,7	0,5
		20.09	11.09		320 ± 40	150 ± 30		120 ± 20	90 ± 20		0,4	0,6
		20.09	15.09		240 ± 30	170 ± 30		80 ± 20	110 ± 20		0,3	0,6
		3.10	15.09		150 ± 20	140 ± 20		170 ± 30	120 ± 20		1,2	0,9
		3.10	15.09		190 ± 30	140 ± 20		230 ± 40	110 ± 20		1,2	0,8
		3.10	15.09		300 ± 50	110 ± 20		100 ± 20	110 ± 20		0,3	1,0
		3.10	15.09		270 ± 30	140 ± 20		110 ± 20	120 ± 20		0,4	0,9
		3.10	15.09		200 ± 30	130 ± 20		120 ± 20	120 ± 20		0,6	0,9
		3.10	15.09		220 ± 30	140 ± 20		130 ± 20	190 ± 30		0,6	1,4
			20.09			170 ± 30			90 ± 20			0,5
			20.09			140 ± 20			80 ± 20			0,6
			20.09			130 ± 20			80 ± 20			0,7
			20.09			150 ± 30			80 ± 20			0,6
			20.09			150 ± 20			60 ± 20			0,4
			20.09			160 ± 30			80 ± 20			0,5
			20.09			160 ± 30			80 ± 20			0,5
	Среднее по району:			310 ± 50	230 ± 30	150 ± 30	80 ± 10	110 ± 20	100 ± 20	0,3	0,5	0,7

ТАБЛИЦА IV. УСРЕДНЕННЫЕ КОНЦЕНТРАЦИИ ^{90}Sr и ^{137}Cs В ВОДАХ ФИНСКОГО ЗАЛИВА В 1971, 1972, 1973 гг.

Год	Концентрация (расп/мин · 100 л)		$^{137}Cs/^{90}Sr$
	^{90}Sr	^{137}Cs	
1971	280 ± 40	70 ± 10	0,3
1972	250 ± 40	120 ± 20	0,5
1973	180 ± 30	80 ± 10	0,4

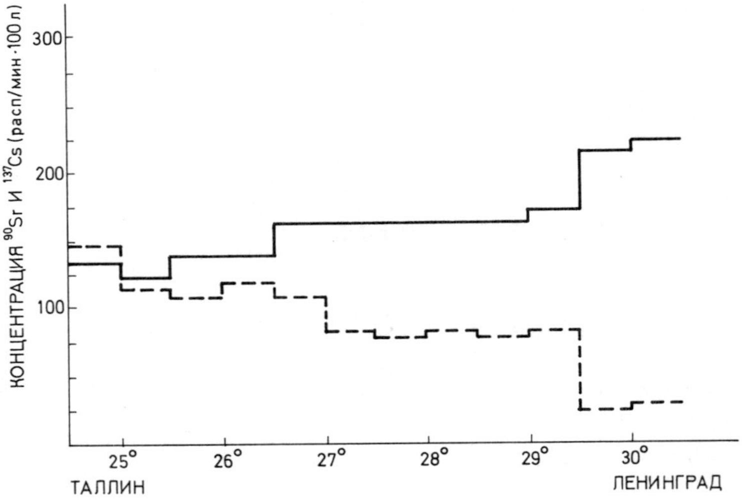

Рис.3. Распределение ^{90}Sr и ^{137}Cs в водах Финского залива в 1973 г.

Средние концентрации ^{90}Sr и ^{137}Cs для вод всего залива в целом составили величины, приведенные в табл.IV.

Рассмотрение данных таблиц III и IV показывает, что содержание ^{90}Sr в Финском заливе имеет тенденцию к снижению. Изменение уровня ^{137}Cs за период наблюдения отчетливо не проявилось.

По отдельным районам Финского залива также наблюдалось снижение концентрации ^{90}Sr от 1971 г. к 1973 г. Увеличение содержания ^{90}Sr было отмечено в 1972 г. в районе Невской губы. Это увеличение совпало с резким обмелением Невской губы летом 1972 г. за счет интенсивного испарения, с одной стороны, и отсутствия осадков, с другой, имевших место в результате необычного для Ленинграда и всего северо-запада СССР жаркого и засушливого лета 1972 г. В этих условиях могло иметь место увеличение концентраций всех видов загрязнений, в том числе и радиоактивных. ^{137}Cs, у которого способность сорбироваться илами и донными осадками значительно выше, чем у ^{90}Sr, мог интенсивно увлекаться взвесями на дно, в результате чего возник дефицит ^{137}Cs при одновременном накоплении ^{90}Sr.

Рассмотрение соотношения $^{137}Cs/^{90}Sr$ (табл.III) показывает, что в районах восточной части Финского залива это соотношение меньше по сравнению с его западной частью. Это с достаточной наглядностью подтверждает рис.3, на котором показано содержание ^{90}Sr и ^{137}Cs в поверхностной воде разных участков Финского залива, следуя с запада на восток. Восточные районы Финского залива наиболее удалены от вод собственно Балтийского моря; это районы, водосодержание которых во многом определяется речными водами. Отсюда, как следствие, малая величина соотношения $^{137}Cs/^{90}Sr$ в восточной части Финского залива. В западной части концентрация ^{137}Cs практически равна концентрации ^{90}Sr. Это выравнивание происходит за счет обмена вод Финского залива с водами открытого Балтийского моря.

В целом наблюдавшаяся в 1972 и 1973 гг. картина изменения содержания в водах Финского залива ^{90}Sr и ^{137}Cs являлась результатом глобальных радиоактивных загрязнений. Затрудненность обмена вод с собственно Балтийским морем и далее с океаном препятствует быстрому снижению концентраций радионуклидов. При имеющей место тенденции к снижению эти концентрации продолжали оставаться достаточно высокими. Кроме того, убыль радионуклидов путем обмена вод пополняется за счет продолжающихся глобальных выпадений, а также за счет речных стоков, вымывающих радионуклиды с земной поверхности.

ЛИТЕРАТУРА

[1] ИВАНОВА, Л.М., Метод одновременного определения стронция-90, цезия-137 и церия-144 в морской воде, Радиохимия 9 5 (1967) 622.

[2] ИВАНОВА, Л.М., ГЕДЕОНОВ, Л.И., ТРУСОВ, А.Г., Разработка методов определения радионуклидов в пробах воды и донных отложений, Научно-техническая конференция СЭВ "Исследования в области обезвреживания жидких, твердых и газообразных радиоактивных отходов и дезактивации загрязненных поверхностей", Колобжег, Польша, 2-7 октября 1972 г.

[3] Reference Methods for Marine Radioactivity Studies, Technical Reports Series, № 118, IAEA, Vienna, 1970.

[4] GYLLANDER, Ch., "Water exchange and diffusion processes in Tvären, a Baltic Bay", Disposal of Radioactive Wastes into Seas, Oceans and Surface Waters (Proc. Symp. Vienna, 1966) IAEA, Vienna (1966) 207.

[5] VOIPIO, A., "Some notes on recent studies of chemical oceanography in Finland", 1st National Conf. Coastal Shallow Water Research, (NSF – ONR), (1962) 785-800.

[6] VOIPIO, A., PAAKOLA, O., Strontium-90 in the Baltic Sea, ICES Hydrographical Committee, № 147 (1964).

[7] SALO, A., VOIPIO, A., "On the ^{137}Cs and ^{90}Sr ratio in Baltic waters". В сборн.: B.Aberg and F.P. Hungate (eds) Radio Ecological Concentration Processes, Pergamon Press, New York, (1966) 827-833.

[8] Environmental Radioactivity in Denmark in 1972, Risø, Rep. № 291 (1973).

[9] Radioactivity in the Marine Environment, prepared by the Panel on Radioactivity in the Marine Environment of the Committee on Oceanography, National Research Council, Washington (1971).

[10] ИВАНОВА, Л.М., ГЕДЕОНОВ, Л.И., МАРКЕЛОВ, В.Н., ПЕТРОВ, Ю.Г., ТРУСОВ, А.Г., ШЛЯМИН, Э.А., Содержание ^{90}Sr и ^{137}Cs в водах Балтийского моря в 1970 г., Атомная Энергия 33 4 (1972) 835.

[11] РАКОВ, Н.А., Исследования по миграции радиоактивного загрязнения, Советско-бельгийский-голландский семинар по переработке и захоронению радиоактивных отходов, Атомная энергия 37 4, (1974) 373.
[12] HARLEY, J., et al., Fission yield and fission product decay, HASL-300, Procedures Manual New York (1972) page F-02-01.
[13] ТРУСОВ, А.Г., ИВАНОВА, Л.М., ГЕДЕОНОВ, Л.И., Соотношение ^{137}Cs-^{90}Sr в океанических и морских водах, Атомная энергия 34 5 (1973) 391.

DISCUSSION

A. PRESTON: Are any data similar to those displayed in Table III for ^{90}Sr and ^{137}Cs available for deep (as distinct from surface) water? If so, do they show a similar pattern of concentrations?

Yu.V. SIVINTSEV: No data of that kind were obtained during these surface-water studies.

PLUTONIUM AND OTHER TRANSURANIUM NUCLIDES IN THE AQUATIC ENVIRONMENT

(Sessions IV and V)

Chairmen

K.C. PILLAI (India)

A. PRESTON (United Kingdom)

PLUTONIUM ISOTOPES IN AQUATIC FOODCHAINS IN THE BALTIC SEA

J.K. MIETTINEN, T. JAAKKOLA, Marja JÄRVINEN
Department of Radiochemistry,
University of Helsinki,
Helsinki, Finland

Abstract

PLUTONIUM ISOTOPES IN AQUATIC FOODCHAINS IN THE BALTIC SEA.

A large number of nuclear power plants will be built along the Baltic coasts in the coming years. Some release of transuranium elements can be expected from this nuclear activity. Since the bays of the Baltic and especially the estuarine and coastal areas are very shallow, even small releases can cause noticeable local concentrations in coastal waters and sediments. It is important to determine the present levels of plutonium in the aquatic biota, before the power reactor activity really begins. For this purpose, plutonium isotopes in sea water, bottom sediment, algae, benthic fauna, and fish were and are being analysed and the results compared with those reported from other marine environments. The dried samples were ashed or wet-ashed and plutonium was separated by ion-exchange and electrodeposited on a platinum disc. The α-activity was determined using a surface barrier semiconductor-detector. Study of coastal marine samples from the Gulf of Finland shows that fish (ruff, flounder, cod) contain 0.04-0.14 pCi ($^{239}Pu+^{240}Pu$)/kg fresh wt, blue mussel 0.6 pCi/kg fresh wt, brown alga 5 pCi/kg fresh wt and sediment 0.18 nCi/kg (1.68 mCi/km^2) in the top centimetre. These values are comparable with those in the Great Lakes of the USA.

INTRODUCTION

The task of determining present levels of plutonium in the aquatic biota of the Baltic Sea was undertaken so that one will be able to observe any changes in plutonium concentration that may occur when the nuclear power plants currently under construction come into operation. Samples of sea-water, bottom sediment, algae, benthic fauna, and fish are being analysed and the results compared with those reported from other marine environments. No results on plankton and seawater are yet ready, but the first samples of the other categories have been analysed.

EXPERIMENTAL

Collection of samples

The samples to be reported now were collected in October 1974 at the Tvärminne Zoological Station of the University of Helsinki on the coast of the Gulf of Finland (59°50' N; 23°15' E). The sediment (ca. 8 kg) was collected by using a seabed 'plane' with an adjustable aluminium blade, which removes the top centimetre when drawn along the bottom. In the laboratory the sample was dried and homogenized. Its mineral content was found to be 85.5% by heating at 700°C for 48 hours. The isopoda, ruff and cod were obtained by a bottom trawl. The samples were deep-frozen, transported to the laboratory, thawed and sorted into groups according to the size. They were then dried overnight at 110°C, placed in plastic bags, crushed by hammer and either dry-ashed or wet-ashed as described later.

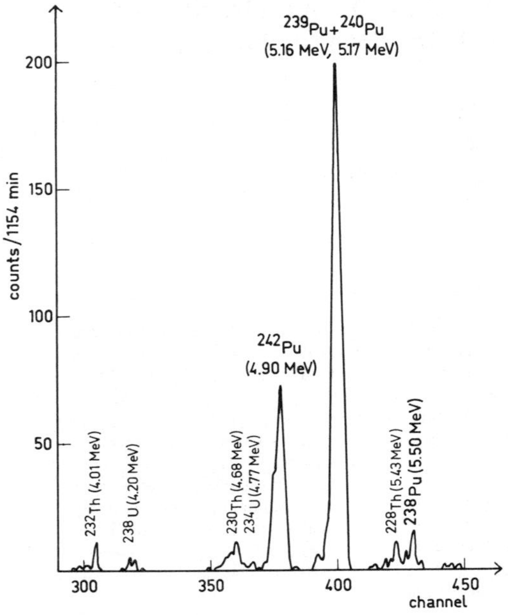

FIG. 1. The plutonium α-spectrum of bottom sediment sampled in the Gulf of Finland in 1974. The ^{242}Pu was added as tracer for yield determination. (Traces of thorium and uranium can be seen).

The flounder was an old sample caught by net in 1970 and kept deep-frozen since then. It was dried and crushed like the ruff and cod. The blue mussels were hand picked from a depth of about 4 m, approximately 300 m from the coast. They were cleaned and sorted according to size, and either separated into soft tissue and shells and analysed separately, or analysed whole. The sorted samples were dried and homogenized as above. The brown alga was collected from the same place as the mussels; it was hand-cleaned, dried, crushed and homogenized.

Analysis of plutonium

The dried tissue samples were mostly wet-ashed with HNO_3-HCl, but some were dry-ashed at 600-700°C. The separation of plutonium was performed according to the HASL Procedures Manual [1] or according to Wong [2]. In both methods the separation of plutonium is based on an anion-exchange technique. After the ion-exchange steps plutonium is electrodeposited on a platinum disc and its α-activity is determined using a surface barrier semiconductor detector and a 512-channel pulse-height analyser. Plutonium-242 isotope was used as tracer for yield determination. In Fig. 1 the α-spectrum of a sediment sample is given. In addition to the plutonium, small α-peaks due to uranium and thorium can be seen. This indicates that complete separation of plutonium from uranium and thorium has not been achieved. However, the small uranium and thorium concentrations do not interfere with the determination of $^{239}Pu + ^{240}Pu$.

RESULTS AND DISCUSSION

The results for the top centimetre of bottom sediment from the Gulf of Finland are presented in Table 1. The values per km^2 were determined by measuring the density of the wet sediment assuming that the sediment sample was precisely the top centimetre. The mean result is 0.178 nCi $^{239,240}Pu$ or 390 (disint/min)/kg or 1.68 mCi/km^2. Bowen [3] reports in 1973 160 (disint/min)/kg dry sediment for a deep North Atlantic ocean sediment and 1.5, 4.2 and 2.4 mCi/km^2 for Atlantic coastal sediments in 1964, 1968 and 1970, respectively. All these values are of the same order of magnitude. The ^{137}Cs content of the sediment sample was 2.05 nCi/kg dry sediment, giving a $^{137}Cs/^{239}Pu + ^{240}Pu$ ratio of 11.5.

Results on some marine organisms are presented in Table 2. Our value for the blue mussel, mean 0.6 pCi/kg wet wt for the whole animal, is about twice as high as those reported by Noshkin from Cape Cod in 1970 [4]. We could not measure ^{238}Pu; Noshkin reports a $^{238}Pu/^{239}Pu$ ratio of 0.099. We have not yet measured seawater. Noshkin reports for mussel a concentration factor of 300 in relation to seawater. Our concentration factor, when the value in fresh mussel (0.6 pCi/kg wet wt) is compared with that of the sediment (178 pCi/kg dry wt), is 0.003. Comparison with sediment may be meaningful, since blue mussel lives on and within the sediment. $^{239,240}Pu$ concentration in isopoda, 0.7 pCi/kg fresh wt, is about the same as in blue mussel. Reduction instead of concentration evidently takes place in this foodchain. Aarkrog [5] reports the ^{239}Pu content of 1.5 pCi/kg fresh wt for shrimps sampled on the Danish coast in 1968.

In the fish the values varied from 0.04 to 0.14 pCi/kg wet wt. We have not yet analysed plankton, but the brown alga (Fucus vesiculosus) contained 5.4 pCi $^{239,240}Pu$/kg wet wt (14.4 pCi/kg dry wt). This value is about 10 times as high as that reported by Noshkin [4] for brown alga from Cape Cod in 1970-1971, namely 0.53, and from the Pacific coast in 1964, 0.47 pCi/kg wet wt. Since alga is the first trophic level and seems to have a remarkable capacity to concentrate plutonium from water, it may be a useful indicator species for the plutonium level of the marine biota. Higher values could be expected from the Baltic Sea than the Atlantic, because the Baltic is shallow and receives much river water. Yet the sediment values in the Baltic and Atlantic are not much different. The higher value in alga may be more due to the surface contamination, because it is difficult to clean up the sediment on the plant if sampled near the bottom.

Our Baltic values are in general in conformity with those of Yaguchi et al. [6] from Lake Michigan. These authors obtained in Mysis relicta 0.57 and Ponterporeia affinis about 1.2 pCi/kg wet wt. Waller et al. [7] reported the following values for Lake Michigan fish: Slimy sculpin (Cottus cognatus) 0.205, chub, alewife and smelt about 0.01 - 0.02 pCi/kg wet wt. These compare well with our fish values 0.04 - 0.14 pCi/kg wet wt. Plutonium concentration decreases in higher trophic levels being lowest in the fish of prey [7]. Bennett [8] reports for edible fresh fish and shellfish from New York in 1972 0.0016 and 0.011 pCi/kg wet wt, respectively. The values refer to muscle.

We have not yet analysed water but Wahlgren and Nelson [9] report for water in Lake Michigan in spring 1975 0.75 fCi/litre. If water in the Baltic were similar, concentration in the top layer of sediment on the weight basis would be 2.4×10^5 times as high as in water. Therefore comparison of

Table 1. $^{239}Pu + ^{240}Pu$ and ^{238}Pu concentrations in marine sediment, Gulf of Finland, 1974. The counting error (one δ) indicated.

Analysis	Dry weight (g)	$^{239}Pu + ^{240}Pu$ nCi/kg dry	$^{239}Pu + ^{240}Pu$ dpm/kg dry	$^{239}Pu + ^{240}Pu$ mCi/km^2	^{238}Pu nCi/kg	^{238}Pu mCi/km^2	Ratio $^{238}Pu/^{239,240}Pu$
No 1	99.9	0.163±0.011	360±30	1.54±0.11	0.008±0.001	0.076±0.009	0.049
2	100.1	0.171±0.007	380±15	1.62±0.06	-	-	-
3	20.1	0.188±0.006	418±13	1.79±0.06	0.015±0.001	0.138±0.009	0.077
4	20.4	0.19 ±0.02	410±40	1.76±0.15	0.016±0.002	0.150±0.02	0.087
Mean		0.178±0.006	392±13	1.68±0.05	0.013±0.001	0.121±0.008	0.071

Table 2. $^{239}Pu+^{240}Pu$ concentrations in some marine organisms in Gulf of Finland, 1974. The standard deviation of the radioassay (one δ) indicated.

Organism	Tissue	Dry wt of sample (g)	$^{239}Pu+^{240}Pu$ pCi/kg dry	$^{239}Pu+^{240}Pu$ pCi/kg wet
Blue Mussel				
(*Mytilus edulis*)	body	40.2	3.0±0.7	0.17±0.04
	shell	67.2	1.6±0.8	1.4 ±0.7
	whole animal	51.1	2.0±0.8	0.6 ±0.2
Isopoda				
(*Mesidotea entomon*)	whole animal	59.6	2.8±0.4	0.7 ±0.1
Ruff				
(*Acerina cernus*)	whole fish	38.0	0.32±0.14	0.09±0.04
Flounder*)				
(*Platichthys flesus*)	whole fish	360.3	0.14±0.04	0.04±0.01
Cod				
(*Gadus morrhua*)	skin	34.4	0.5±0.2	0.14±0.06
	liver	50.3	0.20±0.15	0.14±0.10
Brown alga				
(*Fucus vesiculosus*)		35.0	14.4±2.0	5.4 ±0.8

*) sampled 1970

$^{239,240}Pu$ values of animals and plants living in contact with sediment to $^{239,240}Pu$ concentration in the sediment may be more meaningful than reporting concentration factors in relation with water.

As a conclusion of these preliminary results it can be said that the $^{239,240}Pu$ level in samples from the Gulf of Finland seems to be comparable with those from the Great Lakes in the USA. Forms of the lower trophic levels contain more plutonium than the higher trophic levels since the food-chains seem to lead to reduction. The values of sediment and mussel from the Baltic are twice as high as those from the North Atlantic. A health hazard to humans, if any, could arise from consumption as staple food of bottom animals such as mussel or oysters from shallow bay or estuarine areas in which the sediment contains high concentrations of plutonium.

ACKNOWLEDGEMENT

This investigation was supported by the US Energy Research and Development Administration (ERDA) by Contract CH E(11-1)-3011.

REFERENCES

[1] US AEC, Health and Safety Laboratory: Procedures Manual, HASL-300, E-Pu-06-01 (1972).
[2] WONG, K.M., Anal. Chim. Acta 56 (1971) 355.
[3] BOWEN, V.T., Transuranics in the Marine Environment, HASL-291, I-57 (1975).
[4] NOSHKIN, V.E., Health Physics 22 (1972) 537-549.
[5] AARKROG, A., Health Physics 20 (1971) 31.
[6] YAGUCHI, E.M., NELSON, D.M., MARSHALL, J.S., Plutonium in Lake Michigan Plankton and Benthos, ANL-8060, Pt III, 18 (1973). Argonne National Laboratory, Argonne, Ill., USA.
[7] WALLER, B.J., NELSON, D.M., MARSHALL, J.S., Plutonium in Lake Michigan Fish, ANL-8060, Pt III, 27 (1973).
[8] BENNETT, B.G., Fallout ^{239,240}Pu in Diet, HASL-286, I-36 and HASL-278, I-41 (1974).
[9] WAHLGREN, M.A., NELSON, D.M., Residence Times for ^{239}Pu and ^{137}Cs in Lake Michigan Water, ANL-8060, Pt III, 85 (1973).

DISCUSSION

V.V. PISAREV: At what depth are the plutonium samples taken in the bottom sediments of the Baltic Sea?

J.K. MIETTINEN: The muscles and algae were hand-picked at a depth of 4 metres, 300 metres off-shore. The sediment, isopoda and fish were from a depth of 30 metres, 1 kilometre off-shore.

A.A. FRAIZIER: At what stage in the chemical treatment do you introduce the radioactive tracer (^{242}Pu), and in what chemical form? What do you think of this method of determining the plutonium yield?

J.K. MIETTINEN: We usually add the ^{242}Pu tracer after digestion, but before any chemical treatment is carried out. We consider this method of yield determination reliable.

T.V.A. JAAKKOLA: Perhaps I can add that the ^{242}Pu added as tracer for yield determination was in a 1N HNO_3 solution.

K.E. WHITE: Your sediment samples were taken using a plane with the blade set so as to remove the top 10 mm when drawn along the bottom. Although you selected flat areas, you may not have obtained a true top-layer sample. In my experience, satisfactory sampling of river-, lake- or sea-beds by remote means is very difficult indeed. It is difficult to retain the fine material and to prevent the sample from being disturbed while it is being lifted to the surface. Could you give some more details about the device and the technique of using it, particularly the speed at which it is towed along the bottom? How do you prevent the device from sinking in soft sediments and how is the sample retained? The device has certain advantages over a corer, but there must be problems of sampling accuracy where shell debris or stones are present and where the bed is rippled. Perhaps the Baltic Sea has a more ideal form of bed than those we are familiar with. Do you consider that a surface sample of arbitrary thickness is adequate for calculating the radioactivity per unit area of sea-bed, in the light of concentration-against-depth data given both in Hetherington's and in Thompson's papers at this Symposium (IAEA-SM-198/29 and 198/38 respectively), as well as similar depth distributions quoted in the literature?

J.K. MIETTINEN: Your comments are well founded. High accuracy cannot always be achieved with a bottom plane. It has an aluminium blade located between two metal skis, which slide along the bottom. Heavy weights guarantee good contact with the sediment. When towed slowly by a rowing boat the plane cuts a thin layer (of about 10 mm in our case) from the surface of the sediment, which is then collected in a bag. The thickness of the surface layer can be verified by observation in shallow water. The work of other authors (e.g. Edgington) shows that for 2-3 cm at the surface the concentrations of heavy metals, e.g. plutonium, are fairly uniform. Our purpose was to collect one large sample (8 kg), which was homogenized and can now be used as an intercomparison sample.

J.A. HETHERINGTON: I notice that your values for the ^{238}Pu : ^{239}Pu ratio are somewhat higher than those reported by Noshkin for Buzzards Bay sediments and those that we find in areas not affected by plutonium derived from the nuclear power programme. Could you comment on possible reasons for this apparent discrepancy?

J.K. MIETTINEN: We know that the ^{238}Pu : ^{239}Pu ratio in fall-out was about 3% before the SNAP IX accident in 1964. From 1966 onwards the ratio began to rise in Finland, reaching the value of 8% in 1970-72. We collected only the surface layer (top 10 mm) of sediment, where the ratio must be higher than in deeper layers.

Y. NISHIWAKI: I should like to add a few comments in this connection. The Pu isotope ratio of nuclear debris may differ considerably from one nuclear detonation to another, depending on the conditions of the tests and the materials used. Furthermore, the Pu ratios for the soluble fraction of rainwater are reported to be different from those of insoluble hot particles. Measurements of the ^{238}Pu : (^{239}Pu + ^{240}Pu) activity ratio conducted in Japan about 10 years ago showed that the ratios for rainwater and radioactive hot particles collected in Japan a few days after nuclear detonations were different. The Pu isotope activity ratios for radioactive hot particles after some of the Soviet nuclear tests in 1961 and 1962 were observed to be relatively high (0.1 - 0.3), the ratios of the radioactive hot particles from Chinese nuclear tests in 1964-66 being considerably lower (0.04 - 0.01). Pu isotopes detected in radioactive hot particles (5 - 30 μm) collected a few days after a nuclear detonation may be ascribed to that particular test. However, airborne dust collected after nuclear detonations, as well as rain-water samples, are both likely to contain high proportions of old debris originating from past nuclear tests. Pu ratios in rainwater samples were observed to decrease in Japan during 1963 (0.05 - 0.01), but began to increase from late 1964 onwards until late 1966 (0.01 - 0.1). According to Mamuro et al. of the Radiation Centre of Osaka Prefecture, a ratio as high as 0.5 was observed in 1967. This increase may be considered to be at least partly attributable to the release of ^{238}Pu in the SNAP-IXA accident over the Indian Ocean in April 1964.

The estimation of ^{238}Pu and (^{239}Pu + ^{240}Pu) is generally performed by alpha-ray spectrometry using solid-state detectors. In this case, great care must be taken to eliminate any contamination of ^{228}Th. Otherwise ^{238}Pu may be overestimated. There have also been some attempts to determine ^{239}Pu and ^{240}Pu with X-ray spectrometry using a low-energy photon spectrometer or with micro-mass-spectrometric analysis, since these two isotopes cannot be discriminated by alpha spectrometry. The plutonium isotopes ^{239}Pu and ^{240}Pu have different branching ratios for alpha decays, and decays to the excited states of daughter nuclides are accompanied by gamma transitions with high conversion coefficients and fluorescence yields. Sakanoue and his associates at Kanazawa University in Japan used low-energy γ-ray spectrometry on purified and electro-deposited plutonium to determine ^{241}Pu by measuring the 60 keV γ-ray of ^{241}Am produced by decay of ^{241}Pu. In the 'Bikini ash' collected from the Japanese fishing boat 'Lucky Dragon No.5', which was covered by highly radioactive fall-out at Bikini in March 1954, 26 (disint/min)/mg of (^{239}Pu + ^{240}Pu) were found, and in the calcareous sea-sediments from near Bikini atoll 3 - 10 (disint/min)/g of (^{239}Pu + ^{240}Pu) were detected recently. The $^{241}Am/^{239}Pu$ activity ratio in these samples was estimated to be about 0.3, and the $^{238}Pu/(^{239}Pu + ^{240}Pu)$ activity ratio about 0.4. The $^{240}Pu/^{239}Pu$ activity ratio may be estimated at about 0.7.

Gwyneth D. HOWELLS: Are there any data on the variation of sediment plutonium levels as a function of geographical areas in the Baltic? Could the fish species sampled have been exposed to different levels by migration?

J.K. MIETTINEN: The number of analyses is too small for conclusions to be drawn about possible differences between geographical areas.

C.N. MURRAY: What correction have you made to your ^{238}Pu results for contamination by ^{228}Th, which has a similar energy spectrum? Underestimation of this contamination would tend to cause overestimation of the $^{238}Pu/^{239}Pu$ ratio.

J.K. MIETTINEN: Our samples normally have no thorium-228 contamination. We deliberately showed a slide of an anomalous sample, in order to illustrate the location of possible impurities in the spectrum.

T.V.A. JAAKKOLA: I don't think that thorium has interfered with our ^{238}Pu determinations on the sediment sample. The possible error due to thorium is smaller than the standard deviation in counting.

D.N. EDGINGTON: I should just like to comment that we have made measurements of $^{238}Pu/^{239}Pu$ ratios in many sectioned cores from Lake Michigan. The values of the ratio vary between 0.035 and 0.08 but there is no clear pattern of correlation either with depth (age) in the core or with sampling location.

A. PRESTON: You stated that mussels could constitute an important health hazard to human consumers but that none are eaten in the area studied. I take it you mean that this is only a potential route of human exposure and that, even if mussels were consumed at the present time, levels would need to be very much higher before any risk would be incurred.

J.K. MIETTINEN: That is correct. Even if the levels were many orders of magnitude higher, the mussels could be safely consumed as human food. However, in highly contaminated areas mussels and crayfish would be the prime source of plutonium, particularly if the liver were not removed. In all animals the liver usually contains 2 to 3 orders of magnitude more plutonium than the muscle.

Y. NISHIWAKI: In this same connection I should like to mention that Japanese fishermen sometimes eat the livers of freshly caught fish on the boat. The oil produced from fish liver is often used also as a medicine for growing children.

S. FOWLER: In the case of filter feeders like Mytilus do you think it is entirely valid to base concentration factors on the levels of Pu in sediments? With sediment ingesters like worms this is understandable, but one usually assumes that filter feeders obtain radionuclides from water and associated particulate matter.

J.K. MIETTINEN: This is true. On the other hand, molluscs live within the sediment and, if this contains 250 000 times more plutonium than the water, it can also have an effect.

EXPERIMENTAL STUDIES ON PLUTONIUM KINETICS IN MARINE BIOTA

S. FOWLER, M. HEYRAUD, T.M. BEASLEY
International Laboratory of Marine Radioactivity,
Oceanographic Museum,
Principality of Monaco

Abstract

EXPERIMENTAL STUDIES ON PLUTONIUM KINETICS IN MARINE BIOTA.

Laboratory experiments were undertaken to measure plutonium flux through marine organisms and to clarify the pathways by which this important element is cycled in the marine environment. The use of a specially prepared isotope, plutonium-237, allowed measurements to be made with standard NaI(T1) scintillation techniques. Mussels, shrimp and worms were allowed to accumulate plutonium-237 from seawater for up to 25 days. Accumulation by shrimp was relatively slow and the degree of uptake was strongly influenced by moulting. Cast moults contained large fractions of the shrimps' plutonium content,indicating the high affinity of plutonium for surface areas. Only small amounts of the isotope in the moult are lost to water; hence, moulting is considered to be an important biological parameter in the biogeochemical cycling of plutonium. Mussels attained higher concentration factors than shrimp with most of the accumulated isotope (>80%) located in the shell. Byssus threads often contained large fractions of the mussels' plutonium-237 content and reached concentration factors as high as 4100. Worms readily accumulated plutonium-237 in either the +4 or +6 state, reaching concentration factors of approximately 200. Retention studies indicated a relatively slow loss of plutonium-237 from all animals studied. In the case of mussels, a computed half-time for a large fraction of the animals' plutonium content was of the order of 2 years. The more rapid loss from shrimp ($Tb_{1/2}$ = 1.5 months) was due principally to the large fraction of plutonium lost at moult. Food chain studies with shrimp indicated that tissue build-up via plutonium ingestion would be a slow process. Total excretion was not entirely a result of passing contaminated food through the gut; approximately 15% of the ingested plutonium was removed from the contaminated food and subsequently excreted by processes other than defaecation of labelled food. Ratios of four different plutonium isotopes used in the experiments showed that when the isotopes were present together in the same physico-chemical form, no differences in isotopic behaviour were evident even in cases in which total mass of certain isotopes in the system differed by 10^3. It is concluded that plutonium-237, because of its relatively high specific activity, offers the best possible means for tracing plutonium biokinetics in aquatic systems in which the experimental design dictates that total plutonium concentrations approximate to those currently found in the environment.

INTRODUCTION

During the past 30 years artificially produced plutonium isotopes have entered the oceans principally by way of atmospheric fallout from nuclear testing and controlled release from nuclear reprocessing plants. Additional quantities of plutonium-238 were added in 1964 after the burn-up upon re-entry of the SNAP-9A satellite generator. In addition, the growing use of plutonium in the nuclear industry may increase plutonium inventories in marine ecosystems. The fact that plutonium isotopes are readily concentrated by marine biota is now well documented [1-4]. It is therefore imperative to construct an information base on the behaviour and fate of plutonium in marine ecosystems which will aid in evaluating the impact of future contamination of these products on the aquatic environment.

Although information is accruing on the existing levels of plutonium in water, sediment and aquatic biota [5], knowledge of the kinetics of plutonium in aquatic organisms is scant. Numerous laboratory and field studies on the behaviour of this element in mammals have been reported;

however, it has been pointed out that much of these data may not be easily extrapolated to marine organisms, especially invertebrates [4, 6]. To help fill this need we have begun laboratory studies on the flux of plutonium through selected marine organisms. We report here the results of experiments on mussels, benthic crustacea and marine worms, representatives of three phyla which have been shown to accumulate plutonium to relatively high levels in nature [3]. Our results deal principally with the bio-availability of plutonium from seawater, although some preliminary data on food chain uptake are also presented.

To facilitate the experimental work we have made use of a specially prepared gamma emitting isotope, plutonium-237, which permits direct counting of live animals by standard NaI(Tl) scintillation techniques. The usefulness of ^{237}Pu as a tracer for ^{239}Pu in metabolic studies was first reported by Todd and Logan [7, 8]; however, recent work by Bair et al. [9] has cast some doubt on the applicability of this isotope to adequately trace ^{239}Pu kinetics in certain mammalian systems. We have attempted to address this problem by trying, wherever possible, to compare tissue concentrations of ^{237}Pu with those of ^{236}Pu and ^{238}Pu, two alpha emitters that were present in the ^{237}Pu tracer solution as contaminants.

There is some evidence from recent field studies that ^{238}Pu and ^{239}Pu may behave differently in the environment [4, 10, 11]. Since the original source for certain amounts of these two isotopes is different, it is difficult, if not impossible, to determine whether the observed discrepancy in behaviour is due to inherent nuclear characteristics or merely to the fact that certain proportions of the two isotopes are present in different chemical forms. We have approached this question by performing some double tracer experiments in which we followed the bioaccumulation of ^{238}Pu and ^{239}Pu with time by means of alpha spectrometric techniques.

MATERIALS AND METHODS

The common Mediterranean mussel, *Mytilus galloprovincialis*, the benthic shrimp, *Lysmata seticaudata*, and the marine worm, *Nereis diversicolor*, were selected as the experimental animals. Mussels were collected along the Monaco shoreline and shrimp were trapped by means of placing baited cages near the Monaco harbour. Worms were purchased from a firm on the north coast of France which raises *Nereis* commercially for fish bait. All animals were maintained in the laboratory flowing seawater system.

Isotopes

The ^{237}Pu isotope (46-day half-life) used in our studies was obtained from two different sources. Approximately 2 μCi was received as a gift from the Japanese Atomic Energy Research Institute, Tokai-mura, Japan. Data on the preparation, separation and purity of this isotope have been reported by Murray and Fukai [12]. An additional 10 μCi were purchased from the Oak Ridge National Laboratory, USA. The isotope was produced by irradiating a ^{235}U foil and separated by sorbing ^{237}Pu (+4) on an anion exchanger, followed by stripping with weak HNO_3 and a further purification by extraction into 0.5M TTA (2-thenoyltrifluoroacetone) in xylene and back-extracting with 10M HNO_3. The isotope was finally converted to the chloride form by evaporation and fuming with HCl. As of the production date the alpha content of the radioisotope amounted to 4 nCi per microcurie of ^{237}Pu. Alpha spectrometry showed that 97% and 3% of the alpha activity was due to ^{236}Pu and ^{238}Pu, respectively.

One set of experiments was performed using ^{239}Pu which was supplied by the IAEA Seibersdorf Laboratory. A small quantity of ^{238}Pu (< 1%) was present as a contaminant in this solution.

Our experiments used plutonium in either the +4 or +6 oxidation state. The chemical form of the isotopes was adjusted as follows:

Pu (+4) – Aliquots were reduced to Pu (+3) using 2M $NH_2OH \cdot HCl$ followed by oxidation to the +4 state by addition of 2M $NaNO_2$ solution.

Pu (+6) – Aliquots were taken to near dryness with concentrated $HClO_4$ acid and brought to volume with 1.5M HCl.

Bioaccumulation experiments

Mussels and shrimp. Sufficient ^{237}Pu (+6) was added to a plastic basin containing 3 litres of cotton-filtered seawater to bring the activity level to 60 cpm/ml. The pH was adjusted to 8.0 with NaOH, an air stone added and the solution allowed to equilibrate for 24 hours at 13°C. After this time shrimp (4 large and 4 small) and mussels (4 large and 4 small) were introduced into the radioactive seawater. All animals were held in individual perforated plastic tubes which ensured that the mussels remained submerged and allowed identification of shrimp that moulted. During a 25-day uptake time the animals were periodically rinsed in fresh seawater, counted in a well-type scintillation detector and replaced in the radioactive seawater for further uptake. Any shrimp moults found were counted along with the animal. At each radioanalysis 10 ml of radioactive seawater was also monitored for ^{237}Pu activity. The particulate ^{237}Pu fraction measured in selected aliquots of the uptake medium was defined as that fraction retained on 0.45-μm membrane filters.

Occasionally, before being returned to the tracer solution, the mussels ingested phytoplankton and the shrimp were fed *Artemia* in clean seawater for about one hour. Since the time out of the tracer solution was short compared to the total uptake time, it is probable that these periodic feedings did not significantly alter the uptake pattern. After uptake two mussels and two shrimp were dissected to assess relative ^{237}Pu tissue distribution in shell and soft parts and the remaining individuals were placed in flowing seawater to measure loss of the isotope.

A second experiment which differed somewhat in design from the first, utilized only mussels. In this case two small mussels were added to a 50-ml suspension of mixed phytoplankton containing 75 cpm/ml ^{237}Pu (+4) and allowed to accumulate the isotope for 7 days at 21°C. After uptake these individuals were also placed in flowing seawater to measure ^{237}Pu loss.

In both experiments described above, lack of radiotracer precluded changing the labelled seawater media; thus concentration factors ((cpm/g wet animal)/(cpm/g water)) were computed by taking into account the decreasing amount of isotope in the seawater.

Worms. Five similar-sized Nereis (aver. wet weight = 390 mg) were placed in individual plastic beakers containing 100 ml of ^{237}Pu (+6) labelled seawater at a concentration of 185 cpm/ml. The solution was changed daily and an attempt was made to maintain the ^{237}Pu concentration in the medium relatively constant by periodically adding some tracer at each change. After an 8-day uptake period one individual was sacrificed for ^{236}Pu and ^{238}Pu analysis and the remainder were transferred to a sand-filled basin which received running seawater to follow loss of ^{237}Pu.

To assess whether or not excreted metabolites in the seawater might alter ^{237}Pu uptake kinetics, four worms (aver. wet weight = 386 mg) were placed in cotton-filtered aliquots of the radioactive solution which was used for the first worm experiment outlined above. The radioactive seawater was changed daily; however, no tracer additions were made. Other than the fact that worms from the first experiment had pre-conditioned the radioactive seawater, the two experiments were identical. Uptake was followed for 7 days at which time two worms were sacrificed for ^{236}Pu and ^{238}Pu analysis.

In the final experiment, 10 similar-sized worms (aver. wet weight = 503 mg) were placed individually in 100 ml seawater containing approximately 2 dpm/ml ^{239}Pu (+4). A new solution was prepared daily so that the activity level remained essentially constant throughout the experiment. Four worms were sacrificed on day 3, and three each on days 7 and 15 of uptake; ^{238}Pu and ^{239}Pu activities in both organisms and their medium were determined by alpha spectrometric techniques. The temperature for the three experiments was 13°C.

Retention experiments

Mussels and shrimp. Following the 25-day Pu (+6) uptake period, animals were transferred to clean perforated tubes and held in the laboratory flowing seawater system. It immediately became apparent that owing to the relatively low absolute activity of the isotope in the shrimp, loss could not be monitored for a sufficiently long period of time to be meaningful; hence, another group of shrimp and mussels were allowed to accumulate ^{237}Pu (+6) from seawater of a higher initial activity (350 cpm/ml) for 32 days. These shrimp which contained higher levels of ^{237}Pu than those in the first uptake experiment, were then used to monitor isotope loss to flowing seawater over a 5-week period. The mussels were periodically sacrificed during loss and their ^{236}Pu and ^{238}Pu activity measured. The two mussels which accumulated ^{237}Pu (+4) from water and phytoplankton were transferred to running seawater after a 7-day uptake period and ^{237}Pu loss was followed for 5 weeks. All mussels and shrimp were fed ad libitum during the loss phase.

Worms. Four worms from the first accumulation experiment which utilized the freshly prepared +6 tracer solution were placed in flowing seawater, monitored for ^{237}Pu loss and periodically sacrificed for ^{236}Pu and ^{238}Pu analysis. The animals were not fed during loss.

In all retention experiments the temperature of the running seawater varied between 16° and 21°C depending upon the time of year. The average flow rate was approximately 1 litre/min.

Food chain experiments

To assess the degree to which ingested plutonium might be assimilated by marine invertebrates, some preliminary food chain experiments were designed utilizing benthic shrimp and Artemia brine shrimp. Artemia were placed in seawater containing ^{237}Pu at activity levels near 1000 cpm/ml; this seawater also contained mixed phytoplankton serving as food source for Artemia. After several days of accumulation, the Artemia were removed from the labelling medium and were rinsed in clean seawater for about 1 hour. The labelled Artemia were fed to Lysmata on the following schedule:

(1) One group of four individuals received a single ration of ^{237}Pu-labelled Artemia followed by transfer to running seawater with rations of non-labelled Artemia administered daily throughout the excretion phase.

(2) Two shrimp received a single ration of labelled Artemia and were starved for 8 days in running seawater before receiving non-labelled Artemia.

(3) Two shrimp received daily rations of labelled Artemia for 15 days followed by frequent rations of non-labelled Artemia during the excretion phase.

Radiochemical analyses

^{237}Pu measurements. Measurements of the 100-keV X-ray which results from the electron capture decay of ^{237}Pu to ^{237}Np were made with a multichannel analyser (Intertechnique, Didac 800) coupled to a well-type 7.6 × 7.6 cm NaI(Tl) scintillation detector. Samples were counted and compared with an internal standard to correct for physical decay of the isotope as well as any slight changes in calibration of the instrument. The counting errors associated with samples, backgrounds and standard were propagated and in most cases were < 5% at the 1σ level. On the basis of the information furnished by the two producers, we have estimated the counting efficiency of our system to be approximately 45%; however, as no calibrated ^{237}Pu source was available, the results of the radioactivity measurements are reported as cpm.

Alpha measurements. From time to time selected samples from the ^{237}Pu accumulation and loss experiments were taken to determine ratios of the three plutonium isotopes present. The fact that the ^{237}Pu tracer solution contained virtually no ^{239}Pu permitted the use of the latter as a yield determinant for ^{236}Pu and ^{238}Pu. After establishing the ^{237}Pu content, samples were transferred to new, pre-cleaned 250-ml Pyrex beakers and stored for radiochemical separation.

In addition to samples taken from the ^{237}Pu studies, worms and water from the $^{238/239}Pu$ (+4) uptake experiment were also analysed radiochemically. In these samples ^{236}Pu supplied by the IAEA Seibersdorf Laboratory was used as a yield determinant.

The radiochemical separation of plutonium from all samples followed that reported by Moore and Hudgens [13] with certain modifications which were dictated by the type of sample analysed. Biological samples were wet-ashed with a 1 : 1 mixture of reagent grade concentrated HNO_3 and $HClO_4$ after the addition of the appropriate yield determinant and 10 mg of neodymium carrier. Both water samples and biological specimens were ashed to fumes of $HClO_4$, after which the solutions were transferred to polyethylene centrifuge tubes and the resulting plutonium (+6) was carried on $Nd(OH)_3$ by making the solution basic by addition of NH_4OH. After centrifugation and a single water wash, the precipitate was dissolved in 2$\underline{M}$ HNO_3 and the valence state of the plutonium was adjusted to +3 by reduction with 2$\underline{M}$ $NH_2OH \cdot HCl$. Neodymium fluoride was then precipitated by the addition of concentrated HF. The subsequent chemical purification followed that described by Moore and Hudgens [13] with the purified plutonium fraction being plated onto stainless steel discs after the method of Talvitie [14].

The alpha spectrometry system consisted of 4 solid-state silicon barrier detectors connected to a 1024 channel Ortec analyzer. Counting times varied according to the activity of the sample and ranged from several hours to several days. Counting errors associated with all measurements were propagated and reported at the 1σ level.

RESULTS

Bioaccumulation experiments

The results for ^{237}Pu (+6) uptake by shrimp are presented in Fig. 1. Shrimp slowly accumulated ^{237}Pu over the 25-day period and the degree of accumulation was strongly influenced by moulting. One individual which did not moult during the experiment reached a concentration factor of 19. Two shrimp, dissected at the end of the experiment, contained approximately equal percentages of ^{237}Pu in exoskeleton and soft parts (Table I). Uptake did not appear to be strongly dependent upon shrimp size although the results are difficult to interpret since some individuals moulted during the experiment. Moults cast by three individuals during the first 18 days of uptake contained from 92 to 100% of the shrimps' ^{237}Pu content, indicating a high affinity of plutonium for surface areas (Table II). These percentages may be somewhat of an overestimate since cast moults can further adsorb isotope before they are removed from the radioactive seawater. Retention experiments with moults showed that ^{237}Pu desorption from exuviae was relatively slow; on average about 30% was lost to seawater after 1 week.

Mytilus steadily accumulated ^{237}Pu (+6) from seawater reaching concentration factors ranging from 27 to 70 after 25 days (Fig. 1). The dashed line in this figure indicates the relative amounts of ^{237}Pu that were lost when the byssus was removed. Byssus threads were found to contain as much as 63% of the mussels' total ^{237}Pu content (Table III). Concentration factors ranging from 1860 to 4100 attest to the ability of this tissue to accumulate plutonium. Although the loss of byssus threads by some individuals confounded the results, it appeared that the smaller mussels reached higher concentration factors.

Mussels without byssus, dissected after uptake, contained 80-89% of the ^{237}Pu in their shell, again reflecting the strong affinity of plutonium for

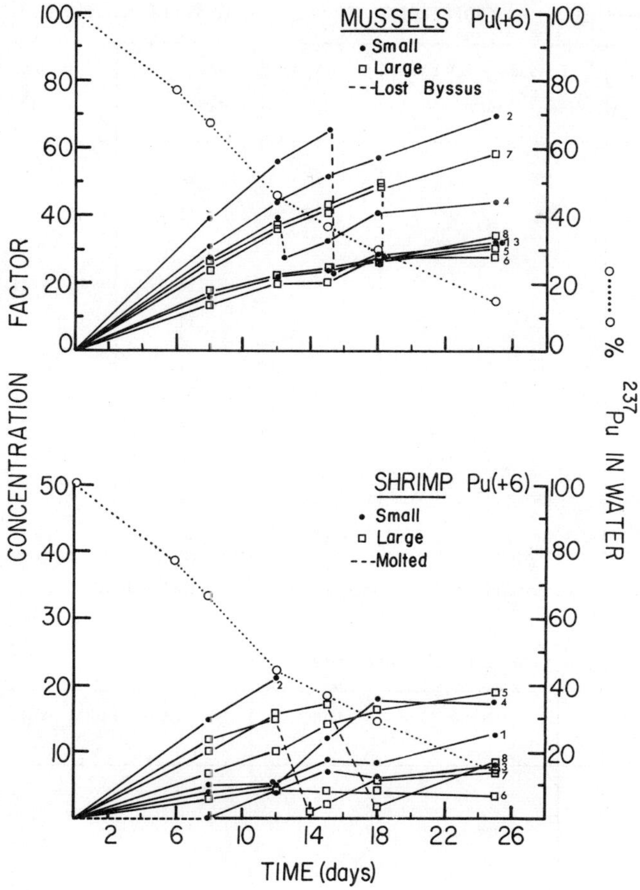

FIG.1. Uptake of plutonium-237 from seawater by mussels (Mytilus galloprovincialis and shrimp Lysmata seticaudata. Points on the uptake curves represent individual organisms; Nos.1-4 are small individuals, 5-8 large individuals.

exposed surface areas. Furthermore, much of the ^{237}Pu in the shell appeared firmly bound as evidenced by the fact that only 25% of the sorbed isotope was removed after 2 hours' rinsing in 0.1N HCl; continued rinsing failed to remove the residual ^{237}Pu (Fig. 2a).

During the mussel and shrimp experiment the ^{237}Pu (+6) particulate fraction remained relatively constant at approximately 20% despite the continued drop in ^{237}Pu concentration in the water with time. Because of the decreasing isotope concentration in the medium, concentration factors were based on average ^{237}Pu water concentrations for given time intervals.

Figure 3 presents data on the uptake of ^{237}Pu (+4) from phytoplankton and water by Mytilus. Concentration factors reached after 7 days were quite similar to those in the ^{237}Pu (+6) experiment, although the two

TABLE I. ^{237}Pu (+6) CONTENT IN DISSECTED TISSUES

Animal wet wt. (g)		Uptake (days)	Loss (days)	Distribution (%) Shell [a]	Distribution (%) Soft parts	Byssus (% of soft parts)
Mussel	2.5	25	0	80	20	
	4.1	25	0	89	11	
Mussel	2.9		73	80	20	64
	5.3		73	89	11	25
Mussel	2.0		178	65	35	
	5.2		178	71	29	
Shrimp	0.5	25		58	42	
	1.1	25		52	48	

[a] Shell represents about 31% and 27% of total wet weight of mussel and shrimp, respectively.

TABLE II. ^{237}Pu (+6) IN SHRIMP MOULTS [a]

Individual No.	Uptake (days)	Loss (days)	Fraction of total body ^{237}Pu (%)	CF
Uptake				
3	7		100	120
7	14		95	160
8	17		92	94
Loss				
1 1st moult		2	42	-
2nd moult		18	0	-
5 1st moult		4	36	-
2nd moult		21	0	-
6 1st moult		11	39	-
2nd moult		31	0	-
2 1st moult		11	19	-
2nd moult		27	0	-
4 1st moult		16	13	-
2nd moult		35	0	-

[a] Moults average 19% of total shrimp wet weight.

TABLE III. ^{237}Pu (+6) IN BYSSAL THREADS OF MUSSELS

Mussel wet wt. (g)	Day uptake	Byssus fraction of total wet wt. (%)	Fraction of whole body ^{237}Pu (%)	CF
1.8	15	1.0	63	4100
5.3	18	0.9	43	2330
2.1	25	0.9	38	2900
6.0	25	1.0	30	1860

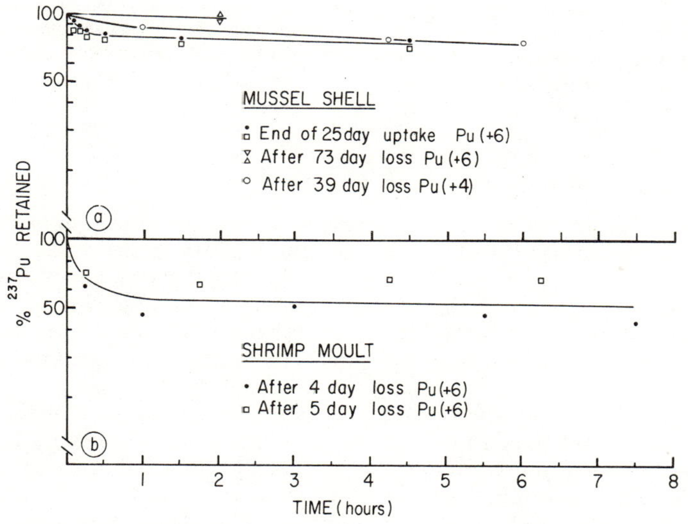

FIG.2. Removal of plutonium-237 from (a) mussel shell and (b) shrimp moult by rinsing with 0.1N HCl.

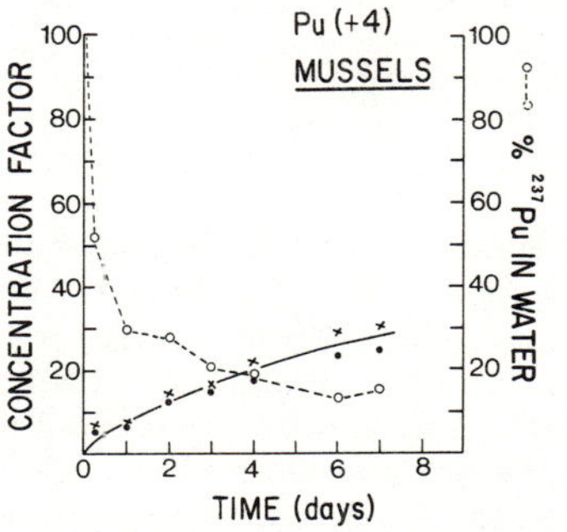

FIG.3. Uptake of plutonium-237 from phytoplankton and seawater by *Mytilus galloprovincialis*. Points on the uptake curves represent individual organisms.

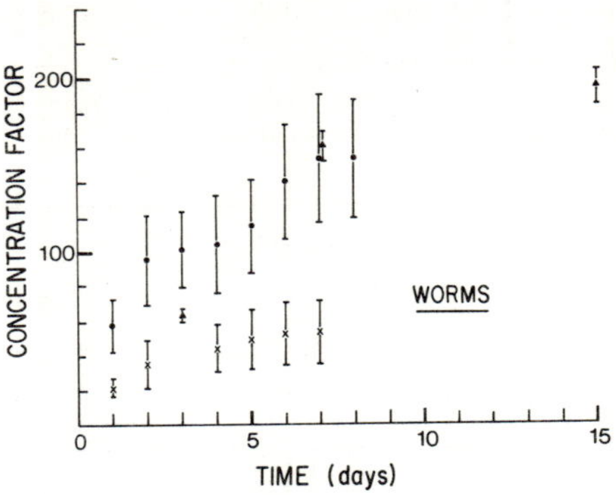

FIG.4. Plutonium uptake from water by Nereis diversicolor. Isotopic forms were (●) plutonium-237 (+6); (▲) plutonium-239 (+4); (×) plutonium-237 (+6) in water previously conditioned by worms. Points on curve indicate mean values; bars indicate 1σ of the mean. The bars for the plutonium-239 experiment represent 1σ based on counting statistics only.

experiments might not be comparable due to differences in experimental design. The relatively rapid drop in ^{237}Pu activity in the water was partially due to the fact that approximately 60% of the isotope was associated with algal cells and other particles at the beginning of uptake, which, with time, were rapidly filtered out of solution by Mytilus. The rapid depletion of particulate ^{237}Pu resulted in an 85% drop in available plutonium during the 7-day period.

Plutonium accumulation by worms is shown in Fig. 4. In experiments utilizing freshly prepared radioactive seawater, no differences in uptake were noted between the two chemical forms of plutonium. Concentration factors for both chemical forms of plutonium were higher than those reached by either mussels or shrimp over corresponding time intervals. However, a marked reduction in ^{237}Pu (+6) uptake was noted in worms which accumulated the isotope from water which had been used for the first worm experiment. This finding suggests that ^{237}Pu might have become associated with excreted metabolites, e.g. mucus, and was thus rendered less available for bioaccumulation.

Retention studies

Loss of ^{237}Pu (+6) was followed in mussels and shrimp which had accumulated the isotope from water for 25 and 32 days, respectively (Fig. 5). The relatively rapid loss of ^{237}Pu from shrimp was, in part, due to a substantial portion of the isotope being lost with the first moult (Table II). It is evident from Table II that smaller fractions of plutonium were associated with moults the longer the shrimp remained in non-contaminated water; this is most likely a result of continual desorption from the outer surface of the exoskeleton which originally held a relatively large fraction of the shrimps' plutonium content. Approximately 50% of the plutonium associated

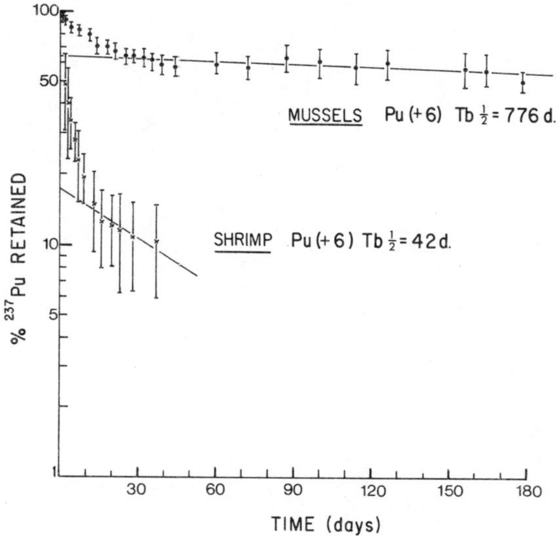

FIG.5. Plutonium loss from mussels and shrimp after one month uptake from seawater. Points and bars represent mean value and 1σ, respectively. $Tb_{\frac{1}{2}}$ calculations are based on a least squares fit of the curve between day 25 and 178 for mussels and day 13 and 28 for shrimp.

with the moult was rapidly leached by rinsing in weak HCl (Fig. 2b). No measurable ^{237}Pu was found in the animals' second moult which indicates that some of the isotope had been incorporated into internal tissues.

Loss of ^{237}Pu from mussels, on the other hand, was quite slow. No difference in loss between small and large individuals was noted, therefore the data for all mussels were combined and plotted in Fig. 5. Graphical analysis showed that total loss was best represented as excretion from two distinct compartments, one fast and one very slow. The biological half-life ($Tb_{\frac{1}{2}}$) for the fast pool which contained 35% of the accumulated plutonium was 7 days; that for the slow-turnover compartment containing 65% of the total plutonium was 776 days.

In all cases byssus threads were radioanalysed intact with the mussels; this precaution was necessary since byssus could account for large fractions of the total plutonium activity in mussels (Tables I and III). The relative distribution of ^{237}Pu between shell and soft parts after 6 months' rinsing suggested that the isotope was being lost slightly faster from shell than from soft parts (Table I). The residual plutonium in the shell appeared to be tightly bound as evidenced by the insignificant amounts that were leached with 0.1$\underline{N}$ HCl rinses (Fig. 2).

Loss of ^{237}Pu (+4) from mussels over 39 days is presented in Fig. 6. Loss from the slow component was approximately exponential from day 5 onwards and resulted in a $Tb_{\frac{1}{2}}$ of only 39 days. The turnover of ^{237}Pu (+4) appeared much faster than in mussels which had accumulated the +6 form; however, direct comparisons are difficult to make since the designs for the two experiments were quite different. More rapid loss can be expected in animals which have had a total accumulation time of short duration [15].

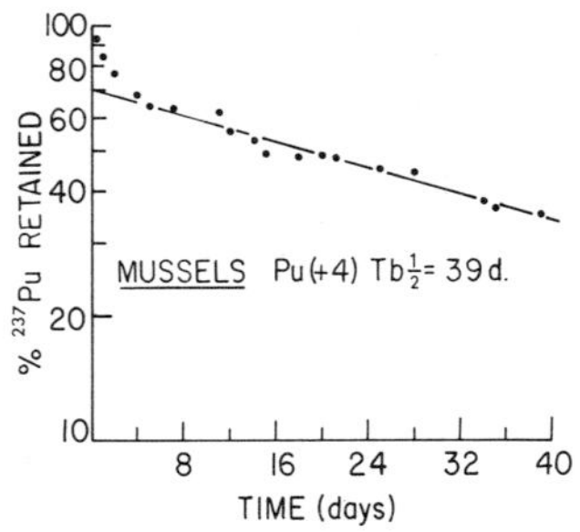

FIG.6. Plutonium loss from mussels (n = 2) after 7 day uptake from a phytoplankton suspension. After day 11 only one individual remained.

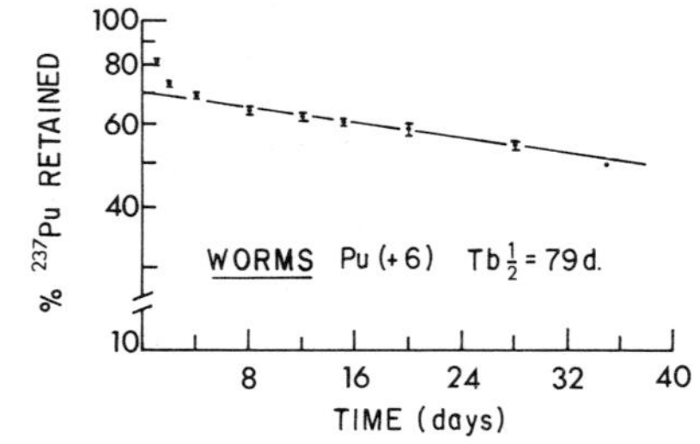

FIG.7. Plutonium loss from worms after 8 day uptake from seawater. Where applicable, points and bars represent mean and 1σ, respectively.

Moreover, low assimilation of plutonium from labelled phytoplankton could also enhance the loss rate due to voiding of this material as faeces. Dissection of one individual after 11 days of loss showed about 80% of the plutonium was located in the shell. This distribution closely parallels that of the + 6 experiments and suggests that uptake from the residual activity in the water occurred despite the large percentage of ^{237}Pu associated with the phytoplankton.

A similar excretion pattern was found with Nereis which had accumulated ^{237}Pu (+ 6) from water for 8 days (Fig. 7). The $Tb_{\frac{1}{2}}$ computed between days 4 and 35 was 79 days.

Food chain experiments

Loss of ^{237}Pu (+ 6) from shrimp after a single ingestion of labelled Artemia was relatively rapid over the first 3 days, after which time the excretion rate decreased dramatically and continued at a nearly exponential rate for 1 month until only 1% remained (Fig. 8). On the other hand, a similar group of shrimp which was not fed for the first 8 days of loss retained 40% of the ingested dose. When feeding was resumed excretion

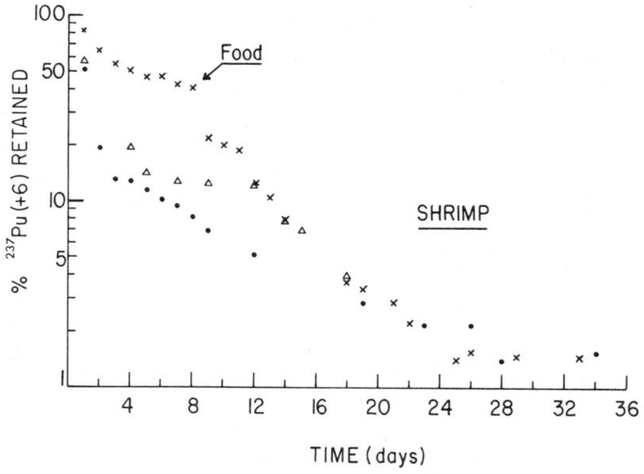

FIG.8. Plutonium excretion from shrimp following ingestion of labelled brine shrimp. The experimental conditions were (●) single ration, (x) single ration followed by starvation, (△) continuous feeding for 15 days. Data points are means of 2 to 4 individuals.

rapidly increased, probably as a result of faecal production and, with time, assumed a rate paralleling that of the other group. A third group receiving daily rations of labelled food for 15 days did not surpass plutonium levels in those animals which ingested only one ration. Furthermore, the ^{237}Pu excretion rates were similar in these groups. This similarity suggests that a temporal build-up of food-chain-derived plutonium into tissues displaying long plutonium residence times would be a very slow process in these organisms.

Plutonium ratio analyses

Ratios for the three plutonium isotopes measured in the two separate mussel and worm experiments are given in Fig. 9. As the experiments spanned several months, all isotope activities have been corrected to a common date for ease in comparison. Counting errors have been propagated and are quoted at the 1σ level.

With few exceptions, within the two separate experiments, the agreement between animal and water ratios was relatively good. Most ratios overlap at 1σ and almost all overlap at 2σ. When the two experiments are compared, a trend is evident for slightly higher 236/238 ratios in worms than in mussels. The difference between the mean ratios for each experiment was tested using a Student's T-test and was found to be significant at the 5% level. The same test applied to the 237/236 and 237/238 ratios indicated there was no significant difference between ratios in mussels and worms. Tracer solution, seawater and worms from the ^{239}Pu (+4) water uptake experiment were also analysed for 238/239 ratios. The data presented in Table IV clearly show no differences in bioavailability for these two plutonium isotopes.

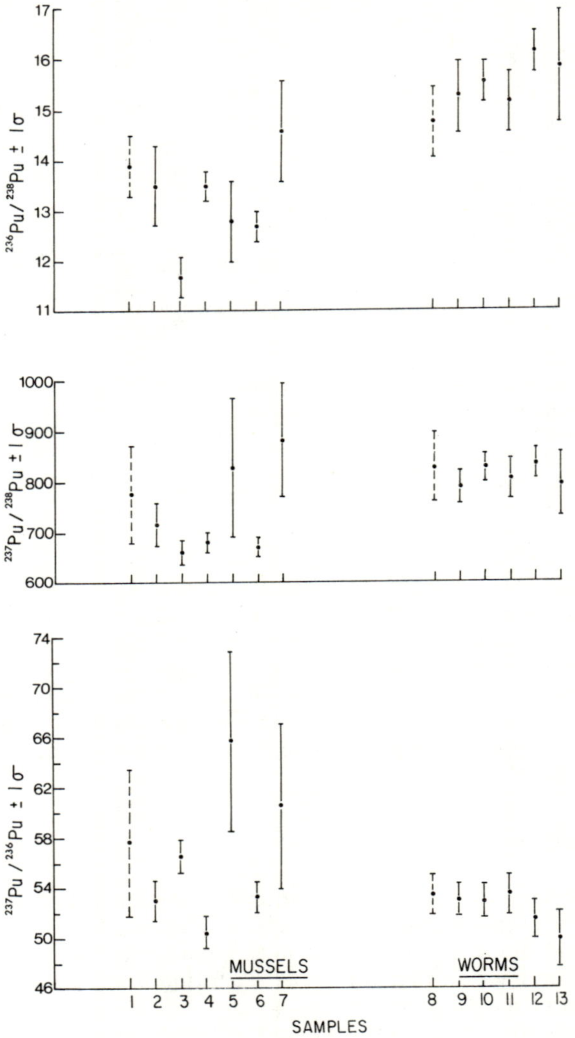

FIG.9. Plutonium isotopic ratios in seawater and organisms.

Mussels: 1, seawater; 2, soft parts 32d uptake; 3, shell 32d uptake; 4, shell 34d loss; 5, soft parts 34d loss; 6, shell 76d loss; 7, soft parts 76d loss.

Worms: 8, seawater; 9, 7d uptake; 10, 8d uptake; 11, 15d loss; 12, 29d loss; 13, 34d loss. Bars represent 1σ of propagated errors.

TABLE IV. $^{238/239}$Pu (+4) RATIOS IN COMPONENTS OF THE WORM-WATER UPTAKE EXPERIMENT

Sample	Uptake time (days)	$\frac{^{238}Pu}{^{239}Pu} \pm 1\sigma$
Spike	-	0.0088 ± 0.0002
Seawater	-	0.0090 ± 0.0006
Worms	3	0.0095 ± 0.0006
Worms	5	0.0089 ± 0.0009
Worms	7	0.0077 ± 0.0006
Worms	15	0.0085 ± 0.0003

DISCUSSION

It is noteworthy that concentration factors in our shrimp experiments agree well with those measured in laboratory studies with lobsters [16] but are an order of magnitude lower than those reported for shrimp which accumulated fallout plutonium in the natural environment [17]. This difference is most probably a reflection of the short uptake time in laboratory studies compared to the field situation in which animals may achieve isotopic equilibrium within their life span. Also, it should not be overlooked that accumulation through the food chain might lead to higher plutonium concentration factors in nature.

Plutonium uptake from water by shrimp can be characterized as a relatively slow process with the major portion of the accumulated isotope sorbed to and firmly bound with the outer portion of the animals' exoskeleton. This pattern of plutonium distribution has been noted in shrimp sampled off the coast of Greenland [17]. Our data also confirm the results of an earlier study on ^{239}Pu accumulation by lobsters in which 90% of the total plutonium content was situated in calcified exoskeleton which accounted for only 43% of total wet weight [16].

Moulting as a mechanism for transporting trace metals and radionuclides has been proposed by several authors [18-21]. In the light of our results, cast crustacean moults, which display high concentration factors and lose plutonium to seawater only very slowly, would also appear to be an important biological mechanism for transporting plutonium to sediments as well as detrital feeders. The potential importance of moulting as a biological parameter in plutonium biogeochemical cycling in the sea becomes evident when one considers that often the bulk of marine animal biomass is composed of small crustacea which moult frequently.

Noshkin et al. [2] have suggested that Mytilus, which appears to derive its plutonium content in relation to water rather than sediment concentrations of the isotope, would be an excellent biological indicator for plutonium in the marine environment. The results from our water uptake studies with

Mytilus support this contention. Concentration factors after 25 days of uptake were somewhat lower than those which have been reported for environmental samples of Mytilus [1, 2]. As in the case with shrimp, this is most probably a result of the mussels' failure to reach equilibrium with plutonium in their medium after such a short uptake time. Nevertheless, since the animals' medium was not changed during the experiment, one cannot overlook the possibility of plutonium complexation with metabolic excreta which might render a fraction of the isotope unavailable for uptake (see Fig. 4).

The remarkable accumulation of plutonium by byssus threads was not entirely unexpected. There are indications in the literature [22] and from some of our own work [23] that byssus threads can accumulate several radionuclides and trace metals to relatively high levels. It is known that byssal matrix is composed of fibrous proteins like collagen [24], substances which play a prominent role in trace element binding. Furthermore, recent microscopic studies have confirmed the presence of fungal hyphae in the byssus threads of Mytilus [25]. Either or both of these factors might act to enhance elemental uptake by this tissue. It has been suggested that certain radionuclides can be excreted by the byssus gland [22, 26]; although experimental evidence is lacking in our study, a similar mechanism might be operative in the case of plutonium. To our knowledge no measurements of environmental levels of plutonium in mussel byssus have been made; however, if natural plutonium tissue distribution patterns follow those seen in our experiments, care should be taken not to contaminate samples of mussel soft parts with attached byssus threads when using Mytilus as an indicator organism for plutonium.

Although it is clear from laboratory and field studies that plutonium is accumulated to a much greater degree in shell than in soft parts of molluscs and crustacea, the exact mechanism of accumulation remains unknown. A recent report proposes that high plutonium concentrations in mollusc and crustacea shell can be explained by attached shell-periphyton uptake rather than by actual accumulation into calcified tissues of these invertebrates [4]. The mussels and shrimp in our studies were either pre-cleaned or visibly free of periphyton. Furthermore, acid rinses which actually attacked the shell and moult tissue failed to remove large fractions of the accumulated plutonium. Thus, mechanisms other than associated epiphytic action were responsible for the observed uptake and relatively firm binding of plutonium in shells noted in our experiments.

Noshkin et al. [2] found that Nereis contained the highest plutonium concentrations of all the invertebrates which they analysed. From considerations of $^{238/239}$Pu ratios in water, sediment and worms, they concluded that most of the plutonium in Nereis was derived from material deposited in the sediments. Our results show that Nereis can effectively accumulate plutonium directly from water and we contend that this process will undoubtedly occur in nature when worms are in contact with seawater. However, data on the ability of worms to accumulate plutonium from sediment and/or detritus will be needed before the relative importance of the two uptake routes can be determined.

It has been assumed from mammalian studies that once plutonium is taken up by marine organisms, the retention times will be long. Our data on mussels (Pu + 6) indicate that a $Tb_{\frac{1}{2}}$ of the order of 2 years may be typical for molluscs which have remained in contact with plutonium-contaminated waters for relatively short periods of time. This value is in the same range as those measured for cadmium [21] and methyl mercury [27]; but is quite long relative to biological half-lives of other radionuclides in Mytilus [15,28].

Although uptake time was short in the worm experiments, indications are that plutonium turnover will also be slow in these organisms. On the other hand shrimp, and presumably other crustacea, can be expected to lose plutonium very rapidly due to the large fractions of the animals' plutonium content associated with moults.

Loss of plutonium from shrimp is most rapid when the isotope is ingested with the food (Fig. 8). Nevertheless, it is evident that total plutonium excretion following a single ingested dose can not be attributed solely to the time needed to clear the gut of contaminated food. If we assume that the rapid loss noted during the first three days represents gut clearance through defaecation, slower turnover thereafter represents excretion due to processes other than passage of contaminated food. To estimate the fraction of the total plutonium pool which contributes to this slower excretion phase, the exponential portion of the curve is extrapolated to time zero, giving a value of roughly 15%. Unfortunately, our data do not permit an assessment of the portion of this 15% which was actually absorbed across the gut and assimilated into tissues; nevertheless, the fact that a fair fraction of ingested plutonium can be removed from food by digestive processes and remain within the shrimp suggests that assimilation efficiencies in some marine invertebrates may be much higher than the tenth to hundredths of one percent reported for mammals.

Recent reports have appeared which show that plutonium isotopes can behave differently in certain ecological systems [9-11, 29]. In the environmental studies, the discrepancies in behaviour can most likely be explained by differing bioavailabilities due to different physico-chemical forms of the isotopes in the original source or to different sources themselves. However, laboratory experiments in which different plutonium isotopes in the same chemical form are separately injected into mammals have led to the conclusion that dissimilar metabolic behaviour is due to the differences in total weight of the isotopes injected into the systems. For example, Bair et al. [9], studying the comparative metabolism of ^{237}Pu and ^{239}Pu in dogs, found that excretion from the group containing ^{237}Pu was 20-50 times higher than excretion from dogs labelled with ^{239}Pu. Owing to differing specific activities, the weights of the isotopes injected differed by 6 orders of magnitude. It was suggested that the large concentration of ^{239}Pu injected formed colloidal particles, thereby behaving differently from the ^{237}Pu which was in much lower concentration. Similar differences in metabolic behaviour between ^{238}Pu and ^{239}Pu in rats have been attributed to the different specific activities, i.e. different concentrations of the two isotopes [29].

Comparison of the plutonium isotope ratios in Fig. 9 and Table IV indicates that if the different plutonium isotopes found in nature are present in the same physico-chemical form, aquatic organisms will be unable to discriminate between them during either accumulation or subsequent excretion phases. In the ^{237}Pu tracer solution, the ^{237}Pu : ^{236}Pu: ^{238}Pu atom ratio was approximately unity and all three isotopes were accumulated and lost in the same relative proportion in all organisms tested (Fig. 9). In the ^{238}Pu and ^{239}Pu experiment with worms, a 239/238 atom ratio of 3×10^4 was used. This differs by roughly a factor of 10 from the ratio in seawater ($\simeq 2 \times 10^3$) computed from the data of Noshkin et al. [2]. Similarly, the two isotopes were accumulated by the worms in the same atom ratio which existed in the labelled seawater (Table IV). Although not tested directly, based on these data we conclude that ^{237}Pu will adequately trace ^{239}Pu in aquatic experiments provided the two isotopes are present together in the same chemical form.

Although ratios in source and organism compared favourably for any one experiment, a discrepancy between 236/238 ratios was apparent in the worm and mussel experiments. The significance of this discrepancy is dubious; however, we believe that the reason for the observed differences is related to physico-chemical changes in the tracer solution rather than to differential bioavailability of the two isotopes in worms and mussels. In fact, plutonium measurements in two samples of the ^{237}Pu tracer itself, taken one month apart and some weeks after both seawater media had been spiked, gave 236/238 ratios higher than those measured in either of the two previously labelled seawaters. Although the reasons for the apparent change in the tracer solution are not known, it is probable that the difference between the 236/238 ratios in the two seawater media is real since the worm experiment was begun two months after the mussel experiment (see Fig. 9).

On the basis of recent measurements of seawater, there are approximately a total of 2.8×10^7 plutonium atoms/litre due principally to fallout [2, 30]. At the ^{237}Pu radioactivity concentrations used in our uptake studies, there were some 3×10^{10} plutonium atoms/litre, or about 10^3 more plutonium atoms than is normally found in seawater. These levels of ^{237}Pu were needed to assure good counting statistics over relatively short counting times. In the case of the $^{238/239}Pu$ experiment with worms, approximately 3×10^{13} plutonium atoms/litre were used. As we are unaware of any existing data on the physico-chemical behaviour of different concentrations of plutonium in seawater, we can only assume that accumulation kinetics measured in our experiments utilizing elevated plutonium levels will adequately reflect what takes place at the natural levels in the environment. The fact that no large difference in uptake was evident between worms accumulating either ^{237}Pu (pg/l) or ^{239}Pu (ng/l) from seawater supports this contention (Fig. 4), although comparison may not be entirely valid due to the different oxidation states used. To answer the question of a possible "concentration effect" on biokinetics, a biological experiment is being designed which employs very low activity levels of ^{237}Pu in seawater in order to more closely approximate the natural plutonium mass levels found in the marine ecosystems.

The low specific activities of ^{238}Pu and ^{239}Pu as well as the limitations of current alpha detection techniques precludes using these isotopes in experimental aquatic studies at concentrations normally found in nature. Plutonium concentration effects have been noted in mammalian systems and could conceivably occur in aquatic ecosystems which contain very low amounts of this element. For these reasons, we concur with the recent suggestion of Moghissi and Carter [31] that in environmental studies, ^{237}Pu is most likely the best tracer for measuring plutonium kinetics in biological systems.

ACKNOWLEDGEMENTS

The Laboratory of Marine Radioactivity operates under a tripartite agreement between the International Atomic Energy Agency, the Government of the Principality of Monaco and the Oceanographic Institute at Monaco. Support for the present work is gratefully acknowledged. We thank Dr. R. Fukai for critical comments on the manuscript. Special thanks are also due to Mr. J. La Rosa and Mr. S. Ballestra for technical assistance in certain phases of the study.

REFERENCES

[1] PILLAI, K.C., SMITH, R.C., FOLSOM, T.R., Plutonium in the marine environment, Nature (London) 203 (1964) 568.

[2] NOSHKIN, V.E., BOWEN, V.T., WONG, K.M., BURKE, C., Radionuclides in Ecosystems (Proc. 3rd Nat. Symp. Radioecol. Oak Ridge 1971), USAEC CONF-710501 P2, 681.

[3] CHERRY, R.D., SHANNON, L.V., The alpha radioactivity of marine organisms, At. Energy Rev. 12 (1974) 1.

[4] ANONYMOUS "Transuranic elements", Ch.3, Assessing Potential Ocean Pollutants (National Academy of Sciences) Washington, D.C. (1975).

[5] NOSHKIN, V.E., Ecological aspects of plutonium dissemination in aquatic environments, Health Phys. 22 (1972) 537.

[6] BOWEN, V.T., Transuranic elements and nuclear wastes, Oceanus 18 (1974) 43.

[7] TODD, R., LOGAN, R., Plutonium-237 labelling: a new technique for plutonium metabolic studies in animals, Int. J. Appl. Radiat. Isotopes 17 (1966) 253.

[8] TODD, R., LOGAN, R., Plutonium-237 for metabolic studies: its preparation and use, Int. J. Appl. Radiat. Isotopes 19 (1968) 141.

[9] BAIR, W.J., WILLARD, D.H., NELSON, I.C., CASE, A.C., Comparative distribution and excretion of ^{237}Pu and ^{239}Pu nitrates in beagle dogs, Health Phys. 27 (1974) 392.

[10] HAKONSON, T.E., JOHNSON, L.J., LA-UR-73-1291, Conf. 730907-14. Los Alamos Scientific Lab., Univ. of California, Los Alamos, New Mexico (1973) 6pp.

[11] EMERY, R.M., KLOPFER, D.C., WEIMER, W.C., BNWL-1950 PT2 (1974) 96.

[12] MURRAY, C.N., FUKAI, R., Paper IAEA-SM-198/25, these Proceedings.

[13] MOORE, F.L., HUDGENS, J.E., Jr., Separation and determination of plutonium by liquid-liquid extraction, Anal. Chem. 29 (1957) 1767.

[14] TALVITIE, N.A., Electrodeposition of actinides for alpha spectrometric determination, Anal. Chem. 44 (1972) 280.

[15] VAN WEERS, A.W., in Radioactive Contamination of the Marine Environment (Proc. Symp. Seattle 1972), IAEA, Vienna (1973) 385.

[16] WARD, E.E., Uptake of plutonium by the lobster, *Homarus vulgaris*, Nature (London) 209 (1966) 625.

[17] AARKROG, A., Radioecological investigations of plutonium in an arctic marine environment, Health Phys. 20 (1971) 31.

[18] FOWLER, S.W., SMALL, L.F., Moulting of *Euphausia pacifica* as a possible mechanism for vertical transport of zinc-65 in the sea. Intern. J. Oceanol. Limnol. 1 (1967) 237.

[19] MARTIN, J.H., The possible transport of trace metals via moulted copepod exoskeletons, Limnol. Oceanogr. 15 (1970) 756.

[20] BERTINE, K.K., GOLDBERG, E.D., Trace elements in clams, mussels and shrimp, Limnol. Oceanogr. 17 (1972) 877.

[21] FOWLER, S.W., BENAYOUN, G., in Comparative Studies of Food and Environmental Contamination (Proc. Symp. Otaniemi, 1973), IAEA, Vienna (1974) 159.

[22] PENTREATH, R.J., The accumulation from water of ^{65}Zn, ^{54}Mn, ^{58}Co and ^{59}Fe by the mussel, *Mytilus edulis*, J. Mar. Biol. Ass. U.K. 53 (1973) 127.

[23] FOWLER, S.W., BENAYOUN, G., "Influence of environmental factors on selenium flux in marine organisms" (in preparation).

[24] BROWN, C.H., Some structural proteins of *Mytilus edulis*, Quart. J. Microsc. Science 93 (1952) 487.

[25] VITELLARO-ZUCCARELLO, L., Ultrastructure of the byssal apparatus of *Mytilus galloprovincialis*. I. Associated fungal hyphae, Mar. Biol. 22 (1973) 225.

[26] FRETTER, V., Experiments with radioactive strontium (^{90}Sr) on certain molluscs and polychaetes. J. Mar. Biol. Ass. U.K. 32 (1953) 367.

[27] MIETTINEN, J.K., HEYRAUD, M., KECKES, S., "Mercury as a hydrospheric pollutant. II. Biological half-time of methyl mercury in four Mediterranean species: a fish, a crab and two molluscs", in: Marine Pollution and Sea Life (M. Ruivo, Ed.), FAO Fishing News (Books) Ltd., London (1972) 295.

[28] YOUNG, D.R., FOLSOM, T.R., Loss of ^{65}Zn from the California Sea Mussel Mytilus californianus, Biol. Bull. 133 (1967) 438.

[29] MORIN, M., NENOT, J.C., LAFUMA, J., Metabolic and therapeutic study following administration to rats of ^{238}Pu nitrate. A comparison with ^{239}Pu, Health Phys. 23 (1972) 475.

[30] NOSHKIN, V.E., Jr., GATROUSIS, C., Fallout ^{240}Pu and ^{239}Pu in Atlantic marine samples, Earth Planet. Sci. Lett. 22 (1974) 111.

[31] MOGHISSI, A.A., CARTER, M.W., Comments on "Comparative distribution and excretion of ^{237}Pu and ^{239}Pu nitrates in beagle dogs", Health Phys. 28 (1975) 825.

DISCUSSION

V.V. PISAREV: Why does the rate of Pu loss from the different organisms vary (Figs 2, 5, 6, 7), and why is the loss curve exponential for all the organisms?

S. FOWLER: The changes in isotope excretion rate seen in biological systems reflect loss from different isotope pools in the organisms. When the pools with fast Pu turnover rates are exhausted, loss from pools with slower turnover becomes evident. In our experiment these slow turnover pools showed similar rates of Pu loss and they were exponential over the time studied.

K.C. PILLAI: Are any data available for the precipitation of Pu in seawater and for sorption on container walls?

S. FOWLER: We have not considered this problem, but I am sure that the chemists in our laboratory will have some information on the subject of Pu precipitation in seawater. As far as sorption on container walls is concerned, we have noted that this occurs after spiking seawater. The rate of adsorption to the walls depends, of course, on the amount of particulate plutonium present.

A.R. BERG: In calculating the ^{237}Pu concentration factor, what kind of correction did you apply in order to take into account the decrease in water radioactivity with time?

S. FOWLER: To calculate a concentration factor at a given point in time, I used a ^{237}Pu concentration for water that reflected the overall average concentration experienced by the organism over the time period in question. In other words, it was a weighted average. This is more representative of the actual ^{237}Pu concentrations available to the animals than when concentration factors are computed using absolute activity levels in both organism and water at any one time.

A.R. BERG: Why is the biological half-life apparently shorter in the accumulation phase than in the excretion phase? This difference seems to indicate that the physico-chemical processes involved in accumulation are different from those in excretion.

S. FOWLER: Yes, I agree. There are numerous examples in the literature showing slower turnover during loss than during uptake. Once isotopes like Pu are incorporated, their release from biota is probably dictated by factors other than simple desorption. Fixation in tissue, especially in hard tissue like shell, may greatly retard the loss process.

J.A. HETHERINGTON: In your uptake studies you appear to have run into the recurrent problem of maintaining the concentration of plutonium in the water at the sort of pH conditions required by your experimental animals. In these circumstances it is difficult to make a correction to your water concentrations in order to calculate the concentration factor, because you cannot assume that the residual activity remaining in the water after, say, 10 days is still available to the animals in the same way as was the activity initially added. Have you studied the chemical form of the plutonium remaining in the water phase during your uptake experiment?

S. FOWLER: No, we have not. This is indeed an important point and I alluded to it in the paper, when discussing the uptake experiments.

ADSORPTION-DESORPTION CHARACTERISTICS OF PLUTONIUM AND AMERICIUM WITH SEDIMENT PARTICLES IN THE ESTUARINE ENVIRONMENT

Studies using plutonium-237 and americium-241

C.N. MURRAY,* R. FUKAI
International Laboratory of Marine Radioactivity,
Oceanographic Museum,
Principality of Monaco

Abstract

ABSORPTION-DESORPTION CHARACTERISTICS OF PLUTONIUM AND AMERICIUM WITH SEDIMENT PARTICLES IN THE ESTUARINE ENVIRONMENT: STUDIES USING PLUTONIUM-237 AND AMERICIUM-241.

The particle formation of plutonium and americium, their adsorption onto fresh water sediments and the desorption from the sediments in sea water were studied in the Laboratory under simulated river-estuary conditions, using γ-emitting plutonium-237 and americium-241. The results of the experiments show that the particle formation of plutonium depends on its valence states, on pH and on the salinity of the medium. For river water at pH4, some 25%, 20% and 30% of the added ^{237}Pu was in particulate form, larger than 0.45 μm, for Pu (III), Pu (IV) and Pu (VI), respectively, while 65%, 90% and 50% of the respective valence states was associated with particles at pH 8. In sea water the general pattern remains similar, although Pu (VI) is more soluble in sea water owing to higher ligand concentrations for carbonate and bicarbonate complexes. The pH-dependency of particle formation of Am (III) is more steep than that of plutonium and seems to be influenced by colloidal substances occurring in the experimental media. The adsorption-desorption characteristics of plutonium and americium with the sediment in river water as well as sea water reflect the characteristics of their particle formation, being dependent upon such properties as valence states, the pH and salinity of the medium. A sewage effluent added to the media has small but measurable effects on the adsorption-desorption processes of plutonium.

1. INTRODUCTION

The present levels of plutonium and transplutonic elements in the marine environment are, in general, still extremely low and the radiological impact of these isotopes on the marine ecosystems, as well as on human health through the consumption of marine food, is considered to be minor in the foreseeable future. Nevertheless, concern has been expressed about future releases of actinide radionuclides from nuclear installations into the marine environment. This concern is based on increasing amounts of highly radiotoxic actinide wastes with long half-lives resulting from the projected development of breeder technology in the coming decades, while the behaviour of actinides in the sea is still inadequately understood. Thus, it is important to deepen one's understanding of the biogeochemical behaviour of actinide radionuclides in the marine environment not only with the aim of proving the usefulness of some of these isotopes as tracers of geochemical processes taking place in the sea, but also to enable one to assess the impact of environment releases of these isotopes on a sound scientific basis.

---

* Present address: Deutsches Hydrographisches Institut, 2000 Hamburg 4, Bernhard-Nocht-Strasse 78, Federal Republic of Germany.

The knowledge so far obtained on the behaviour of actinides in the sea is limited. The data available for plutonium on marine environmental samples such as sea water, sediments and marine organisms, up to 1971, have been compiled and reviewed, with extensive discussion, by Noshkin [1]. Studying the radioecological behaviour of plutonium released from the Thule accident in Greenland in 1968, Aarkrog [2] estimated that 85-95% of the plutonium oxide released to the marine environment rapidly deposited on the sea bed with the debris of the accident and only 1% was left suspended as fine particles in the water phase. Concerning fallout plutonium, Miyake and Sugimura [3] noted that $^{239}Pu/^{90}Sr$ ratios in Pacific surface water are lower than those in ground fallout, and Bowen et al. [4] also showed that the upper layers of the Atlantic are deficient in ^{239}Pu compared to ^{90}Sr and ^{137}Cs. On the basis of sediment inventory and the vertical distribution in the water column of plutonium, Noshkin and Bowen [5] estimated sinking rates of 70 to 392 m/a for plutonium associated with a mixed population of particles. In the discharge area around Windscale, Templeton and Preston [6] found that about 80% of plutonium in sea water was associated with particles being retained by 1.2 μm pore size filter and that over 80% of plutonium was contained in the top 5 cm of the sediment, tending to penetrate further than ^{137}Cs, ^{90}Sr or ^{106}Ru.

All these findings based on in-situ data obtained for various sources of plutonium inputs into the sea indicate that plutonium easily forms particles and/or tends to associate itself with particulate matter occurring in natural sea water. Since particulate matter tends to sink in the sea, the downward vector is predominant in the transport of plutonium in natural conditions. Thus the marine sediment represents eventually a sink of plutonium, although the route through which plutonium reaches the sediment is likely to depend on various factors, such as the chemical forms of plutonium, the sorption characteristics of particulate matter, the intensity of advection and vertical mixing, etc. To date, no systematic study on the sorption characteristics of actinides with sediments in relation to their chemical forms has been carried out.

On the other hand, laboratory studies by Kharkar et al. [7], Murray and Murray [8] and Robertson et al. [9] showed that, although many trace components are firmly associated with suspended sediment particles in the river environment, significant fractions of components such as manganese, cobalt, zinc and silver are likely to desorb when the suspensions are brought into the seawater medium. Fukai et al. [10] demonstrated that such desorption indeed occurs for zinc in the Var Estuary. No such information is available for actinides.

In the present study, laboratory experiments on the adsorption-desorption characteristics of plutonium and americium with sediment particles under simulated estuarine conditions were conducted, advantage being taken of the easy measurability of gamma-radiation of plutonium-237 and americium-241. As has been pointed out by Todd and Logan [11,12] the application of plutonium-237 to these experiments proved the utility of this isotope also in environmental plutonium studies by eliminating complications in laboratory safety measures and tedious radiochemical procedures during the experiments, unavoidable when using other plutonium isotopes for the same purpose. The generalization of the results obtained by using ^{237}Pu to the behaviour of other plutonium isotopes in the environment, however, should be made carefully, since different plutonium isotopes might behave differently in ecosystems [13].

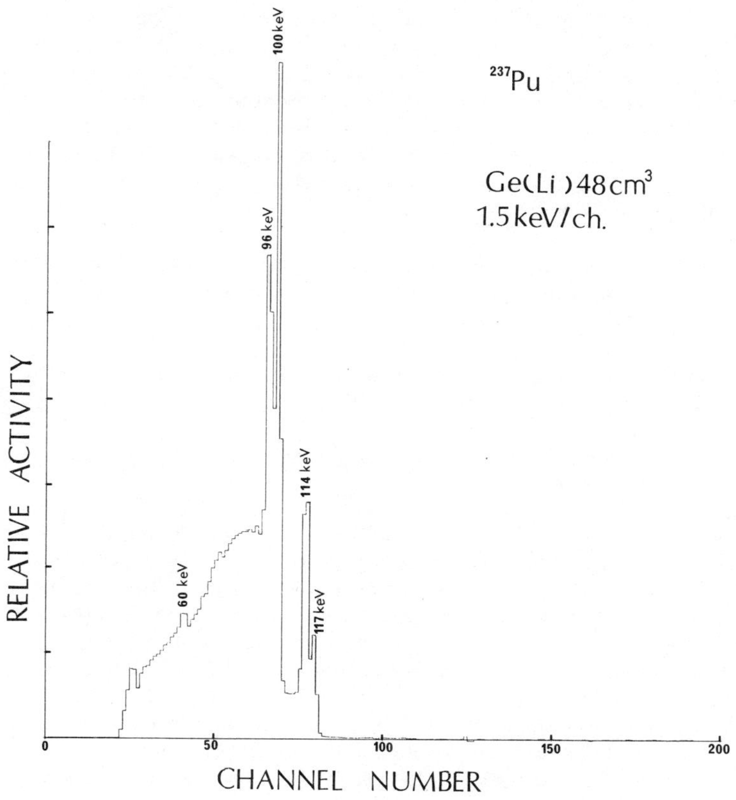

FIG. 1. A gamma spectrogram of plutonium-237.

2. MATERIALS AND EXPERIMENTAL METHODS

2.1. Materials

Plutonium-237: Around 2 μCi of the isotope was produced by the Institute of Physical and Chemical Research, Tokyo, using natural uranium as target material through the following nuclear reaction:

$^{238}U\ (^3He,\ 4n)^{237}Pu$

The isolation of plutonium from the irradiated target and the purification of the isolated isotope were carried out at the Japanese Atomic Energy Research Institute, Tokai-mura, Japan. Figure 1 shows a typical gamma-spectrogram of the ^{237}Pu obtained, measured by a Ge(Li) detector. As can be seen, major peaks occur around 100 keV; they represent X-rays in K-series of $K\alpha_2$ (97 keV), $K\alpha_1$ (101 keV), $K\beta_1$ (114 keV) and $K\beta_2$ (117 keV), which are due to transitions in the de-excitation of ^{237}Np; 33 keV and 60 keV γ-peaks of ^{237}Pu are also shown, but their abundances are not large. The spectrogram and the measured half-life of 46.2 days against the cited value 45.6 days [14] indicate that the radiochemical purity of the ^{237}Pu obtained is satisfactory for gamma measurements. The overall alpha impurities (^{236}Pu, ^{238}Pu) were estimated to be less than 1%. To ensure the chemical forms of plutonium being utilized, aliquot portions of the ^{237}Pu solution were treated as follows:

Pu (III) – An aliquot portion was evaporated to dryness with a 2 ml of hydroxylamine solution (50 mg $NH_2OH \cdot HCl$) and the residue was taken up with 3 ml of 1.5<u>N</u> HCl.

Pu(IV) – Instead of hydroxylamine solution, 2 ml conc. HNO_3 was used in which solid $NaNO_2$ was dissolved.

Pu (VI) – 2 ml conc. $HClO_4$ was used in place of hydroxylamine solution for Pu (III).

<u>Americium-241:</u> Around 2.5 μCi was obtained from the CEA, France. The chemical form Am (III) was stabilized as was done for Pu (III).

<u>River sediment:</u> This was collected from the vicinity of the Var River mouth and stored wet in a refrigerator at a temperature below 4°C. Mineral composition, exchange capacity, organic carbon content (= 1%) and other chemical characteristics of the sediment have already been described by Duursma and Eisma [15].

<u>River water:</u> This was collected from the upper reaches of the Var River, immediately filtered through a 0.45 μm filter and stored in a similar way to the sediment. The pH of the water was ≃ 8.1 and Ca and Mg concentrations were 50 and 10 mg/l, respectively.

<u>Sea water:</u> This was collected from a station off Monaco and treated in a similar way to the river water. The chlorinity of the sea water was ≃ 20‰ .

<u>Sewage:</u> This was collected directly from the sewage line at Monaco, filtered through a 0.45-μm filter, oxygenated for 24 hours by having air bubbled through it, and stored in a refrigerator at a temperature below 4°C until used. The organic carbon content measured by permanganate titration against a glucose standard was 13.4 mg/l (as glucose) and the phosphate concentration was 3.3 mg PO_4-P/l. The dissolved oxygen concentration of the sewage effluent did not fall below 5.8 ml/l during the experiments.

2.2. Experimental procedures adopted

The procedures adopted for the experiments on particle formation of the isotopes added, adsorption by sediment from the river water medium and desorption from the sediment in the sea water medium, have been described in detail by Murray and Murray [8] and for the sewage experiments by Murray and Meinke [16]. Briefly, the river-estuary conditions were simulated by continuous mechanical shaking of 100-ml water samples with or without the addition of 100 mg sediment at a given pH for a duration of several days; after the expected equilibria of the reactions have been

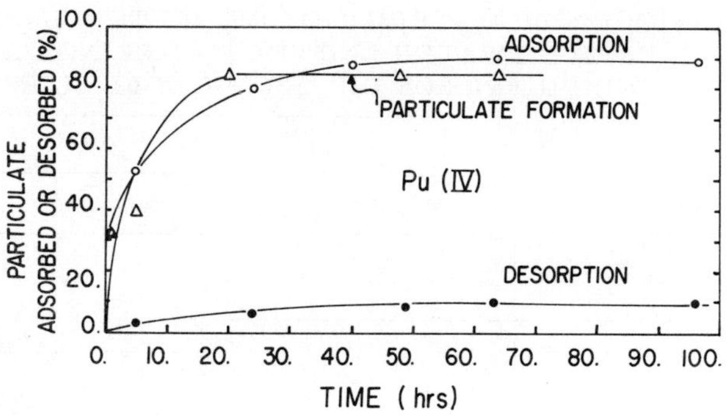

FIG. 2. Equilibrium time for particle formation and adsorption on sediments of Pu (IV) in Var River water (pH ≃8) and its desorption from sediments in the sea water medium.

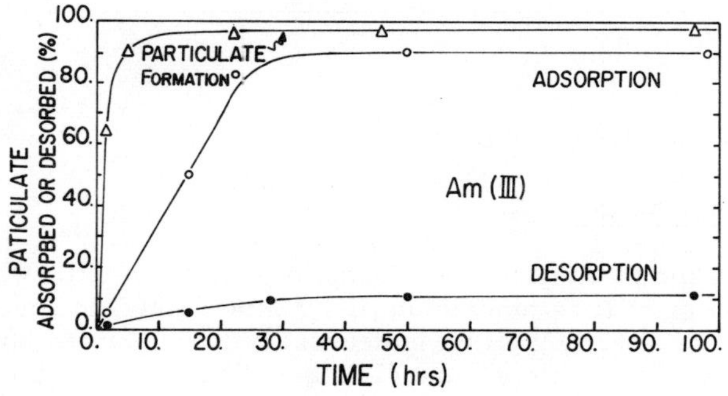

FIG. 3. Equilibrium time for particle formation and adsorption on sediments of Am (III) in Var River water (pH ≃ 8) and its desorption from sediments in the sea water medium.

reached, an aliquot portion of the samples (normally 5 ml) was taken and filtered (usually 0.45-μm filter); both the filter and filtrate were counted by an Intertechnique 200 channel γ-analyser connected to a 7.5 cm × 7.5 cm NaI(Tl) well crystal, using a γ-energy range of 0 to 200 keV. On the basis of these γ measurements, the percentages of the isotopes in particulate state, adsorbed or desorbed fraction were computed. The overall reproducibility of these results was estimated to be within ±5%. It should be kept in mind, however, that the particulate fraction separated by the filtration would miss some particles which pass through the filters of chosen pore size.

In the experiments on particle formation and adsorption, the radioisotopes were added to the sample waters, while the release of the radioisotopes from the labelled sediments obtained in the adsorption experiments into the sea water medium was measured in the desorption experiments.

As shown in Figs 2 and 3, although the equilibrium time with Pu (IV) and Am (III) for particle formation as well as adsorption and desorption can be reached within 50 hours, the experiments were carried out, in any case, over a 96-hour period to ensure equilibrium conditions.

TABLE I. PARTICLE SIZE DISTRIBUTION OF PLUTONIUM (IV) AND AMERICIUM (III) IN RIVER WATER AND SEA WATER INITIALLY FILTERED THROUGH FILTERS OF DIFFERENT PORE SIZES

Isotope	^{237}Pu (IV)				^{241}Am (III)			
Medium	River water		Sea water		River water		Sea water	
Pore size of initial filter	0.45 μm	0.10 μm	0.45 μm	0.10 μm	0.45 μm	0.10 μm	0.45 μm	0.10 μm
Retained by	(%)	(%)	(%)	(%)	(%)	(%)	(%)	(%)
1.2 μm filter	87	85	92	84	88	64	82	62
0.45 μm filter	-	-	89	85	97	71	89	68
0.22 μm filter	87	85	87	86	99	83	89	73
0.10 μm filter	96	93	87	85	100	93	95	73

3. RESULTS AND DISCUSSIONS

3.1. Particle formation

Since it is known that all three forms of plutonium (III, IV and VI valence states) as well as americium (III valence state) in aqueous solution tend to hydrolyse at higher pH, it is necessary to determine how these chemical forms of the isotopes are distributed between soluble and particulate fractions in the simulated natural conditions, before attempting sorption-desorption experiments.

To examine the sizes of the particles being formed in river water and sea water, percentage retention of Pu (IV) and Am (III) on filters of various pore sizes (0.10, 0.22, 0.45 or 1.2 μm) was measured in the media filtered initially through 0.10-μm or 0.45-μm filters. The shaking was continued for 96 hours as described above and the pH of the media was kept around 8. The adsorption on blank filters was tested by double-layer filtration and was estimated as being less than 1%. The results are presented in Table I. As shown in the table, Pu (IV) retention by the filters of various pore sizes does not differ greatly between the media initially filtered through 0.45-μm and 0.10-μm filters for both river water and sea water; a slightly increased retention by 0.10-μm filter compared with that by other filters was found in river water for both initial filtrations. Except for this increase, no systematic difference was found among the values for Pu (IV) even between those for river water and sea water. This indicates that the size of the particles formed of Pu (IV) depends neither on the pore size of the initial filtration nor on salinity of the media; the sizes of the particles formed are comparatively homogeneous.

The picture is different for Am (III). The retention of Am (III) on the filters of various pore sizes from the media initially filtered through 0.10 μm is definitely lower than that from the 0.45-μm filtered media for river water as well as sea water, and tends to increase with decreasing

TABLE II. PARTICLE SIZE DISTRIBUTION OF PLUTONIUM (VI) IN RIVER WATER AND SEA WATER INITIALLY FILTERED THROUGH 0.45-μm FILTERS

Medium	River water	Sea water
Retained by:	(%)	(%)
1.2 μm filter	34	34
0.45 μm filter	48	40
0.22 μm filter	57	41
0.10 μm filter	56	40

pore size in all cases. The gradient of the increase is largest in the case of the river water initially filtered through 0.10-μm filters, where the retention by 1.2-μm filter, 64%, increased to 93% by 0.10-μm filter. In the sea water medium the corresponding increase was from 62% to 73%. These observations indicate that the particle formation of Am (III) depends on the pore size of the initial filtration and there is an appreciable size distribution of the particles formed. This suggests that particle formation of Am (III) in river water as well as sea water is much more influenced by the occurrence of substances which pass through the filters than that of Pu (IV).

In Table II the results of similar experiments with Pu (VI) are given, but in the media initially filtered only through 0.45-μm filters. The results show that the particle formation of Pu (VI) is much less pronounced in both river water and sea water than that of Pu (IV). Especially in sea-water, percent retention of Pu (VI) by the filters of different pore sizes ranged only from 30% to 40%. This lower particle formation of Pu (VI) may indicate the stabilization of soluble species of Pu (VI) by complex formation with carbonate and bicarbonate [17]. The carbonate content of river water is estimated to be high because of its high calcium and magnesium concentrations and high pH, due to the calcareous terrain of the water-shed.

To examine the variation of particle formation as a function of changes in pH of the media, ^{237}Pu (III, IV, VI) and ^{241}Am (III) were added to river water and sea water filtered through 0.45-μm filters, and the percent retention of these isotopes was measured by filtration also through 0.45-μm filters. The pH of the media varied between 2.5 and 11. In Figs 4 and 5, the results obtained in the river water and sea water are respectively presented. These figures indicate that even in the pH range below 4, there is a fraction of Pu (III, IV and VI) associated with particles larger than 0.45 μm in size. For river water at pH 4 approximately 25%, 20% and 30% of the added ^{237}Pu was in particulate form for the III, IV, and VI valence states, respectively. As pH was increased, the particulate fraction increased. Thus at pH 8 some 65% (Pu-III), 90% (Pu-IV) and 50% (Pu-VI) was respectively in particulate fraction. However, a plateau of the curve exists for Pu (III) between 5.5 and 8. Although the general shape of the

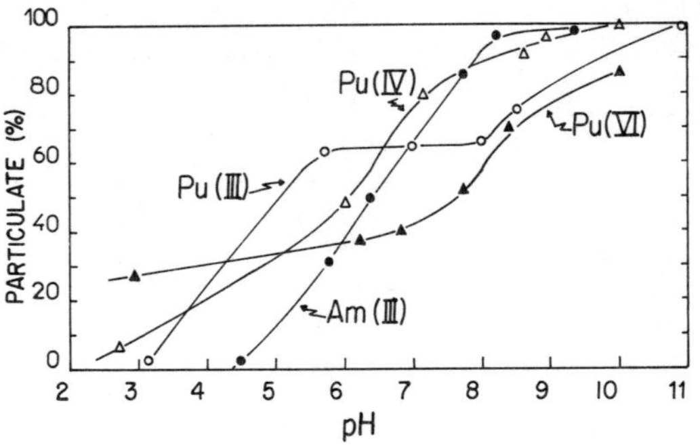

FIG. 4. Particle formation of Pu (III, IV, VI) and Am (III) in Var River water as a function of changes in pH.

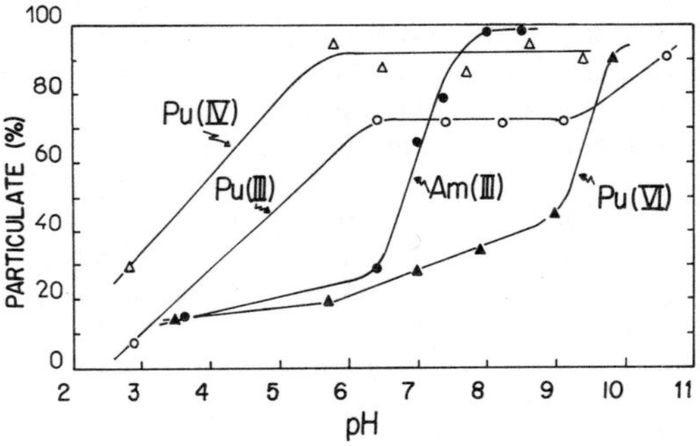

FIG. 5. Particle formation of Pu (III, IV, VI) and Am (III) in sea water as a function of changes in pH.

curves for particle formation of the three forms of plutonium in sea water is similar to that in river water, the curve for Pu (IV) is displaced towards lower pH and those for Pu (III) and Pu (VI) towards higher pH. Thus the plateau of the curve for Pu (III) occurs between pH 6.5 and 9 and only 35% of Pu (VI) is in particulate form at pH 8 in sea water. As has been previously mentioned, the lower particle formation of Pu (VI) in both river and sea water is considered to be due, at least partially, to complex formation with carbonate. It is also worthwhile to note the different shapes of the pH dependency curves on the particle formation for the three valence states of plutonium, since these differences might be related to specific colloidal properties of the particle formed.

The pH dependency curve for particle formation of Am (III) in river water is similar to that for Pu (IV), although Am (III) seems slightly more pH-sensitive. Moreover, the curves for Am (III) for both river water and sea water are similar, except for a slight displacement of the curve for sea water towards lower pH and for the sign of some particle formation in sea water below pH 4.

On the basis of these results, it is estimated that in natural conditions such as in the Var River or in sea water, the major fraction of Pu (III), Pu (IV) and Am (III) is likely to become particles, larger than 0.45 μm, while a greater part of Pu (VI) may stay in solution.

3.2. Adsorption in fresh water

As has been described, the experimental method utilized for measuring the adsorption of plutonium or americium onto sediment particles by filtration cannot differentiate ions, complexes or particles attached to the sediment surface from particles not attached to the sediment. That is why, as shown in Fig.2, the adsorption curve for Pu (IV) coincides virtually with that of particle formation. The same situation was not, however, observed for Am (III). As shown in Fig.3, the adsorption curve for Am (III) shows a slower rate of adsorption than that of particle formation, although equilibrium values of both do not differ greatly. This indicates that, in the river water medium, the particle formation of Am (III) was prevented by the addition of sediment particles in the adsorption medium, especially in the initial period of the experiments. Although it is difficult to estimate the mechanism of the delayed adsorption, a clue may be obtained from particle size distribution of Am (III) in river water, presented in Table I. The values for Am (III) in Table I show that not only the particle formation depended on the pore size of the filters used for the initial filtration, but there was also a considerable size difference among the particles formed even after a 96-hour shaking period. If the added sediment contained appreciable amounts of colloidal fraction, which can pass through a 0.45-μm filter, Am (III) would immediately have associated with the colloid fraction, passed through the filter and been measured as 'non-adsorbed fraction'; later, in the course of the 96-hour period, these colloidal particles would have grown to larger size and eventually been measured as adsorbed fraction. Since the pH-dependency of Am (III) for the adsorption is steep in a pH range of 6.5-8.5, a slight change in the pH of the experimental system might have accentuated the retarded adsorption.

The variations of adsorption on sediment particles of Pu (III, IV and VI) and Am (III) as a function of change in pH of the river water medium, are shown in Fig.6. The adsorption curves for three forms of plutonium are not significantly different from each other, except for lower adsorption of Pu (IV) and Pu (VI) at around pH 3, compared with that of Pu (III); at around pH 8 all three forms of plutonium are well asssociated with the sediment particles (85-95%). The curve for Am (III) is markedly different from those of plutonium; the adsorption of Am (III) changes sharply between pH 6.5 and 8.5, being 30% at pH 7 up to 90% at pH 8.

These results indicate that Pu (III, IV, VI) as well as Am (III) released into a fresh water system like the Var River (pH $\simeq$ 8) in calcareous terrain are well associated with sediment particles, while the adsorption of Am (III) in an average fresh water system (pH $\simeq$ 6.5) may be considerably lower.

Since the sediment used in the adsorption experiments was low in organic content (organic C $\simeq$ 1%) the adsorption characteristics studied are considered to represent chiefly inorganic interactions. In the natural condition, there are indications that organic particles play an important role in the vertical transport of plutonium, as pointed out by Bowen et al.[4]. Further studies on the adsorption characteristics of actinides on organic particles are apparently needed.

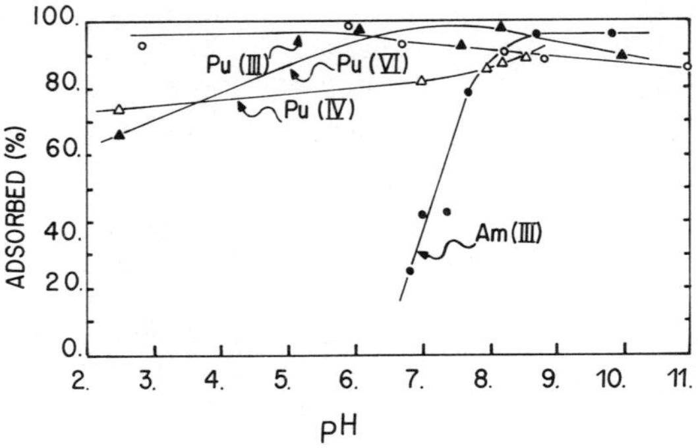

FIG. 6. Adsorption of Pu (III, IV, VI) and Am (III) on sediments in Var River water at different pH values.

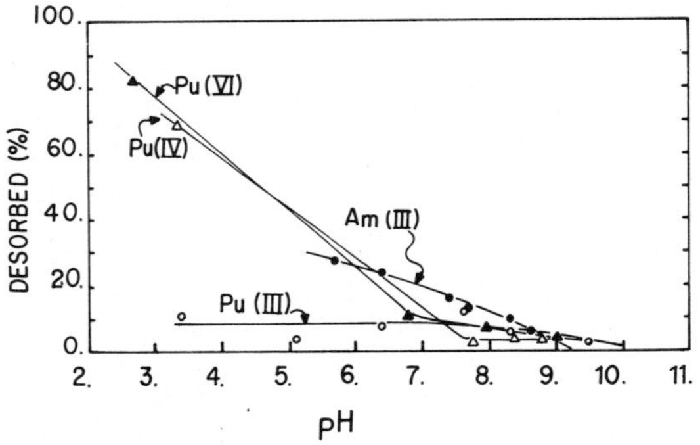

FIG. 7. Desorption of Pu (III, IV, VI) and Am (III) from sediments in the sea water medium at different pH levels.

3.3. Desorption under contact with sea water

The pH-dependency of the desorption of adsorbed fraction of ^{237}Pu (originally in III, IV and VI states) and ^{241}Am (originally in III state) from the sediment under contact with sea water was studied, the results of experiments being shown in Fig.7. As illustrated in the figure, in a low pH range (pH 3-4) the Pu (IV) and Pu (VI) desorb much more (60-70%) than Pu (III) does (10%), although the desorption of all three forms in a pH range of 7-8 is low (5-10%). The desorption of Am (III) seems more pH-sensitive in the range of 7-8; systematic increase of the desorption was observed with lowering pH, being $\simeq$ 15% at pH 8 and $\simeq$25% at pH 7. Thus, with the normal sea water (pH $\simeq$ 8), between 5-10% of Pu (III, IV and VI) and 10-15% of Am (III) would be desorbed from fresh water sediments in contact with sea water. If the pH of sea water is lowered to pH 7 by some

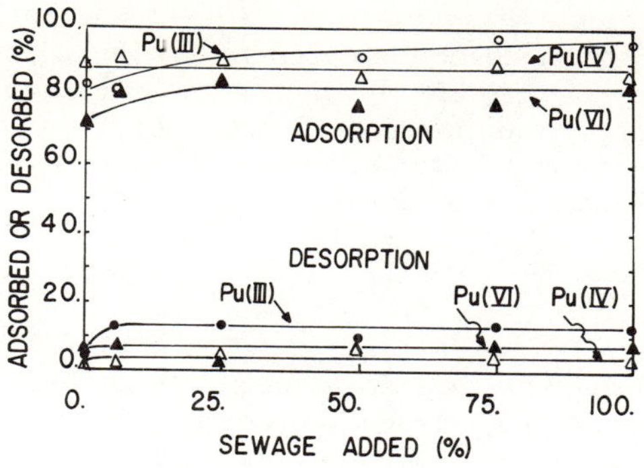

FIG. 8. Effects of sewage effluent on the adsorption-desorption processes of Pu (III, IV, VI).

cause, the americium desorption could double, while the plutonium desorption would remain more or less unchanged. As mentioned for the adsorption, the organic content of sediments may also play an important role in the desorption process of actinides from sediments.

3.4. Effects of sewage effluent

To examine the effects of some soluble organic substances on the adsorption-desorption processes of plutonium with sediments, a series of experiments were conducted using a sewage effluent. The chemical characteristics of the sewage solution used has been described in the previous section. The adsorption experiments were carried out with Pu (III, IV and VI) in mixtures of 5, 25, 50 and 75 vol/vol percent sewage solution with the Var River water; undiluted sewage effluent and the Var River water bracketed both ends of this concentration range. For the desorption experiments, the Monaco sea water described previously was used. The pH of all media was kept at ≃ 8. The rest of the experimental procedures were the same as the normal adsorption-desorption measurements [16].

In Fig.8 the percentage adsorption or desorption of the three forms of plutonium is illustrated as a function of volume percent of sewage effluent in the adsorption media. These results show that the adsorption of Pu (IV) does not depend on the sewage concentration, while that of Pu (III) and Pu (IV) tends to show a slight increase up to a sewage concentration of 25% and no further increase of adsorption was observed with increasing sewage concentration. In the absence of sewage effluent, the adsorption of Pu (III) was 85%, while at 100% vol/vol sewage, this was increased to more than 95%. In any case it appears that the net increase of the adsorption due to the addition of sewage effluent does not exceed ≃ 10%.

As to the desorption of plutonium from the sediment labelled initially with Pu (III), Pu (IV) and Pu (VI), no significant effect was observed with the increasing concentrations of sewage effluent in the labelling medium, except for a possible increase in the desorption of Pu (III) between 0-5% vol/vol sewage concentration.

The results indicate that the sewage effluent has small but measurable effects on the adsorption-desorption process for Pu (III) and Pu (VI) with sediments, but it is not possible to attribute the effects to complexation of plutonium with organic or inorganic ligands, or to complicated surface phenomena at the sediment-water interface. Further studies are required to elucidate the mechanisms involved.

4. SUMMARY

The results of the experiments demonstrate the following characteristics of particle formation, adsorption and desorption of plutonium and americium in simulated river-estuary conditions.

(1) The particle formation of Pu (IV) in river water and sea water depends neither on the pore size (0.10 μm and 0.45 μm) of filters used for the initial filtration, nor on the salinity of the medium; the sizes of the particles formed are comparatively homogeneous.

(2) The particle formation of Am (III) in river water and sea water seems to be influenced by colloidal substances occurring in the medium; there is an appreciable size distribution of the particles formed.

(3) The particle formation of Pu (IV) in river water and sea water is lower, suggesting possible complex formations with carbonate and biocarbonate.

(4) In natural conditions, such as the Var River or the sea (pH ≃ 8), the major fractions of Pu (III), Pu (IV) and Am (III) are likely to become particles larger than 0.45 μm, while a greater part of Pu (VI) may stay in solution.

(5) The major part of Pu (III, IV and VI) and Am (III) in a fresh water system with a high pH value (≃ 8) such as the Var River is associated with sediment particles, while in an average fresh water system (pH ≃ 6.5) the adsorption of Am (III) may be considerably lower.

(6) The desorption of Pu (originally III, IV and VI states) and Am (originally III state) from fresh water sediment in normal sea water (pH ≃ 8) is low (5-15%). If the pH of sea water is lowered to 7, the americium desorption could double, while the plutonium desorption would remain more or less unchanged.

(7) The sewage effluent shows small but measurable effects on the adsorption-desorption characteristics of Pu (III) and Pu (VI) in simulated river-estuary conditions.

ACKNOWLEDGEMENTS

The authors wish to express their gratitude to Prof. K. Kimura of the University of Tokyo, Japan, who organized the production of the ^{237}Pu, and to Dr. T. Nozaki of the Institute of Physical and Chemical Research, Tokyo, for his successful efforts in specially producing it for us.

The authors are also indebted to Dr. T.M. Beasley for his valuable advice, Mrs. M. Blessington and Mrs. E. Illes for assistance in the manuscript preparation and Mr. L. Huynh-Ngoc for the drawing.

The Laboratory of Marine Radioactivity operates under a tripartite agreement between the International Atomic Energy Agency, the Government of the Principality of Monaco and the Oceanographic Institute at Monaco. Support for the present work is gratefully acknowledged.

REFERENCES

[1] NOSHKIN, V.E., Ecological aspects of plutonium dissemination in aquatic environment, Health Phys. 22 (1972) 537.

[2] AARKROG, Radioecological investigations of plutonium in an Arctic marine environment, Health Phys. 20 (1971) 31.

[3] MIYAKE, Y., SUGIMURA, Y., Plutonium content in the western North Pacific waters, Pap. Meteorol. Geophys. (Tokyo) 19 (1968) 481.

[4] BOWEN, V.T., WONG, K.M., NOSHKIN, V.E., Plutonium 239 in and over the Atlantic Ocean, J. Mar. Res. 29 (1971) 1.

[5] NOSHKIN, V.E., BOWEN, V.T., "Concentrations and distributions of long-lived fallout radionuclides in open ocean sediments", Radioactive Contamination of the Marine Environment (Proc. Symp. Seattle, 1972) IAEA, Vienna (1973) 671.

[6] TEMPLETON, W.L., PRESTON, A., "Transport and distribution of radioactive effluents in coastal and estuarine waters of the United Kingdom", Disposal of Radioactive Wastes into Seas, Oceans and Surface Waters (Proc. Symp. Vienna, 1966) IAEA, Vienna (1966) 267.

[7] KHARKAR, D.P., TUREKIAN, K.K., BERTINE, K.K., Stream supply of dissolved silver, molybdenum, antimony, selenium, chromium, cobalt, rubidium and cesium to the oceans, Geochim. Cosmochim. Acta 32 (1968) 285.

[8] MURRAY, C.N., MURRAY, L., "Adsorption-desorption equilibria of some radionuclides in sediment-fresh-water and sediment-sea water systems", Radioactive Contamination of the Marine Environment (Proc. Symp. Seattle, 1972) IAEA, Vienna (1973) 105.

[9] ROBERTSON, D.E., SILKER, W.B., LANGFORD, J.C., PETERSEN, M.R., PERKINS, R.W., "Transport and depletion of radionuclides in the Columbia River", Radioactive Contamination of the Marine Environment (Proc. Symp. Seattle, 1972) IAEA, Vienna (1973) 141.

[10] FUKAI, R., MURRAY, C.N., HUYNH-NGOC, L., Variations of soluble zinc in the Var River and its estuary, Estuar. Coast. Mar. Sci. 3 (1975) 177.

[11] TODD, R., LOGAN, R., Plutonium-237 labelling: A new technique for plutonium metabolic studies in animals, Int. J. Appl. Radiat. Isot. 17 (1966) 253.

[12] TODD, R., LOGAN, R., Plutonium-237 for metabolic studies: Its preparation and use, Int. J. Appl. Radiat. Isot. 19 (1968) 141.

[13] HAKONSON, T.E., JOHNSON, L.J., "Distribution of environmental plutonium in the Trinity Site ecosystem after 27 years", Los Alamos Sci. Lab., Univ. Calif., Los Alamos, New Mexico, Rep. Conf. 730907-14, LA-UR-73-1291 (1973).

[14] STROMINGER, D., HOLLANDER, J.M., SEABORG, G.T., Table of isotopes, Rev. Modern Phys. 30 (1958) Part 2, 585.

[15] DUURSMA, E.K., EISMA, D., Theoretical, experimental and field studies concerning reactions of radioisotopes with sediments and suspended particles of the sea. Part C. Applications to field studies, Neth. J. Sea Res. 6 (1973) 265.

[16] MURRAY, C.N., MEINKE, S., Influence of soluble sewage material on adsorption and desorption behaviour of cadmium, cobalt, silver and zinc in sediment-fresh water, sediment-sea water systems, J. Oceanogr. Soc. Japan 30 (1974) 216.

[17] CLEVELAND, J.M., "The chemistry of plutonium", Gordon and Breach Science Publishers Inc., New York (1970) 133, 395-396.

DISCUSSION

J.K. MIETTINEN: In your river sediments there are organic and inorganic components. What percentage of your adsorbed ^{237}Pu was in the organic component, and what in the inorganic component?

C.N. MURRAY: Mr. Duursma of the Monaco laboratory, who has details of the mineralogy of this sediment, established that the Var river sediment used in these experiments contained about 0.5% organic material. It is a high-carbonate sediment with a clay fraction of only about 30%. We have no values for the proportion of ^{237}Pu on the organic, carbonate and clay fractions, but we felt that the inorganic carbonate fraction would be the most important.

R. BITTEL: You have been concerned mainly with the plutonium present in valence states III and IV. In the event of an accident involving sodium-cooled fast reactors, the plutonium would primarily be in valence state VII. Have you any experimental knowledge on the behaviour of plutonium of this valence state in the environment?

C.N. MURRAY: We also studied the VI valence state in detail, as well as the III and IV states. We felt that the VI form would normally be the highest state of Pu capable of existing for any length of time in the environment. The formation of Pu (VII) valence species, as discussed by Métivier in 1974 (Radioprotection), would require fairly extreme conditions, such as you mention in the case of an accident to a sodium-cooled reactor, but the instability of such a species would most likely be very great and its disproportionation in the environment to lower plutonium valence states would be very rapid indeed. Thus the longer-term environmental impact on a river estuarine situation is similar to that in the laboratory work presented in the paper. The very short-term impact, close to an accident area, where Pu (VIII) species might be found could possibly be somewhat different initially.

K.C. PILLAI: I should just like to stress that percentage particulate activity is a relative term, especially in tracer experiments. It all depends how much tracer you add, since the solubility product of Pu in sea water is very low.

C.N. MURRAY: Under the conditions used in the experiments, the present work would seem to show that, of the three forms of Pu investigated (+3, +4, +6) the +6 state is most soluble (as measured by a 0.45-μm filter), both in fresh water and sea water. Data show that all three forms are strongly adsorbed on fresh-water sediments, irrespective of their apparent solubility.

SOME INVESTIGATIONS INTO THE BEHAVIOUR OF PLUTONIUM IN THE MARINE ENVIRONMENT

J.A. HETHERINGTON, D.F. JEFFERIES, M.B. LOVETT
Ministry of Agriculture, Fisheries and Food,
Fisheries Radiobiological Laboratory,
Hamilton Dock, Lowestoft, Suffolk,
United Kingdom

Abstract

SOME INVESTIGATIONS INTO THE BEHAVIOUR OF PLUTONIUM IN THE MARINE ENVIRONMENT.

Plutonium and other transuranic elements are produced during the operation of magnox reactors, although they do not usually present any problems of waste management until the fuel reaches the reprocessing stage. At this stage one of the chief aims is to separate and recover the plutonium from the depleted uranium and the fission products, the incentive being the value of plutonium itself as a nuclear fuel particularly for the fast breeder reactors. For this reason the amounts of plutonium appearing in the low activity waste streams for discharge to the environment are usually small, and their radiological impact insignificant compared with that of the more abundant fission products. However, with the prospect of use of large quantities of plutonium in the fast reactors, considerable interest surrounds all aspects of the properties of plutonium, and in an attempt to give a fuller understanding of the behaviour of plutonium in the marine environment use has been made of the small amounts which have been discharged in recent years to the north-east Irish Sea from the fuel reprocessing plant at Windscale in Cumbria. Details of the amounts of ^{239}Pu and ^{240}Pu discharged to sea are given and the resultant measured concentrations of the isotopes in the water throughout the Irish Sea are shown. The distribution of plutonium in sea water is shown to follow the same basic pattern as the fission products, details of which have been reported previously. By comparing the concentration gradients with distance from the point of discharge for plutonium and caesium, a value for the rate of loss of plutonium from the water to the other important compartment, namely the sediment, has been deduced. Measurements have been made of ^{239}Pu and ^{240}Pu in seabed and estuarine sediments and by combining these with the water data values of the concentration factor for the nuclide in sediment have been calculated. The results of a number of core samples have been used to calculate an approximate inventory of the amount of plutonium at present residing in the sea bed. The concentrations of plutonium in samples of seaweed, fish and shellfish are reported. Comparison of these results with those in water gives an indication of the concentration factors for the nuclide in these materials.

1. INTRODUCTION

Plutonium and other transuranic elements are produced during the operation of magnox reactors, although they do not usually present any problems of waste management until the fuel reaches the reprocessing phase. At this stage one of the chief aims is to separate and recover the plutonium from the depleted uranium and fission products, the incentive being the value of plutonium itself as a nuclear fuel. For this reason alone the amounts of plutonium reaching the waste streams for discharge to the environment are usually small and their radiological impact insignificant compared with that of the more abundant fission products. However, with the prospect of use of increasing quantities of plutonium in fast reactors, considerable interest surrounds all aspects of the properties of this element and, in an effort to obtain a better understanding of its behaviour in the marine environment, use has been made of its presence in the discharges to the north-east Irish Sea from the fuel reprocessing plant at Windscale. The location of the plant and the important features of the area referred to in this paper are shown in Fig. 1.

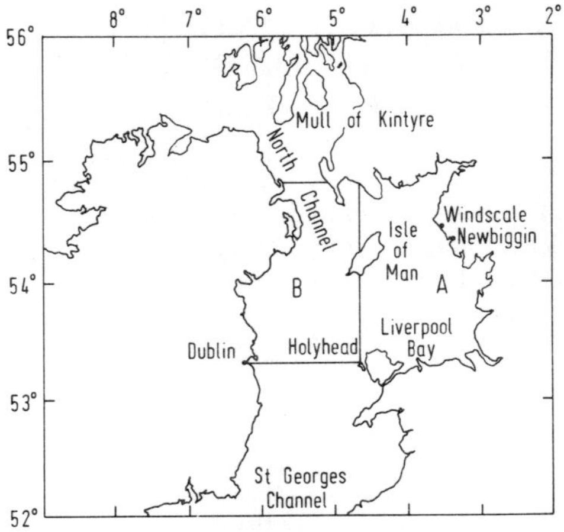

FIG. 1. Map of the north-east Irish Sea area showing the location of the Windscale reprocessing plant.

Plutonium nuclides have been present in the effluent from Windscale for many years and they have formed a substantial part of the total α component of the discharges reported previously [1, 2]. For this reason, and in accordance with the UK policy for the control of radioactive waste, measurements have been made of the plutonium present in marine samples from the vicinity of Windscale. This paper is concerned with studies which have been carried out by the Fisheries Radiobiological Laboratory (FRL). Materials measured have included the important elements of human food chains such as fish, shellfish and the seaweed Porphyra umbilicalis, as well as sea water and sediment, whose role as a "sink" for plutonium introduced into the sea is well recognized [3, 4]. Table I shows the discharge rates of plutonium during the past 8 years and the resultant concentrations in two critical food-chains, fish and edible seaweed, expressed as a percentage of the derived working limit (DWL) for these chains over the period 1968-74. It may be seen that levels of contamination have remained extremely low and the safety of discharges from the point of view of public radiation exposure is clearly demonstrated.

Table I also shows how the amounts of plutonium discharged as waste have increased by a factor of 3-4 over the past 8 years. Concurrent with this increase, which was anticipated as the first phase of the UK nuclear power programme came into operation, the programme of this laboratory has been widened to provide more information of a research nature about the partition of plutonium amongst the major compartments of sea water, sediment and biota, and this paper describes some of the first results of this work. The object of the study, which should be seen in the context of the potential increases in the uses of plutonium as a nuclear fuel in the future, aims to answer some of the questions posed by the introduction of a long-lived radionuclide into the marine environment. These include the potential for build-up in various compartments and the hazards associated with the solubilization and mobilization in the environment of compounds which begin as insoluble, biologically unavailable forms [3].

TABLE I. Average monthly discharge rate (Ci) of plutonium nuclides from Windscale to the Irish Sea and the resultant levels of contamination of edible seaweed and fish, expressed as a percentage of the derived working limits (DWL)

Year	Average discharge (Ci/month)	*Porphyra* spp. (% DWL)*	Fish	
			Caught within 5 km of outfall (% DWL)	Commercial landings from NE Irish Sea (% DWL)
1967	41	-	-	-
1968	69	2.0	-	-
1969	68	1.0	-	-
1970	78	0.4	0.009	< 0.0007
1971	94	0.8	0.10	-
1972	129	0.8	0.08	-
1973	148	0.3	0.15	0.005
1974	104	< 0.05	0.08	0.002

*DWL calculated on the assumption that laverbread is made entirely from Cumbrian weed.

2. SAMPLING PROCEDURES AND METHODS OF ANALYSIS

2.1. Sea water

Samples of 25 litres of sea water have been collected from a grid of stations throughout the Irish Sea during research vessel cruises in 1973 and 1974. The samples have been collected from the vessel's clean salt water system, which draws water from a depth of approximately 3 metres below the surface, and the samples are representative therefore of the surface profile. Each sample was filtered through a 0.22 μm millipore filter immediately after collection and the filtrate stored in a polythene container.

For the analysis of plutonium the millipore-filtered sample was acidified and ^{236}Pu tracer added. The Pu was coprecipitated with calcium oxalate under reducing conditions at pH 2 provided by the addition of sodium sulphite. After filtration the oxalate was converted to the carbonate by ignition at 600°C and dissolved in 6$\underline{M}$ HCl/H_2O_2. To this solution 5 mg of neodymium carrier was added and the plutonium coprecipitated with neodymium hydroxide. The precipitate was redissolved in approximately 5 ml of HCl, giving a solution which was 9$\underline{M}$ with respect of HCl, and to this 10 drops of 16$\underline{M}$ HNO_3 were added. This solution was passed through an ion-exchange column [0.7 cm x 5 cm Bio-rad AG1 - X4, 100-200 mesh]

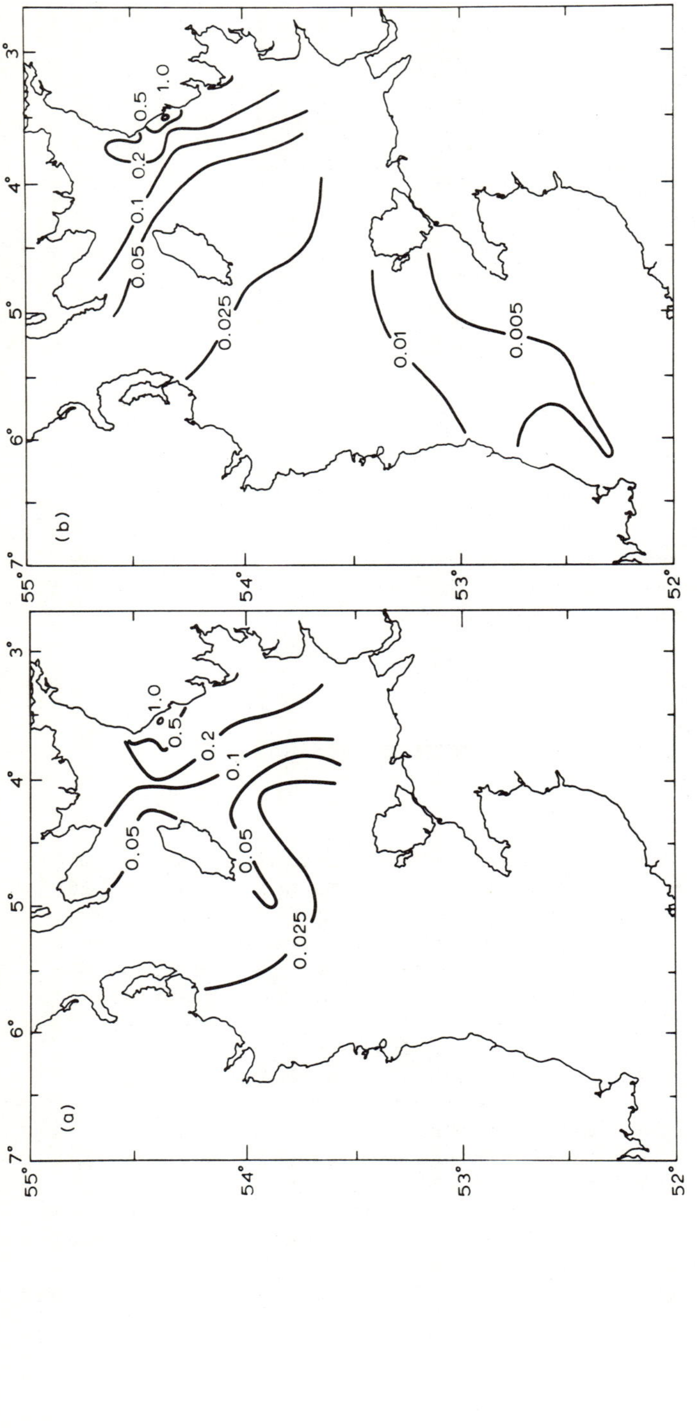

FIG.2. The contours of concentration of ^{239}Pu (pCi/l) in filtered sea water: (a) July 1973, and (b) July 1974.

which retains the Pu (in the Pu^{VI} form) while allowing the lanthanides and trivalent actinides to pass through. The column was washed with 8M HNO_3 followed by 11M HCl to remove the bulk of other α-emitting nuclides and then a second wash with 11M HCl/0.1M HI solution eluted the Pu in the Pu^{III} form. This was evaporated to dryness, the halides removed by the addition of HNO_3 followed by H_2SO_4, and the Pu electroplated on to a stainless steel disc from a slightly acidic ammonium sulphate solution. The disc was rinsed with distilled water, heated to redness and the $^{239+240}Pu$ was determined by α-spectrometry using a silicon surface barrier detector. It is not possible to differentiate between ^{239}Pu and ^{240}Pu by radiometric methods, but mass spectrometric measurements on some of the environmental samples has shown that the ratio of ^{239}Pu:^{240}Pu is not less than 10:1. Throughout this paper the results are presented as ^{239}Pu and it should be taken that up to 10% of this is due to ^{240}Pu, on a weight basis.

2.2. Sediment and biological materials

Sediment samples have been collected in the form of 10-cm-diameter core samples. In the intertidal zone these have been obtained by driving a thin-walled PVC tube into the sediment directly, whilst for the seabed samples the cores have been obtained as a sub-sample of a larger section collected with a Reineck box corer.

For the analysis of the sediment and biological materials, the organic content was destroyed by ashing at 450°C. ^{236}Pu tracer was added to the residue, which was then digested with either HF/HNO_3 followed by 3M HCl/H_2O_2 or by a single 6M HCl/H_2O_2 treatment. Any insoluble matter was filtered off and discarded and after the oxalate precipitation on the filtrate the procedure was the same as for sea water.

3. DISTRIBUTION AND DISPERSION OF PLUTONIUM IN SEA WATER

The results of the analyses of plutonium in surface seawater samples collected during the cruises in July 1973 and July 1974 are reproduced in contour form in Fig. 2 (a and b). The contours, which were interpolated from 26 sampling points, represent the concentrations of ^{239}Pu remaining in water after passing it through a 0.22 μm filter. The data show that detectable levels of plutonium due to Windscale discharges are present throughout the Irish Sea. In the immediate vicinity of the Windscale pipeline, the concentrations are between 500 and 1000 times greater than those reported due to weapons fallout levels in the Atlantic water entering the Irish Sea[4], while concentrations of an order of magnitude greater than fallout are detectable at distances in excess of 100 km from the outfall.

The shape of the contours of concentration of ^{106}Ru in filtered sea water within approximately 50 km of the outfall has been reported earlier[5, 6], and it is noteworthy that the contours for ^{239}Pu within this area show the same southward displacement. Over the past four years, and in parallel with the work with plutonium, the measurements have been extended throughout the Irish Sea and similar measurements have been obtained for ^{137}Cs. The distributions of these nuclides for July 1973 are given in Fig. 3 (a and b) for the purposes of comparison with the plutonium picture, and it may be seen that the behaviour of all three nuclides is qualitatively similar.

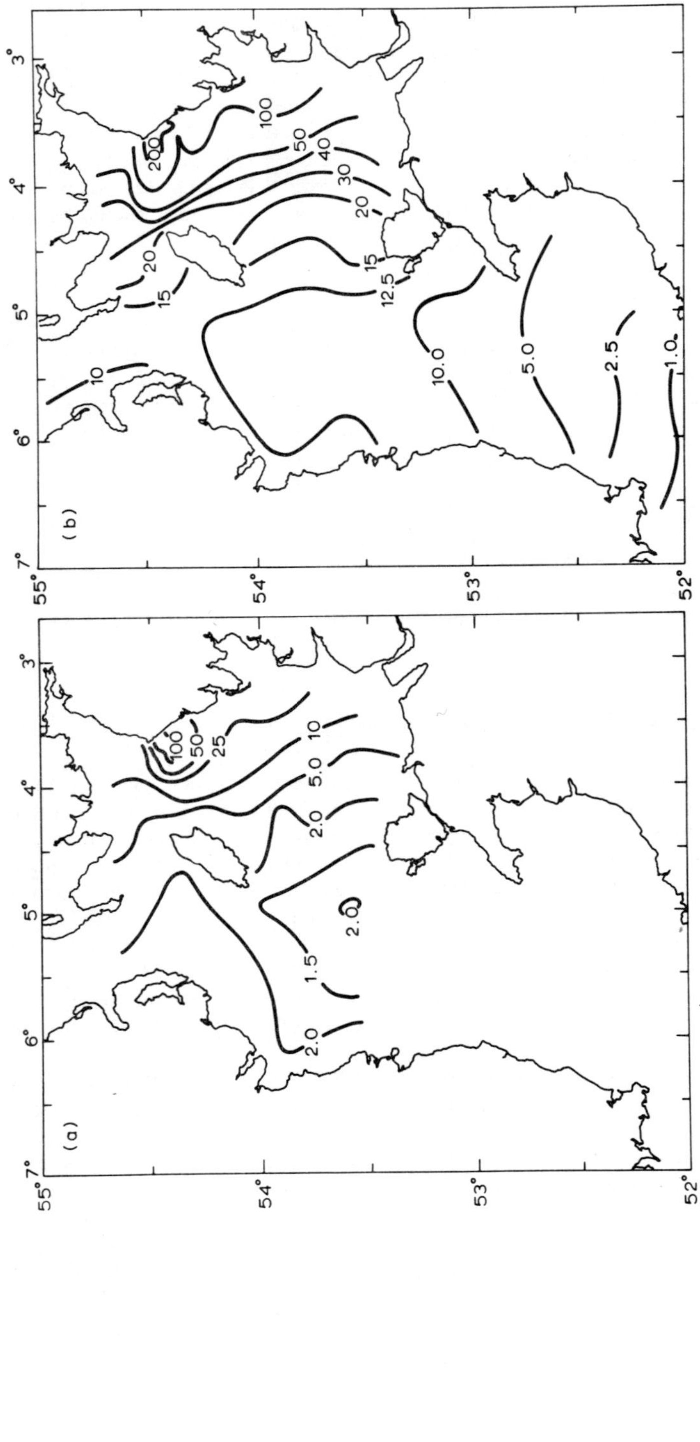

FIG.3. The contours of concentration (pCi/l) of (a) ^{106}Ru and (b) ^{137}Cs in filtered sea water in July 1973.

A full discussion of the way in which the contaminants released from Windscale move through the Irish Sea, based on the ^{137}Cs evidence, is given in papers by Jefferies [7, in prep.]. It has been found that immediately after discharge, the effluent remains stratified because of its lower density relative to sea water and is moved to and fro with the tidal plug for a short time - typically 3 to 4 days, but very much a function of the prevailing weather conditions. It is gradually removed from the plug, both by diffusion and by a southerly residual drift following the Cumbrian coast. Thereafter it joins the circulation of water in Liverpool Bay, eventually to be flushed from the area by the general northward movement of Atlantic water which enters via St George's Channel in the south and leaves again by the channel between the Mull of Kintyre and Ireland in the north (see Fig. 1). The contours for plutonium show that this nuclide follows these same general pathways within the Irish Sea.

TABLE II. Concentrations of ^{239}Pu, ^{144}Ce, ^{137}Cs and ^{106}Ru (pCi/l per Ci/day) discharged in filtered sea water (i) within 10 km of the outfall and (ii) at 75-110 km from the outfall

Nuclide	Seawater concentrations		(i)/(ii)
	(i) within 10 km of outfall	(ii) at 75-110 km from outfall	
^{239}Pu	0.150	0.012	12
^{144}Ce	0.063	0.006	11
^{137}Cs	2.7	0.24	11
^{106}Ru	0.57	0.038	15

A comparison of the ways in which the concentrations of ^{239}Pu, ^{144}Ce, ^{137}Cs and ^{106}Ru in filtered sea water change with distance from Windscale is given in Table II. This shows the average concentrations of each nuclide at 10 stations within 10 km of the outfall and at a similar number of stations at 75-100 km distance, all the concentrations having been normalized to a discharge rate of each nuclide of 1 Ci/day. Two points of interest emerge from the table. First it is seen that the concentrations of ^{239}Pu and ^{144}Ce in filtered water per unit rate of discharge are of the order of 30 times less than that of ^{137}Cs, indicating that there is a substantial loss of the former nuclides from the water phase very soon after their injection into the sea compared with caesium. Second it is seen that once this initial loss has taken place the residual fractions of both ^{239}Pu and ^{144}Ce remaining in the water phase behave thereafter in a fashion which is conservative and very similar to ^{137}Cs.

TABLE III. Total inventories (Ci) of ^{239}Pu and ^{137}Cs in the sea water in the area A + B as defined in Fig. 1

	^{239}Pu		^{137}Cs	
	1973	1974	1973	1974
Total in sea water in Area A + B (Ci)	60	56	29 000	60 000
Total as % of preceding year's discharge	4.4	4.0	94	136

An estimate of the magnitude of the fraction which is removed from the water phase has been obtained by computing the total amounts of ^{239}Pu and ^{137}Cs present in the sea water in the area of 2.5 x 10^4 km^2 between latitude 54°50' in the north and the line joining Dublin and Holyhead in the south (the area defined by the shaded areas A and B on Fig. 1). The results of this calculation based on the data for July 1973 and July 1974 are given in Table III, and the figures are expressed also as percentages of the amount of each nuclide discharged during the 12 months before the date of sampling. The results again emphasize that the fraction of the plutonium entering the sea which is lost from the water mass is very large compared with caesium, and it is seen that the inventory of plutonium in the water of the area did not change significantly between July 1973 and July 1974, despite the injection of some 1800 Ci during that period. The existence of this equilibrium situation leads to the conclusion that at least 96% of the plutonium introduced during the year 1973-74 disappeared from the water mass by some means.

A fraction of the activity discharged is flushed from the area by the general water circulation, as described earlier, and an estimate of this factor has been made from the observations on the behaviour of radiocaesium isotopes. These observations have suggested that the water in this area is turned over with a half period of approximately 1 year (Jefferies, in prep.). Thus, with an equilibrium inventory of 60 Ci of plutonium, the amount lost per year via water transport is not greater than 40 Ci and it is concluded that the greater part of the plutonium - of the order of 96% - entering the water is removed to the sediments of the area and that this loss occurs in the immediate vicinity of the outfall.

This rapid loss of plutonium from the water is in keeping with the conclusions of Noshkin[4]. In a review of the information available on the behaviour of plutonium in sea water from a number of sources, including nuclear weapons fallout, satellite debris, and nuclear fuel reprocessing operations, he reports that in all cases a fraction of the order of 90% of the injected plutonium is lost from the water phase.

In view of the speed with which the plutonium disappears from the water, it is not considered likely that the primary mechanism for the removal is of a biological nature as proposed by Bowen et al.[8] on the basis of their own work and that of Pillai[9]. Rather, it is assumed that the loss is due to the hydrolysis and consequent precipitation of the plutonium, present initially in the effluent in a soluble phase, either as a hydroxyl complex of the element itself or in association with suspended material already present in the sea water. Whatever the mechanism, the process removes the discharged plutonium from both solution and suspension in the water mass. Measurements carried out on the residue remaining after filtration of the seawater samples have shown that 70% of the total activity in the unfiltered sea water is present either in solution or in association with suspended material with a size < 0.22 μm, whilst only 30% was found with the suspended material size > 0.22 μm. The value of the suspended sediment load with size > 0.22 μm in these samples was between 0.5 and 10 ppm, with a mean of 2.5 ppm.

3. THE INTERACTION OF PLUTONIUM WITH SEDIMENTS

3.1. Surface sediments

The foregoing section has reinforced the importance of the sediments in any study of the behaviour of plutonium in the sea and this present work has included measurements on both marine and estuarine core samples.

The results of the plutonium analyses on a series of seabed surface samples collected during July 1974 are shown in Table IV. The samples were the top 20 mm profiles of cores taken with minimum disturbance from a Reineck box sample. Also given in the table are the plutonium concentrations in the surface water at each of the stations where sediment has been measured, and these data have been used to calculate the concentration factors at each point. At first sight it appears that there is a very large spread in the values of the concentration factor, especially between the near and distant stations.

This conclusion neglects, however, to take account of the difference in the particle size distribution of the various sediment samples. Work by Hetherington and Jefferies[6] has demonstrated that the uptake of fission products by sediment in the area is a surface adsorption phenomenon, and similar measurements of the distribution of α-emitting contaminants among the various sizes of particles in samples from the Newbiggin area have shown a similar relationship, with the specific activity increasing steadily as one goes from sand towards clay (see Fig. 4). When account is taken of this it is found that the high values of the observed concentration factors are associated with the muddier stations (such as No. 44), while the sandier stations (such as No. 61) are the ones with low values for the concentration factor. The values of the concentration factor of a typical sediment from this area containing approximately 5% clay, 50% silt and 45% sand is 2×10^4.

Furthermore, when sites of a similar mineralogical nature are compared no significant difference is found between the concentration factors at stations near to the outfall and those at distance. The sediment found at station 30 for example is very similar in terms of particle size to that found at stations 42 to 48 and, with the exception of station 44, which had a uniquely high concentration of ^{239}Pu in the sediment, there is little difference in the concentration factors with distance.

The levels of plutonium in the superficial sediments throughout the area would appear therefore to be related by a constant factor to the levels in the sea water at

TABLE IV. Concentration of ^{239}Pu in sediment (pCi/g dry) and filtered sea water (pCi/l) and the concentration factor (concentration per unit weight of dry sediment/concentration per unit weight of sea water) at a number of stations in the Irish Sea

Station	Distance from outfall (km)	Concentration of ^{239}Pu†		Concentration factor
		Water (pCi/l)	Sediment (pCi/g dry)	
45	1	0.48 ± 0.01	36 ± 3	7×10^4
44	4.5	0.37 ± 0.01	80 ± 6	22×10^4
47	4.5	0.65 ± 0.01	23 ± 2	3.5×10^4
41	9	0.31 ± 0.01	9 ± 1	3×10^4
42	9	0.46 ± 0.01	12 ± 0.4	3×10^4
43	9	0.31 ± 0.01	29 ± 2	9×10^4
48	9	1.02 ± 0.01	17 ± 1	2.0×10^4
60	75	0.035 ± 0.016	0.22 ± 0.02	0.6×10^4
59	90	0.05*	0.37 ± 0.05	0.7×10^4
30	110	0.03*	0.6 ± 0.05	2×10^4
61	110	0.040*	0.04 ± 0.004	0.1×10^4

† Errors quoted are those due to counting statistics.
* Interpolated values.

the same point, a fact which is perhaps surprising in view of the evidence from the water studies which suggests that large amounts of plutonium should reach the sea bed very soon after injection. The failure to find any sign of an enhanced concentration factor of even a local nature in the superficial sediment in the area of the outfall implies that a large fraction of the plutonium is being removed, not only from the water mass but from the vicinity altogether. In view of the speed of this removal it is most probably because the precipitating plutonium is being dispersed from the area in association with the general sediment transport. It is known from seabed drifter experiments [10] that there is a residual current at the sea bed which will tend to move sediment from the area of the outfall in a southerly direction, and the hydrographic work carried out in the area by Ramster and Hill [11] has shown that the magnitude of this current is of the order of 3 km/day. Radioecological studies [6, 12] have also shown that there is a movement of sediment, showing the same degree of fission product contamination as the sediment found in the immediate vicinity of the Windscale outfall, into estuaries such as Ravenglass to the south. It is concluded therefore that the fraction of the plutonium which leaves the water does so because it joins the particulate phase, which in turn is being dispersed with the general sediment load, and in this respect plutonium is again found

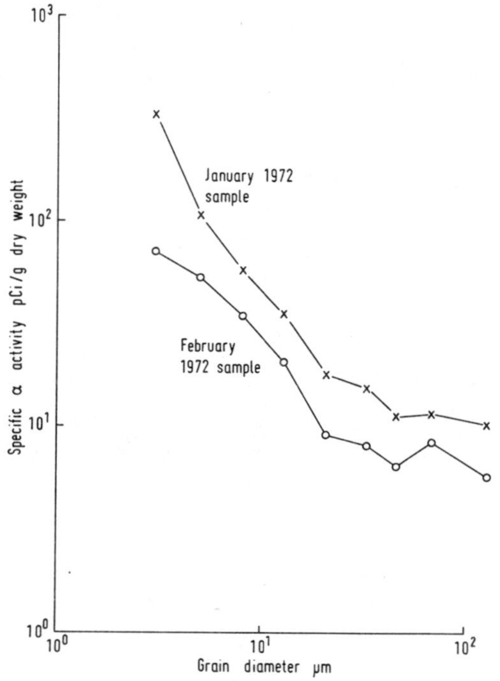

FIG.4. Variation of the α-emitting contamination (pCi/g dry weight) with grain diameter (μm) in a sample of surface sediment from the Ravenglass Estuary.

to behave in a similar fashion to other nuclides such as ^{106}Ru and ^{144}Ce. Following their discharge to sea, these nuclides, like plutonium, are lost rapidly from the water phase, but their concentration in sediment relative to the prevailing water concentrations shows very little enhancement in the vicinity of the outfall compared with the situation at distance.

The one piece of evidence contrary to this interpretation of the plutonium measurements concerns the small proportion of the nuclide found in association with the suspended solid component of the sea water. If the sediment movement is the primary vehicle for the transport of the plutonium, then the ratio of the plutonium in the suspended fraction to that in solution might be expected to be much larger than the value of 30:70 observed. It will be recalled, however, that this figure refers to the measurements made on samples of surface sea water and it is not necessarily relevant to the situation nearer to the sea bed. Since the bed load transport is governed chiefly by the current regime in the region immediately above the bed, and the southerly residual currents occur in the deeper layers, measurements of the plutonium in suspension and solution in these profiles are required to complete the picture. Work to do this is now in hand.

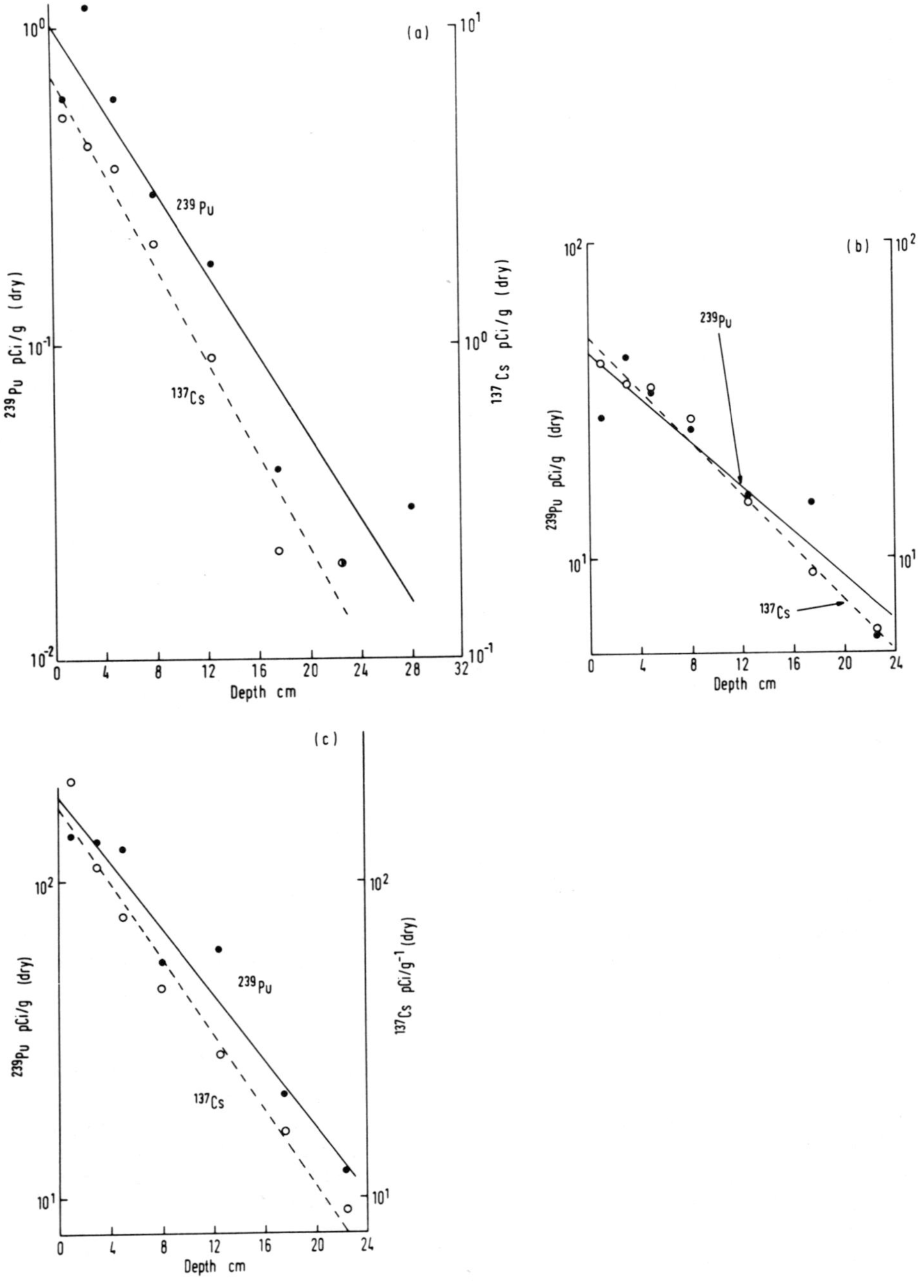

FIG. 5. The distributions of ^{239}Pu (●) and ^{137}Cs (o) with depth in three sediment core samples: (a) station 30, (b) station 43, and (c) Newbiggin.

3.2. Deeper sediments

In order to investigate the behaviour of plutonium in the deeper sediment layers, analyses of all the profiles of three cores have been completed. The cores selected were from stations 43 and 30 which are taken as being representative of the sea bed both close to and distant from the outfall, together with a core from an intertidal zone at Newbiggin in the Ravenglass Estuary. The distributions of ^{239}Pu and also ^{137}Cs with depth in these cores is shown in Fig. 5 (a, b and c), and it is seen that in all three cases the concentrations of both nuclides decline approximately exponentially with depth. Lines have been fitted to each set of data by the method of least squares and from the slopes of the lines the depths for the concentrations of each nuclide to reduce by a factor of 2 have been calculated. These values are given in Table V.

TABLE V. Depths (cm) for the concentrations of ^{239}Pu and ^{137}Cs to fall by a factor of 2 in three core samples

Station number	Half depths (cm)	
	^{239}Pu	^{137}Cs
30	4.6	4.1
43	8.6	7.2
Newbiggin	5.8	5.1

From the values of these half depths it is seen that in each case the penetration of ^{239}Pu is slightly greater than that of ^{137}Cs and this is in agreement with the observations made by Templeton and Preston during the early 1960s[13]. The differences between the distributions of the two nuclides is not large however and, in view of the vastly different chemical properties of caesium and plutonium, the fact that they appear to behave so similarly in sediments is regarded as much more significant and is taken as evidence that sedimentation of contaminated material is the primary mechanism by which these nuclides are being incorporated into the sediments of this area.

It is of interest to note that Noshkin[4] found a similar result for the patterns of ^{137}Cs and ^{239}Pu contamination due to nuclear weapons fallout in the sediment of Buzzard's Bay. In that case it was claimed that the sedimentation rate in the area was too low to account for the presence of ^{137}Cs and ^{239}Pu in the deeper layers and it was proposed that the observed distributions were due either to molecular diffusion and sorption/desorption taking place in the bulk sediment or to biological reworking of the deposits or to a combination of these processes. Presumably similar processes are at work in the sediments of the Irish Sea but, whereas the similarity between the distributions of nuclides with unlike chemical properties is compatible with biological reworking of the sediments, it is not quite so easy to reconcile the identical behaviour of chemically different species under the influence of processes such as molecular diffusion and sorption/desorption reactions. If, however, it can be demonstrated that molecular diffusion and related processes are playing a significant part in the redistribution of both plutonium and caesium in these sediments and that both nuclides are behaving similarly under their influence, it will be important evidence in relation to the chemical form assumed by the plutonium discharged to the marine environment.

From the results of the measurements of ^{239}Pu in sediment completed so far, an estimate of the total amount residing in the sea bed has been made. By dividing the bed into a series of concentric rings, centred on the outfall, and by multiplying the mean surface concentration in each area by the mean depth derived from the core sample data for each area, it is estimated that between 10^3 and 10^4 Ci of ^{239}Pu are residing in the bed of the Irish Sea to the east of the Isle of Man. The total amount discharged from Windscale during the past 15 years is approximately 10^4 Ci and the estimated amount in the sediment is therefore of the same order as this and compatible with the belief that a major fraction of the plutonium reaches the sea bed and remains there.

This conclusion raises immediately questions concerning the long-term consequences of the introduction of long-lived nuclides into this water mass. If the plutonium which joins the sediment phase becomes permanently incorporated into it, the equilibrium inventory of plutonium in the area will be dictated essentially by the rate of removal of the sediment itself. Belderson[14] and Belderson and Stride[15] have concluded that there is a general transport of bed material into the area to the east of the Isle of Man, although no detailed information has been found about its behaviour after it has arrived in the area. The Solway Firth has been receding throughout recorded history with all the channels being slowly filled with marine-derived material[16], and there are several readily identifiable areas on the Cumbrian coast, such as Maryport Harbour and Ravenglass, where sediment is accumulating. These factors, together with the similarity of order between the estimated plutonium burden of the sediments to the east of the Isle of Man and the integrated amount of the nuclide discharged from Windscale, suggest that the rate of loss of the plutonium from the area is small and until further information becomes available it seems only reasonable to assume that the rate is zero. This is the basis which is being taken for an assessment of the longer-term radiological consequences of this discharge. This will include the consequences to man via the established marine food chains and also an assessment of the potential significance of the plutonium to the benthic communities in the same way as has been done already for the major fission product nuclides[17].

4. PLUTONIUM IN BIOLOGICAL MATERIALS

Measurements of ^{239}Pu have been made on samples of edible algae, of fish flesh and of the soft parts of shellfish taken from the Irish Sea. By comparing these results with the measured values of the seawater concentrations, values of the concentration factor of plutonium in each material have been obtained. In this context

TABLE VI. Concentration factors (concentration per unit wet weight/concentration per unit weight of sea water) of ^{239}Pu in biological materials

Material	^{239}Pu concentration factor
Edible algae	
Porphyra spp.	3×10^3
Fish flesh	
Plaice (Pleuronectes platessa)	3×10^1
Dab (Limanda limanda)	3×10^1
Salmon (Salmo salar)	1×10^0
Molluscan soft parts	
Mussels (Mytilus edulis)	2×10^3
Winkles (Littorina littorea)	2×10^3

the concentration factor is defined as ratio of the radioactivity per gramme wet weight of material to the radioactivity per gramme of filtered sea water, and the results are shown in Table VI.

The concentration factor of 3×10^3 for the edible algae Porphyra spp. falls just within the upper limit of the range of values quoted for this type of material by Noshkin[4]. For the two demersal fish species, dab and plaice, the factor is 3×10^1 in each case and is greater than the concentration factors reported for plutonium in the muscle of any of the species considered by Noshkin[4]. This value of 3×10^1 refers however to fish caught within 5 km from the Windscale discharge point, and consideration of the situation at distance suggests that the concentration factor falls as the distance from the outfall increases. Reference to Table I shows that the mean concentration of ^{239}Pu in the flesh of fish caught throughout the area of radius 100 km from Windscale is some 30 times less than the concentration in the fish taken from the immediate area of the discharge point. The data in Table II, however, show that the concentration in the water falls only by a factor of 12 between the area of the outfall and the region 75-100 km distant. Thus, the evidence suggests that the concentration factor in fish flesh at distance may be lower than 3×10^1 by a factor of at least 3. A full explanation for this is being sought, but meanwhile it is suggested that it may be a reflection of the variations in the nature of the sea bed from one area to another, and particularly of the predominance of mud in the vicinity of the Windscale pipeline. Pentreath[18] has shown that, with the exception of elements such as caesium, the fission and neutron activation products which reach fish muscle do so via the food chain. It is suggested that the same may be true for plutonium and that fish deriving their food from muddier areas of the sea bed will show a relatively higher level of this nuclide than the fish feeding from the sandier areas.

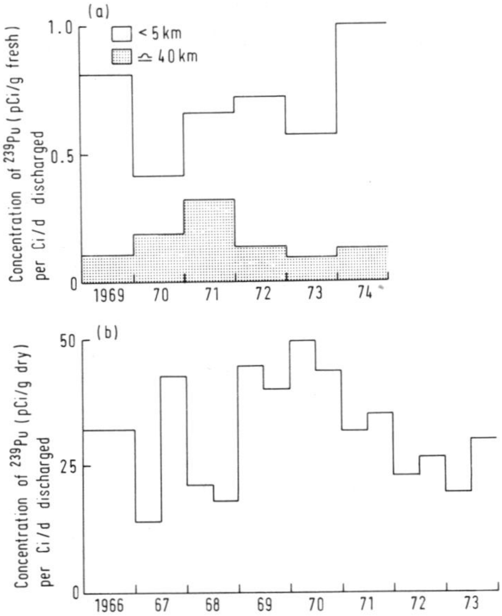

FIG. 6. Concentrations of ^{239}Pu (a) in edible seaweed (pCi/g wet weight) at two stations on the Cumbrian coast and (b) in surface sediment (pCi/g dry weight) at Newbiggin as a function of time. The concentrations are corrected for a 1 Ci/day discharge rate.

The concentration factor of 2×10^3 for the soft parts of molluscan shellfish, collected within 5 km of the Windscale outfall, is almost two orders of magnitude greater than that for fish flesh and is similar to the factor of $1\text{-}2 \times 10^3$ for both ^{106}Ru and ^{95}Zr+^{95}Nb in these species reported by Preston and Jefferies [19]. The factor of 1 for salmon is consistent with the fact that this is a migratory species which is not likely to have been in the area for very long.

In Fig. 6 the concentrations of ^{239}Pu in edible seaweed collected on the Cumbrian coast and of surface sediment from Newbiggin in the Ravenglass Estuary have been plotted as a function of time. The data have been normalized to a release rate of 1 Ci per day of ^{239}Pu from Windscale, and it is seen that there has been no build-up of the nuclide in the weed or surface sediments per unit discharge rate during this period despite a fourfold increase in the discharge rate.

5. CONCLUSIONS

A preliminary survey of the behaviour of ^{239}Pu following its discharge from Windscale to the Irish Sea has been made. The results show that a large fraction of the plutonium discharged is lost very rapidly from the water phase, but that the residual few percent remaining in the water behaves thereafter conservatively and in a similar fashion to ^{137}Cs. Measurements of ^{239}Pu in seabed sediment have failed to show any accumulation in the vicinity of the outfall and it is concluded that once the plutonium is lost from the water it is dispersed in association with the general sediment transport processes in the area. An interesting similarity between the ^{239}Pu and ^{137}Cs distributions with depth in core samples has been confirmed, and it is suggested that this implies that sedimentation is the primary mechanism by which these nuclides are being incorporated into the sediments, although the similarity of behaviour of the two nuclides in this respect as well as in the sea water may be important from the point of view of understanding the chemical form of plutonium in sea water. Measurement of ^{239}Pu in important biological species has shown that the concentration factors are similar to those reported by other workers, with the exception of the factor for inshore demersal fish muscle. An explanation for the relatively high value found for this material has been suggested. The concentrations of plutonium in the important food materials have been summarized and the safety of past discharges in a public health context has been clearly demonstrated. No evidence of build-up of plutonium in any material has been found and concentrations in sea water, sediment and biological materials have been related by a constant factor to the rate of discharge of the nuclide.

REFERENCES

[1] HOWELLS, H., "Discharge of low-activity radioactive effluent from the Windscale works into the Irish Sea", Disposal of Radioactive Wastes into Seas, Oceans and Surface Waters (Proc. Symp. Vienna, 1966), IAEA, Vienna (1966) 769.

[2] MITCHELL, N. T., Radioactivity in Surface and Coastal Waters of the British Isles (1971), Tech. Rep. Fish. radiobiol. Lab., Lowestoft (1973) 9.

[3] STANNARD, J. N., "Plutonium in the environment", Ch. 15, Handbook of Experimental Pharmacology (HEFFTER, K. W. A., Ed.), Springer-Verlag, Berlin (1973) 36.

[4] NOSHKIN, V. E., Ecological aspects of plutonium dissemination in aquatic environment, Hlth Phys. 22 6 (1972) 537.

[5] PRESTON, A., JEFFERIES, D. F., MITCHELL, N. T., "Experience gained from the controlled introduction of liquid radioactive waste to coastal waters", Nuclear Techniques in Environmental Pollution (Proc. Symp. Salzburg, 1970), IAEA, Vienna (1971) 629.

[6] HETHERINGTON, J. A., JEFFERIES, D. F., The distribution of some fission product radionuclides in sea and estuarine sediment, Netherlands J. Sea Res. 8 4 (1974) 319.

[7] JEFFERIES, D. F., PRESTON, A., STEELE, A. K., Distribution of caesium-137 in British coastal waters, Marine Pollution Bulletin 4 8 (1973) 118.

[8] BOWEN, V. T., WONG, K. M., NOSHKIN, V. E., Plutonium-239 in and over the Atlantic Ocean, J. Mar. Res. 29 1 (1971) 1.

[9] PILLAI, K. C., SMITH, R. C., FOLSOM, T. R., Plutonium in the marine environment, Nature, Lond. 203 (1964) 568.

[10] PERKINS, E. J., BAILEY, M., WILLIAMS, B. R. H., The biology of the Solway Firth in relation to the movement and accumulation of radioactive materials, Pt IX, UKAEA PG Rep. 605 (CC) (1964).

[11] RAMSTER, J. W., HILL, H. W., Current systems in the northern Irish Sea, Nature, Lond. 224 (1969) 59.

[12] LOWTON, R. J., MARTIN, J. H., TALBOT, J. W., "Dilution, dispersion and sedimentation in some British estuaries", Disposal of Radioactive Wastes into Seas, Oceans and Surface Waters (Proc. Symp. Vienna, 1966), IAEA, Vienna (1966) 189.

[13] TEMPLETON, W., PRESTON, A., "Transport and distribution of radioactive effluents in coastal and estuarine waters of the United Kingdom", Disposal of Radioactive Wastes into Seas, Oceans and Surface Waters (Proc. Symp. Vienna, 1966), IAEA, Vienna (1966) 267.

[14] BELDERSON, R. H., Holocene sedimentation in the western half of the Irish Sea, Marine Geol. 2 (1964) 147.

[15] BELDERSON, R. H., STRIDE, A. H., Tidal currents and sand wave profiles in the north-eastern Irish Sea, Nature, Lond. 222 (1969) 74.

[16] PERKINS, E. J., WILLIAMS, B. R. H., The biology of the Solway Firth in relation to the movement and accumulation of radioactive materials, Pt II, UKAEA PG Rep. 587 (CC) (1966).

[17] WOODHEAD, D. S., "Levels of radioactivity in the marine environment and the dose commitment to marine organisms", Radioactive Contamination of the Marine Environment (Proc. Symp. Seattle, 1972), IAEA, Vienna (1973) 499.

[18] PENTREATH, R. J., "The roles of food and water in the accumulation of radionuclides by marine teleost and elasmobranch fish", Radioactive Contamination of the Marine Environment (Proc. Symp. Seattle, 1972), IAEA, Vienna (1973) 421.

[19] PRESTON, A., JEFFERIES, D. F., "Aquatic aspects in chronic and acute contamination situations", Environmental Contamination by Radioactive Materials (Proc. Symp. Vienna, 1969), IAEA FAO/WHO, Vienna (1969) 183.

DISCUSSION

S. FOWLER: I was interested to note that your computed concentration factors for mussels are about an order of magnitude higher than those measured in environmental samples of mussels collected from other parts of the world. Your ^{239}Pu water concentrations were based on the amounts of ^{239}Pu in filtered sea water, however, and not on the amounts really available for uptake. Had you used total ^{239}Pu concentrations in water, your concentration factors would have been lower. Mr. Pillai has also measured ^{239}Pu in mussels, and perhaps he would like to comment.

J.A. HETHERINGTON: Yes, our values for the concentration factor of ^{239}Pu in the soft part of mussels are, in fact, higher than those reported for samples collected from other areas by a factor of between 5 and 10. The concentration factors I have quoted are referred to ^{239}Pu concentrations in filtered sea water and you are correct in saying that the factors would be lower if the values for unfiltered water were used. You will see from the paper, however, that the fraction of the sea-water plutonium which was retained on a 0.22 μm filter was of the order of 30% of the total in unfiltered water. Thus, even if this fraction were taken into account, the concentration factor would be reduced by less than a factor of 2 and would still be higher than those reported by other workers.

It is not certain to what extent the availability of plutonium to this or any other species varies with the size of the particles with which the nuclide is associated in sea water. Nevertheless, this information is important and we hope that an answer may be obtained in the course of the work which is being done to learn more about the chemical and physical nature of plutonium in sea water.

K.C. PILLAI: The Pu concentration factors for mussels referred to by Mr. Fowler apply to fall-out. We have observed higher Pu concentrations in certain benthic organisms collected from discharge locations, for example in Arca from Bombay harbour. The levels are 10-150 times higher than fall-out levels observed in similar organisms in other areas.

J.A. HETHERINGTON: Thank you for making that valuable point. I stressed the fact that the plutonium we are measuring was derived from nuclear fuel reprocessing and that there is no a priori reason why its behaviour should be identical to that of plutonium originating from nuclear weapons testing in the atmosphere.

It may be worth while adding that we have found some evidence that the concentration factor for ^{137}Cs in demersal fish species is lower by a factor of about 2 for caesium derived from the nuclear fuel cycle than for weapon fall-out caesium. If the behaviour of a simple element such as caesium depends on its source, then it is almost inevitable that the behaviour of more complicated polyvalent metals such as plutonium will also depend on their origin.

R. BITTEL: Has any mineralogical analysis been performed on suspended matter and bottom sediments in the Irish Sea with a view to predicting the possible consequences of contamination of the oceans by plutonium?

J.A. HETHERINGTON: We have examined the relationship between grain size and uptake of plutonium by sediments, and the provisional results show that this nuclide follows a very similar pattern to the gross alpha contamination shown in Fig. 4. During research cruises in the Irish Sea this summer we shall be extending this work to the suspended-sediment phase. In view of the importance of sediments in relation to the behaviour of plutonium in sea water we regard this work as vital and we are particularly interested in the work being done in the vicinity of Cap de la Hague, where we understand the sediment régime is very different from that in the north-east Irish Sea.

Y. NISHIWAKI: What is the chemical form or valence state of the plutonium released at the discharge point? Are any chelating agents discharged at the same time?

J.A. HETHERINGTON: I have no details about the chemical form of plutonium in these effluents. However, we do know that the streams which reach the final low-activity waste tanks are strongly acid (pH = 2-3) and that plutonium is present up to this stage in soluble form. Before discharge to the sea the effluent is neutralized by addition of ammonium hydroxide, and we may expect that this adjustment of the pH will predispose the plutonium to assume less soluble forms, i.e. forms which would possibly precipitate were it not for the turbulence in the neutralizing tanks and the final discharge tanks caused by the pumping operations. I believe therefore that the plutonium reaches the sea either as a hydrous complex of a colloidal nature or in a form which will show a strong affinity for solid surfaces, particularly those provided by the suspended sediment phase of the sea water.

No chelating agents, as such, are added to the effluents but, as may be expected with a chemical plant constructed largely from steel-based components, there are substantial amounts of iron present in the liquors and this may well cause some chelation of plutonium.

IAEA-SM-198/38

PLUTONIUM IN THE AQUATIC ENVIRONMENT AROUND THE ROCKY FLATS FACILITY

M.A. THOMPSON
The Dow Chemical Company,*
Rocky Flats Division,
Golden, Colorado,
United States of America

Abstract

PLUTONIUM IN THE AQUATIC ENVIRONMENT AROUND THE ROCKY FLATS FACILITY.

The Rocky Flats Plant of the United States Energy Research and Development Administration has been fabricating and chemically recovering plutonium for over 20 years. During that time, small amounts of plutonium have been released with liquid process and sanitary waste discharges. The liquid waste flows through a series of holding ponds from which it is discharged into a creek that is part of a municipal drinking water supply. The water flows for about 1.5 km between the last holding pond and the municipal drinking water reservoir. In addition, liquid wastes containing high levels of chemical contaminants and plutonium concentrations less than allowable drinking water standards have been discharged to large evaporation ponds. The fate of the plutonium in both the surface and subsurface aquatic environment has been extensively monitored and studied. It has been found that plutonium does not move very far or very rapidly through subsurface water. The majority of the plutonium released through surface water has been contained in the sediments of the plant holding ponds. Small amounts of plutonium have also been found in the sediments of the draining creek and in the sediments of the receiving reservoir. Higher than normal amounts of plutonium were released from the waste treatment plants during times when suspended solids were high. Various biological species have been examined and plutonium concentration factors determined. Considerably less than 1% of the 210 mCi of plutonium released has been detected in biological systems including man. After more than 20 years of large scale operations, no health or environmental hazard has been identified due to the release of small amounts of plutonium.

Introduction

The Rocky Flats Plant is a prime production facility of the United States Energy Research and Development Administration (formerly the Atomic Energy Commission). The plant began operation in 1953 with a primary mission of fabricating and recovering nuclear materials including plutonium-239. Typically 2400 kg of plutonium are recovered and recycled each year. During the 22 years of plant operation, small amounts of plutonium have been released with process and sanitary waste discharges. All waste water is treated in a sanitary or process treatment facility. After treatment, all water discharged from the plant flows through a series of holding ponds, where settling of solids can occur prior to final release of the water to a stream which flows into a municipal drinking water reservoir. Other aqueous process wastes containing high levels of chemical contaminants and plutonium concentrations less than allowable drinking water standards have been discharged into large evaporation ponds.

Since the start of plant operations, all liquid discharges from the plant have been extensively monitored. It has been known that small amounts of plutonium have been released and have accumulated in the discharge paths

* A prime contractor for the United States Energy Research and Development Administration CONTRACT AT(29-1)-1106.

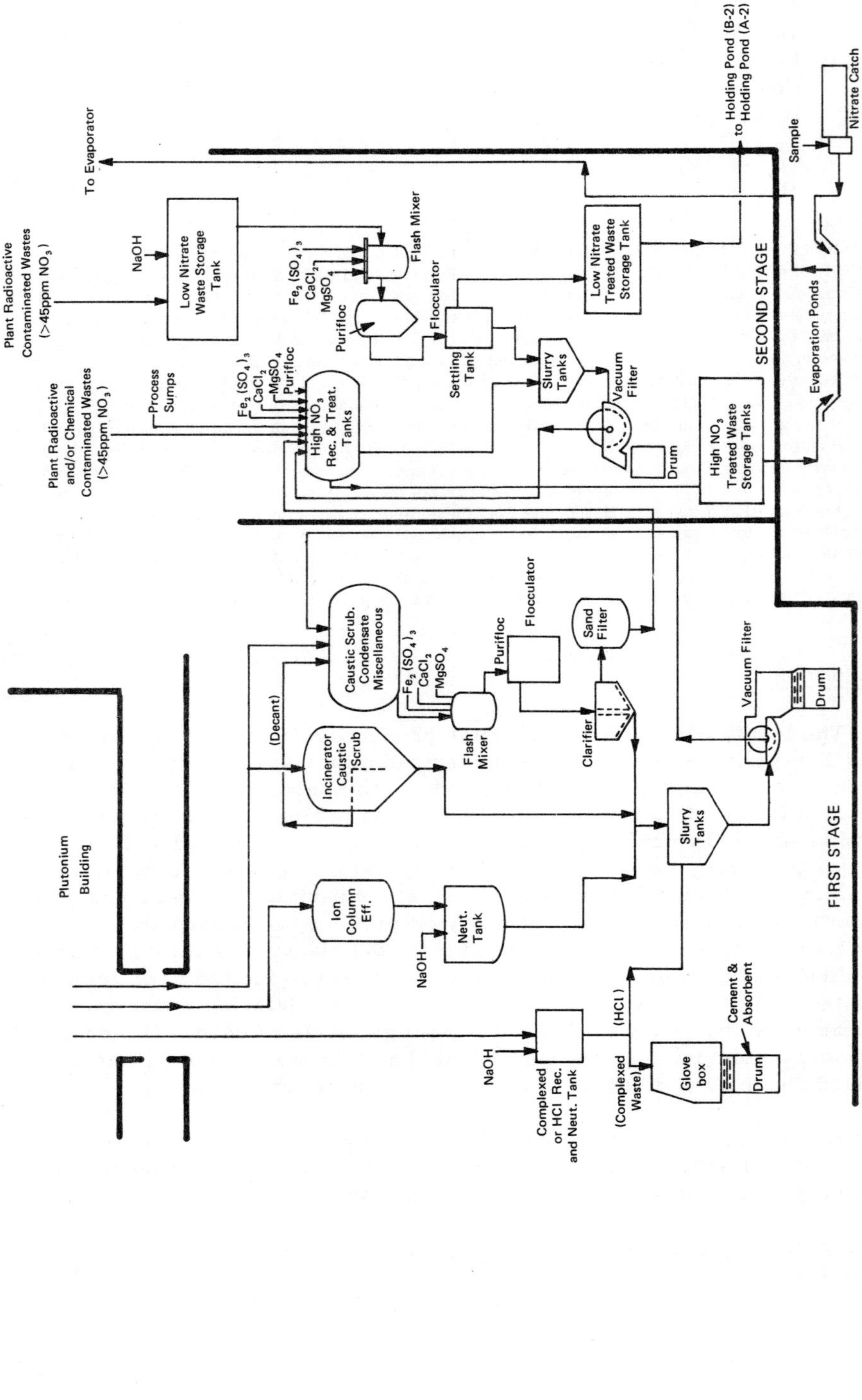

FIG. 1. Process waste treatment facility – Rocky Flats Plant.

of the aquatic environment around the plant. During the past several years, extensive studies have been undertaken to measure the amounts of plutonium in the various compartments of the aquatic system and to study the movement and behavior of the plutonium in the environment. The studies have been carried out by plant personnel as well as by independent organizations such as Colorado State University [1, 2] and the United States Environmental Protection Agency [3]. Following is a summary of the plant waste treatment methods, their efficiency for removing plutonium and a summary of the current location of all the plutonium released from the plant to the aquatic environment.

Waste Water Treatment Facilities

The Rocky Flats Plant has two separate waste water treatment facilities. One treats process liquid wastes containing plutonium, uranium and chemicals in concentrations above allowable drinking water standards. The second system handles sanitary waste liquids. During approximately a year and a half of plant operation, all treated process wastes were discharged through the sanitary waste treatment facility prior to discharge off the plant site.

The process waste treatment consists of a two-stage ferric hydroxide carrier precipitation process as shown schematically in Figure 1. The process removes over 99% of the plutonium and americium from the waste stream. Solids, containing most of the plutonium and uranium, are dried and packaged in special containers for storage at a USERDA waste storage site. The effluent liquid, containing residual amounts of plutonium and americium are currently stored in large ponds on the plant site. Prior to December 1973, the treated process liquids were released off site through Walnut Creek,which flows into a municipal drinking water supply. Between July 1972 and December 1973, the liquids were routed through the sanitary waste treatment facility for additional treatment prior to release off site via Walnut Creek. Typically the process waste treatment facility processes 3 280 000 liters of liquid per month with the discharged water containing significantly less than the currently accepted standard for plutonium in drinking water of 1 667 pCi/l.

The sanitary waste water treatment system consists of a tertiary process as shown in Figure 2. Prior to December 1974, the process consisted of only secondary treatment. Currently, sanitary waste is being given tertiary treatment. Approximately 20×10^6 liters of sanitary waste water per month are treated and released. During 1974, the average plutonium concentration in the water released from the treatment plant was 1.78 pCi/l. Sludge from the treatment plant is currently packaged as non-retrievable waste and shipped to a U.S. Energy Research and Development Administration waste storage site.

Figure 3 shows a flow chart for the entire aqueous waste system at Rocky Flats. During the 22-year history of the plant, an estimated 75 mCi of plutonium-239 have been released with water discharges. An additional 135 mCi are estimated to be in the aqueous environment within the plant boundary.

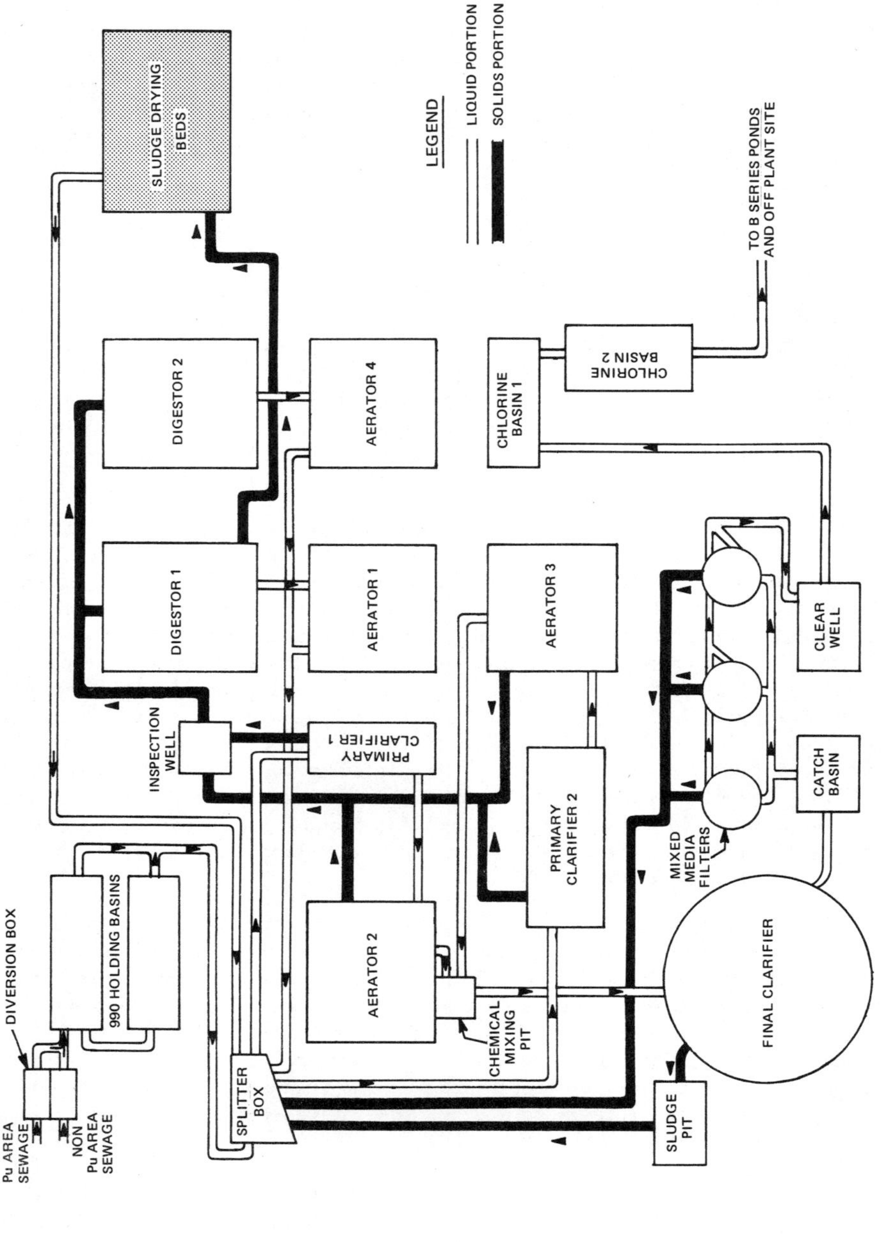

FIG. 2. Sewage treatment facility – Rocky Flats Plant.

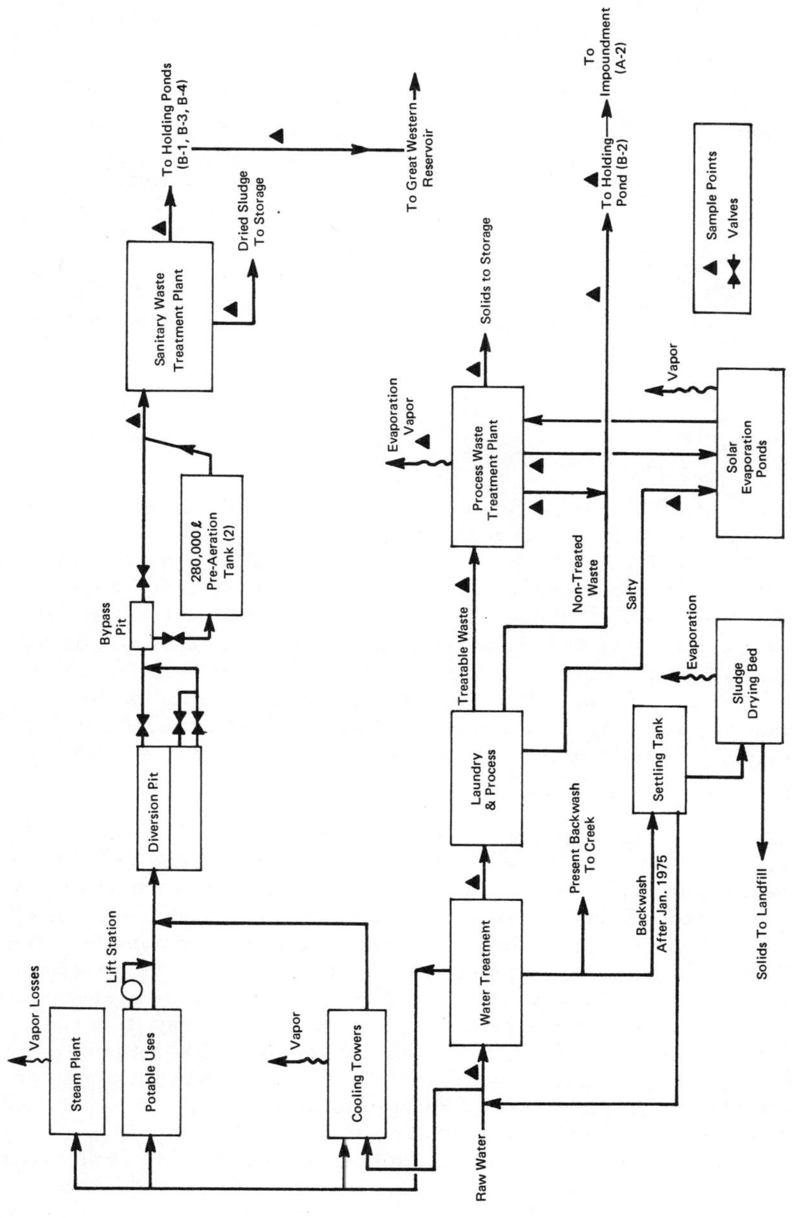

FIG. 3. Aqueous waste flow — Rocky Flats Plant.

TABLE I. TYPICAL DECONTAMINATION FACTORS

	Plutonium Concentration		Decontamination
	Influent	Effluent	Factor
Process Waste Treatment (g/l)			
1st Stage	10^{-3}	10^{-4} - 10^{-5}	10 - 100
2nd Stage	10^{-4}	10^{-8}	10^{4}
Sanitary Waste Treatment (pCi/l)			
Tertiary Treatment	2.4	0.92	2.6

Process Decontamination Factors

Between July 1972 and December 1973, process waste water containing plutonium was decontaminated by first the process waste treatment facility and then by the sanitary waste water treatment plant. Currently, process waste is decontaminated by only the process waste treatment facility. Table I shows typical influent and effluent values for the two processes and the decontamination achieved. The Decontamination Factor is defined as the plutonium concentration in the influent divided by the plutonium concentration in the effluent. The concentration of the second stage influent can be greater than the first stage effluent because certain wastes are added directly to the second stage without prior first stage treatment.

Sanitary waste water passes from the primary clarifier to an aerator to a final clarifier and sand filters. The residence time of the water in the sanitary treatment facility is about 24 hours. Sludge from the primary clarifier and aerator go to a digestor with the supernate from the digestor recycling to the aerator and primary clarifier. As the plutonium concentration builds up in the digestor sludges, it also builds up in the supernate that is being recycled to the primary clarifier. The plutonium concentration in the supernate may actually exceed the plutonium concentration in the plant influent. Therefore, the decontamination factor will become zero as the sludge and supernate become increasingly contaminated with plutonium. For example, after the plutonium concentration increased in the sludge to 50 000 d/m/g (dry weight) and the influent water contained 60 pCi/l plutonium or less, the effluent water had higher concentrations of plutonium than the influent. Plutonium concentrations increased by factors ranging from 2 to 30. This was caused by the recycle of the digestor supernate and also because the digestors became overloaded with sludge and the sludge and associated plutonium were washed out of the system. It was found that over 95% of the plutonium was associated with the solids, suggesting that sanitary sewage sludge is a good medium for decontamination of waste water if the sludge does not become overloaded with plutonium. Current operation of the tertiary sanitary treatment facility shows incoming water with typical plutonium concentration of 2.4 pCi/l and the effluent with a concentration of 0.92 pCi/l.

Holding Ponds

Water to be released from the plant after treatment by the process waste facility or the sanitary waste facility, is passed through a series of on-site holding ponds. Until December 1973, both treated process and sanitary waste water passed through four ponds with a residence time of about ten days. Currently the system transfers all process water to one pond for initial settling and then to a second pond for storage and evaporation. All sanitary water is passed through three holding ponds prior to release off site. The current water handling system is shown schematically in Figure 4.

During the years in which plutonium-containing process water was released off site, after passing through the four holding ponds, additional decontamination of the process water was achieved due to the settling out of the plutonium in the ponds. Table II shows typical decontamination factors achieved.

As the plutonium slowly built up in the bottom of the ponds, to a maximum average concentration in the first pond of 1302 pCi/g, the decontamination factor did not significantly change, indicating that the pond sludge has not become saturated with plutonium and that the plutonium is tightly bound to the sludge particles. Laboratory studies [1] showed no detectable transfer of plutonium from pond sediments containing 1000 d/m/g plutonium to water after one week. After approximately two months, the water showed an activity of 0.01 d/m/ml. Laboratory experiments [1], which measured the decrease in plutonium activity from water in contact with sediment showed that the transfer kinetics of plutonium from water to sediment can be expressed by a function consisting of two exponential terms as follows:

$$C(t) = C_o (Ae^{-xt} + Be^{-yt})$$

where C_o = initial Pu concentration in water

$C(t)$ = Pu concentration in water at any time

t = time in days

x = 0.92

y = 0.058

The study and equation predict that 75% of the plutonium in the water delivered to the first holding pond should be transferred to the sediment in an average time of 3.1 days. This is in reasonably good agreement with actual pond data.

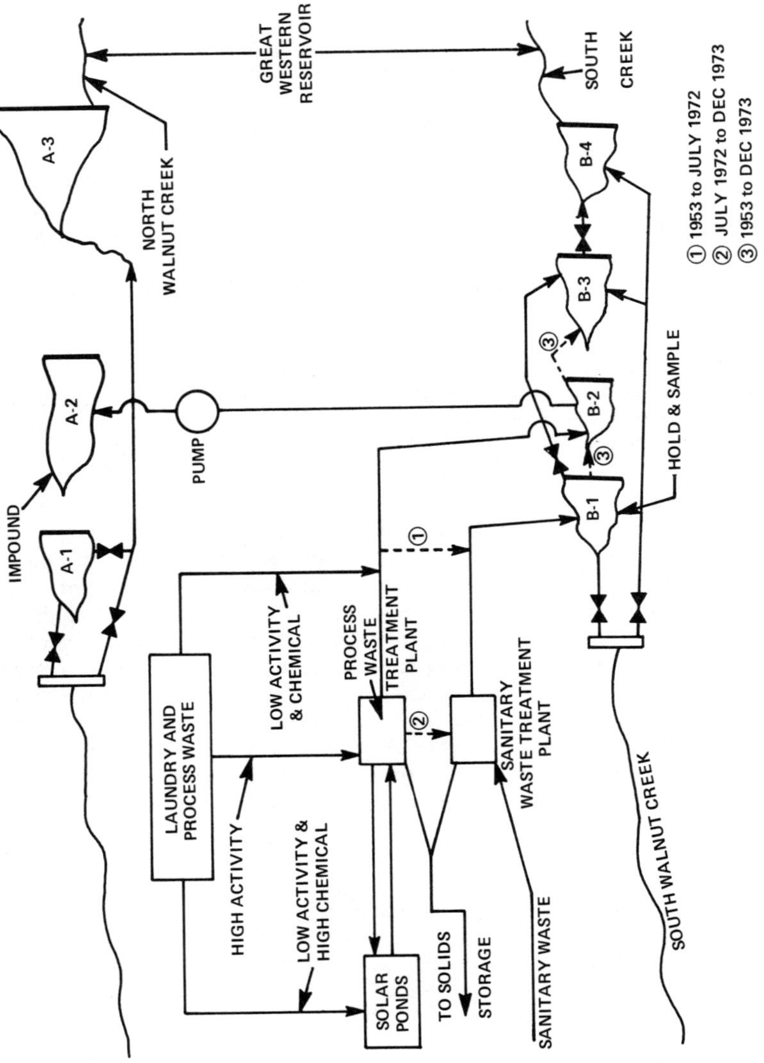

FIG. 4. Liquid waste flow – Rocky Flats Plant.

TABLE II. HOLDING POND DECONTAMINATION FACTORS

Average Plutonium Concentration (pCi/l)			
1st Pond Inlet	4th Pond Outlet		Decontamination Factor
157	15.3		10.2
85.4	9.2		9.2
49.2	3.3		14.9
22.1	1.5		14.7
8.1	1.1		7.3
2.9	0.87		3.3
		Average	9.9

TABLE III. PLUTONIUM INVENTORY IN HOLDING PONDS

Pond	Pu Inventory (mCi)	%	Average Plutonium Concentration (pCi/g)
1	84.5	62.4	1302
2	27.0	19.9	204
3	19.4	14.3	203
4	4.6	3.4	55

Coring and analysis of the sludge from the pond bottoms has been carried out over several years. The measured inventory is shown in Table III. In general, highest plutonium concentrations were found in the sediments near the pond inlet. Above-average concentrations were also inconsistently found at other pond locations such as near the pond outlet or in the original stream channel prior to the pond construction. There was no evidence to suggest that the plutonium in the sediment was being consistently moved toward the pond outlet. Samples as deep as 80 cm were taken and all showed above-background amounts of plutonium. The plutonium showed a high affinity for the clay fraction of the sediments.

Creeks and Reservoirs

After discharge from the last holding pond, the water flows relatively slowly down a small creek (Walnut Creek) for about 1.5 km and into a reservoir which is a drinking water supply for a city. The elevation drop from the pond to the reservoir is about 52 m. Additional plutonium is removed from the water during its flow down the creek as shown in Table IV. Approximate flow rate is 0.1 m/sec. Sediment samples to a depth of 15 cm

TABLE IV. AVERAGE DECONTAMINATION DURING FLOW IN CREEK

Year	Average Plutonium Concentration (pCi/l) Last Pond	1.5 km downstream	Decontamination Factor
1974	1.78	0.84	2.1
1973	7.37	3.11	2.4

have been taken from the creek bottom and analyzed for plutonium. Plutonium concentrations varied from 0.3 to 4.5 d/m/g with the higher concentrations being found in the deeper sediments and farther down stream (closer to the reservoir). Assuming an average plutonium concentration in the sediment of 2 d/m/g, approximately 0.2 mCi of plutonium is contained in the sediment to a depth of 15 cm.

Walnut Creek empties into Great Western Reservoir, which is the drinking water supply for a city. At full capacity the reservoir is a maximum of 19 m deep and covers 668 000 m^2 with a volume of 3 430 000 m^3. Except during periods of heavy rain and runoff, the reservoir is not filled to capacity. More typically, the reservoir is maintained at about 13 m depth, covering an area of 294 500 m^2 with a volume of 1 162 000 m^3. Water samples are taken routinely from the reservoir and analyzed for plutonium. Periodically, sediment and core samples are taken from the reservoir and also analyzed.

Sediment and core samples have shown plutonium concentrations as high as 4 pCi/g (dry weight) which is about 40 times the plutonium concentration attributable to world-wide fallout of 0.1 pCi/g (dry weight). The average concentration to a depth of 5 cm was 1.4 pCi/g (dry weight). In general, samples at a depth greater than 5 cm showed no plutonium concentrations greater than world-wide fallout. In a few cases, the plutonium concentration at the 2.5 to 5 cm depth was greater than in the 0 to 1 cm depth. The highest concentrations of plutonium are found near the inlet of the reservoir and along the dam where the greatest sedimentation rate has also been found. There is an estimated total inventory of 75 mCi of plutonium-239 and -240 in the reservoir with between 5 and 10% of the total due to fallout. Water from the reservoir was passed through a 0.3-μm filter followed by passage through anion, cation and aluminum oxide sorption bed in order to determine the particulate, anionic, cationic and non-ionic fractions of the plutonium in the water. The results showed approximately 5% of the plutonium to be particulate, 35% cationic, 54% anionic and 6% non-ionic. The water in the reservoir is routinely analyzed for plutonium as is the domestic water in the city served by the reservoir. Table V shows the annual results of these analyses as compared to background plutonium concentrations in other nearby lakes and reservoirs due to fallout.

TABLE V. PLUTONIUM CONCENTRATIONS IN RESERVOIR AND DOMESTIC WATER

Year	Average Plutonium Concentration (pCi/l) Great Western Reservoir	Domestic Water	Background Water
1974	<0.03	< 0.02	< 0.02
1973	<0.08	< 0.11	< 0.31
1972	<0.67	< 0.46	< 0.15

All of the reservoir, domestic water and background results are essentially the same within the limits of analytical and sampling variations. Background concentrations of plutonium in water varied from < 0.02 to 1.25 pCi/l. The results indicate that the sediment in the reservoir is effectively holding the plutonium and preventing its movement into the municipal drinking water. The reservoir water passes through a filter plant prior to domestic consumption and a small amount of plutonium is removed with the suspended silt. Based on an assumed average concentration of plutonium in the silt, which is backwashed from the filters, of 2.3 d/m/g and 1000 kg of silt removed per month, approximately 0.25 mCi of plutonium would have been removed after 20 years. The plutonium concentrations in both the reservoir water and domestic water is less than 0.01% of the currently accepted limit for plutonium in drinking water.

Ground Water

A series of 33 wells have been drilled at various locations around the plant site to depths varying between 30 and 130 feet. The well locations were selected to intercept ground water in areas where potential contamination might be expected. Well locations are near holding ponds, evaporation ponds and the creek bed. Some wells have been monitored for the entire 22-year history of the plant. Although the evaporation ponds have been known to leak, in no case has plutonium above background levels been detected in any of the wells, suggesting that soil is a good medium for removing plutonium from an aqueous media.

Biological Systems

Various biological systems in and along the discharge water have been studied and analyzed for effects of plutonium exposure and possible bioconcentration of plutonium. Table VI is a summary of the plutonium concentration factor found for various species examined. Concentration factor is defined as the plutonium concentration in the biosystem divided by the concentration in the water associated with the biosystem. It is not clearly defined if the plutonium is on the surface of the various species or taken into the structure; however, careful washing and preliminary studies indicate that appreciable amounts of the plutonium detected were within the structure of the system.

TABLE VI. CONCENTRATION FACTORS IN VARIOUS BIOLOGICAL SPECIES

Species	Concentration Factor
Bullhead minnow	.89 - 7.5
Seston[a]	10^4 - 10^5
Vascular plants	10^{-1} - 10^{-2}
Zooplankton	10^3
Crayfish	10^3
Algae	10^4

a) Seston is primary phytoplankton, detritus and small zooplankton

TABLE VII. PLUTONIUM LOCATION

Location	Amount (mCi)	%
Plant Holding Pond Sediments	135.5	64.2
Creek Bed Sediments	0.2	.09
Reservoir Water[a]	0.002	<0.001
Water Treatment Plant Silt	0.25	0.1
Reservoir Sediments	75	35.5
Ground Water	0	0
Biological Systems	< .001	<0.001
Man[a] (10 000 people - from drinking for 20 years)	0.03	0.01
TOTAL	210.98	

a) Based on an estimated 0.01 pCi/l concentration of plutonium due to Rocky Flats operations. For man, assumption is made that all plutonium is retained in the system.

Summary

An estimate has been made of the location of the plutonium released with waste water during the 22 years of operation of the Rocky Flats Plant. Table VII summarizes the estimated amounts and percentages in the various systems. It can be seen that only a small fraction of the released plutonium concentrates in biological systems or in man. This is true even considering the very conservative assumption that the plutonium concentration in water due to Rocky Flats operations is 0.01 pCi/l. There is no evidence to suggest that exposure to such low levels of plutonium present any health hazard to man or cause any damage to biological systems.

REFERENCES

[1] JOHNSON, J.E., SVALBERG, S., PAINE, D., The study of plutonium in the aquatic systems of the Rocky Flats environs, USAEC Contract No. 41493-F, June 1974.

[2] WHICKER, F.W., Radioecology of some natural organisms and systems in Colorado. Colorado State University Report COO-1156-70, May 1, 1974.

[3] LAMMERING, M.W., Plutonium levels in the sediment of area impoundments environs of the Rocky Flats plutonium plant — Colorado. US Environmental Protection Agency EPA-908/2-75-001, February 1975.

DISCUSSION

J.A. HETHERINGTON: I believe you said that higher concentrations of plutonium were detected in the deeper sediment profiles at some sampling stations. If this is so, is it because the concentrations of the nuclide in the water entering the pond decrease with time, or is it evidence of a redistribution of plutonium after its incorporation into sediments?

M.A. THOMPSON: In the stream flowing from the holding ponds higher plutonium concentrations were detected at lower sediment levels, and this might indicate that additional sediment is being washed into the stream. In the receiving reservoir there were a few cases where the plutonium concentrations were found to be higher at greater depths, but this was not general.

J.A. HETHERINGTON: Your reply shows that fresh sediment is being transported into the ponds, and this is significant in relation to your other observation that the plutonium sink constituted by these sediments does not appear to saturate. This may also be due to the fact that there is a consistent supply of uncontaminated sediment.

K.E. WHITE: How do you dispose of the plutonium-contaminated sewage sludge from the drying beds?

M.A. THOMPSON: The plutonium-contaminated sewage sludge was dried, put into waste storage containers, and is now being stored at a radioactive-waste storage site.

K.E. WHITE: You use a number of purification ponds before the effluent is released to the reservoir via a creek. May I sound a note of warning regarding the nominal retention times quoted (capacity divided by throughput). The mixing expected — and the decay of short-lived nuclides — may be hindered by river-like currents round the banks and consequent short-circuiting between inlet and outlet of each lagoon. Where the influent is warmer than the water in the pond, this can be particularly serious because of density currents across the surface. A similar process is also possible in the main reservoir, and with systems of this kind we have found the degree of short-circuiting to vary on a day-by-day basis, depending on the mixing action caused by the wind.

I should also like to stress that, although samples taken from the boreholes around your site did not indicate any plutonium and you conclude that soil is a good medium for removing such contamination, it does not follow that groundwater transport is not possible at other sites — particularly if fissure flow exists. This raises the general question of thoroughness of environmental monitoring around nuclear sites. According to one school of thought, detailed surveys in the surrounding environment may not be necessary in view of the possibility of controlling at source by merely monitoring the main liquid-effluent stream, or stack release in the case of gaseous effluents. I suggest caution with regard to such a philosophy. Most nuclear sites are complex assemblies of buildings, many of which — in addition to reactor buildings — contain considerable quantities of radioactive materials. Contamination can be removed from these sites by wind, rain or percolation from areas below ground. Would you agree that water authorities abstracting water from either boreholes or open waters in areas adjacent to nuclear sites would be wise to undertake a monitoring programme of their own completely independent of those conducted by the owners of the sites?

M.A. THOMPSON: Yes. An independent monitoring programme provides additional assurance to the public that discharges are within permissible limits and also provides a check on the monitoring programme of the site owner.

THE BEHAVIOUR OF PLUTONIUM AND OTHER LONG-LIVED RADIONUCLIDES IN LAKE MICHIGAN
I. Biological transport, seasonal cycling, and residence times in the water column*

M.A. WAHLGREN, J.S. MARSHALL
Radiological and Environmental Research Division,
Argonne National Laboratory,
Argonne, Illinois,
United States of America

Abstract

THE BEHAVIOUR OF PLUTONIUM AND OTHER LONG-LIVED RADIONUCLIDES IN LAKE MICHIGAN: I. BIOLOGICAL TRANSPORT, SEASONAL CYCLING, AND RESIDENCE TIMES IN THE WATER COLUMN.

Eight operating nuclear reactors are situated on the shores of Lake Michigan, but their releases of radioactivity have been much less than that entering the lake from stratospheric fallout. Measurements of $^{239,240}Pu$, ^{241}Am, and ^{137}Cs from the latter source have been made in order to study biological transport, seasonal cycling, and residence times of long-lived radionuclides in the lake. The apparent turnover times for the residual fallout $^{239,240}Pu$ and ^{137}Cs, which are present as nonfilterable, ionic forms, are about 3 to 4 years. Resuspension may be occurring at a low rate, probably through the feeding activities of benthic organisms. Transport by settling of phytodetritus and zooplankton fecal pellets is postulated to be the cause of the rapid decline of the concentration of $^{239,240}Pu$ in surface waters observed during summer thermal stratification of the lake, while the concentration of ^{137}Cs remained almost constant. Concentration factors for fallout $^{239,240}Pu$, ^{137}Cs, and ^{90}Sr at various trophic levels in the food chain in Lake Michigan have been measured. Analyses of biological samples taken at various distances from the Big Rock Point Nuclear Power Plant and of plant waste discharge show that any plutonium possibly released from the recycle plutonium test fuel is too low to be detectable in the presence of fallout plutonium. Measurements of $^{239,240}Pu$, ^{137}Cs, and ^{90}Sr on a comparison set of surface water and net plankton samples from all five Great Lakes indicate generally consistent behaviour patterns in these lakes.

1. INTRODUCTION

Approximately 40% of the population of the United States lives in states bordering the Laurentian Great Lakes. The economic advantage to the electrical generating industry of utilizing these lakes for once-through cooling has long been recognized, both in the United States and in Canada. The advent of large multiple-unit nuclear plants has led to the operation of eight such reactors on Lake Michigan, and a total of sixteen on the four lower Great Lakes. The rapid growth of the nuclear power industry has generated considerable public concern about possible environmental effects of radioactive discharges, whether routine or accidental, and this concern has been directed primarily toward plutonium. The emphasis in the Argonne Radioecology Program has, therefore, been concentrated on $^{239,240}Pu$, ^{241}Am, ^{137}Cs, and ^{90}Sr, which are each sufficiently long-lived to be potentially available to man for hundreds or thousands of years. The discharges of radioactivity from

* Work performed under the auspices of the USERDA.

nuclear plants to Lake Michigan up to the present have been insignificant relative to the inputs which resulted from atmospheric nuclear weapons testing. The residual fallout radionuclides, therefore, have been utilized to study the biogeochemical properties of these radioelements (Pu, Am, Cs, Sr) in Lake Michigan.

2. EXPERIMENTAL METHODS

2a. Sample Collection

Water samples ranging in volume from 50—1000 ℓ are collected by means of a submersible pump (to about 40 meters) or by use of 30 ℓ Niskin bottles. The samples are screened through 85 and 35 μm Nitex nets, followed by filtration through 3 and 0.45 μm Millipore filters. A limited number of these samples are then further fractionated into a submicron fraction with an ultrafilter having a molecular weight cutoff ~30 000 followed by passage through anion, cation, and charcoal beds. Tripton samples have been collected in a series of sediment traps of 15 cm diameter which readily collect sufficient material for radiochemical analysis on a monthly retrieval basis. Benthic floc is collected using the gravity corer. Plankton tows are taken using #25 and #6 nets which generally yield samples enriched in phytoplankton and zooplankton, respectively. Benthic invertebrates are collected with a benthic sled towed at minimum ship speed (~6 km/h). The fish samples have been provided by various government agencies and commercial fishermen.

2b. Sample Analysis

A relatively detailed description of radiochemical methods for Pu, Cs, and Sr has been given [1] and for Am [2]. Through the use of large samples and long counting times, counting errors ≦ 10% are attained for Pu, Sr, and Cs. Due to the extremely low concentrations of ^{241}Am in samples other than sediments, generally only upper limit values are obtained for this radioelement.

3. EXPERIMENTAL RESULTS

3a. Turnover Times for ^{239,240}Pu, ^{241}Am, ^{137}Cs, and ^{90}Sr

A synoptic lake-wide sampling cruise was carried out in early June 1972, at a time when the lake has been reported to be well-mixed, following the late winter wind-driven and spring convection mixing periods [3]. The results of the study demonstrated a high degree of homogeneity at this season [4], and permitted estimation of the 1972 inventory of ^{239,240}Pu, ^{137}Cs, and ^{90}Sr in the water of Lake Michigan. The fallout inputs to the lake can be approximated by three methods, i.e., ^{90}Sr monitoring at nearby land stations [5], our measurements of ^{90}Sr in Lake Michigan water and sediments, and world-wide studies of ^{239,240}Pu fallout in the 40—50 °N latitude band [6]. By utilization of ^{239,240}Pu/^{90}Sr ratios reported from many years of stratospheric monitoring [7], we find agreement within 20% for these estimates. The ^{90}Sr data indicate a ^{239,240}Pu input to the lake of 1.8 mCi·km^{-2} versus the 2.2 mCi·km^{-2} reported from soil measurements.

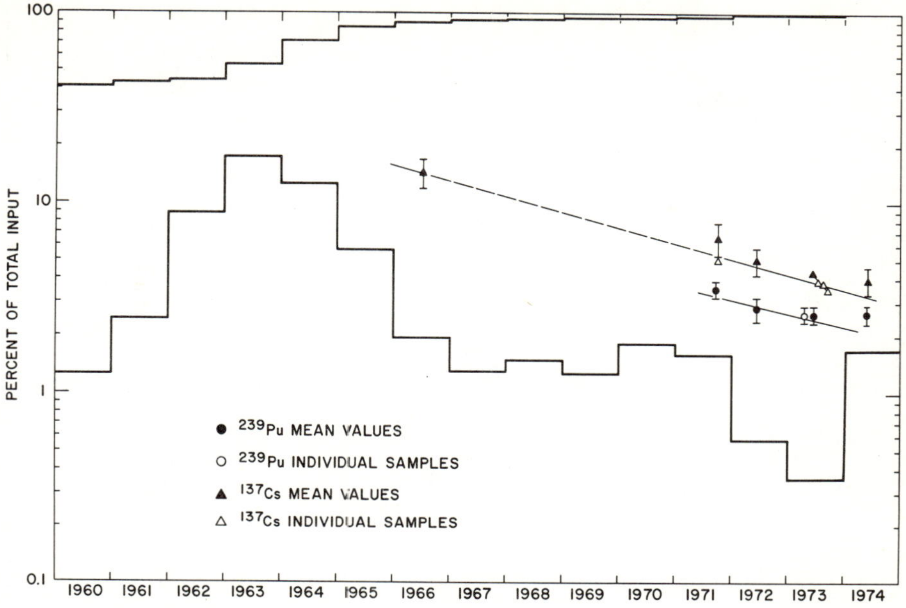

FIG. 1. Relative annual fallout radionuclide inputs (lower histogram) and cumulative total input (upper histogram) for Lake Michigan, 1960-1974, and fractions of total ^{239}Pu and ^{137}Cs inputs remaining in the water during 1966-1974 (data points and straight lines).

In Figure 1, the relative annual inputs of fallout radioactivity (based on ^{90}Sr monitoring on land) to Lake Michigan are shown. The integral of this curve (the upper histogram) approximates the build-up of the ^{90}Sr concentration in Lakes Michigan (and Superior), which have long water residence times (100 y and 190 y, respectively). The experimental measurements of Lake Michigan water concentrations of ^{239,240}Pu and of ^{137}Cs during 1971—1974 are shown in the figure, and demonstrate an apparent turnover [8] time of 3—4 yrs. We note that in order to have transported ~95% of the ^{239,240}Pu and ^{137}Cs inputs to the lake to the sediments, turnover times of ~1 y are required. We note also that no significant change in water concentrations occurred during or following the low fallout inputs in 1972 and 1973, nor was there a marked increase following a greater input in 1974. Measurements of ^{239,240}Pu and ^{137}Cs in unfiltered water samples from a major Lake Michigan tributary indicate that leaching from the watershed contributes an insignificant input of either radioelement to Lake Michigan. These data suggest that a steady-state condition exists, where the annual sedimentation of ^{239,240}Pu and ^{137}Cs is equal to the rate of return from the dissolution of the benthic floc layer. An alternative explanation is that both plutonium and cesium exist in at least two physicochemical forms with widely differing turnover times.

Analysis of sediment (and soil) samples yield ^{239,240}Pu/^{241}Am ratios ~4, while analysis of a limited number of Lake Michigan water samples yields ratios > 15—> 40. Better measurements of the ratio are being obtained by

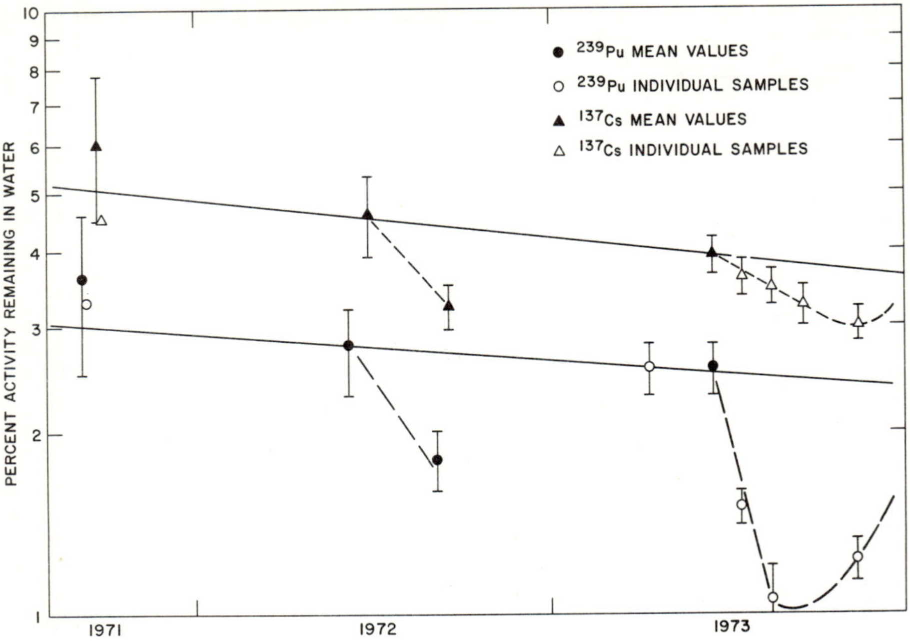

FIG. 2. Percentages of input ^{239}Pu and ^{137}Cs activities remaining in Lake Michigan water, 1971-1973. Solid lines show apparent turnover times of 3 to 4 years. Dashed lines show seasonal cycling and rapid temporary depletions of activities in surface waters during summer thermal stratification.

compositing samples and making longer counts. The turnover time of ^{241}Am in Lake Michigan cannot be estimated from available data, since the primary source of ^{241}Am is radiogenic, i.e., the decay of 13 y ^{241}Pu, and the nuclear transformation occurs primarily in the surficial sediments. If the postulate of cycling of plutonium and cesium activities from the benthic floc layer is correct, the americium results indicate that either the binding of americium to sediments is much greater than plutonium or cesium, or that the sedimentation process in the water column is much more efficient. Measurements of ^{90}Sr in Lake Michigan sediments shows that the sediments contain no more than 1—2% of the lake inventory of this radioelement and indicate that the removal of ^{90}Sr from Great Lakes waters will be governed by the time required for radioactive decay and displacement from the lake by flushing.

3b. Annual Cycling Between Surface and Deeper Waters

During the field season in 1973, it was observed that the concentration of $^{239,240}Pu$ (and to a much lesser extent, ^{137}Cs) in surface waters decreased markedly during July and August following the formation of a stable thermocline, and by November was returning toward the early season concentration, as the fall mixing of surface and deeper waters occurred (see Figure 2). The decrease in plutonium concentration parallelled, in time and in degree, that reported for soluble Si in offshore Lake Michigan waters [9]. This seasonal cycling of plutonium was observed again in 1974. The removal mechanism is postulated to involve sorption by phytoplankton (primarily diatoms) and subsequent settling of phytodetritus and zooplankton fecal pellets from the epilimnion.

3c. Distribution of $^{239,240}Pu$ and ^{137}Cs Between Particulate, Colloidal and Soluble Fractions

Filtration experiments on 50 ℓ of lake water during 1972 and 1973 demonstrated that the concentrations of $^{239,240}Pu$ and of ^{137}Cs in filtered (0.45 μm) and unfiltered water were generally equivalent within the analytical error (± 10%). During 1974 and 1975, filtration of large volume samples (300—1000 ℓ) has allowed direct measurement of the radioactivity retained with the 0.45—3 μm and 3—35 μm fractions. The data from 1974 are summarized in Table I. The small particle fraction which would be expected to retain < 2 μm clay particles is found to be insignificant in terms of sorbed $^{239,240}Pu$ or ^{137}Cs. The 3—35 μm fraction, generally diatom-dominated phytoplankton, contains only 5—10% of either radioelement at various depths from surface to 50 meters, except during the summer decline of plutonium concentration in surface waters when the 3—35 μm fraction may contain up to 30% of the $^{239,240}Pu$ and 10% of the ^{137}Cs. A number of the filtered water samples were then passed through anion, cation, and charcoal beds. Only the anion bed retains a readily measurable amount of $^{239,240}Pu$, having ≧ 90% of the nonfilterable activity in the water. During the fall and winter of 1974, a few lake water and precipitation (snow and rain) samples were processed through the extensive fractionation using filters, ultrafilters and resin beds. The cumulative analytical errors resulting from analysis of small amounts of activity in the fractions result in less than satisfactory material balances, but these preliminary results [9] show definite trends which are summarized in Table II. The particle size associations of activity observed in precipitation samples are consistent with previously reported stratospheric particle size distributions [10] and demonstrate that only a small fraction of the fallout input to the lake is associated with particles large enough to be sedimented rapidly in the water column by gravitational force. The fallout inputs during 1974 are estimated to have added an amount of $^{239,240}Pu$ (and ^{137}Cs) equivalent to approximately 30% of the previous amount contained in the water; in the lake water samples taken during the fall season neither the 30 Å — 0.4 μm nor the cation fractions contain detectable activities as observed in the precipitation samples. This indicates a relatively rapid removal of the new $^{239,240}Pu$ and ^{137}Cs inputs from the water column by processes that involve physicochemical transformations and association with particles that settle rapidly. The association of $^{239,240}Pu$ with chemically defined phases of the benthic floc has been studied by Alberts et al. [11] using extraction techniques developed for soils [12], and adapted for study of the associations of trace elements with stream particulates [13]. These results demonstrate that the $^{239,240}Pu$ in floc is associated only with the reductant-soluble (metal oxides) fraction. These data suggest that iron and manganese oxides may act as carriers during the latter stages of the sedimentation process or even during its association with settling phytodetritus and fecal pellets.

TABLE I. Concentrations[a] of $^{239,240}Pu$ and ^{137}Cs in filterable particulates at an offshore station.

Date	Depth (meters)	0.45 μm Filter		3 μm Filter	
		$^{239,240}Pu$	^{137}Cs	$^{239,240}Pu$	^{137}Cs
18 June 74	3	< 0.01 (< 1.5)	0.27 (0.6)	0.036 (5.5)	2.2 (5.1)
16 July 74	3	0.096 (20)	0.64 (1.5)	0.034 (7.2)	2.0 (5.0)
	20	< 0.01 (< 2)	0.31 (0.7)	0.054 (9.0)	2.6 (5.5)
6 Aug 74	3	< 0.005 (< 1)	0.20 (0.5)	0.025 (5.7)	0.78 (1.9)
	33	< 0.007 (< 1)	0.38 (0.9)	0.101 (28)	8.83 (20)
17 Sept 74	3	0.018 (6)	0.34 (0.7)	0.130 (31)	0.68 (1.6)
	40	0.014 (3)	0.24 (0.6)	0.122 (25)	4.92 (11)
18 Oct 74	3	---	---	0.083 (25)	1.47 (3.4)
	45	---	---	0.098 (25)	2.27 (5.1)
19 Nov 74	3	---	---	0.080 (25)	5.89 (13)

(a) The concentration listed is expressed as pCi/m^3, the number following in parentheses expresses this concentration as percent of the radionuclide collected by filtration compared to total filterable and nonfilterable present.

TABLE II. Concentrations of $^{239,240}Pu$ (fCi/ℓ) in various fractions[a] of Lake Michigan water from an offshore station, 1974.

Date	Sampling Depth in meters	Activity Retained by 3 μm Filter	Activity Retained by 0.45 μm Filter	Activity Retained by Anion Resin	Activity Retained by Cation Resin
17 Sept	3	0.13	0.02	0.42	0.01
	33	0.12	0.01	0.48	0.01
18 Oct	3	0.08	--	0.14	0.07
	40	0.10	--	0.46	0.03
	60	--	--	0.43	0.04

(a) The efficiency of backwashing of the fiber tube bundle used as an ultrafilter is uncertain; however, the large fraction which passes through and is retained by anion resin indicates that little $^{239,240}Pu$ can be in the 30 Å — 0.45 μm size range. An improved ultrafiltration system which permits direct analysis of the retained material is being used in ongoing studies.

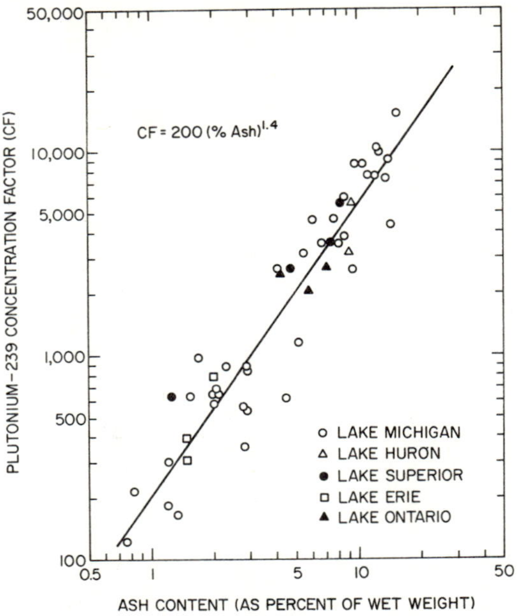

FIG. 3. Relationship between ^{239}Pu concentration factor and ash content in Great Lakes net plankton, including mixing plankton dominated by diatomaceous phytoplankton (high ash) and zooplankton (low ash).

3d. Relationship of Plutonium to Ash Content of Plankton Samples

The relationship between ^{239}Pu concentration factor (C.F.) and ash content in Great Lakes net plankton samples, including many diatomaceous phytoplankton-dominated (high ash) samples and zooplankton-dominated (low ash) samples from Lake Michigan and a few comparison samples from the other lakes, is shown in Figure 3. The use of concentration factors (i.e., ratio of ^{239,240}Pu concentration in biota on a wet weight basis to that in water) is convenient for comparison of lakes having varying plutonium concentrations. The use of the logarithmic plot permits observation of variability of samples about the curve. The data indicate that a generally consistent relationship exists between the ash content and the observed plutonium concentration factors regardless of species composition or water quality differences. Note that the plutonium concentration factor for samples with high ash content approaches 10 000. Comparisons of inshore and offshore plankton samples suggested that for diatoms, the concentration factor is dependent on the soluble silica/plutonium ratio. This potential relationship is under study during the ongoing field season. Studies are also underway to determine whether these and other particulate samples contain significant amounts of mineral material, and if so, to evaluate the role of these materials in radionuclide transport.

TABLE III. Comparison of Lake Michigan particulate fractions.

Fraction	Number of Samples	Ash/Dry Ratio Mean	Range	Pu Content ($fCi \cdot g^{-1}$ ash) Mean	Range
Phytoplankton	12	0.53	0.39-0.66	55	42- 73
3—35 μm Seston					
June-Sept[a]	5	0.36	0.30-0.42	200	188-212
Oct-Nov	3	0.75	0.65-0.85	126	112-136
Tripton	4	0.90[b]	0.88-0.92	156	140-197
Floc	12	0.90[c]		220	180-290

(a) This set excludes 2 August samples taken during a transient bloom of the gelatinous zooplankter, Holopedium.

(b) The ash/dry ratio of the settling material, before decomposition while in the submerged trap, is uncertain; a comparison of 3 day versus 30 day collections during the spring of 1975 shows values of 0.84 and 0.85, respectively. This indicates that either the decomposition is extremely rapid or is insignificant.

(c) These samples were not ashed prior to analysis; the ash/dry ratio is estimated from organic carbon measurements of the upper few centimeters of surficial sediment.

3e. Comparison Studies of Plankton, Seston, Tripton and Floc

During 1974, preliminary comparisons were made of the plutonium and ash contents of particulate fractions (defined by sampling method). These results are summarized in Table III. The plutonium concentrations of these fractions show a considerable similarity, again indicating a relationship between ash and plutonium contents. We note that the spring and fall seston samples appear to fall into two distinct groups, in terms of both ash and plutonium content, which are probably related to major periods of production and decomposition of biogenic material. Insufficient data are available to define the relationships of the several types of particulate material in the transport of plutonium. Additional studies are underway during the 1975 field season using improved sample collection and processing techniques to further develop these relationships. ^{137}Cs and a limited number of ^{241}Am analyses are also being carried out on these samples.

TABLE IV. Summary of radionuclide analyses of Lake Michigan biota.

Sample Type	Number of Samples[a]	pCi/kg wet weight			
		$^{239,240}Pu$	^{137}Cs	^{90}Sr	^{241}Am
Water	14	0.00076	0.061	0.810 (4)	≦ 0.00002 (5)
Cladophora	16	2.8	62	39 (6)	0.30 (2)
Mixed Plankton[b]	22	4.3	163	30 (12)	0,29 (2)
Zooplankton	9	0.26	34	49 (5)	
Mysis	7	0.57	230	--	0.045 (2)
Pontoporeia	2	1.23	172	--	0.35 (1)
Slimy Sculpin	7	0.18	216	85 (1)	
Bloater	8	0.028	270	32 (3)	
Alewife	7	0.019	165	23 (2)	
Smelt	6	0.015	225	54 (2)	
Whitefish	5	0.024	250	--	0.003 (2)
Coho Salmon	2	0.005	250	13 (1)	≦ 0.0001 (1)
Chinook Salmon	2	0.002	290	11 (1)	
Lake Trout	7	0.003	420	--	≦ 0.001 (2)

(a) The number of samples analyzed for ^{90}Sr and ^{241}Am are given in parentheses following the ^{90}Sr and ^{241}Am values.

(b) Diatomaceous phytoplankton-dominated net plankton samples containing relatively little zooplankton or detritus.

3f. Food Chain Behavior

The results from analysis of Lake Michigan biota are summarized in Table IV. These data show that $^{239,240}Pu$ and ^{137}Cs are highly concentrated by phytoplankton (average C.F.'s = 5700 and 3700, respectively), and that ^{90}Sr is concentrated to a much lesser extent (C.F. = 40). Comparison of the concentrations of radioelements observed at the highest trophic level, trout and salmon, to those in phytoplankton shows ratios of ~3 for ^{137}Cs, ~0.3 for ^{90}Sr, and $\sim 10^{-3}$ for $^{239,240}Pu$. The ^{90}Sr uptake into aquatic food chains has been shown to be inversely dependent on the calcium concentration in the water [14,15]; the Lake Michigan calcium concentration is relatively high (~30 ppm) among the Great Lakes [16]. The relationship between ^{137}Cs and potassium is more complex [17]. The increased ^{137}Cs concentration at higher trophic levels in Lake Michigan is similar to the increase in the ^{137}Cs/potassium ratio in the same levels, in agreement with the results observed in other ^{137}Cs food chain studies [18]. The behavior of plutonium in the food chain is not inconsistent with the indirect evidence cited in previous sections showing a possible relationship with soluble silica, which is not a significant nutrient element at higher levels in the food chain.

The concentration factors for these radionuclides in edible fish from Lake Michigan are 4—40 for $^{239,240}Pu$, 160—420 for ^{137}Cs, and 15—67 for ^{90}Sr. The concentration factor for ^{241}Am cannot be determined from these

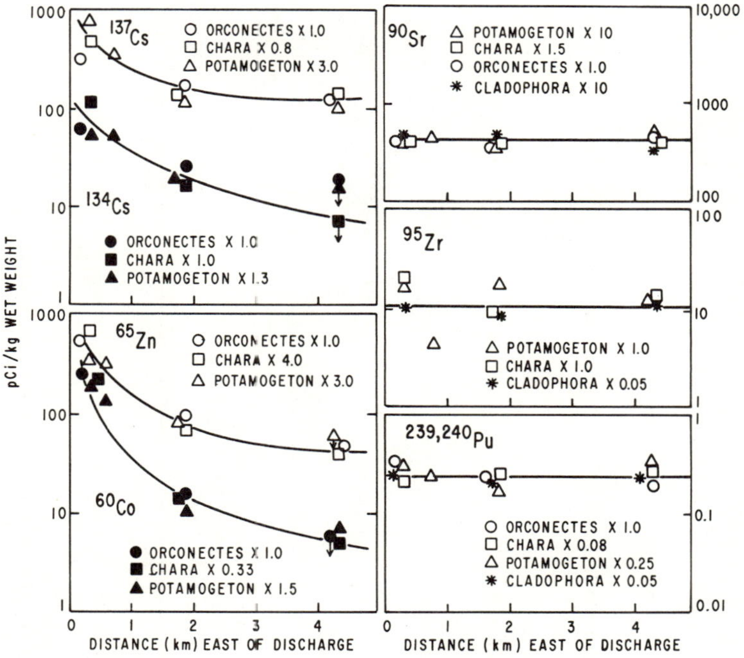

FIG. 4. Normalized radionuclide concentrations in littoral attached algae, rooted plants and crayfish collected at different distances from the Big Rock Point Nuclear Plant on Lake Michigan.

TABLE V. Radionuclide concentrations[a] in Big Rock waste releases.

Sample Number	Date of Release	Concentration, pCi/ℓ[b]						
		$^{239,240}Pu$	^{238}Pu	^{137}Cs	^{134}Cs	^{90}Sr	^{65}Zn	^{60}Co
48 - 70	19 Oct 70	17.4	7.2	730 000	273 000	---	265 000	43 000
20 - 71	14 Jun 71	83.3	36.6	556 000	255 000	---	1 010 000	357 000
22 - 74	26 Nov 74	20.5	9.4	512 000	139 000	---	18 800	73 000
24 - 74	28 Dec 74	70.5	34.3	477 000	144 000		165 000	519 000
Mean Concentration in waste solutions		48	22	570 000	200 000	~10 000[c]	365 000	250 000
Mean Concentration in incoming cooling water		.0007	0.00004	.050	< .001	0.8	< 0.010	< 0.010
Mean Concentration added by waste dilution in plant cooling water[d]		0.0002	0.00010	2.7	1	~0.05	1.8	1.2

(a) For ^{95}Zr only an upper limit ≦ 1% of the $^{137,134}Cs$ could be determined.

(b) The 1 σ counting error is ~5%.

(c) Axtell, 1974.

(d) Total annual discharge/total annual volume of water used in cooling.

data. Small but detectable levels of ^{241}Am are observed in the benthic invertebrates, Mysis and Pontoporeia, but may represent contamination from the surficial sediments and may not be biologically available to predators. In contrast to the plankton-plutonium data, concentrations of plutonium in fish show only a weak correlation with ash content.

3g. Radionuclide Concentrations in Biota near the Big Rock Nuclear Plant

Previous studies utilizing measurements by gamma-ray spectrometry have shown a highly localized increase in activity of several radionuclides in biological samples taken near the Big Rock Plant [19]. For a number of years, the fuel loading of this plant has included a limited number of mixed oxide fuel elements containing recycled plutonium for testing purposes. Yaguchi et al. [20] have compared the concentrations of $^{239,240}Pu$, ^{90}Sr, and gamma emitting radionuclides in littoral organisms near this reactor and at control sites elsewhere in Lake Michigan. The results of the study, shown in Figure 4, demonstrate that ^{134}Cs, ^{65}Zn, and ^{60}Co concentrations, which are not detectable in control samples from other Lake Michigan sites, are high near the plant cooling water discharge. In the same samples, however, the concentrations of ^{90}Sr, $^{239,240}Pu$, and ^{95}Zr show no dependence on distance from the plant discharge, suggesting that either the plant discharges were insignificant relative to the fallout background in lake water or they are released in biologically inert form. To evaluate the former possibility, samples of Big Rock Plant waste discharge solutions from 1970, 1971, and 1974 were radiochemically analyzed. The results of these analyses (see Table V) indicate that the biological radionuclide concentration data are consistent with those which would be predicted from the composition of the waste solutions. Comparison of the isotopic ratios of $^{238}Pu/^{239,240}Pu$ in the plant waste (45%) with those of biological samples (7%) taken nearby and at remote sites further demonstrates that the contribution of Big Rock Plant wastes to the lake biota is undetectable.

3h. Comparison of the Five Great Lakes

In order to establish whether the results of the detailed study of Lake Michigan radioecology might be applicable to the other Great Lakes, a comparison set of water, plankton, and fish samples was obtained during the spring season of 1973. The results of this study are summarized in Table VI. While the number of samples of various types is limited, the results of the study are encouraging in that similar concentration factors are observed, where comparison between lakes is possible. The plutonium data have been evaluated in terms of inventory estimates and trends of biogeochemical behavior [21], and of food chain relationships [22].

The ^{90}Sr concentration observed in each of the Great Lakes during 1973 shows excellent agreement (5—15%) with calculated concentrations obtained from input estimates and a chain of lakes model [23], when losses to the sediments and inputs from watersheds are both assumed to be small. Comparisons of $^{239,240}Pu/^{137}Cs$ ratios from the 1973 water concentrations showed that the ratios in Lake Michigan and Lake Huron appear to be higher than for the other 3 lakes, and the same trend is observed in plankton ratios.

TABLE VI. Mean $^{239,240}Pu$, ^{137}Cs, and ^{90}Sr concentrations in Great Lakes water and biota, June 1972 to November 1973.

Lake	Sample	Number[a] Analyzed	pCi/kg Wet Weight $^{239,240}Pu$	pCi/kg Wet Weight ^{137}Cs	pCi/kg Wet Weight ^{90}Sr
Superior	Water	3	0.00057	0.072	0.530
	Mixed Plankton	3	2.5	200	23
	Zooplankton	1	0.40	190	13
	Smelt	1	0.004	240	--
Michigan	Water	14	0.00076	0.058	0.820 (4)
	Mixed Plankton	22	4.3	150	30 (12)
	Zooplankton	9	0.26	34	49 (5)
	Smelt	6	0.015	220	54 (2)
	Perch	2	0.013	325	--
	Alewife	7	0.019	165	23 (2)
Huron	Water	3	0.00062	0.042	0.960
	Mixed Plankton	2	2.8	54	21
	Smelt	1	0.008	150	--
	Perch	1	0.015	260	--
	Alewife	2	0.10	120	--
Erie	Water	3	0.00021	0.019	1.06
	Zooplankton	3	0.085	34	11
	Smelt	1	0.040	15	--
	Perch	1	0.002	23	--
Ontario	Water	3	0.00026	0.025	1.26
	Mixed Plankton	3	0.61	34	21
	Alewife	1	0.044	46	--

(a) The number of Lake Michigan samples analyzed for ^{90}Sr is given in parentheses following the ^{90}Sr values.

For ^{137}Cs, the fraction [21] of the fallout input present in the water in 1973 is generally inversely proportional to the mean depth squared, suggesting that the turnover rate is determined by factors such as the fraction of the lake volume that is biologically productive. The plutonium data show a similar trend except that the Lake Michigan and Lake Huron values lie considerably above the line. Anomalous plutonium behavior in these two lakes is not related directly to the pH or concentrations of the major chemical constituents of the lakes.

4. SUMMARY

a. $^{239,240}Pu$ and ^{137}Cs show remarkably similar biogeochemical properties, e.g. turnover times and associations with seston fractions.

b. There is a strong inverse relationship between concentration factors for $^{239,240}Pu$ in organisms surveyed and their trophic level. For ^{137}Cs the opposite behavior is noted, with a small increase in concentration going from lower to higher trophic levels.

c. The concentration of plutonium in plankton samples is found to be strongly correlated with ash content. Sorption of plutonium by phytoplankton, and subsequent settling of phytodetritus, and zooplankton fecal pellets is postulated to account for the rapid removal of plutonium from the epilimnion during July and August.

d. $^{239,240}Pu$, ^{137}Cs, and ^{90}Sr show comparable biogeochemical behavior in all five Great Lakes, based on a limited set of comparison samples. The fraction of $^{239,240}Pu$ and ^{137}Cs present in the water after a number of years can be related to the mean depth of the lake.

ACKNOWLEDGEMENTS

We are indebted to the many people and institutions who have collaborated on sample collection and other portions of this study. We are especially indebted to D. M. Nelson, E. M. Yaguchi, B. J. Waller, and K. A. Orlandini for their invaluable help in the field and in the laboratory.

REFERENCES

[1] NELSON, D.M., YAGUCHI, E.M., WALLER, B.J., WAHLGREN, M.A., Radiochemical methods, Argonne National Laboratory Radiological and Environmental Research Division Annual Report, Ecology Part III, January—December 1973, ANL-8060 (1974) 6—17.

[2] ORLANDINI, K.A., WAHLGREN, M.A., ^{241}Am in Lake Michigan environmental samples, Argonne National Laboratory Radiological and Environmental Research Division Annual Report, Ecology Part III, January—December 1974, ANL-74-3 (1975).

[3] CHURCH, P.E., Convection in the annual temperature cycle of Lake Michigan, Ann. N.Y. Acad. Sci. 48 (1947) 789—800.

[4] WAHLGREN, M.A., NELSON, D.M., Plutonium in Lake Michigan water, Argonne National Laboratory Radiological and Environmental Research Division Annual Report, Ecology Part III, January—December 1972, ANL-7980 (1973) 7—14.

[5] HARDY, E.P., Fallout program quarterly summary report, U.S. Atomic Energy Comm. Rep., HASL-281 (1974).

[6] HARDY, E.P., KREY, P.W., VOLCHOK, H.L., Global inventory and distribution of fallout plutonium, Nature 241 (1973) 444—445.

[7] KREY, P.W., SCHONBERG, M., TOONKEL, L., Updating stratospheric inventories to January 1973, "Fallout program quarterly summary report", U.S. Atomic Energy Comm. Rep., HASL-281 (1974).

[8] BROWNELL, G. L., BERMAN, M., ROBERTSON, J.S., Nomenclature for tracer kinetics, Int. J. Appl. Radiation Isotopes 19 (1968) 249—262.

[9] HOLLAND, R.E., BEETON, A. M., Significance to eutrophication of spatial differences in nutrients and diatoms in Lake Michigan, Limnol. and Oceanogr. 17 (1972) 88—96.

[10] LOYSEN, P., Particle size distribution of stratospheric aerosols at 110,000 feet, Proc. Conf. on Radioactive Fallout from Nuclear Weapons Tests, USAEC Report CONF-765 (1965) 135—143.

[11] ALBERTS, J.J., WAHLGREN, M.A., REEVE, C.A., JEHN, P.J., Sedimentary $^{239,240}Pu$ phase distributions, Argonne National Laboratory Radiological and Environmental Research Division Annual Report, Ecology Part III, January—December 1974, ANL-74-3 (1975).

[12] CHANG, C.S., JACKSON, M.L., Fractionation of soil phosphorus, Soil Sci. 84 (1957) 133—144.

[13] GIBBS, R. J., Mechanisms of trace metal transport in rivers, Science 180 (1973) 71—73.

[14] TEMPLETON, W.L., BROWN, V.M., The relationship between the concentrations of calcium, strontium, and strontium-90 in wild brown trout, Salmo trutta L., and the concentrations of the stable elements in some waters of the United Kingdom, and the implications in radiological health studies, Int. J. Air Wat. Poll. 8 (1964) 49—75.

[15] OPHEL, I.L., FRASER, C.D., JUDD, J. M., Strontium concentration factors in biota and bottom sediments of a freshwater lake, Proc. Int. Symp. Radioecology Applied to the Protection of Man and his Environment, Rome, Italy (1971) 509—530.

[16] BEETON, A.M., CHANDLER, D.C., The St. Lawrence Great Lakes, Chapter 19, "Limnology in North America", FREY, D.G., Ed., University of Wisconsin Press (1963).

[17] SPIGARELLI, S.A., Effects of biogeochemical factors on the accumulation of ^{137}Cs fallout by largemouth bass, Ph.D. thesis, Michigan State University (1970).

[18] PENDLETON, R.C., MAYS, C.W., LLOYD, R.D., CHURCH, B.W., A trophic level effect on ^{137}Cs concentration, Health Phys. 16 (1969) 167—183.

[19] NELSON, D.M., ROMBERG, G.P., PREPEJCHAL, W., SPIGARELLI, S.A., Effects of a routine radioactive release on aquatic radionuclide concentrations, Proc. 15th Conf. Great Lakes Res. (1972) 260—269.

[20] YAGUCHI, E.M., NELSON, D.M., MARSHALL, J.S., Plutonium and other radionuclides in biota near the Big Rock Nuclear Plant, Argonne National Laboratory Radiological and Environmental Research Division Annual Report, Ecology Part III, January—December 1973, ANL-8060 (1974) 32—35.

[21] WAHLGREN, M.A., NELSON, D.M., Plutonium in the Laurentian Great Lakes: comparison of surface waters, Verh. Internat. Verein. Limnol. 19 (in press).
[22] MARSHALL, J.S., WALLER, B.J., YAGUCHI, E.M., Plutonium in the Laurentian Great Lakes: food chain relationships, Verh. Internat. Verein. Limnol. 19 (in press).
[23] LERMAN, A., Strontium-90 in the Great Lakes: concentration-time model, J. Geophys. Res. 77 (1972) 3256—3264.

THE BEHAVIOUR OF PLUTONIUM AND OTHER LONG-LIVED RADIONUCLIDES IN LAKE MICHIGAN
II. Patterns of deposition in the sediments*

D.N. EDGINGTON
Radiological and Environmental Research Division,
Argonne National Laboratory,
Argonne, Illinois

J.A. ROBBINS
Great Lakes Research Division
The University of Michigan,
Ann Arbor, Michigan,
United States of America

Abstract

THE BEHAVIOUR OF PLUTONIUM AND OTHER LONG-LIVED RADIONUCLIDES IN LAKE MICHIGAN: II. PATTERNS OF DEPOSITION IN THE SEDIMENTS.

At present eight operating nuclear power reactors, and one more which will be in operation by 1980, are situated along the shores of Lake Michigan. Since their releases of radioactivity have been much less than that entering the lake from stratospheric fallout, $^{239,240}Pu$ and ^{137}Cs from the latter source have been measured in order to attempt to predict the depositional patterns of long-lived radionuclides in the lake.

Ninety-seven percent of the $^{239,240}Pu$ and ninety-five percent of the ^{137}Cs that has entered the lake by wet and dry deposition of atmospheric fallout now reside in the sediments. The vertical distribution of ^{137}Cs has been measured in 60 sediment cores, and $^{239,240}Pu$ in 14 cores taken from all over the lake. The sampling locations were chosen to be representative of the various types of sedimentation, namely areas with high sedimentation associated with river inputs, offshore locations where organic-rich, very fine-grained material is settling out, and areas of low sedimentation where erosion may be occurring.

The results of these studies indicate that the scavenging of $^{239,240}Pu$ and ^{137}Cs appears to occur at the same overall rate but controlled by different processes in the water column. Analysis of the profiles shows that the sedimentation rate varies by more than an order of magnitude and that the total content of these two radionuclides is strongly correlated with the mass-sedimentation rate. As a result, there are large areas of the lake where there is virtually no deposition, and other areas where the total content of $^{239,240}Pu$ and ^{137}Cs per unit area is several times greater than would be expected. This depositional pattern is probably related to the hydrodynamic properties of the lake. The shape of the profiles in areas of high and low sedimentation also suggests that there is a redistribution of the radioactivity which has already been deposited in the sediments. In areas of high sedimentation, the concentration in the upper few centimetres of the core is much higher than can be explained in terms of new deposition or erosion from the watershed.

Therefore, any monitoring scheme for the long-term effects of releases from nuclear power plants must take account of the massive redistribution of sediment that may occur and must recognize that the final point of deposition may be far removed from the site of release.

1. INTRODUCTION

As one of the requirements for the granting of a construction permit or an operating license for a nuclear powered generating station in the United States, the applicant must plan and carry out environmental studies in relation

* This work was performed under the auspices of the USERDA.

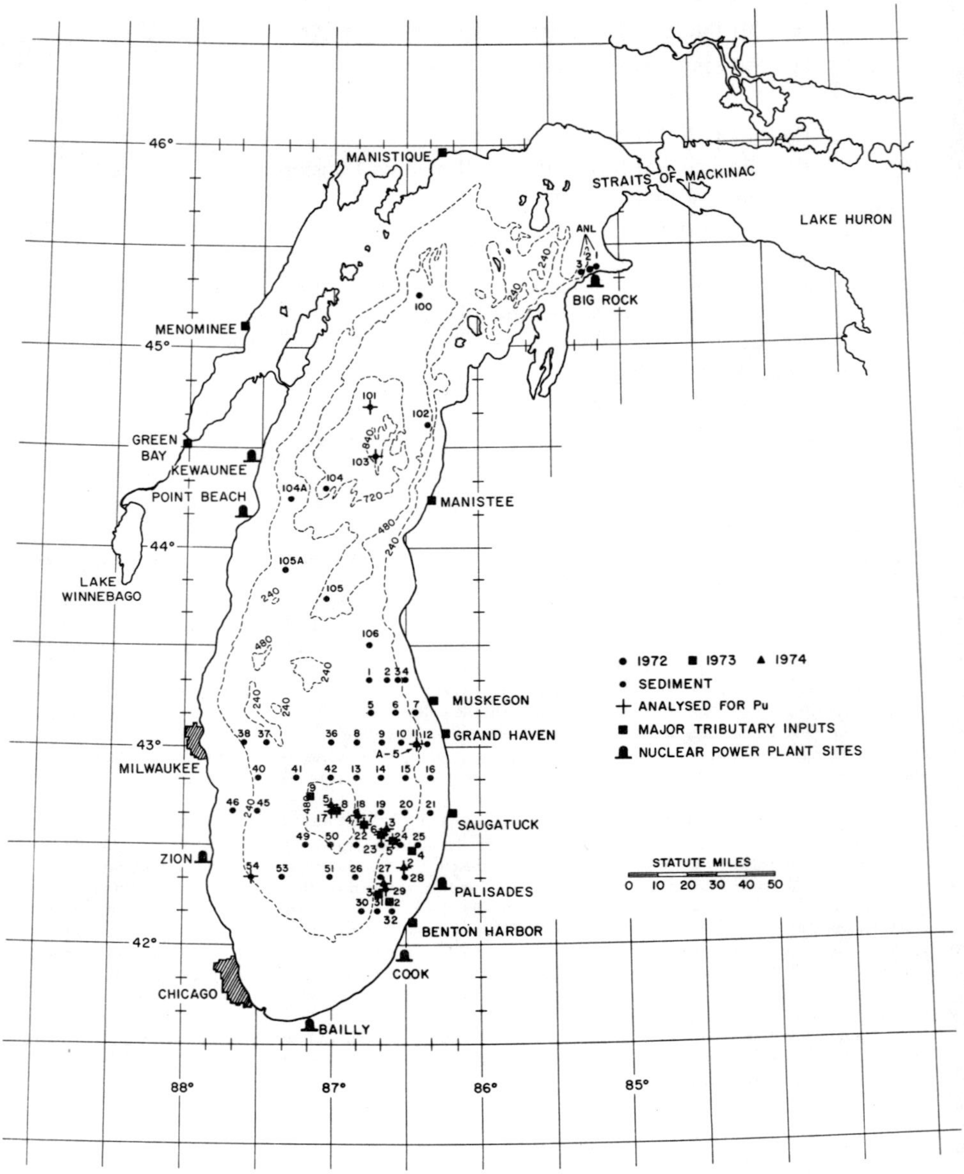

FIG. 1. Stations occupied in Lake Michigan for the sampling of sediments in 1972, 1973 and 1974.

to the operation of the plant. Since the sediments will be the ultimate sink for most of the long-lived radionuclides which may be discharged into the environment as a result of the operation of nuclear reactors, an effective monitoring scheme for this part of the ecosystem is of prime importance. This is particularly true for those plants that will be built on the shores of lakes or coastal enbayments where little or no loss of sediment will be associated with water exchange.

The Laurentian Great Lakes are one of the largest freshwater resources in the world, and a significant fraction of the total population of Canada and the United States live within their basins. As of May 1975, there were a total of seventeen operating nuclear power plants located in the Great Lakes Basin, which means that more than 30% of the total nuclear power capacity of North America was located in this region [1]. Of these plants, eight are located along the shores of Lake Michigan (Figure 1), which is the only one of the lakes totally contained within the continental United States. Lake Michigan is particularly vulnerable to the effects of the addition of undesirable pollutants because of the long residence time (> 100 years) for conservative materials in the water column [2]. There is an almost infinite residence time for pollutants that are readily transported to the sediments. This arises because Lake Michigan is virtually disconnected from the main flow pattern of the other lakes and there are no possible flushing mechanisms.

Since the releases of radioactivity from these operating nuclear plants, which have to be as low as feasible in order to meet the federal guidelines [3], have been much lower than the amount entering the lake from stratospheric fallout, $^{239,240}Pu$ and ^{137}Cs from the latter source have been measured in order to attempt to predict the patterns of deposition of long-lived radionuclides in Lake Michigan. This information will be important in evaluating the suitability of additional sites for power plants, for the prediction of the effects of nonroutine releases, and for the implementation of meaningful routine surveys.

The intent of this paper is to summarize the results obtained from the analysis of a large number of sediment cores for ^{137}Cs, and a lesser number for $^{239,240}Pu$, taken from Lake Michigan in 1972, 1973 and 1974. An attempt will be made to relate the observed distributions to limnological factors and to possible releases of radioactivity from nuclear power plants.

2. METHODS

2a. Sample Collection

Multiple 3"-diameter cores were taken from mainly the soft clays and muds at the stations shown in Figure 1. Coring operations were carried out on the University of Michigan research vessels, Inland Seas, Laurentian and Mysis, during several cruises between 1972 and 1974. The cores were characterized in nearly all instances by the presence of an extremely fluid brown flocculent material up to 1 cm thick at the sediment/water interface overlying either more consolidated black sediment or, particularly on the western side of the lake, pink glacial till. The gravity coring unit was lowered as slowly as possible into the sediments to prevent lateral motion of this flocculent material. The plastic core liners, containing 2—3 ft of sediment with overlying water, were removed from the corer barrel and stored in an upright position. Cores that showed the minimum of disturbance were carefully sectioned, using a hydraulic extruder, after the floc had been allowed to settle. Sections were taken at 0.5-, 1.0-, or 2.0-cm intervals, depending upon the depth in the core, by extruding the sediment into a scribed collar (the liquid floc was generally removed by use of a syringe). The sectioned material was stored in preweighed plastic bottles and frozen for return to the laboratory. The bulk density was measured before air drying at 90°C for 48 hr before analysis so that the porosity of each section could be determined.

2b. Radioactivity Measurements

The ^{137}Cs activity in each section was determined by gamma counting, in standard glass scintillation vials, ≦ 20 g of the oven-dried samples for up to 800 min by use of a 10 x 10 cm NaI detector multichannel-analyzer system. Standards of ^{137}Cs, uranium ore, thorium ore, and potassium having the same density and shape as the unknowns were used and the ^{137}Cs concentrations were determined by least-squares analysis of the resulting gamma spectra [4].

These samples were analyzed for plutonium by use of techniques similar to those described by Golchert [5], Talavitie [6], and Wong [7]. Up to 5-g aliquots of the ground and well-mixed air-dried material, with 1—2 pCi of ^{242}Pu as an isotopic dilutant, were digested in concentrated nitric acid to destroy organic material and then in aqua regia. If the sample is to be analyzed for ^{210}Pb [8], the sample is treated with concentrated hydrochloric acid until all the nitric acid has been destroyed. The sample is then centrifuged and filtered to remove all solids that are resistant to this digestion procedure and the supernatant taken down almost to dryness. This residue is taken up in 20 ml of 8 M nitric acid which is then passed through a column of Dovex 1 x 8 (100—200 mesh) which has been preconditioned with nitric acid. The column is washed with 75 ml of 8 M nitric acid to remove all traces of uranium and almost all other heavy elements. In order to remove thorium, the column is then washed with 100 ml of 12 M hydrochloric acid. Plutonium is then eluted with 50 ml of a mixture of 0.1 M hydrochloric acid and 0.01 M hydrofluoric acid.

The eluent is taken to dryness, redissolved in 5 ml of concentrated HCl, taken to dryness again. This step is repeated twice. Finally, the residue is taken up in a few drops of 6 M HCl and then 5 ml of a mixture of 1 M NH_4Cl and 0.01 M oxalic acid with warming. The plutonium is plated out of this solution onto pre-electropolished stainless steel discs as described by Talavitie [6].

The plutonium is finally determined by alpha spectroscopy by use of silicon barrier detectors whose background in the region of interest 4.5—5.5 MeV is ≦ 0.001 cpm. Activities of ^{238}Pu and $^{239,240}Pu$ are determined by comparing the integrated count in their respective peaks with that in the ^{242}Pu peak.

3. RESULTS AND DISCUSSION

The vertical distribution of ^{137}Cs has been measured in sixty sediment cores, and $^{239,240}Pu$ in fourteen. A large proportion of these cores were taken from the southern basin of the lake (i.e., south of latitude 43°N, Figure 1).

Since there have been only minor releases of radioactivity from nuclear power plants around the lake, effectively all the $^{239,240}Pu$ and ^{137}Cs found in the lake results from nuclear testing in the 1950's and early 1960's. As only a minor contribution to the total activity in the lake has come from runoff into tributary rivers and streams [9], the activity is a result of wet and dry deposition of radioactive fallout over the lake.

In all the cores studied, the $^{239,240}Pu$ and ^{137}Cs activity was confined to the uppermost 6 cm of the sediment, and in many cores no activity was found below 3 cm. These findings indicate that surveys of radionuclides performed using less exact sampling techniques, such as taking grab samples, are probably quite unreliable because of the likelihood of losing the uppermost flocculent layer during sampling. Typical $^{239,240}Pu$ and ^{137}Cs profiles are shown in Figures 2—5. These profiles illustrate the different observed types of distribution of these two radionuclides with depth in sediment cores taken from Lake Michigan. The shapes of the profiles will depend upon the time variability of the input of $^{239,240}Pu$ or ^{137}Cs to the lake, their residence times in the water column, the sedimentation rate, the effects of vertical mixing due to biological or physical processes in the uppermost sediment layers, and the finite thickness of the sampling interval [8].

Wahlgren and Nelson [10] have shown from measurements of $^{239,240}Pu$ and ^{137}Cs in samples of water from the lake during the period 1966—1973, that the residence time of these radionuclides in the water column is of the order of 1 yr (in contrast to $\geqq$ 100 yr for conservative elements). The result is that less than 5% of the total activity which entered the lake over the last 20 yr remains in the water column, and also, since the annual flux to the lake has been extremely variable, two horizons should be preserved in the sediments. One corresponding to the onset of nuclear testing around 1950 and the other to the maximum in activity corresponding to the 1963 input. Present-day inputs are more than an order of magnitude lower than this latter value [11].

The total concentration of $^{239,240}Pu$ or ^{137}Cs at any location in the lake will depend upon three factors: the flux P_t (in pCi/cm^2 per yr) at the air/water interface, the sedimentation rate R (in cm/yr) or ω (in mg/cm^2 per yr), and a factor f by which the flux P_t' (in pCi/cm^2 per yr) measured at the sediment/water interface is normalized to the flux P_t at the air/water interface, where $P_t' = fP_t$. Here the time t, measured backward from the present, is in years. Therefore, the activity in sediment at a depth z in a core corresponding to t years ago is

$$A_t = (fP_t/\omega)\, e^{-\lambda t}$$

where λ is the radioactive decay constant. As the sedimentation rate in Lake Michigan (between 0.01 and 0.3 cm/yr) results in an annual deposit that is thinner than the thinnest section that can be reliably produced, the activity in a layer between t and t+δt years, corresponding to a depth between z and z+δz is

$$A_{t+\delta t} = \int_t^{t+\delta t} (fP_t/\omega)\, e^{-\lambda t}\, dt$$

Here the relationship between t and z is given by

$$t = (\rho_s/\omega) \int_o^z (1-\phi_z)\, dz$$

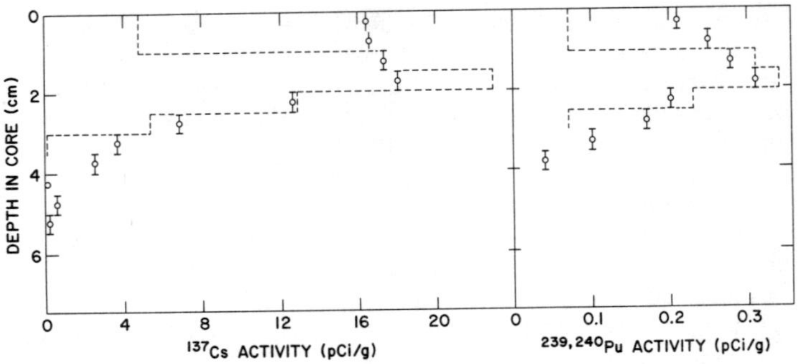

FIG. 2. Measured and calculated ^{137}Cs and $^{239,240}Pu$ profiles in the sedimentary column at station 73-5. The sedimentation rate is 34 mg/cm^2 • yr or 0.17 cm/yr at the surface and there is no surficial mixing.

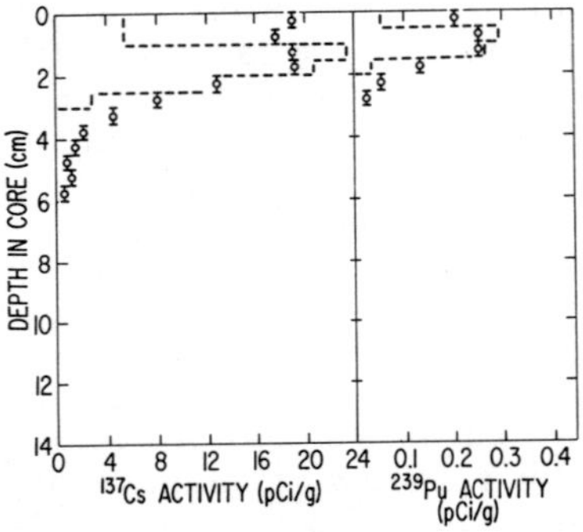

FIG. 3. Measured and calculated ^{137}Cs and $^{239,240}Pu$ profiles in the sedimentary column at station 73-6. The sedimentation rate is 32 mg/cm^2 • yr or 0.17 cm/yr at the surface and there is no surficial mixing.

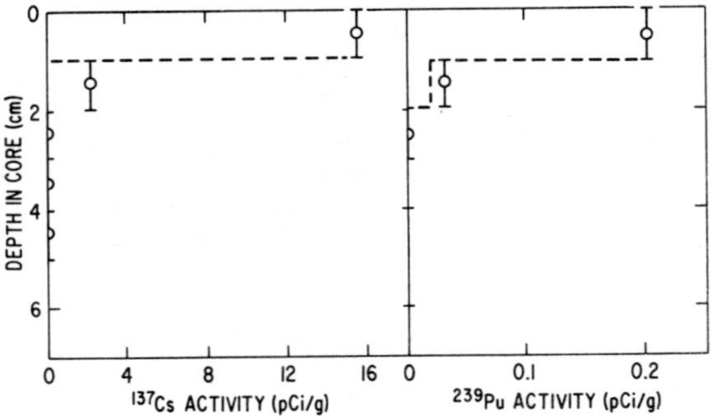

FIG. 4. Measured and calculated ^{137}Cs and $^{239,240}Pu$ profiles in the sedimentary column at station 72-54. The sedimentation rate is 4.8 mg/cm^2 • yr or 0.015 cm/yr at the surface and there is mixing down to a depth equivalent to 82 years of deposition.

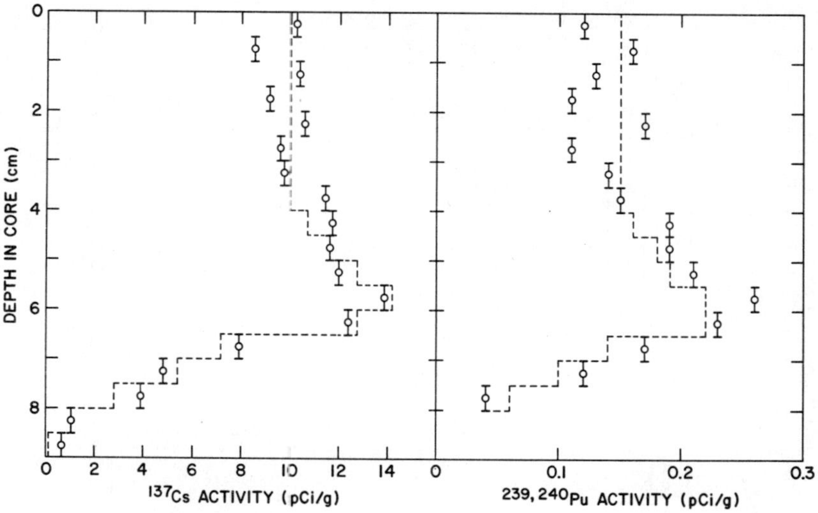

FIG. 5. Measured and calculated ^{137}Cs and $^{239,240}Pu$ profiles in the sedimentary column at station 74-1. The sedimentation rate is 96.3 mg/cm^2 · yr or 0.50 cm/yr at the surface and there is mixing to a depth equivalent to 12 years of deposition.

where ρ_S is the density of the sediment particles and ϕ_z is the porosity at depth z. These equations provide a simple model for the distribution of radioactivity in sediments.

However, in several cores it has been observed that the two horizons appeared to be deeper in the core than would be expected on the basis of sedimentation rates from the ^{210}Pb measurements, and that there was a corresponding interval of constant activity of ^{210}Pb at the top of the sediment cores. These observations have been interpreted in terms of a homogenous layer at the surface of the sediment column due to either physical or biological turbation, as has been postulated elsewhere [12]. The depositional model can be altered to take account of this process. It is assumed that this homogenous layer is of constant thickness, all new inputs are instantaneously mixed within it, and a constant fraction of the total mass of sediment contained within it (equivalent to the annual deposition) is permanently buried each year. Thus, if the first introduction of plutonium or cesium into the sediments occurred k years ago and the thickness of the homogenous layer is equivalent to N years of sedimentation, it can readily be shown that the activity measured at present in sediment taken from a depth corresponding to jth year of deposition is

$$A_{j+N} = \frac{f}{N\omega}\left[\sum_{n=j}^{k}\left(\frac{N-1}{N}\right)^{n-j} \cdot e^{(n-j)\lambda} \cdot P_n\right] e^{(1-j)\lambda}$$

where $j \leq n \leq k$. The best values for the sedimentation rate, mixing depth, and the normalization factor f are obtained from the distribution of each radionuclide in the core by means of an iterative least-squares method.

TABLE I. Summary of measured and calculated parameters from the distribution of $^{239,240}Pu$, ^{137}Cs, and ^{210}Pb in sediment cores taken from Lake Michigan in 1972, 1973, and 1974.

Station	Sedimentation Rate						Mixing Depth						Total Activity pCi/cm²		Flux Normalization factor			Flux Ratios		
	$^{239,240}Pu$		^{137}Cs		^{210}Pb		$^{239,240}Pu$		^{137}Cs		^{210}Pb									
	(a)*	(b)*	(a)*	(b)*	(a)*	(b)*	cm	yr	cm	yr	cm	yr	$^{239,240}Pu$	^{137}Cs	f_{Pu}	f_{Cs}	f_{Pb}	Pu/Cs	Pb/Pu	Pb/Cs
72-1	---	---	0.14	1.90	0.15	2.10	---	--	1.6	110	1.4	110	---	0.38	---	0.05	0.21	---	---	4.20
72-11	0.07	21.8	0.08	23.0	0.08	23.0	0.5	7	0.5	7	0.0	0	0.068	3.81	0.46	0.50	1.04	0.92	2.26	2.08
72-17(2)	0.07	15.1	0.05	11.1	0.08	16.9	0.5	7	1.0	18	1.0	14	0.05	2.32	0.30	0.37	2.46	0.81	8.20	6.64
72-29	0.32	113.6	0.26	92.3	0.24	85.2	2.5	8	3.2	12	3.8	16	0.347	19.37	2.18	2.39	4.51	0.91	2.06	1.88
72-31	---	---	0.06	16.1	0.08	23.0	---	--	2.0	49	1.0	15	---	7.77	---	0.88	1.11	---	---	1.26
72-54	0.02	5.9	0.013	3.8	0.02	5.9	1.0	68	1.0	104	1.0	68	0.08	7.45	0.58	0.59	0.91	0.98	1.57	1.54
72-100	---	---	0.09	20.0	0.14	29.5	---	--	1.5	17	1.0	8	---	8.92	---	1.07	3.19	---	---	2.92
72-101'	0.05	11.0	0.05	13.0	0.06	14.0	1.0	24	1.0	20	2.0	36	0.168	7.55	0.91	0.86	2.69	1.06	2.95	3.13
72-102'	---	---	0.01	6.9	0.02	11.1	---	--	1.0	108	1.0	68	---	1.75	---	0.34	0.94	---	---	2.94
72-103'	0.05	10.5	0.07	12.8	0.08	15.2	1.0	20	0.5	8	0.0	0	0.020	0.715	0.31	0.080	1.58	3.87	5.09	19.75
72-104'	---	---	0.09	16.7	0.06	11.4	---	--	0.0	0	0.0	0	---	1.61	---	0.20	1.18	---	---	5.90
72-105'	---	---	0.09	15.8	0.09	17.5	---	--	0.5	7	0.0	0	---	5.2	---	0.52	1.58	---	---	3.04
72-106'	---	---	0.06	13.3	0.04	8.3	---	--	0.0	0	1.0	28	---	1.89	---	0.27	0.70	---	---	2.59
73-2	---	---	0.08	68.0	0.08	68.0	---	--	1.5	22	1.0	14	---	5.84	---	0.57	1.22	---	---	2.14
73-3	---	---	0.24	70.3	0.30	87.9	---	--	7.0	43	7.0	35	---	31.60	---	3.78	5.05	---	---	1.33
73-4	---	---	0.16	36.9	0.18	41.5	---	--	2.0	15	2.0	10	---	12.61	---	1.41	2.75	---	---	1.95
73-5	0.17	34.3	0.17	34.3	0.20	40.4	0.0	0	0.0	0	0.0	0	0.195	11.92	0.80	1.03	2.94	0.77	3.67	2.85
73-6	0.17	32.0	0.17	32.0	0.23	43.4	0.0	0	0.0	0	0.0	0	0.117	13.10	0.65	1.04	3.80	0.62	5.84	3.65
73-7	0.15	28.5	0.15	28.5	0.18	32.1	1.0	10	1.5	10	0.0	0	0.170	11.00	0.95	1.15	3.32	0.83	3.49	2.89
73-8	0.06	8.7	0.07	10.0	0.10	14.5	2.0	35	2.0	35	2.0	25	0.081	7.61	0.48	0.88	2.43	0.54	5.06	2.76
73-9	---	---	0.12	17.2	0.14	20.0	---	--	2.0	20	2.0	18	---	9.12	---	1.04	2.96	---	---	2.85
74-1	0.50	96.3	0.50	96.3	0.48	92.8	4.0	12	4.0	12	5.2	16	0.427	27.12	2.50	3.14	5.26	0.80	2.10	1.68
74-2	0.10	23.8	0.08	19.0			2.0	18	2.0	29			0.115	8.73	0.61	0.89		0.69		
74-3	0.12	18.8	0.10	15.8	0.28	44.2	2.0	25	2.0	25	0.0	0	0.140	10.94	0.93	1.15	3.30	0.81	3.55	2.86
74-4			0.04	6.5	0.14	22.7			1.5	60	1.5	13		4.59		0.63	2.19			3.48
74-5																				
A-74-5	0.14	31.0			0.19	42.0	2.0	18			2.5	16	0.083		0.77		2.09		2.71	

*(a) cm/yr
*(b) $mg/cm^2/yr$

TABLE II. Summary of measured and calculated parameters from the distribution of ^{137}Cs in sediment cores taken from Lake Michigan in 1972.

Station	Sedimentation Rate cm/yr	Sedimentation Rate mg/cm²/yr	Mixing cm	Mixing yrs	Total Activity pCi/cm²	f
2	0.30	59.1	2.2	8	21.81	2.16
3	0.36	83.1	3.0		23.68	2.57
4	0.05	52.9	1.0		1.76	0.22
6	0.10	15.7	0.5		5.37	0.60
7	0.10	64.8	0.5		2.68	0.34
9	0.10	15.4	0.5		4.86	0.57
10	0.05	14.3	2.0		10.42	1.37
13	0.05	5.7	1.0	33	3.46	0.34
14	0.10	1.2	0.0	0	6.80	
15	0.10	14.3	2.0	32	6.60	0.80
16	0.32	101.5	3.2	12	17.25	2.04
17	0.10	27.1	0.5	5	6.26	0.85
18	0.10	13.0	2.0	29	6.49	0.77
19	0.10	13.9	1.0	15	5.33	0.57
20	0.10	16.3	2.0	30	7.51	0.97
21	0.20	102.3	1.5	9	7.91	0.93
22	0.05	9.1	3.0	78	9.84	1.09
23	0.05	11.0	3.0	75	11.57	1.08
24	0.05	11.4	3.0	71	11.30	1.33
25	0.10	26.8	3.0	42	8.56	1.06
26	0.10	17.0	2.0	43	7.69	0.85
27	0.25	53.3	2.5	13	17.80	1.90
28	0.05	13.8	2.5	75	6.09	0.74
33	0.05	17.8	1.0	22	1.67	0.37
34	0.05	28.7	1.0	21	2.32	0.36
37	0.10	25.2	0.5	6	8.39	1.05
38	0.05	32.5	0.0	0	5.85	0.67
41	0.10	15.4	1.5	20	5.62	0.73
42	0.05	12.4	0.0	0	1.32	0.11
49	0.05	8.3	1.0	24	0.375	0.12
50	0.05	8.8	1.0	23	2.94	0.43
51	0.05	8.7	1.0	26	3.00	0.47
53	0.15	35.5	1.0	9	10.33	1.63
61	0.15	67.2	1.5	14	7.64	0.92
67	0.10	22.7	1.5	17	7.55	0.88
69	0.05	21.9	0.5	11	4.85	0.55
100	0.10	11.2	0.5	6	1.53	0.20

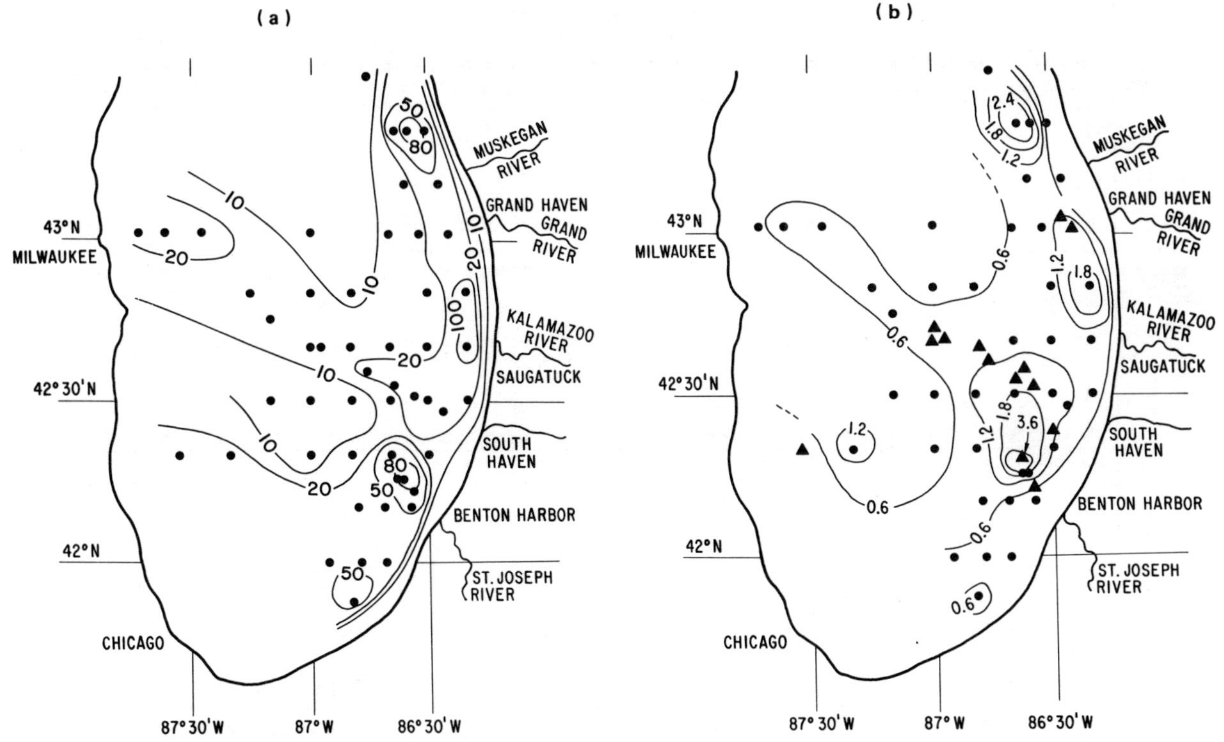

FIG. 6. Comparison of the mass sedimentation rate and total deposition of ^{137}Cs or $^{239,240}Pu$ in southern Lake Michigan.

(a) The variation in mass sedimentation rate ($mg/cm^2 \cdot yr$).

(b) The variation in the flux normalization factor for ^{137}Cs. Sampling locations where $^{239,240}Pu$ have been measured are identified by a triangle.

The dashed lines in Figures 2—5 represent the best fits to the data calculated from this model. Values of the sedimentation rate, mixing depth, normalization factor, and the total activity per unit area of the lake bottom are given in Tables I and II. In those cores for which the sedimentation rate has been determined radiometrically from the ^{210}Pb profiling [8,13] (Table I), the depths to which $^{239,240}Pu$ and ^{137}Cs activities are found are consistent with the known history of deposition from the atmosphere as a result of nuclear testing. The agreement between the values of the sedimentation rates calculated from the measured profiles of these three radionuclides indicates that this method of sampling resulted in no significant losses of material from the sediment/water interface. If significant losses of material had occurred during sampling, the sedimentation rates calculated from the $^{239,240}Pu$ and ^{137}Cs profiles would be lower than those calculated from the ^{210}Pb profiles. This arises because the flux of ^{210}Pb to the lake is constant (P_t = const = 0.2 pCi/cm^2 yr)[14], and the sedimentation rate is calculated solely from its rate of decay, while P_t is variable for the other two nuclides and the estimate of the sedimentation rate is almost entirely based on the position of two horizons as described earlier.

As a result of the seismic profiling experiments conducted in southern Lake Michigan, it is known that the distribution of sediments is extremely variable [15]. Large areas of the western side of the lake exhibit virtually zero sedimentation; but in other areas close to the eastern shore, where there are several large rivers (Figure 6a), the post-glacial sediment layer is up to 10 m thick. The $^{239,240}Pu$ and ^{137}Cs profiles shown in Figures 2 and 3 (station 73-5 and 73-6) are for areas where there is an intermediate rate of sedimentation, and the profiles shown in Figures 4 and 5 are for areas of low (72-54) and high (74-1) sedimentation, respectively.

The profile shown in Figure 4 is representative of many areas on the western side of the lake where there is a thin layer of brown floc (1—2 cm thick) overlying glacial till or clean sand. Since detectable concentrations of $^{239,240}Pu$ and ^{137}Cs have been found within the sampling interval that includes the interface between the floc and the till, it is reasonable to suppose that this floc layer is of a transitory nature and that the sediments as represented by this core are undergoing episodic periods of deposition and erosion. As a result, there is a large probability that this flocculent material, which contains relatively high concentrations of $^{239,240}Pu$ or ^{137}Cs, is undergoing resuspension and transport to other locations in the lake.

Such a remobilization of sedimentary material is required in interpreting the two profiles representative of regions of intermediate sedimentation (Figures 2 and 3). The profiles calculated on the basis of the least-squares fits indicate a much greater drop in the expected concentration in the upper few centimeters (as a result of the large decrease in atmospheric flux since 1963) than is reflected in either the $^{239,240}Pu$ or ^{137}Cs data. This excess plutonium or cesium can originate from two possible sources: runoff from the watershed or remobilization of surficial sedimentary material. The present estimate of the contribution of cesium (or plutonium) from the watershed due to erosional losses from the soil is still less than the atmospheric inputs [9]. Therefore, it would appear that the excess radioactivity in these two cores is due to remobilization of surface sediments, probably from those areas where little permanent sedimentation occurs. The final profile (Figure 5) is representative of an area where there is high sedimentation and considerable mixing. At this station, the sedimentation rate is 0.5 cm/yr and the mixing depth is 4 cm.

The effect of mixing is to produce a region of nearly constant activity at the top of the core and to broaden the maximum in concentration corresponding to the 1963 maximum in input.

The resuspension and transport of sediments will have far-reaching effects on the long-term distribution of long-lived radionuclides such as plutonium in Lake Michigan (and other lakes). They will be particularly important in terms of the fate of radionuclides discharged by nuclear power plants and in determining the validity of present and proposed monitoring schemes. From the data shown in Tables I and II, it is clear that there are significant variations in the distribution of $^{239,240}Pu$ and ^{137}Cs in the sediments of the lake. The total deposition of plutonium varies between 0.02 and 0.427 pCi/cm^2 and that for cesium between 0.38 and 31.6 pCi/cm^2. The average deposition per unit area has been 0.2 pCi/cm^2 and 8.0 pCi/cm^2 for Pu and Cs, respectively, and therefore, the flux normalization factors f vary between 0.05 and 3.1. These results suggest that there are large differences in the fluxes of plutonium or cesium arriving at the sediment/water interface. These differences in concentration at the surface, which the data presented in Tables I and II show to be very local, could arise from either a large variability in the flux from the atmosphere to the lake surface or from differences associated with the hydrodynamic and scavenging processes removing plutonium and cesium from the water column. Since it is well known, from the monitoring of world-wide fallout for ^{90}Sr, that there are only small variations in flux within any latitude band [11], the observed differences in deposition probably arise for the latter reason.

If radioactivity remained unassociated with bulk sedimentary material in the water column and arrived independently at the bottom, or if the sedimentation rate and the flux normalization factor f were constant over the whole area of the lake, the total activity per unit area should have the constant value = $f\sum P_t$ for all t. However, since the sedimentation rate is extremely variable (Figure 6a) and the values of f vary from station to station as well as from nuclide to nuclide (Table I), the removal of radioactivity from the water column is more than likely controlled by the composition of the particulates in the water column. Therefore, in order to compare the nature and magnitude of this scavenging process, ratios of the flux normalization factors for the three radionuclides were calculated. The values are shown in Table I.

From these values it will be noted that, in general, ^{210}Pb is scavenged more effectively at deeper water stations ($\geqq$ 300 ft) where the $^{210}Pb/^{137}Cs$ and $^{210}Pb/Pu$ ratios are $\geqq 2$, and cesium and plutonium are more effectively scavenged at the shallower water stations. The values of the $Pu/^{137}Cs$ ratios are, with one exception (station 73-103'), very close to unity and therefore appear to be removed to the sediments with the same efficiency regardless of sampling location.

These observations can be accounted for in terms of their known behavior in the environment. Shukla and Leland [16] have concluded that the scavenging of lead from the water column is governed primarily by its complexation with organic matter rather than by absorption on clays or hydrous oxides. Since the fraction of organic carbon in the sediments increases with increasing water depth [17], the preferential scavenging of ^{210}Pb offshore would not be unexpected. On the other hand, ^{137}Cs is very strongly adsorbed by the clay component of soils [18,19] and, consequently, a comparatively greater scavenging in the inshore regions where there may be a large proportion of fine-grained clays in the water column is not unexpected. The strong association between plutonium and ^{137}Cs is rather unexpected because of their widely differing chemical properties. However, Alberts and Wahlgren [20], have shown that the plutonium is associated primarily with the hydrous oxide fractions of the sediment from Lake Michigan. Therefore, since Shimp et al. [17] have shown that there is a strong correlation between hydrous oxides (such as Fe_2O_3) and the < 2 μm clay fraction, the observed similarity between the behavior of plutonium and cesium in the lake can readily be understood.

The apparent similarity between the behavior of plutonium and cesium in Lake Michigan is fortunate because it means that measurements of ^{137}Cs, which are far less costly and time consuming to perform than those for $^{239,240}Pu$, can provide the necessary information to predict the long-term behavior of plutonium in the sediments. The two maps shown in Figure 6 compare the deposition of ^{137}Cs and, by inference, $^{239,240}Pu$ (presented in terms of the flux normalization factor f) and the mass sedimentation rate ω. The triangles denote stations where plutonium has been measured. It will be noted that the distribution of radioactivity parallels that of the sediments. Particularly high depositions occur to the north and west of the major river inputs and the greater part of the radioactivity is concentrated within the sediments of the eastern half of the lake. The pattern of sedimentation in the southern half of the lake is very interesting in that whereas the bottom topography indicates an almost continuous slope down to the maximum depth of 540 ft, the maximum depth of sediment occurs at a water depth of only 240 ft (80 m). In addition, no sedimentation occurs at water depths less than 150 ft (50 m) even on the east side of the lake near the rivers, and therefore very little radioactivity is being deposited in the nearshore zone. From this one concludes that the deposition of radioactivity in the lake is being controlled mainly by lake-wide processes such as current structure and sources of the sediment. Therefore, because of the very short residence times in the water column, radioactivity which is released to the lake along the western shore will be rapidly transported to areas where there is a significant deposition of sediment; and for Lake Michigan, these areas are the band of sedimentation parallel to the eastern shore. Discharges of radioactivity along the eastern shore would also be found predominantly in this band of high sedimentation.

These results are important in that they suggest that any monitoring schemes to determine the long-term fate of effluents from nuclear power reactors must take account of the limnological properties of the receiving body of water. For Lake Michigan, a fixed sampling grid for sediments in the immediate environs of the plant would provide no significant information because of the physical and current structure of the lake.

ACKNOWLEDGEMENTS

We would like to express our appreciation to Captain R. Thibault and the crews of the University of Michigan research vessels Inland Seas and Laurentian for their invaluable help in obtaining the cores for this study; to M. M. Thommes and K. Johansen for their assistance in the field work; to J. Karttunen for his infinite patience in the analyses of the samples; to J. J. Alberts and M. A. Wahlgren for the use of their unpublished data.

REFERENCES

[1] "Operating experience of nuclear generating facilities in the United States", Nuclear Industry 22 (1975) 22.

[2] BEETON, A.M., CHANDLER, D.C., The St. Lawrence Great Lakes, p. 535—558, "Limnology in North America", FREY, D. G., Ed., University of Wisconsin Press.

[3] U. S. Nuclear Regulatory Commission Rules and Regulations, Title 10, Part 20, Code of Federal Regulations, "Standards for protection against radiation".

[4] LUCAS, H.F., EDGINGTON, D.N., Computer analysis of gamma-ray spectra: validity of the results, p. 1207—1214, "Modern trends of activation analysis", DeVOE, J.R., Ed., Nat. Bur. Stand. Spec. Publ. 312 (1969).

[5] GOLCHERT, N.W., SEDLET, J., Radiochemical determination of plutonium in environmental water samples, Radiochem. Radioanal. Letters 12 (1972) 215—221.

[6] TALAVITIE, N.A., Radiochemical determination of plutonium in environmental and biological samples by ion exchange, Anal. Chem. 43 (1971) 1827—1830.

[7] WONG, K. M., Radiochemical determination of plutonium in sea water, sediments, and marine organisms, Anal. Chim. Acta 56 (1971) 355—364.

[8] ROBBINS, J.A., EDGINGTON, D.N., Determination of recent sedimentation rates in Lake Michigan using Pb-210 and Cs-137, Geochim. Cosmochim. Acta 39 (1975) 285—304.

[9] EDGINGTON, D.N., RITCHIE, J.C., ROBBINS, J.A., Comments on paper, "Use of rivers to predict accumulation in sediments of radionuclides discharged from nuclear power plants" by PLATO, P., Health Physics (1975) in press.

[10] WAHLGREN, M.A., NELSON, D.M., Residence times for ^{239}Pu and ^{137}Cs in Lake Michigan waters, Argonne National Laboratory Radiological and Environmental Research Division, Ecology Part III, January—December 1973, ANL-8060 (1973) 85—89.

[11] Health and Safety Laboratory, Environmental Quarterly, Appendix, April 1, 1975, U.S. ERDA, HASL-291, (1975) A-43—45, A-101—102.

[12] DAVIS, R.B., Pollen studies of the near surface sediments in Maine lakes, Quarternary Paleoecology, CUSHING, E.J., WRIGHT, H.E., Eds., Yale University Press, New Haven (1967) 153—157.

[13] EDGINGTON, D.N., ROBBINS, J.A., Unpublished data.

[14] JAWOROWSKI, Z., Temporal and geographical distribution of Radium D (Lead-210), Nature 212 (1966) 886—889.

[15] LINEBACK, J.A., GROSS, D.L., Depositional pattersn, facies, and trace element accumulation in the Waukegan member of the Lake Pleistocene-Lake Michigan formation in southern Lake Michigan, Ill. State Geol. Survey, Environ. Geol. Notes, EGN-58 (1972).

[16] LELAND, H.V., SHUKLA, S.S., Factors affecting distribution of lead and other trace elements in sediments of southern Lake Michigan, p. 89—129, "Trace metals and metal-organic interactions in natural waters", SINGER, P.C., Ed., Ann Arbor Science, Ann Arbor, Michigan (1973).

[17] SHIMP, N.F., SCHLEICHER, J.A., RUCH, R.R., HECK, D.B., LELAND, H.V., Trace element and organic carbon accumulation in the most recent sediments of southern Lake Michigan, Ill. State Geol. Survey, Environ. Geol. Notes No. 41 (1971).
[18] JENNE, E.A., WAHLBERG, J.S., Role of certain stream-sediment components of radio-ion sorption, U.S. Geol. Surv. Prof. Paper 443-F (1968) 16 p.
[19] McHENRY, J.R., RITCHIE, J.C., GILL, A.C., Accumulation of fallout cesium-137 in soils and sediments in selected watersheds, Water Resources Research 9 (1973) 676—686.
[20] ALBERTS, J.J., WAHLGREN, M.A., Unpublished data.

DISCUSSION

J.A. HETHERINGTON: After you have made the correction to your sedimentation rates, as deduced from the observations of fall-out Cs and Pu, to take account of the mixing layer, are you confident that the values you then obtain are compatible with the belief that both of these nuclides are reaching these sediments mainly via sedimentation and that once they are incorporated into the sediments they are fixed there?

D.N. EDGINGTON: I am confident that Pu, Cs and Pb are reaching the sediments mainly in association with settling solids. In the mixing model there is no correction applied to the sedimentation rate. If the assumptions made are correct, the value of the sedimentation rate calculated from the model will depend on the difference between the depths of the two Cs or Pu horizons corresponding to the onset of nuclear testing and the maximum in stratospheric fall-out in 1963, and for ^{210}Pb on the slope of the non-constant portion of the profile. The sediment rates calculated by these means are independent of the magnitude of the mixing depths.

S. BONOTTO: Could you explain why the deposition of sediments is more pronounced on one side of Lake Michigan than on the other?

D.N. EDGINGTON: The patterns of sediment deposition in Lake Michigan are determined by the sources of the sedimentary material and by the current structure. There are large deposits of sediments on the eastern side because this is where the major tributary streams enter the lake and also because there is considerable erosion of large sand dunes and cliffs. That the major deposition of sediment does not move to the centre of the basin is presumably due to the presence of a current gyre close to the eastern shore.

A. PRESTON (Chairman): Perhaps I can make a few comments at this point. I have been interested to note that nothing said or reported upon during this session on plutonium disturbs the geometry of the environmental problems of transuranic elements as outlined in the paper I presented in Session I (SM-198/57). The behaviour of plutonium when introduced to aquatic environment is sediment-orientated, whether that plutonium is of weapon origin or results from fuel reprocessing, and whether the environment is fresh water or saline. Furthermore, the biological availability of such material – once removed to sediments – is very low, if it exists at all.

Our future attention needs to be directed towards the long-term consequences of this removal to sediments, i.e. the fate of the sediments, the possibilities of remobilization from sediment, and the biological consequences, if any, of remobilization.

Present rates of introduction of plutonium to aquatic environment are causing no contemporary radiological problems.

IDENTIFICATION OF CRITICAL PATHWAYS RELATIVE TO THE PHYSICAL, CHEMICAL AND BIOLOGICAL SYSTEMS

(Sessions VI and VII)

Chairmen

Yu. V. SIVINTSEV (USSR)

W. FELDT (Germany, Fed. Rep.)

CONSEQUENCES DU DEPLACEMENT DES SEDIMENTS SUR LA DISPERSION DES RADIONUCLEIDES

R. SCHAEFFER*
Electricité de France,
Paris, France

Abstract–Résumé

THE EFFECT OF SEDIMENT DISPLACEMENT ON THE DISPERSAL OF RADIONUCLIDES.

The bulk of discharged radionuclides are adsorbed by the sediments in water courses. It follows from this that the volumetric activity of the water decreases with distance from the disposal point in an approximately exponential manner. However, the activity accumulated by the sediments deposited on the river bed does not follow this law: the bottom transport to which these sediments are subjected reverses the direction of the variation for them so that their activity increases progressively in the downstream direction, reaching a maximum which may be situated a great distance away after a long period of disposal. In the long term, therefore, the sediments could constitute the main source of irradiation of populations living on river banks fairly far from a disposal point. In view of this variation of sediment activity in time and in space, the paper draws attention to the fact that a very limited number of measurements repeated at a single point is not representative.

CONSEQUENCES DU DEPLACEMENT DES SEDIMENTS SUR LA DISPERSION DES RADIONUCLEIDES.

Les sédiments des cours d'eau adsorbent la plus grande partie des radionucléides rejetés. Il en résulte que l'activité volumique de l'eau décroît en fonction de la distance au point de rejet selon une loi approximativement exponentielle. Mais l'activité accumulée par les sédiments déposés sur le fond de la rivière ne suit pas cette loi: le charriage de fond auquel ceux-ci sont soumis inverse pour eux le sens de variation, de sorte que leur activité croît progressivement de l'amont vers l'aval jusqu'à un maximum qui peut se situer à une grande distance après une longue période de rejet. Les sédiments pourraient donc constituer à long terme la principale source d'irradiation pour des populations séjournant sur berges assez loin du point de rejet. Compte tenu de cette évolution de l'activité des sédiments à la fois dans le temps et dans l'espace, l'attention est attirée sur l'absence de représentativité d'un trop petit nombre de mesures répétées en un seul point.

1. INTRODUCTION

La plus grande partie des radionucléides rejetés en rivière est adsorbée soit sur les sédiments en suspension qui après un certain parcours se déposent au fond du lit, soit sur les sédiments déjà en place sur ce fond. On doit donc s'attendre à ce que les sédiments des cours d'eau accumulent au cours du temps une fraction importante des activités qui seront rejetées par l'industrie nucléaire, et il importe de prévoir dès à présent les modalités d'évolution de ce processus.

Nous étudierons d'abord la variation de l'activité adsorbée sur les sédiments en fonction de la distance et de la durée de rejet, puis nous examinerons ses conséquences sur la valeur du facteur de concentration des nucléides dans les sédiments et la dose d'irradiation qui peut en résulter pour les populations.

---

* Depuis mars 1975: Ministère de l'industrie et de la recherche, Service central de sûreté des installations nucléaires, 13 rue de Bourgogne, 75007 Paris.

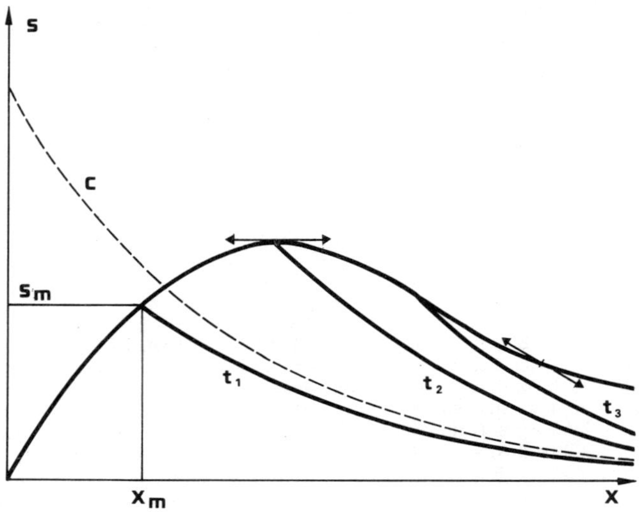

FIG. 1. Variation générale, en fonction de la distance x au point de rejet, de l'activité s des sédiments pour différentes durées t de rejet (en tirets, variation de l'activité volumique C de l'eau).

2. ACTIVITE ADSORBEE SUR LES SEDIMENTS

L'observation montre qu'en raison de l'adsorption des nucléides par les sédiments l'activité volumique de l'eau (y compris les sédiments en suspension qui ne se sont pas encore déposés) décroît en fonction de la distance x au point de rejet selon une loi que l'on peut approximativement traduire par une exponentielle de la forme e^{-kx} (fig. 1).

La valeur de la constante k dépend notamment :

- de la nature du nucléide considéré,
- de la composition chimique et de la granulométrie des sédiments,
- du temps de contact de l'eau contaminée avec les sédiments,

c'est-à-dire de la vitesse moyenne d'écoulement de l'eau.

Si l'on désigne par k_o et k'_o les constantes relatives à l'élément stable (ou à un isotope de longue période) pour deux valeurs w et w' de la vitesse moyenne d'écoulement de l'eau, on a la relation :

$$k_o = k'_o \; w'/w$$

S'il s'agit d'un isotope de période courte, il convient de tenir compte, en outre, de sa constante de décroissance radioactive λ qui entraîne une diminution de son activité sur la distance x :

$$k = k_o + \frac{\lambda}{w}$$

TABLEAU I. VALEURS (km^{-1}) DE LA CONSTANTE k

Nucléide	Canaux d'irrigation du Rhône	Grand fleuve
^{60}Co	0,025	0,005
^{90}Sr	0,01	0,002
^{106}Ru	0,025	0,005
^{137}Cs	0,05	0,01

Pour l'évaluation de cette constante, nous nous sommes référés aux résultats des mesures effectuées sur des canaux d'irrigation du Rhône [1, 2], pour lesquels la vitesse moyenne d'écoulement de l'eau était voisine de 0,4 m/s.[1]

Nous prendrons dans cette étude l'exemple d'un grand fleuve auquel nous attribuerons une vitesse moyenne d'écoulement de l'eau de 2 m/s et nous supposerons que ses sédiments sont de même nature que ceux des canaux ci-dessus.

Le tableau I rapporte pour quelques-uns des principaux nucléides rejetés par l'industrie nucléaire ces valeurs de base et celles que l'on en déduit pour le fleuve en question.

On serait tenté de penser que l'activité accumulée par les sédiments décroît également de l'amont vers l'aval de la même façon que l'activité volumique de l'eau, de sorte que, pour une période de rejet donnée, les deux activités présenteraient entre elles, à toute distance, un rapport constant.

En réalité, la force tractrice, que la composante du poids de l'eau parallèle à la direction du courant principal exerce sur le fond du lit de la rivière, anime les couches superposées de sédiments d'un mouvement de translation dont la vitesse décroît progressivement en profondeur à partir d'un maximum à la surface du tapis alluvionnaire. Nous désignons par v la vitesse moyenne de déplacement de la couche supérieure d'une trentaine de centimètres d'épaisseur [3] où la majeure partie de l'activité adsorbée se trouve concentrée.

Ce charriage de fond a pour conséquence d'inverser la variation de l'activité du sédiment par rapport à celle de l'eau (fig. 1) : l'activité s du sédiment croît de l'amont vers l'aval pour atteindre un maximum s_m à une distance x_m. Les valeurs de ces deux grandeurs augmentent avec la durée t du rejet jusqu'à ce que les pertes par décroissance radioactive équilibrent les apports ; à partir de ce moment, la position du maximum reste inchangée, le front des sédiments contaminés continue de progresser mais son activité va en s'amenuisant du fait que la décroissance radioactive l'emporte.

[1] Pour le cobalt et le ruthénium, nous avons adopté des valeurs intermédiaires entre celles qui ont été trouvées pour le strontium et le césium.

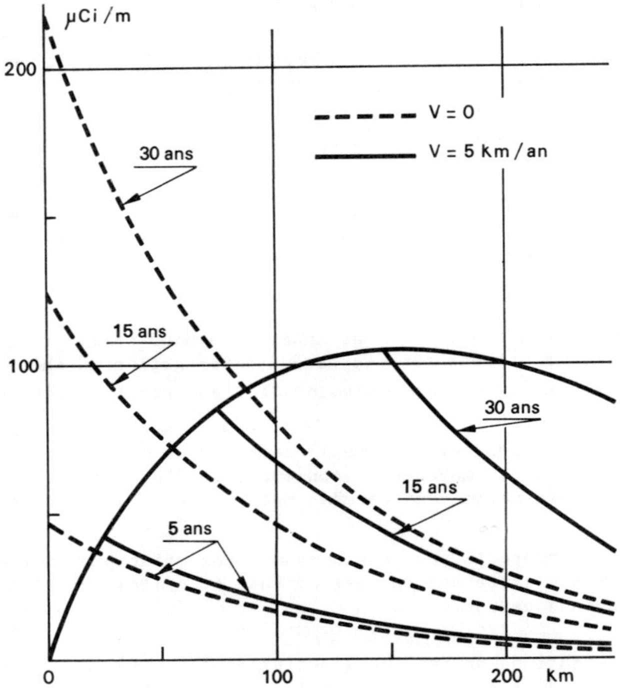

FIG. 2. Variation, en fonction de la distance au point de rejet, de l'activité en césium-137 (microcuries par mètre de longueur de lit) des sédiments du cours d'eau de référence (voir texte) pour différentes durées de rejets effectués à raison de 1 curie par an. En tirets, variation en l'absence de charriage de fond; en traits pleins, variation avec un charriage de fond de 5 km/an.

Nous établissons à l'annexe I les expressions mathématiques qui traduisent ces phénomènes. Pour simplifier, nous supposons que le débit du cours d'eau garde une valeur moyenne constante en aval du point de rejet, que son lit présente une section régulière, que tous les sédiments contaminés sont animés de la même vitesse de translation, qu'il n'y a pas de phénomènes de désorption.

En fait, le débit des cours d'eau est variable dans le temps et augmente avec le bassin versant, les équipements hydrauliques modifient les régimes d'écoulement de l'eau et de transport des sédiments, la vitesse de charriage des sédiments dépend de la granulométrie, des phénomènes de désorption peuvent survenir en période de crues par suite de variations du pH.

La théorie que nous exposons est donc très schématique et il conviendrait d'adapter le calcul à chaque cas d'espèce.

Si l'on adopte pour le fleuve que nous avons choisi comme exemple une vitesse moyenne de charriage de fond de 5 kilomètres par an, l'application numérique de ces expressions mathématiques au cas de rejets de césium-137 effectués à raison de 1 curie par an conduit aux résultats représentés graphiquement par la figure 2. Aux fins de comparaison, nous avons également tracé les courbes qui correspondraient à une absence de charriage de fond.

TABLEAU II. VALEURS DE x_m (km) ET DE s_m (MICROCURIES PAR METRE DE LONGUEUR DE LIT ET PAR CURIE ANNUELLEMENT REJETE) POUR DIFFERENTES DUREES DE REJET

Nucléide	Période (ans)	t = 5 ans		t = 15 ans		t = 50 ans		t = 100 ans	
		x_m	s_m	x_m	s_m	x_m	s_m	x_m	s_m
^{60}Co	5,2	25	17	75	25	80	26	80	26
^{90}Sr	28	25	9	75	23	250	40	310	44
^{106}Ru	1	25	6	25	6	25	6	25	6
^{137}Cs	30	25	40	75	88	150	105	150	105

Nous remarquons que celui-ci abaisse le niveau de contamination maximale des sédiments, mais le déplace progressivement vers l'aval jusqu'à ce qu'il atteigne au bout de 30 années de rejet le point d'équilibre situé à 150 kilomètres. Le charriage de fond augmente donc les possibilités d'irradiation de populations éloignées.

Le tableau II rapporte pour quelques autres nucléides les valeurs de x_m et de s_m. Pour les isotopes de période relativement courte, tel que le ruthénium 106, le maximum s'établit très près du point de rejet.

Nos résultats semblent assez conformes aux observations qui ont pu être faites pour certains cours d'eau en aval d'installations nucléaires exploitées depuis longtemps. C'est ainsi que les sédiments du Rhône à l'aval de l'usine de traitement des combustibles de Marcoule présentaient en 1968 un pic de contamination de 22000 picocuries de césium-137 par kilogramme de sédiments à environ 75 kilomètres de l'usine [4].

3. FACTEUR DE CONCENTRATION DES NUCLEIDES DANS LES SEDIMENTS

Le facteur de concentration est défini comme le rapport des activités massiques présentées au même endroit par les sédiments et l'eau.

Les expressions mathématiques de ce facteur sont établies à l'annexe II.

Nous les avons appliquées au fleuve que nous avons précédemment choisi comme exemple en lui attribuant un débit moyen de 1400 m^3/s, une largeur de 150 m et une masse volumique des sédiments de 1,5.

Les résultats de nos calculs sont représentés graphiquement sur la figure 3 dans le cas de rejets de césium-137 et selon les deux hypothèses : absence de charriage de fond et existence d'un charriage de fond de 5 km/an.

Dans la première hypothèse, la valeur du facteur de concentration est indépendante de la distance au point de rejet, parce que l'activité massique des sédiments et celle de l'eau suivent la même loi de variation. Par contre,

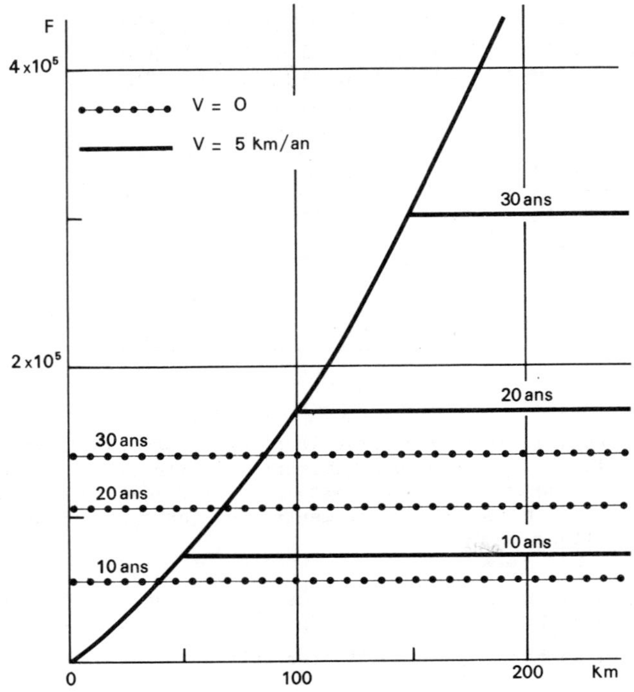

FIG. 3. Variation, en fonction de la distance au point de rejet, du facteur de concentration F du césium-137 dans les sédiments du cours d'eau de référence (voir texte) pour différentes durées de rejet. En pointillés, variation en l'absence de charriage de fond; en traits pleins, variation avec un charriage de fond de 5 km/an.

en présence d'un charriage de fond, le facteur de concentration augmente avec la distance ; cette croissance se poursuit même au-delà du point d'équilibre (150 km), parce que l'activité des sédiments diminue moins vite que celle de l'eau.

Ces variations mettent en évidence le caractère relatif du facteur de concentration.

Mis à part les zones à proximité immédiate du point de rejet, nous trouvons des valeurs en général très supérieures à celles qui ont été mesurées jusqu'à présent en milieu marin [5]. Ces écarts peuvent être attribués aux conditions différentes de dilution, mais aussi à la forte teneur de l'eau de mer en éléments alcalins qui concurrencent le césium dans les phénomènes d'adsorption.

4. DOSES DELIVREES AUX POPULATIONS PAR LES SEDIMENTS

Les expressions mathématiques des doses résultant de l'irradiation externe par les sédiments sont données à l'annexe III.

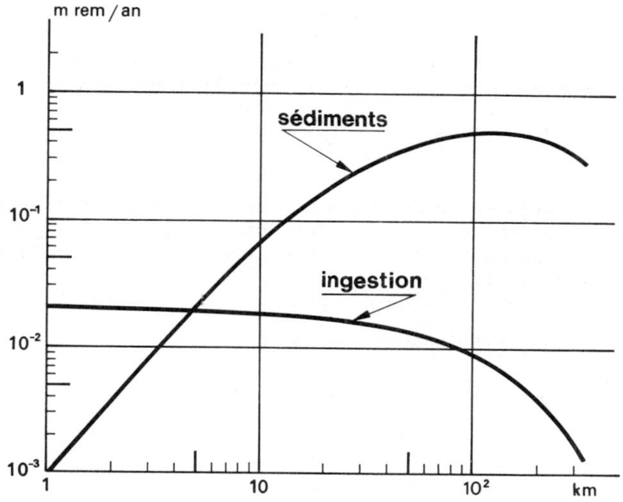

FIG. 4. Variation, en fonction de la distance au point de rejet, de la dose annuelle délivrée à long terme à l'organisme entier des personnes du public par un rejet continu de césium-137 effectué à raison de 1 curie par an dans le cours d'eau de référence (voir texte), selon la voie d'irradiation: irradiation externe par les sédiments, contamination interne par ingestion d'eau et d'aliments.

Nous les avons appliquées au cas de rejets continus de césium-137 effectués pendant plus de 30 ans à raison de 1 curie par an dans le fleuve choisi comme exemple, en admettant qu'une personne du public peut séjourner 500 heures par an au contact des sédiments à l'occasion d'activités récréatives en bordure de cours d'eau [6].

Les valeurs obtenues ne sont guère différentes pour la peau et l'organisme entier : respectivement 0,65 et 0,50 millirem par an au point d'accumulation maximale de la contamination.

La figure 4 représente la variation de la dose annuelle à l'organisme entier en fonction de l'éloignement de la zone récréative du point de rejet. Aux fins de comparaison, nous avons également représenté la variation du débit de dose qui pourrait être dû à la contamination interne par ingestion d'eau et d'aliments, calculé selon la méthode que nous avons déjà exposée ailleurs [2]. Nous remarquons que l'irradiation externe par les sédiments serait capable d'engendrer, en des points très éloignés du point de rejet, des doses plus de 100 fois supérieures à celles qui correspondraient à la voie de contamination par ingestion. A ces distances, l'irradiation externe par les sédiments ne saurait donc être négligée.

5. CONCLUSION

Malgré le caractère théorique de cette étude, nous pouvons conclure à l'évolution de l'activité des sédiments (et des organismes benthiques qui leur sont liés, tels que les algues et les mollusques) à la fois dans le temps et dans l'espace, et nous attirons l'attention sur l'absence de représentativité d'un trop petit nombre de mesures répétées en un seul point.

ANNEXE I

Calcul de l'activité adsorbée sur les sédiments

Désignons par :

k_o et k les constantes de la décroissance exponentielle de l'activité volumique de l'eau en fonction de la distance x au point de rejet, respectivement sans et avec décroissance radioactive (notations déjà explicitées dans le texte, § 2).

λ la constante de décroissance radioactive,

a l'activité rejetée dans le cours d'eau par unité de temps,

t la durée du rejet,

v la vitesse moyenne de translation des sédiments de fond,

s l'activité des sédiments de fond par unité de longueur du lit, à la distance x.

L'activité qui traverse par unité de temps de rejet une section droite de la rivière située à une distance X a pour valeur :

$$Q = ae^{-kX} \qquad (1)$$

Considérons une deuxième section droite distante de la première d'une longueur élémentaire dX. L'activité perdue par l'eau entre ces deux sections droites, par unité de longueur et par unité de temps de rejet, a pour valeur :

$$dQ/dX = - kQ$$

Celle perdue du fait de l'adsorption seule a pour valeur :

$$dQ/dX = - k_oQ$$

Elle est égale à l'activité déposée sur le sédiment, par unité de longueur et par unité de temps de rejet. Comme le sédiment met un temps dt pour franchir la distance dX, l'activité déposée par unité de longueur pendant ce temps a pour expression :

$$ds(X) = k_oQ\ dt,$$

ou, en remplaçant Q par l'expression (1) et dt par dX/v :

$$ds(X) = \frac{ak_o}{v} e^{-kX}\ dX$$

Du fait de la décroissance radioactive du nucléide, cette activité va devenir à la distance x :

$$ds(x) = ds(X)\ e^{-\lambda(x-X)/v}$$

L'intégration s'écrit :

$$s(x) = \frac{ak_o}{v}\ e^{-\lambda x/v} \int_{x_o}^{x} e^{-(kv-\lambda)X/v}\,dX$$

La valeur de la borne inférieure x_o dépend de la position initiale du sédiment, selon qu'il se trouvait en amont ou en aval du point de rejet au temps $t = o$.

a) <u>le sédiment se trouvait en amont</u>, c'est-à-dire $o < x \leq vt$.

Dans ce cas, $x_o = o$

$$s = \frac{ak_o}{kv - \lambda}\ (e^{-\lambda x/v} - e^{-kx}) \qquad (2)$$

La valeur de cette fonction passe par un maximum pour :

$$x_m = \frac{\log_e\ (kv/\lambda)}{k - (\lambda/v)} \qquad (3)$$

b) <u>le sédiment se trouvait en aval</u>, c'est-à-dire $x \geq vt$.

Dans ce cas, $x_o = x - vt$

$$s = \frac{ak_o}{kv - \lambda}\ e^{-kx} \left[e^{(kv-\lambda)t} - 1 \right] \qquad (4)$$

Remarque : si v tend vers zéro, seule l'équation (4) est valable ; elle devient :

$$s = \frac{ak_o}{\lambda}\ e^{-kx}\ (1 - e^{-\lambda t}) \qquad (5)$$

ANNEXE II

Calcul du facteur de concentration

En plus des notations déjà explicitées à l'annexe I, désignons par :

q le débit moyen du cours d'eau,

L la largeur de son lit,

h l'épaisseur de la couche supérieure des sédiments du fond du cours d'eau, où l'activité adsorbée se trouve concentrée,

ρ la masse volumique des sédiments.

L'activité massique des sédiments a pour expression :

$$A = \frac{s}{Lh\rho}$$

Celle de l'eau a pour expression :

$$C = \frac{a}{q} e^{-kx}$$

Le facteur de concentration s'écrit donc :

$$F = \frac{A}{C} = \frac{sq}{aLh\rho}$$

ANNEXE III

Calcul des doses résultant de l'irradiation externe par les sédiments

En plus des notations déjà explicitées à l'annexe II, désignons par :

$\dot{D}$ le débit de dose, exprimé en millirems par heure, qui correspond à une activité volumique des sédiments de 1 curie par mètre cube,

θ la durée de séjour d'un individu en bordure du cours d'eau, exprimée en nombre d'heures par an.

La dose reçue, exprimée en millirems par an, a pour valeur :

$$D = F\, C \rho\, \theta\, \dot{D}$$

L'expression de $\dot{D}$ est différente selon qu'il s'agit de rayonnements bêta ou de rayonnements gamma.

a) Rayonnements bêta

On a l'expression générale suivante, tirée de la référence [7] :

$$\dot{D} = 100\, n\, \eta \left[17{,}8 \left(R_o - h + h \log_e \frac{h}{R_o}\right) + 5{,}85\, R_o^{0{,}5} \left(1 - \frac{h}{R_o}\right)^2\right]$$

où :

h est la profondeur du point considéré dans l'organisme, exprimée en centimètres,

R_o est le parcours maximal de l'électron dans les tissus, exprimé en centimètres,

η est le rendement du rayonnement bêta par désintégration,

n est le pouvoir d'arrêt des tissus relatif aux sédiments.

Pour déterminer la valeur de ce dernier paramètre, nous nous sommes référés à la composition du sable de Seine : la somme pondérée des pouvoirs d'arrêt partiels propres à chacun des éléments constituants de ce sable donne n = 0,77.

b) Rayonnements gamma

L'épaisseur de la couche de sédiments contaminés (quelques dizaines de centimètres) étant très grande par rapport à la longueur de relaxation des rayonnements gamma dans le sable (en général, quelques centimètres), on peut admettre qu'un état d'équilibre s'est établi et que l'énergie absorbée par unité de volume de sable est égale à l'énergie émise dans ce même volume.

Dans ces conditions, le débit de dose dans le sable, exprimé en millirads par heure, qui correspond à une activité volumique de 1 curie par mètre cube, a pour valeur :

$$2{,}12 \times 10^3 \eta \, E/\rho$$

E représentant l'énergie (MeV) du rayonnement gamma et η son rendement par désintégration.

$\dot{D}$ est égal à ce débit de dose divisé par deux (puisque l'organisme est irradié sous un demi-angle solide) et multiplié par le rapport des kerma dans les tissus et le sable :

$$\dot{D} = \frac{1{,}06 \times 10^3 \eta \, E}{\rho} \, \frac{(\mu_K/\rho)\ \text{tissus}}{(\mu_K/\rho)\ \text{sable}}$$

REFERENCES

[1] PICAT, P., GRAUBY, A., «Evolution de la radiocontamination des canaux d'irrigation alimentés par le Rhône», Actes Symp. int. de radioécologie, CEA, CEN de Cadarache, 1969.

[2] SCHAEFFER, R., «Incidence des reconcentrations radioécologiques sur la valeur de la concentration maximale admissible des eaux de rivière», Population Dose Evaluation and Standards for Man and his Environment (C.R. Journées d'études Portoroz, 1974), AIEA, Vienne (1974) 525-38.

[3] BRISBIN, L.I., Jr., BEYERS, R.J., et al., Health Phys. 27 (1974) 19-27.

[4] COMMISSARIAT A L'ENERGIE ATOMIQUE, Radioécologie du bassin rhodanien, CEA, CEN de Cadarache, Service de recherches toxicologiques et écologiques (1974).

[5] PRESTON, A., MITCHELL, N.T., «Evaluation of public radiation exposure from the controlled marine disposal of radioactive waste (with special reference to the United Kingdom)», Radioactive Contamination of the Marine Environment (C.R. Coll. Seattle, 1972), AIEA, Vienne (1973) 575-93.

[6] UNITED STATES ATOMIC ENERGY COMMISSION, Attachment to Concluding Statement of Position of the Regulatory Staff, Draft Regulatory Guides for Implementation, Docket No. RM-50-2, Washington, D.C. 20545 (1974) A-13.

[7] SCHAEFFER, R., «Calculs de dose en irradiation externe par les gaz rares», 3e Congrès de l'Association int. de protection contre les radiations, Washington, 9-14 sept. 1973.

DISCUSSION

V.V. PISAREV: Can your formulae be applied to ^{90}Sr?

R. SCHAEFFER: The formula I gave is general and applies to any nuclide, but the numerical value of the constant k depends on the nature of the particular nuclide. In this connection I would refer you to Table I.

V.V. PISAREV: Why does a peak form in the concentration in the bottom sediments, followed by a drop?

R. SCHAEFFER: The peak corresponds to a state of equilibrium between the contributions from activity adsorbed on the sediments and the losses due to radioactive decay.

S. BONOTTO: You showed that the activity increases progressively in the downstream direction. Is it known what percentage of the activity is fixed by the flora and fauna (macroscopic or microscopic plants, and animals) as one goes downstream?

R. SCHAEFFER: My work dealt only with the contamination of sediments. However, I think it is logical to assume that the benthic organisms living in contact with the sediments fix an amount of radioactivity which is proportional to that of the sediments and that, consequently, for them too the contamination increases with distance from the point of release up to the maximum corresponding to the equilibrium state. This dependence would explain why their concentration factors are sometimes very high.

In other organisms, in particular fish, the activity fixed varies in the opposite manner, and I have dealt with this question elsewhere (see Ref. [2] of the paper).

Yu.V. SIVINTSEV (Chairman): What flow-rate was assumed for the water in your calculations? Is it constant along the course of the river?

R. SCHAEFFER: For simplicity, I assumed that the flow-rate has a constant mean value for a distance of about 200 km downstream of the discharge point. Actually, of course, the flow-rate varies along the course of the river, but this would not significantly affect our calculations.

Yu.V. SIVINTSEV: Can you demonstrate correspondence between the calculated and experimental data - if you have any experimental data?

R. SCHAEFFER: All indications in the literature point to a general increase in the activity of the sediments in the downstream direction for a considerable distance from the point of discharge. There is, therefore, good qualitative agreement between experiment and theory. Quantitatively, the agreement is by no means as good, owing to the bed irregularities and the presence of hydroelectric installations.

P.L. BOVARD: In calculating the adsorbed activity, did you take account of the grain size as well as the time of contact?

R. SCHAEFFER: The sediments were assumed to be of the same nature (grain size and physico-chemical state) as those in the Rhône irrigation canals. This is explained in the paper (section 2).

P.L. BOVARD: As you stated, damming modifies the laws of deposition. Is it possible to take account of damming or dredging, using your theory?

R. SCHAEFFER: Damming or dredging can be taken into account in the calculation, but they complicate the laws of hydraulics by introducing empirical factors. For this reason it is better to use small-scale models for such studies.

P.L. BOVARD: In most cases a river does not behave like a homogeneous liquid flow but like a juxtaposition of flows, especially in the case of confluence with large tributaries. Has this any influence on the use of the formula and on the measurement of the mean flow rate?

R. SCHAEFFER: In a non-linear calculation, replacing the true distribution of particular values by a mean value always introduces an error. However, this can be regarded as acceptable if the aim of the research is to obtain only a very approximate result, as is the case here.

V.V. PISAREV: Your paper presents a simple mathematical model for the accumulation of radionuclides in the bottom sediments of rivers and, although there are only few experimental data, you get good agreement with the model. However, I think that a strict solution of this problem is possible only with the aid of second-order differential equations.

R. SCHAEFFER: It is possible to take account of various characteristics of the watercourse in this calculation: variation in flow-rate, irregularities in the bed downstream of the release point, etc. However, owing to the complexity of the laws of hydraulics and sedimentology governing these phenomena, I think that the quest for greater mathematical accuracy in this field may be fruitless and that a simplified calculation may be preferable.

BEHAVIOUR OF DISCHARGED RADIONUCLIDES FROM FUEL REPROCESSING OPERATIONS IN THE AQUATIC ENVIRONMENT OF BOMBAY HARBOUR BAY

K.C. PILLAI, N.N. DEY, Elizabeth MATHEW, B.U. KOTHARI
Environmental Studies Section,
Health Physics Division,
Bhabha Atomic Research Centre,
Trombay, Bombay, India

Abstract

BEHAVIOUR OF DISCHARGED RADIONUCLIDES FROM FUEL REPROCESSING OPERATIONS IN THE AQUATIC ENVIRONMENT OF BOMBAY HARBOUR BAY.

Bombay Harbour bay has been receiving low-level liquid effluents from the irradiated fuel reprocessing facility of the Research Centre at Trombay for nearly a decade. The paper gives details of the extensive studies carried out in order to assess the environmental impact of such releases. The interaction of the waste effluents with different matrices of the aquatic environment like sea water, sediment and organisms results in fractionation of the waste constituents and selective accumulation. The interaction and distribution of the radionuclides ^{90}Sr, ^{106}Ru, ^{144}Ce, ^{137}Cs and ^{239}Pu were studied in detail.

High accumulation of radionuclides was observed in suspended and bottom sediments. Distribution factors in silt were of the order of 10^3 to 10^5 for Ru, Cs, Ce and Pu. Though the releases of Pu were extremely small, measurable quantities were found to be accumulated near the discharge area. Core samples of bottom sediments near the low water line indicated the pattern of deposition of radionuclides. The nature of the radionuclides in the sediments was investigated.

Studies on the uptake of these radionuclides in bay organisms from areas close to the discharge location showed selective accumulation of radioisotopes of Cs and Ru by benthic organisms. Ruthenium-106 accumulated in the Arca shell was found to be mostly in the outer layers of the shell and a significant portion of it was present in the organically bound form. The uptake of Pu by benthic organisms indicated the possibility of higher accumulation of Pu from waste effluents compared to those from fall-out sources.

Solar salt from salt pans scattered along the coastline of the bay was examined for possible contamination of this marine product. Only ^{90}Sr was found to be accumulated in salt. $^{90}Sr/Sr$ ratios in salt samples from different salt pans along the bay indicated the extent of mixing of effluents with the waters at different locations in the bay.

The possible effect of the accumulation of radionuclides on sediments, fish and salt are evaluated. The capacity of suspended silt for removing radionuclides in coastal waters is computed. An assessment of maximum radiation exposure that can result from waste releases into the bay is made.

INTRODUCTION

Bombay harbour bay has been receiving low-level radioactive liquid effluents after treatment from the Bhabha Atomic Research Centre, Trombay, for more than a decade. In addition, the bay is also a recipient of domestic and industrial wastes from the city of Bombay and adjoining areas. The Research Centre, situated on the west coast of the bay, has four research reactors, thorium and uranium plants, an irradiated fuel reprocessing plant, metallurgy and engineering laboratories, isotope production facilities and other research laboratories.

The bay opens up on the south into the Arabian sea. A part of the Ulhas river flows into the bay through the Thana Creek in the north. Figure 1 is a map of the bay and adjoining areas. The bay is subject to semidiurnal

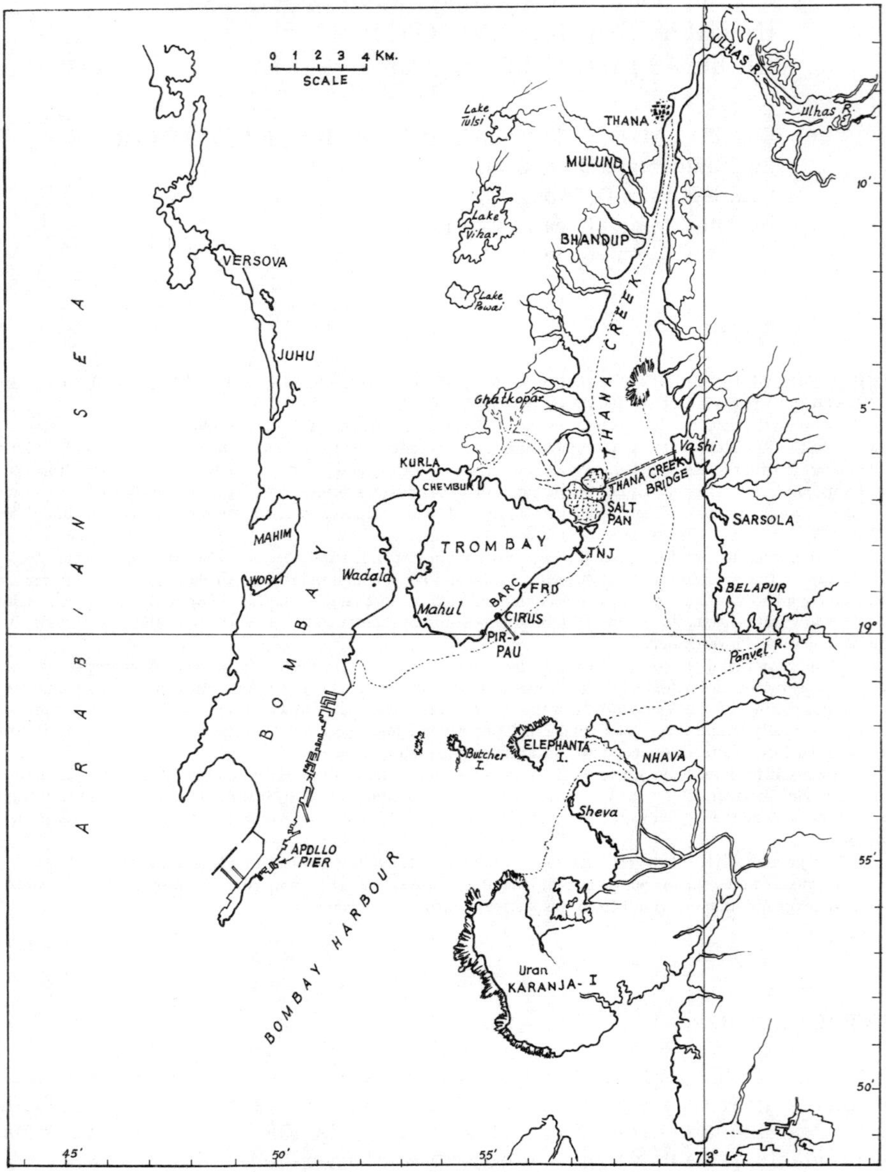

FIG. 1. Bombay Harbour Bay and adjoining areas.

tides. Extensive studies on water movement patterns in the bay, dilution factors available for effluent discharges, possible inflow of river waters into the bay and evaluation of the bay environment as a recipient of low-level radioactive liquid effluents were carried out before commissioning of the CIRUS reactor and other production plants at the Research Centre and are reported elsewhere [1-4].

The irradiated fuel reprocessing facility, with a nominal capacity of 30 t/a of uranium, started operations in 1964. The processes followed for recovery of plutonium and concentration and containment of fission-product wastes are given in detail elsewhere [5, 6]. The low-level liquid effluents are discharged through a pipeline into Bombay harbour bay. The major radionuclides in the low-level liquid effluents released from the plant into the bay are ^{144}Ce, ^{137}Cs, ^{90}Sr, ^{106}Ru and ^{95}Zr. The effluents also contain traces of alpha activity, the major component of which is identified as Pu. Investigations carried out to study the behaviour of discharged radionuclides in the bay environment and to assess the impact of such releases are reported here.

LABORATORY AND FIELD STUDIES

Some experiments conducted to study the nature of the effluents revealed that about 25% of ^{144}Ce in the effluent was in the particulate form (0.22 μm Millipore) and that ^{137}Cs and ^{106}Ru were in the soluble form. Soluble ^{144}Ce and ^{137}Cs were in the cationic form whereas ^{106}Ru was all in the anionic form. However, on mixing the effluents with sea water (1 : 1000) and filtering within an hour through 0.22 μm Millipore filter paper it was observed that all ^{144}Ce was rendered insoluble and was retained on the filter paper, leaving all Ru and Cs in the soluble form. The ionic nature of the dissolved species was studied by passing the filtrate through cation (Na^+) and anion (Cl^-) resins, and it was found that there was no change in the forms of Cs and Ru. These laboratory studies however were not extended further to study changes that might take place with more time of contact with sea water. Earlier [7] it had been found that a maximum of 43 dpm/l of ^{239}Pu only could be present in soluble form in sea water as a result of mixing of the effluents with sea water.

The physico-chemical state of radionuclides in the effluents at the point of introduction into the coastal waters and/or the physico-chemical state into which the radionuclides are transformed soon after introduction, determine their initial fractionation in the neighbourhood of the discharge area. The hydrolysis and precipitation of the introduced compounds on mixing with sea water and sorption of radionuclides by silt and sedimentation determine the extent of relative amounts of discharged radionuclides that can get transported by dispersion and tidal transport from the discharge location. Therefore interactions of released radionuclides with different matrices of the environment were studied by obtaining samples of sea water, suspended silt and sediments from the discharge areas and measuring their radionuclide content.

The nature of radionuclides still remaining in solution after the initial interaction in the receiving medium may further change with time and environmental conditions. The silt that has accumulated radionuclides can get transported, deposited or get resuspended in the bay. Bioaccumulation

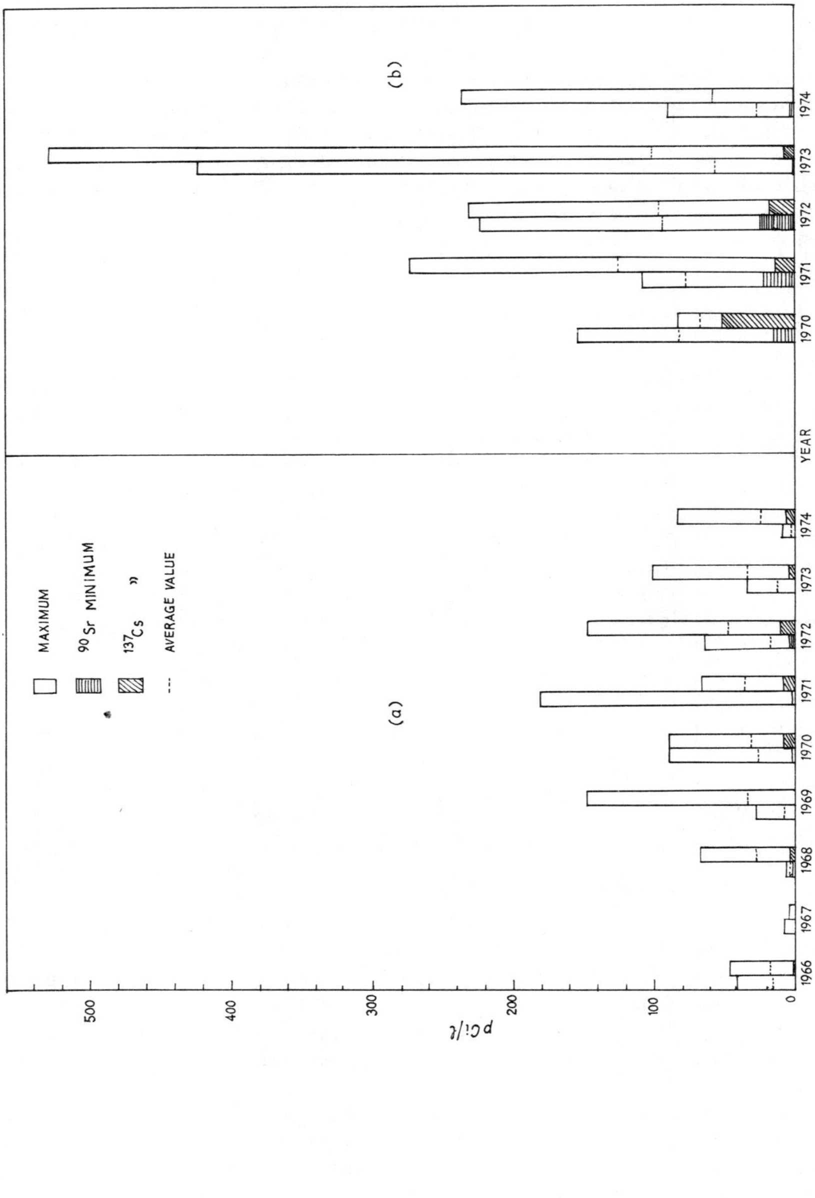

FIG. 2. Range of values of ^{90}Sr and ^{137}Cs in sea water at (a) CIRUS Jetty, (b) Thana Creek Bridge (Trombay end).

TABLE I. AVERAGE CONCENTRATION OF ^{90}Sr AND ^{137}Cs IN SEA WATER SAMPLES FROM SHORE LINE STATIONS IN THE BAY
Values in pCi/l

Year	CIRUS Jetty		Thania Creek Bridge		Belapur		Apollo Pier		Outside bay	
	^{90}Sr	^{137}Cs	^{90}Sr	^{137}Cs	^{90}Sr	^{137}Cs	^{90}Sr	^{137}Cs	^{90}Sr	^{137}Cs
1974	3.2	23.0	26.6	57.9	6.2	14.4	-	-	-	-
1973	12.2	34.4	55.6	102.5	6.7	26.1	1.2	6.8	0.05	2.5
1972	18.3	48.5	94.1	96.1	16.0	31.4	2.7	8.5	1.6	4.8
1971	50.5	35.7	78.6	126.4	34.1	27.5	3.9	6.5	0.7	1.4
1970	25.9	32.5	85.0	68.2	N.D.	1.7	3.3	4.8	1.1	1.1
1969	8.4	33.1	-	-	-	-	-	-	-	-
1968	4.5	28.1	-	-	-	-	0.14	0.4	0.14	1.2
1967	2.4	1.5	-	-	-	-	-	-	-	-
1966	15.5	18.1	-	-	-	-	-	-	N.D.	0.5

N.D. = not detected

of radionuclides can take place either from sea water medium or through the sediments. Marine products like solar salt obtained from the bay waters might also accumulate radionuclides. The distribution and accumulation of radionuclides in different matrices of the bay environment are thus necessary to assess the impact of radioactive effluent releases into it.

Sea water

The sea water samples collected were surface samples and mostly from shore-line stations on either side of the discharge point. The frequency of sampling was once a week or once a fortnight. Occasionally sea water samples from the mouth of the bay, from the east coast of the bay and also from outside the bay were obtained. Suspended silt in sea water was removed by filtration (Whatman No.42) and ^{90}Sr and ^{137}Cs were estimated in sea water by radiochemical methods. A few samples from areas close to the discharge location were analysed also for ^{106}Ru, ^{144}Ce and ^{239}Pu.

As a result of the tidal oscillations the effluent stream may move towards the north or south. The range of concentrations of ^{90}Sr and ^{137}Cs in sea water during 1966-74 from stations near to discharge locations on either side are given in Fig. 2. In many sea water samples it was found that $^{137}Cs/^{90}Sr$ ratios were lower in sea water compared to effluents showing preferential removal of ^{137}Cs by silt. Transient high levels of ^{90}Sr for short durations were observed in sea water during 1971 and 1972 owing to higher discharges in those years.

Table I gives the average concentrations of ^{90}Sr and ^{137}Cs in sea water during the years from the stations observed. The average values for ^{90}Sr and ^{137}Cs at Thana Creek Bridge, CIRUS and at the mouth of the bay

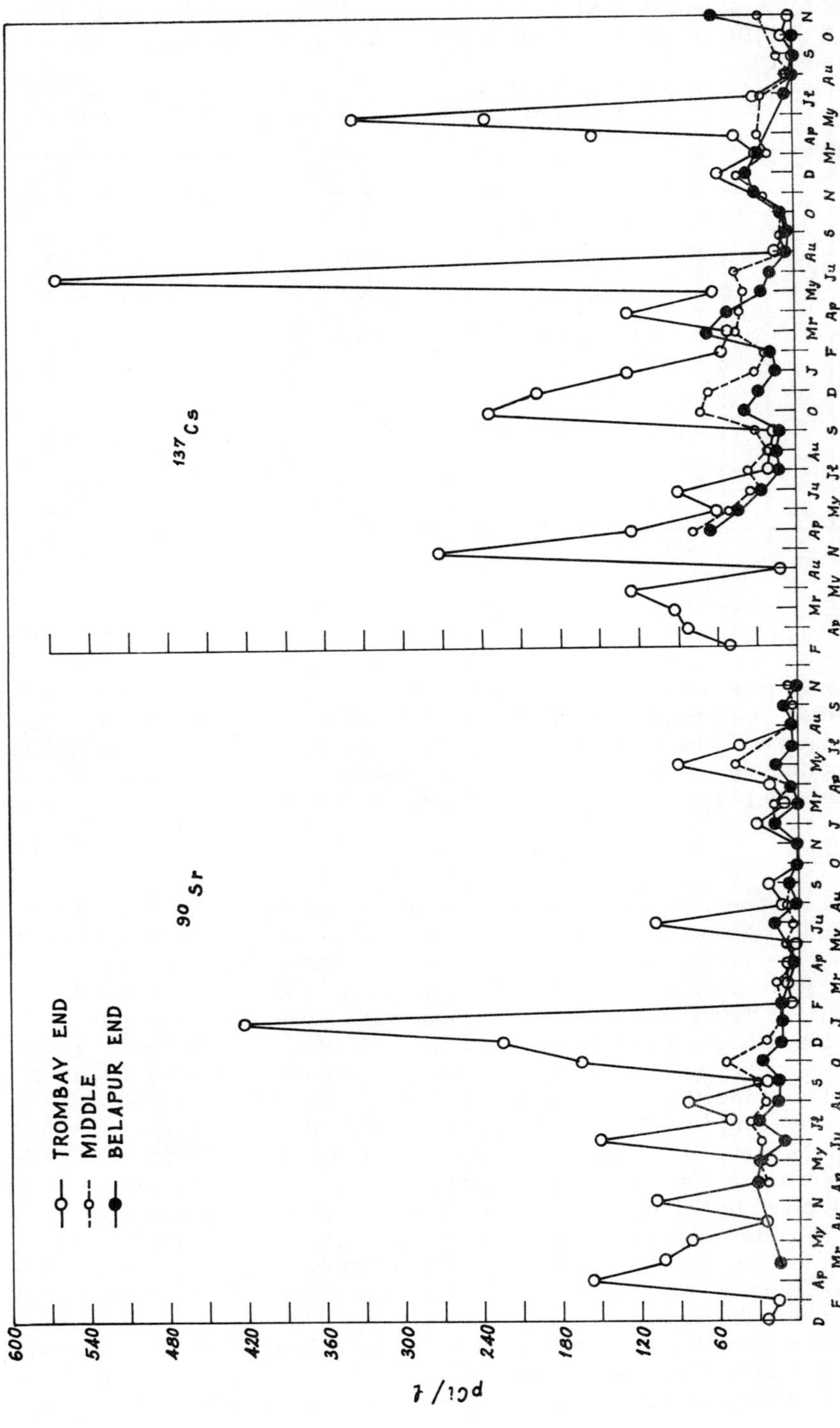

FIG. 3. Concentration of ^{90}Sr and ^{137}Cs in sea water samples (pCi/l) taken from the Thana Creek Bridge.

(Apollo Pier) show the extent of mixing of the effluents with the tidal stream flowing near the Trombay shore. Compared to the average values of ^{90}Sr and ^{137}Cs in sea waters outside the bay (samples from Worli, Mahim and Juhu, see Fig. 1) the average concentrations of these nuclides at Apollo Pier were never more than 5 times as high. The levels of ^{90}Sr and ^{137}Cs in sea water on the eastern part of the bay (Belapur) indicate the extent of lateral mixing of the effluents in the bay and the build-up over the years.

The Thana Creek Bridge spans the entire width of the bay between Trombay and Vashi (see Fig. 1). By making use of the bridge, sea water samples from Trombay and the middle of the bay and the Belapur end were obtained simultaneously. Samples were collected periodically in different tidal conditions and ^{90}Sr and ^{137}Cs levels were determined. Figure 3 gives the levels of these nuclides in sea water in these three locations in the bay for a number of days. The levels of ^{90}Sr and ^{137}Cs showed wide fluctuations. Half the number of observations indicate that at the Belapur end the bay waters had on average more than 20% and 40%, respectively, of the ^{90}Sr and ^{137}Cs concentrations at the Trombay end, which is evidently due to the larger mixing of the effluent-carrying shore stream with the bay waters.

Suspended silt

The suspended silt from sea water which was separated by filtration was dried at 105°C and weighed and gamma counted by means of a NaI(Tl) detector and 512 channel analyser. The dried sample was also counted for beta activity.

The silt load in the bay waters varied with the seasons and tidal conditions. From the data on the sea water samples collected during the last 5 years it is seen that the silt load varied from 9 to 2814 mg/l. Nearly 50% of the samples during this period showed silt concentrations less than 200 mg/l. Higher silt loads were obtained in samples taken near the shore. Suspended silt showed accumulation of mainly ^{137}Cs but in samples collected from areas near discharge locations ^{106}Ru and ^{144}Ce were also observed. Only total beta and ^{137}Cs values are taken for comparing the activity levels at different locations in the bay. Figure 4 shows the range of values of silt load in sea water and total beta and ^{137}Cs in suspended silt from shore line stations on either side of the discharge point. Table II gives the average values of total beta and ^{137}Cs levels in suspended silt obtained from various stations. Maximum activity concentrations in silt were at Trombay Naval Jetty (TNJ), which is close to the discharge location, and the maximum was recorded in 1971. The average values were highest in all stations during this year.

To obtain quantitative data on the extent of accumulation of radionuclides by the suspended silt, samples of sea water taken from areas close to the discharge location, both sea water and silt, were analysed separately for ^{90}Sr, ^{137}Cs, ^{106}Ru, ^{144}Ce and ^{239}Pu. Both Ru and Ce in sea water were separated by coprecipitation with Mn. Plutonium in sea water samples was estimated by coprecipitation with $BiPO_4$, radiochemical separation and alpha spectrometry using an internal tracer ^{236}Pu [7]. Since ^{239}Pu and ^{240}Pu peaks cannot be separated by alpha spectrometry, the Pu concentrations are expressed as ^{239}Pu. Silt samples were treated repeatedly with 8N HNO_3, and ^{90}Sr and ^{239}Pu were estimated in the extract radiochemically.

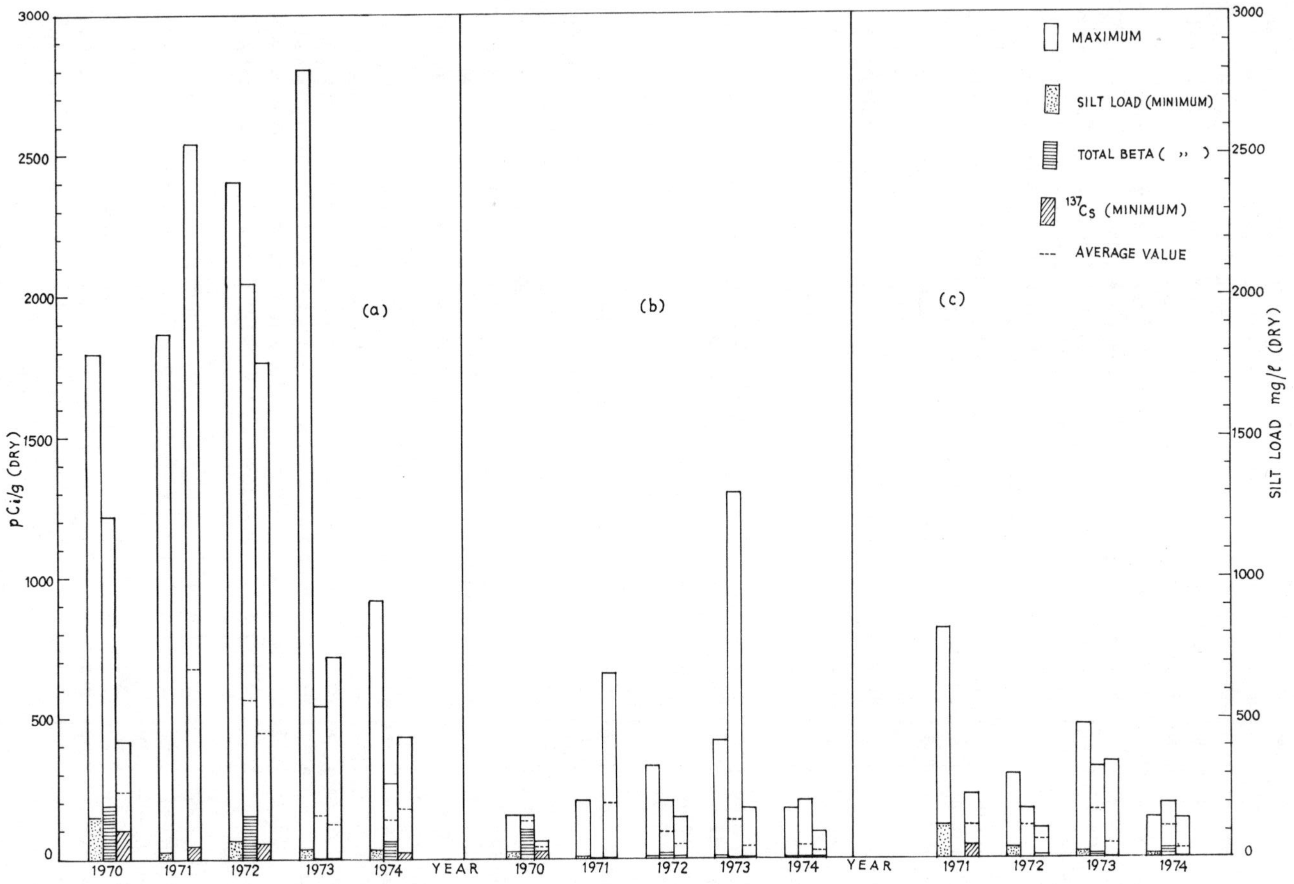

FIG. 4. Range of values for silt load in sea water (mg/l), total beta activity and ^{137}Cs in silt (pCi/g) at (a) Trombay Naval Jetty, (b) CIRUS, (c) Thana Creek Bridge.

TABLE II. AVERAGE VALUES OF TOTAL BETA AND ^{137}Cs IN SUSPENDED SILT SAMPLES FROM SHORE LINE STATIONS IN THE BAY
Values in pCi/g (dry)

Year	Trombay Naval Jetty		Thana Creek Bridge		Belapur		CIRUS		Apollo Pier		Outside bay	
	Total beta	^{137}Cs	Total beta	^{137}Cs	Total beta	^{137}Cs	Total beta	^{137}Cs	Total beta	^{137}Cs	Total beta	^{137}Cs
1974	144.5	175.5	123.7	50.7	31.7	27.4	50.2	31.8	-	-	-	-
1973	162.1	129.0	132	84.0	35.1	14.5	138.7	45.6	34.3	N.D.	30.6	3.2
1972	571.2	445.8	118.6	71.0	51.2	16.7	96.6	56.0	N.D.	N.D.	44.5	8.6
1971	-	679.1	-	126.7	-	42.6	-	195.1	-	N.D.	-	-
1970	619.0	242.0	-	-	-	-	135.5	52.1	-	-	17.6	1.8

N.D. = not detected

TABLE III. K_d FACTORS FOR RADIONUCLIDES IN SUSPENDED SILT

Nuclide	K_d factors		
	Minimum	Maximum	Average
^{90}Sr	4.3	31.0	15.8
^{137}Cs	4.22×10^2	1.57×10^3	8.70×10^2
^{106}Ru	1.75×10^3	8.72×10^3	4.86×10^3
^{144}Ce	1.69×10^4	1.32×10^5	8.67×10^4
^{239}Pu	4.84×10^4	1.31×10^5	8.96×10^4

TABLE IV. AVERAGE VALUES OF ^{137}Cs IN SHORE SEDIMENTS
Values in pCi/g (dry)

Year	Trombay Naval Jetty			Thana Creek Bridge	CIRUS	Belapur
	^{106}Ru	^{137}Cs	^{144}Ce	^{137}Cs	^{137}Cs	^{137}Cs
1974	N.D.	147.8	N.D.	48.7	-	17.1
1973	N.D.	179.3	107.3	-	-	24.6
1972	51.1	237.9	186.3	-	-	-
1971	40.9	347.0	302.1	88.9	1077.1	14.3
1970	7.0	76.9	80.6	54.8	476.6	-
1969	-	-	-	-	-	-
1968	-	11.7	-	11.1	-	-

N.D. = not detected

From the concentration of these nuclides in sea water and suspended silt, and the silt load in the water, K_d factors (distribution coefficient) for the nuclides in silt have been estimated. The range of values and the average K_d factors obtained are given in Table III. The highest K_d factor obtained was for ^{239}Pu, followed by ^{144}Ce, ^{106}Ru, ^{137}Cs and ^{90}Sr.

Shore and bottom sediments

The shore and bottom sediments accumulate radionuclides by sorption from sea water or by sedimentation of suspended radioactive solids. During low tides, large areas of the shore get exposed. The extent of contamination of the shore sediments was investigated by collection of surface grab samples. The sediments, after drying at 105°C, were counted by gamma spectrometry. In a few cases ^{90}Sr and ^{239}Pu were analysed radiochemically. Table IV gives the average concentration of radionuclides in shore sediments

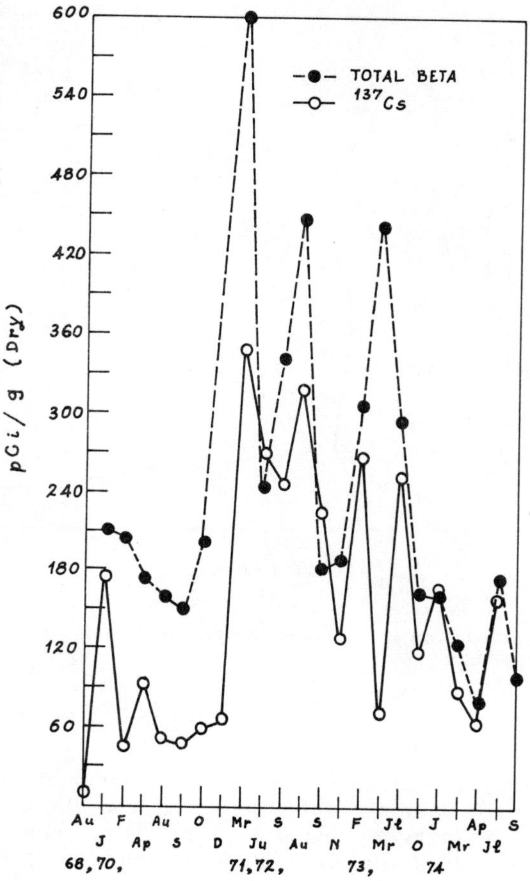

FIG. 5. Build-up of activity on shore sediments at Trombay Naval Jetty from 1968 - 1974. The measurements were not taken regularly, and the specific dates given are not critical; the curves show the build-up over the years.

along the bay from four locations, two each on either side of the discharge point, over a period of 3-4 years. Significant concentrations of ^{144}Ce and ^{106}Ru were found in stations very close to the discharge area only. In areas away from discharge locations, ^{137}Cs was the predominant nuclide in shore sediments.

Figure 5 gives the build-up of activity (total beta and ^{137}Cs) over the years in shore sediments at one station close to the discharge area. The maximum concentration of radionuclides in shore sediments was observed at this location (cf. Table IV). This is due to the greater time of contact of the effluent plume with the shore sediments during high tide and possible higher sedimentation of the suspended matter that gets transported from the discharge area owing to the slowing-down of the tidal currents at this location as a consequence of the land projecting into the bay.

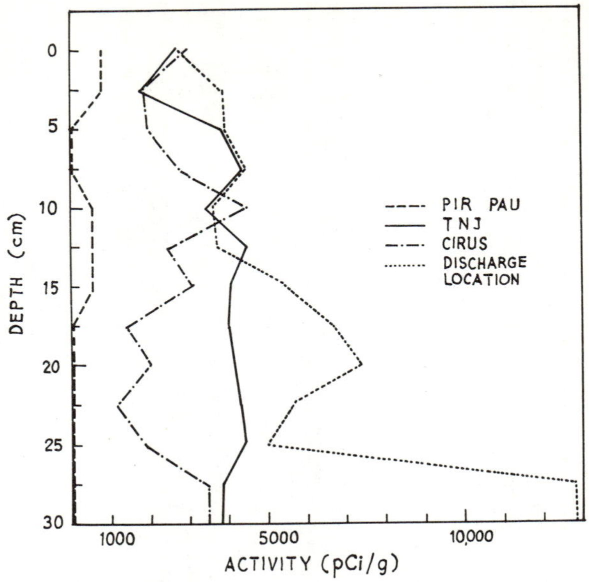

FIG. 6. Distribution of total beta activity in sediment cores collected along the low water line near discharge location (12 Dec. 1972).

The shore sediments in general have shown accumulation lower than that of suspended silt from sea water collected from the location. The accumulation was lower by factors of 2-10 for ^{137}Cs. This is due to the greater time of contact of silt with sea water, fineness of the floating silt and consequent higher sorption capacity and the possibility of suspended silt getting transported from the discharge area. In addition, shore sediments might get depleted in its finer fraction of clay and silt by wave action and tidal currents. The exposed shore areas during low tide can be a source of external exposure. The main contribution is from ^{137}Cs.

Surface sediments collected in 1971 from areas close to the discharge location along the low water line during neap tides had shown more accumulation of radionuclides than the shore sediments. Cerium-144 and ^{106}Ru were found to be more localized than ^{137}Cs [7]. Depth distribution of radionuclides in this area of maximum accumulation were investigated by collecting core samples of 30-cm length. Along the low water line, core samples were collected from the discharge location and on either side of it. The sediments were divided into 12 fractions of 2.5 cm each, cut, dried and weighed. The fractions were counted for beta activity. The depth versus beta activity levels of the sediment cores are shown in Fig.6. No definite pattern of distribution of activity with depth could be observed, possibly because of the non-uniformity of discharges. The distribution of activity in the top portions of the core (0-15 cm) was more or less uniform in the core samples; however the lower halves of the cores in all cases (TNJ, discharge location and CIRUS) show the progressive reduction in

TABLE V. ACCUMULATION OF RADIONUCLIDES IN BOTTOM SEDIMENTS IN THE BAY – RANGE OF CONCENTRATIONS
Values in pCi/g (dry)

	1968	1970
Number of samples:	18	16
Radionuclides		
Pu-239	0.003 - 0.134	0.002 - 0.827
Ce-144	N.D.	N.D. - 102.9
Cs-137	0.9 - 54.7	0.44 - 90.42
Ru-106	N.D.	N.D. - 22

N.D. = not detected

activity with distance. It is apparent that accumulation of activity was mostly between TNJ and CIRUS jetty only and the levels were very much low at Pir Pau.

Bottom sediments obtained from different locations in the bay during the pre-monsoon periods of 1968 and 1970 were examined for radionuclide content. The surface sediments collected were grab samples. The range of accumulation of radionuclides in the sediments is given in Table V. Maximum accumulation was observed in samples from the area opposite the discharge point. During 1968 the levels of ^{144}Ce and ^{106}Ru were very low and could not be estimated by gamma spectrometry. Caesium-137 accumulated in bottom sediments more than any other nuclides discharged. However, in 1970 ^{144}Ce accumulated in sediments to the maximum extent in the area opposite the discharge point. The accumulation of ^{137}Cs in sediments in this location was higher than in 1968 by a factor of 2. However, for ^{239}Pu the factor was about 7, even though there was no proportionate increase in alpha activity in the discharged effluents. Evidently, this was due to the localisation of Pu, like Ce as observed earlier, due to its higher K_d factors. Further, the distribution of activity showed that significant accumulation of activity in bay sediments was restricted to the stream flowing near Trombay. Patel et al. [8] have studied the depth distribution of activity in bottom sediments in the bay and reported ^{137}Cs activity up to a depth of 30 cm in some of the core samples.

The accumulation of radionuclides in bottom sediments is of significance from an ecological point of view since the sediments serve as a source of nutrients especially for benthic organisms and not only do they supply some of the accumulated radioisotopes for uptake but the general contamination level in the sediments determines the chronic external dose to such organisms.

Nature of nuclides in sediments

The radionuclides that remain associated with sediments are influenced greatly by the chemical, biochemical and microbiological changes that take place in the environment. In the dynamic process of exchange of these

TABLE VI. PERCENTAGE OF LEACHABLE NUCLIDES IN SEDIMENTS

Leaching solutions	^{137}Cs	^{106}Ru	^{144}Ce
1. 0.05N HCl + 75 ml H_2O_2 pH 5.8	0.8	3.6	0.6
2. 0.05N HCl + 200 ml H_2O_2 pH 5.8	1.5	5.3	N.D.
3. 1N ammonium acetate pH 7.0	9.3	N.D.	N.D.
4. 0.1N HCl 300 ml	7.3	14.2	50.8
5. 1N HCl 250 ml × 3 times	25.2	25.8	100
6. 4N HCl 250 ml	40.8	20.5	100
7. 5% EDTA pH 7.0 300 ml	7.8	7.0	37
8. 5% EDTA pH 7.0 600 ml	14.4	22.3	67.2
9. 5% EDTA pH 7.0 900 ml	23.5	35.0	77.4
10. Filtered sea water	4.0	N.D.	N.D.
11. Filtered sea water + organic matter	3.5	N.D.	N.D.

N.D. = not detected

nuclides with other matrices of the aquatic environment, it is the labile components that are involved. In sediments, such labile constituents can be composed of the exchangeable or removable ions on the surface, or the hydrous precipitates that undergo dissolution by the complexing action of the chemicals in the environment.

To study the nature of radionuclides sorbed on sediments, contaminated bay sediments from the discharge location were leached with different leaching agents (over night) and the amount of radionuclides ^{137}Cs, ^{106}Ru and ^{144}Ce, leached out was determined [9]. The percentage of leachable radionuclides of the total present in the sediment was estimated (see Table VI). Ammonium acetate did not remove any of these nuclides; this indicates that these nuclides are not present in the sediment in an exchangeable form. It is to be noted that positive values for ^{137}Cs are due to the presence of the nuclide in sea water present in the wet sediment. Ruthenium-106 from H_2O_2-treated sediment indicates the presence of organically bound Ru in sediments. This was also evident from the laboratory experiments reported earlier on the sorption of Ru on sediments and organic-matter-free sediment. The sediment containing organic matter sorbed 6-7 times more Ru than the one which is organic-matter-free. The presence of non-dialysable Ru in sea water as well as in well waters (near fuel reprocessing plants) has also been established; this indicates the existence of Ru in colloidal form or as organic complexes [10]. Hydrochloric acid of different concentrations only partly solubilized Ru from sediments whereas all Ce was solubilized; this indicates that Ce was retained by the sediment as hydroxide or carbonates. Complexing with EDTA removes 35% of Ru and 77% of Ce from sediments, which indicates the possible solubilization of these nuclides with similar complexing agents present in the aquatic environment. Sediments when leached with filtered sea water did not release any activity. Therefore it is unlikely that the sorbed nuclides will get released into the water when the sediments get transported to other areas. Small amounts of organic matter added to sea water also did not release any activity in the sediment on contact for a period of 16 hours. However, it is not possible to rule out the solubilization of nuclides with longer period of contact.

On leaching another batch of sediments with 6M HNO_3, only 23% of ^{137}Cs could be removed from the sediment. Earlier studies have shown the dominant montmorillonitic nature of the clay fractions in the bay sediments, so ^{137}Cs, picked up by mineral interaction and exchange, is not likely to be desorbed easily. Strontium, on the other hand, is present in the sediment in available form and can be leached out easily [9]. In experiments with contaminated sediments it was found possible to remove quantitatively all Pu sorbed in the sediments by repeated leaching with 8M HNO_3 [7]. Pu sorbed on sediments could not be released into sea water on repeated leaching with the same.

Organisms

Organisms (both benthic and non-benthic) collected periodically from the bay were examined for radionuclide content. Most of the organisms (especially the benthic organisms like crab and Arca) analysed were obtained from areas close to the discharge location. The specimens, after initial cleaning and separation into different parts like flesh, bones (shells) and gut, were weighed and dried at 105°C. These dry specimens were used for gamma-spectrometric measurements. In some cases ^{106}Ru, ^{90}Sr and ^{239}Pu were measured by radiochemical methods.

Table VII gives the concentration of various radionuclides in different tissues in the organisms. The levels indicated are in pCi/g wet tissue. The accumulation of radionuclides varied with the species. Benthic organisms like Arca and crab obtained from the discharge area showed accumulation of discharged nuclides ^{106}Ru, ^{144}Ce, ^{137}Cs, ^{90}Sr and ^{239}Pu. However, species like Arius showed mostly ^{137}Cs only in their flesh tissues and ^{106}Ru and ^{144}Ce could not be detected. Though these nuclides were detected in the gut, apparently only ^{137}Cs and ^{90}Sr have been metabolized. On examining the species Arca and catfish obtained at the same period, it was observed that ^{137}Cs levels were higher in catfish than in Arca flesh. With the latter, ^{106}Ru concentrations in flesh were higher than ^{137}Cs. The gamma ray spectra of bottom sediments and the spectra of Arca (flesh and shells) collected from the discharge area are given in Fig. 7. The gamma spectra have been normalized to the same weight of each of the components. As indicated earlier, sediments in the discharge location had accumulation of radionuclides in the order of $^{144}Ce > ^{137}Cs > ^{106}Ru > ^{90}Sr > ^{239}Pu$. However, in the flesh of these organisms, the concentration of radionuclides was in the order of $^{106}Ru > ^{144}Ce > ^{137}Cs > ^{90}Sr > ^{239}Pu$.

Arca samples collected on the same day from the discharge area and about one kilometre on either side of it parallel to the shore and close to the low water line clearly indicate the trend of accumulation of radionuclides in these organisms (see Table VII). The concentration of radionuclides including Pu, in flesh as well as in shells, showed a downward trend in either direction from the discharge point. In this case no attempts were made to correlate the uptake with the age of the organisms or to estimate

TABLE VII. RADIONUCLIDE CONCENTRATION IN BAY ORGANISMS

Serial No.	Date of collection	Location	Species	Part analysed	pCi/g (wet)				
					^{90}Sr	^{106}Ru	^{137}Cs	^{144}Ce	^{239}Pu
(1)	(2)	(3)	(4)	(5)	(6)	(7)	(8)	(9)	(10)
1.	2- 1-70	TNJ	Prawns	Whole	-	N.D.	8.18	N.D.	N.D.
2.	5- 6-70	TNJ	Crab (Scylla serrata)	Flesh Shell	- -	1.44 N.D.	9.24 25.7	1.23 N.D.	N.D. N.D.
3.	18- 9-70	TNJ	Crab (Scylla serrata)	Flesh Bone	9.63 184.0	5.35 N.D.	3.71 15.52	4.54 N.D.	0.0016 0.0025
4.	18- 9-70	TNJ	Catfish (Arius sp.)	Flesh	-	N.D.	10.06	N.D.	-
5.	30- 9-70	TNJ	Catfish (Arius sp.)	Flesh	-	N.D.	2.56	N.D.	-
6.	30-11-70	Offshore	Mackerel	Flesh Bone	0.054 -	N.D. -	0.023 -	N.D. -	0.005 0.0028
7.	7- 5-71	TNJ	Crab (Scylla serrata)	Flesh	-	20.2	20.5	3.66	-
8.	11- 6-71	TNJ	Crab (Scylla serrata)	Shell + Bone	-	85.3	54.7	N.D.	0.0033
9.	11- 6-71	TNJ	Catfish (Arius sp.)	Flesh Bone Gut	- - -	N.D. N.D. 230.5	29.6 62.6 76.2	N.D. N.D. 33.7	0.0004 0.0060 0.0023
10.	25-11-71	Discharge location	Arca (Anadara (granosa)	Flesh Shell	- -	28.76 60.0	13.05 N.D.	75.4 N.D.	0.0672 0.0079
11.	25-11-71	Cirus	Catfish (Arius sp.)	Flesh Bone	1.61 165	N.D. N.D.	29.1 68.6	N.D. N.D.	0.0055 0.024
12.	3- 6-72	TNJ	Arca (Anadara (granosa)	Flesh Shell	0.44 102.9	14.61 25.31	8.91 N.D.	7.41 N.D.	0.0139 0.0133
13.	18- 8-72	TNJ	Mudskipper (Periopthalmus schlosseri)	Flesh Bone Gut	- - -	N.D. 2.2 1.9	38.4 4.8 4.0	4.0 N.D. 1.0	- - -
14.	12-12-72	1.0 km north of discharge point	Arca (Anadara granosa)	Flesh Shell	- -	4.96 1.02	2.26 N.D.	3.52 N.D.	0.0139 0.0232
15.	12-12-72	Discharge location	Arca (Anadara (granosa)	Flesh Shell	- 337.8	19.71 169.2	4.47 N.D.	12.26 N.D.	0.04047 0.1457
16.	12-12-72	1.0 km south of discharge	Arca (Anadara granosa)	Flesh Shell	- -	5.38 6.2	2.77 N.D.	3.65 N.D.	- 0.01481
17.	21- 3-73	TNJ	White sardine (Kowala coval)	Whole	0.063	0.091	0.93	0.03	-
18.	16- 4-73	TNJ	Catfish (Arius sp.)	Flesh Bone Gut	0.74 238.8 -	N.D. 0.23 N.D.	6.9 1.5 7.0	N.D. 2.0 N.D.	- - -
19.	12- 9-73	TNJ	Prawns	Flesh Shell	N.D. 0.03	N.D. N.D.	N.D. N.D.	N.D. N.D.	- -
20.	18- 9-73	Thana Creek Bridge	Tilapia mossambica	Flesh Bone Gut	N.D. 5.0 0.033	N.D. N.D. N.D.	0.63 1.9 5.8	N.D. N.D. N.D.	- - -

TABLE VII. (contd)

(1)	(2)	(3)	(4)	(5)	(6)	(7)	(8)	(9)	(10)
21.	26- 9-73	TNJ	Prawns	Flesh	0.0067	N.D.	0.11	0.08	-
				Shell	0.45	N.D.	N.D.	0.59	-
22.	30-10-73	Sowri	Catfish	Flesh	0.022	N.D.	0.8	N.D.	-
			(Arius sp.)	Bone	3.4	N.D.	0.87	N.D.	-
				Gut	0.12	N.D.	0.91	N.D.	-
23.	7- 1-74	TNJ	Arca	Flesh	0.34	2.55	0.71	N.D.	0.0031
			(Anadara	Shell	0.18	4.72	6.9	N.D.	0.0025
			(granosa)						
24.	5-12-74	Versova	Catfish	Flesh	0.017	-	0.068	N.D.	0.0007
		(Offshore)	(Arius sp.)	Bone	0.02	-	0.068	N.D.	0.0038
25.	26-12-74	Mahul	Tilapia mossambica	Whole	0.004	-	0.105	N.D.	0.0015

N.D. = not detected

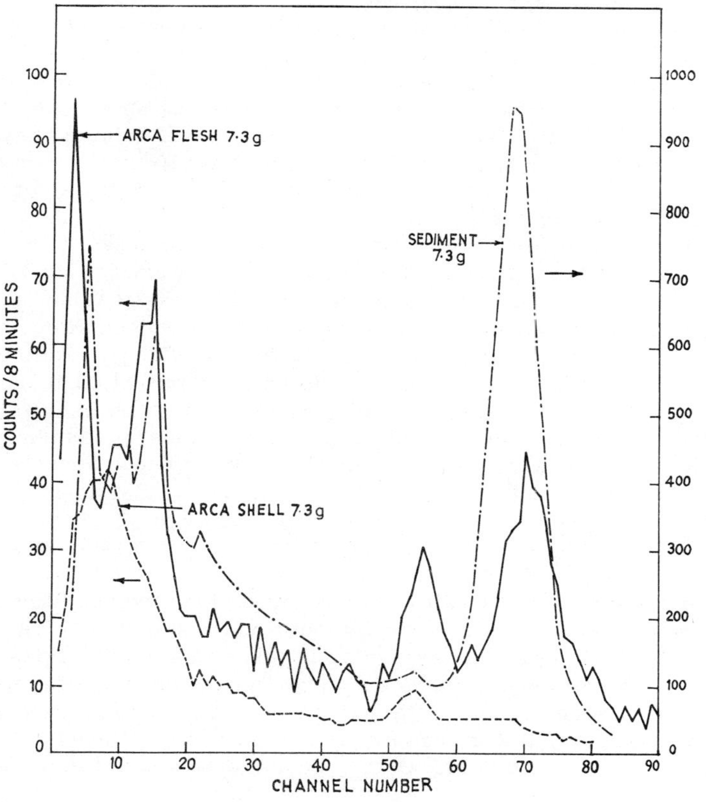

FIG.7. Gamma spectrum of sediment, Arca flesh and shell from discharge location (12 Dec. 1972).

the concentration factors of various nuclides in the organisms. The specimens of Arca collected from Trombay Naval Jetty over the years also reflect the ambient levels in that location. As such, Arca can be used to monitor the radionuclides discharged from fuel reprocessing operations.

The accumulation of ^{106}Ru was higher in shells than in the flesh of Arca. The higher accumulation is due to the high calcium carbonate concentration in the shell. It has been reported [11] that peel of the molluscan shells, which is an extension of the organic matrix of the shell, contains a high concentration of Co and ^{137}Cs, and possibly the concentrations in peel and outer layers are comparable. Chemical analysis of the matrix of the decalcified shell has been shown to contain protein, scleroproteins and polypeptides [12]. The shells were solubilized in 0.1<u>N</u> HCl and the insoluble residue containing mostly the outer coating on the shell were separated out. The insoluble residue amounting to 0.5 and 0.7% (in two cases) of the weight of the shell was found to contain more than 50% of the ^{106}Ru contained in the shell. The residue contained about 27 and 39% of organic matter (loss on ignition at 600°C) respectively in the two cases. After repeated treatment of the residue with 0.2<u>M</u> Na_2CO_3 − NaOH, the organic matter extracted was filtered, purified and dialysed [13]. The dialysed extract was filtered through Millipore (0.22 μm) and counted. The gamma spectrum showed the presence of ^{106}Ru; it is given in Fig.8 along with the gamma spectrum of the residue obtained on acid treatment. ^{106}Ru in the extract accounted for nearly 28% of the total ^{106}Ru in the residue obtained from the shell; this indicates the presence of organic complexes of Ru in the shell of the Arca. The sediment also has shown the presence of Ru in the organically bound form. The presence of soluble organic complexes in sea water has already been reported [10]. Further studies are in progress to understand the nature of these complexes and their significance in the bay environment.

The plutonium levels in flesh and bones/shells of organisms from the bay revealed that Arca flesh had high accumulating capacity for this element. The levels observed compared to flesh of fish (Arius sp.) obtained at the same location were 12 - 160 times as high. Concentrations of fall-out Pu in similar organisms are reported [14] to be lower by a factor of 30 - 150. These results, along with the low $^{90}Sr/^{239}Pu$ ratios found in a few Arca flesh species in contrast to the very large ratios at the discharge location, indicate that discrimination against plutonium may not be as high as that observed for fall-out plutonium. These aspects will have to be examined further [15].

Salt

There are a number of salt pans scattered along the coastline of the bay from which salt is produced during non-monsoon months, i.e. from October to May. In salt pans sea water is taken in at spring tides. Water is allowed to undergo preliminary evaporation and concentration when gypsum separates out. Most of the silt in sea water also settles at this stage. Clean supernatant is then evaporated in pans, bitterns are discarded and the crystals are taken out. Both pure and crude varieties of salt are produced.

Samples of salt from salt pans located within a radius of about 20 km from the Research Centre were periodically monitored. Salt was solubilized

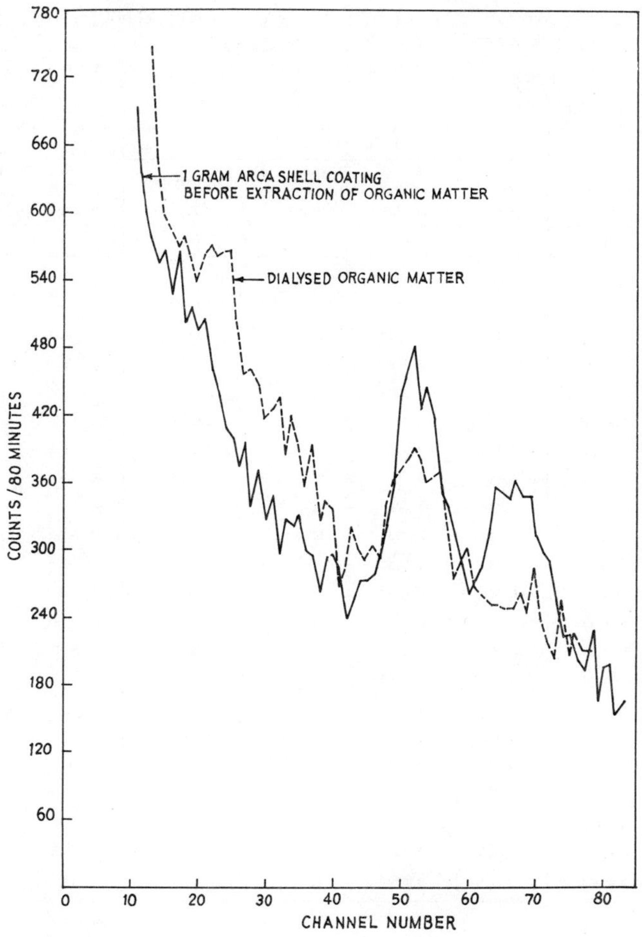

FIG. 8. Gamma spectrum of organically bound ^{106}Ru from Arca shell.

in water and the silt was separated by filtration (Whatman No. 42). The silt activity was measured by gamma spectrometry. Strontium-90 and ^{137}Cs were estimated in the soluble portion of the salt. The stable strontium content in some samples was determined by atomic absorption spectrophotometry. Data on concentrations of other trace elements in salt — Ca, K, Sr, Zn, Mn, Fe and Cu — and their contribution in the daily dietary intake of trace elements have been reported elsewhere [16].

The results are given in Table VIII. The silt content in pure varieties of salt varied from 167 to 1250 ppm whereas in crude salt it varied from 1360 to 6965 ppm. By gamma-ray spectrometry only ^{137}Cs could be estimated in silt; ^{106}Ru and ^{144}Ce could not be detected. Caesium-137 in silt varied with time and distance of the salt pan from the discharge area.

TABLE VIII. ^{90}Sr AND ^{137}Cs IN SOLAR SALT OBTAINED FROM DIFFERENT SALT PANS IN THE BAY REGION
Columns 3, 4 and 7: counting error ± 10%

Date of collection	Location	Soluble portion of salt				Silt ^{137}Cs (pCi/g (dry))	Silt content in salt (ppm)
		^{90}Sr (pCi/g)	^{137}Cs (pCi/g)	Stable Sr (ppm)	Sp. activity (pCi ^{90}Sr/g Sr)		
1	2	3	4	5	6	7	8
10-12-69	Trombay	0.127	0.032	-	-	0.09	-
9- 4-71	Trombay	0.63	0.01	45.7	13800	-	-
15- 1-72	Trombay	1.71	0.0043	39.0	44000	14.3	6965*
5-11-73	Trombay	2.81	0.013	-	-	32.1	1250
24-11-69	Bhandup	0.10	0.0028	-	-	-	-
6- 4-71	Bhandup	0.99	0.004	69	14300	N.D.	270
15- 1-72	Bhandup	0.21	0.005	29.6	7300	3.5	2407*
17-11-72	Bhandup	0.36	0.0076	77	4689	3.3	5620*
16- 4-71	Mulund	0.34	0.009	56	6100	-	-
15- 1-72	Mulund	0.34	0.005	53.2	6400	N.D.	845
17-11-72	Mulund	0.52	0.0047	-	4474	4.8	1360*
24-11-69	Belapur	0.005	0.0023	-	-	-	-
16- 4-71	Belapur	0.006	0.005	19.2	300	-	541
15- 1-72	Belapur	0.10	0.013	32.0	3200	N.D.	3248*
17-11-72	Belapur	0.002	0.0065	59.0	48.97	3.2	4900*
16- 4-71	Sarsola	0.08	0.01	17.2	4700	-	167
15- 1-72	Sarsola	0.193	0.007	45.5	4200	5.3	8330*
5-11-73	Sarsola	0.116	0.002	-	-	N.D.	412
26- 8-68	Wadala	0.058	0.0046	-	-	0.0146	-
10- 4-70	Wadala	0.054	0.0046	-	-	-	-
31- 3-71	Wadala	0.055	0.0013	31.2	1800	N.D.	359
29- 3-72	Wadala	0.24	0.0080	52.8	4500	18.3	428

* Crude salt.
N.D. = not detected.

Strontium-90 gets preferentially concentrated in salt compared to ^{137}Cs, especially in salt from pans nearer to the discharge location. Caesium-137, unlike ^{90}Sr, gets depleted in water. The lower concentration in salt is also possibly due to the higher removal of ^{137}Cs in bitterns.

Strontium-90 levels in salt obtained from salt pans located on the west coast of the bay were higher than those obtained on the eastern coast. This seems clearly to be due to sea water with larger concentrations of radionuclides entering the salt pans due to their proximity to the discharge location. Strontium concentration varied from 17 to 77 ppm in different salts; the average amounted to 44 ppm. Compared to other trace elements the contribution of Sr from salt to the daily intake of the element is significant [16]. The specific activity of salt (pCi ^{90}Sr/g of Sr) from salt pans nearer to the discharge location was considerably higher than those from the farthest locations and possibly these values reflect the extent of mixing and dilution of the effluents with sea water at different locations.

EVALUATION OF EXPOSURE

Extensive studies carried out in the bay environment have shown that suspended silt and shore sediments accumulate radionuclides to a large extent. Though suspended silt picked up radionuclides ^{144}Ce, ^{106}Ru, ^{137}Cs and ^{239}Pu, only the shore sediments near the discharge location showed accumulation of ^{144}Ce and ^{106}Ru. The major contaminant from the point of external exposure in shore sediments is ^{137}Cs. Organisms from the bay, especially the benthic organisms near the discharge location, showed accumulation of discharged nuclides. Salt obtained by evaporation of bay waters are found to accumulate ^{90}Sr. The effect of accumulation of radionuclides in all these matrices of the aquatic environment are to be examined for evaluating the radiation exposure to personnel utilizing the bay.

Suspended silt from sea waters had varying K_d factors ranging from 1.6×10^1 to 8.9×10^4. On the basis of these K_d values and on assuming an average silt load of 1000 mg/l of sea water, decontamination factors obtained were estimated (see Table IX). The removal of released activity due to sorption of radionuclides by suspended silt thus assumes significance in the bay waters. Transport of the contaminants through suspended silt to the shore is of significance in the evaluation of radiation exposure.

TABLE IX. DECONTAMINATION FACTORS OBTAINED WITH SUSPENDED SILT FOR DIFFERENT RADIONUCLIDES

Radionuclide	Sr-90	Cs-137	Ru-106	Ce-144	Pu-239
Decontamination Factor	1	2	9	87	90

Exposure of the population generally does not arise from contaminated shore sediments of the bay since the shore areas where contamination levels are significant are not accessible. Only fishermen who occupy such areas occasionally for fishing might receive external exposure. The area between Trombay Naval Jetty and CIRUS has to some extent shown accumulation of radionuclides. On the basis of the maximum value of ^{137}Cs in shore sediments observed at Trombay Naval Jetty the external exposure from the sediments amounts to only 0.13 mrad/h. It is seen from Table IV that in other areas of the bay the accumulation of radionuclides in sediments is lower by orders of magnitude. Considering the maximum possible time they might spend on the shore [4] this source is most unlikely to give any significant radiation exposure to fishermen. The inhalation hazard that might arise as a result of resuspension of dried shore sediments containing traces of Pu are negligibly small.

The annual fish catch in the bay accounts for only 5% of the fish sold in the Bombay markets. The Arca flesh has shown the maximum concentration of radionuclides. It is most unlikely that the same person may eat the contaminated fish caught from the discharge area throughout the year. Even in such a case the maximum radiation exposure will be of the order of 3% of the dose limit.

In the case of salt, the maximum accumulation of ^{90}Sr was in the salt pan at Trombay. Salt manufactured in these salt pans is for domestic consumption and industrial use locally, and is also exported to other areas. Even in the unlikely case of a person consuming the most contaminated salt throughout the year the maximum exposure possible is only 5% of the dose limit. The average levels in salt in different areas amount to about one-tenth of the maximum obtained at Trombay and consequently the radiation exposure might not exceed 0.5% of the dose limit through this source.

Extensive studies carried out in Bombay Harbour bay have shown the fractionation, distribution and accumulation of released radionuclides in different matrices of the bay. Transient high levels of activity have been observed only at close proximity to discharge locations. The fuel reprocessing operations and the consequent waste releases into the bay have not created any significant environmental radiation problems.

ACKNOWLEDGEMENTS

The authors are grateful to Dr. A.K. Ganguly, Director, Chemical Group, and Mr. S.D. Soman, Head, Health Physics Division, B.A.R.C., for their guidance and keen interest in the work. Mr. A. Kuttappan collected and processed the bay organisms and Mr. S. Ganapathy participated in the salt measurements; the authors wish to acknowledge their assistance. Mr. L.H. Peshori and his group co-operated with the authors in the spectrometric work.

REFERENCES

[1] PILLAI, K.C., SUBBARATNAM, T.S., and GANGULY, A.K., Tidal movement and water renewal rates in Bombay harbour bay, AEET Rep. AEET/HP/RWD-2 (1958).

[2] KAMATH, P.R., PILLAI, K.C., BHAT, I.S., NAIDU, J.R., BORKAR, M.D., BHARATHAN, K.G., GANGULY, A.K., Tidal movement studies at Thana Creek, AEET Rep. AEET/HP/RWD-6 (1961).

[3] PILLAI, K.C., GANGULY, A.K., Evaluation of maximum permissible concentration of radionuclides in sea waters of Bombay, AEET Rep. AEET/HP/R-11 (1961).

[4] PILLAI, K.C., "Aquatic pollution control systematics for discharge of radioactive effluents", Proceedings of the COST Seminar on Pollution and Human Environment, BARC-Bombay (1970) 164.

[5] SETHNA, H.N., SRINIVASAN, N., "Fuel reprocessing plant at Trombay", Int. Conf. Peaceful Uses Atom. Energy (Proc. Conf. Geneva, 1964), 10 (1964) 272.

[6] SETHNA, H.N., FAREEDUDIN, S., KATIYAR, H.C., MECKONI, V.N., SRINIVASAN, N., "Integrated planning of nuclear industry", Int. Conf. Peaceful Uses Atom. Energy (Proc. Conf. Geneva, 1971), 4 (1972) 243.

[7] PILLAI, K.C., "Determination of plutonium in the marine environment", Reference Methods for Marine Radioactivity Studies II, Technical Reports Series No. 169, IAEA, Vienna (1975) 97.

[8] PATEL, B., MULAY, C.D. and GANGULY, A.K., Radioecology of Bombay harbour — a tidal estuary, Estuarine and Coastal Marine Science 3 (1975) 13.
[9] GANAPATHY, S., PILLAI, K.C. and GANGULY, A.K., Adsorption of trace elements by near shore sea bed sediments, BARC Rep. BARC-376 (1968).
[10] PILLAI, K.C., DEY, N.N., "Radioruthenium in coastal waters", Reference Methods for Marine Radioactivity Studies II, Technical Reports Series No. 169, IAEA, Vienna (1975) 147.
[11] HARRISON, F.L., "Accumulation and loss of cobalt and caesium by the marine clam, Mya arenaria, under laboratory and field conditions", Radioactive Contamination of The Marine Environment (Proc. Symp. Seattle, 1972), IAEA, Vienna (1973) 453.
[12] WILBUR, K.M., "Shell formation and regeneration", Ch. 8, Physiology of Mollusca (WILBUR, K.M., YONGE, C.M., Eds) Academic Press, New York & London (1964).
[13] ELIZABETH, K., GANGULY, A.K., Organic Materials in the Marine Environment and Their Interactions with Some Metal Ions, BARC Rep. BARC-402 (1969).
[14] NOSHKIN, V.E., BOWEN, V.T., WONG, K.M., BURKE, J.C., Radionuclides in Ecosystems, Proc. Third Nat. Symp. Radioecology, 2 (1971) 681.
[15] PILLAI, K.C., MATHEW, E., "Plutonium in the aquatic environment: its behaviour, distribution and significance", Paper IAEA-SM-199/27 to be presented at the International Symposium on Transuranium Nuclides in the Environment, San Francisco, 17-21 Nov. 1975.
[16] GANAPATHY, S., KOTHARI, B.U., PILLAI, K.C., Solar salt as a medium for transfer of radionuclides and trace elements to man. Paper presented at the Regional Congress on Radiation Protection, BARC-Bombay, Dec. 1974.

DISCUSSION

G. STOTT: In the United Kingdom there are two spent-fuel reprocessing plants some 500 miles apart. In the low-level liquid effluent streams from these plants ^{95}Zr and ^{95}Nb are important radionuclides and have been considered in the critical path approach. I am puzzled by the absence of ^{95}Zr and ^{95}Nb in the results you presented for Trombay.

K.C. PILLAI: The zirconium concentrations are so low that zirconium can be detected only at the discharge location.

Gwyneth D. HOWELLS: Have you observed seasonal variations in the levels of radionuclides in sediments, and does the seasonal growth of macro-algae influence the sediment uptake? In work on the Hudson River at Indian Point we found that three major factors influenced sediment uptake: the discharges from the reactor, the hydrological regime, and the seasonal growth of aquatic vegetation.

K.C. PILLAI: We did observe seasonal variations, especially during and after the monsoon months, i.e. from June to September. The shore sediments showed lower activity levels after the monsoon. This can be seen from the build-up of activity on shore sediments shown in Fig.5. Even the suspended sediments exhibited lower activity levels during the monsoon, owing to the larger amount of silt coming into the bay through run-off. Aquatic vegetation from the bay has not been studied, but I feel it may be of less significance owing to the large silt load in the bay waters.

J.A. HETHERINGTON: It is of interest to find yet another case in which sediments are important in relation to the behaviour of radionuclides discharged in the course of nuclear power operations. I notice particularly your observation that ^{137}Cs cannot be removed from the sediments even by acid leaching, and your suggestion that this is because the nuclide is being absorbed by montmorillonite clays. Can you say what weight fraction of these sediments is clay and what fraction of the clay is montmorillonite?

K.C. PILLAI: We have observed differences in regard to the sediments at different locations. At Trombay, for example, ^{137}Cs in the sediments cannot be leached out by acid leaching; at a station 100 kilometres away, however, the sediments do not pick up as much ^{137}Cs and here the ^{137}Cs can be leached out with acid. I assume that the montmorillonite in the Trombay waters explains this difference. I think the clay content is about 2%.

Y. NISHIWAKI: Were the distribution coefficients (K_d) calculated with respect to the soluble fraction of radionuclides? Or was the fraction of radionuclides which may be regarded as insoluble at the time of discharge included in the radioactivity of the silt? If the latter were the case, I think it would be more appropriate to say "apparent distribution coefficient".

K.C. PILLAI: The K_d factors were calculated on the basis of the concentration of radionuclide left behind in the sea water after removal of suspended silt. The fraction of the radionuclides removed by silt will include the insoluble radionuclides present in the discharge, those insolubles produced as a result of interaction of effluents with seawater, and also some of the sorbed nuclides present in soluble form in sea water. I suppose one could indeed say "apparent distribution coefficient".

Yu.V. SIVINTSEV: Do I understand correctly that your K_d coefficient is equivalent to the F.C. value? Is it the same as the concentration of radionuclide in the sediments divided by the concentration of radionuclide in the water?

K.C. PILLAI: Yes. It is the ratio of concentration of radionuclide per unit weight of sediment to the concentration of the nuclide in the same weight of water. The activity discharged gets partitioned between the suspended sediment and the water and therefore I prefer to call it K_d (distribution coefficient).

V.V. PISAREV: In this connection I should like to say that the use of different symbols for the same concept is rather confusing, and to put in a plea for the use of standardized symbols.

CONCENTRATION OF RADIOACTIVE COBALT BY SEAWEEDS IN THE FOOD CHAIN

M. NAKAHARA, T. KOYANAGI, M. SAIKI
National Institute of Radiological Sciences,
Chiba-shi, Japan

Abstract

CONCENTRATION OF RADIOACTIVE COBALT BY SEAWEEDS IN THE FOOD CHAIN.

On the pathway of radioactive substances in marine environments, seaweeds play an important role because of their higher concentration factors for many radionuclides and because they constitute a link in the food chain.

In the present work, uptake, distribution and excretion of radioactive cobalt were studied on several kinds of seaweeds by radioisotope tracer experiments under laboratory conditions and concentration factors were calculated. The concentration factors were also estimated from the results of stable cobalt determination by activation analysis or atomic absorption spectrometry on seaweeds and seawater, and compared with the results of tracer experiments.

The seaweeds showed the species specificity for the concentration of stable and radioactive cobalt with diverse values of concentration factors and biological half-lives.

The transfer of radioactive cobalt in the food chain from contaminated seaweeds to mollusca was examined by feeding abalones, *Haliotis discus,* with four kinds of seaweed labelled with ^{60}Co and observing retention. Absorption rate for radioactive cobalt by abalones calculated at two days after feeding showed diverse values depending upon the species of seaweed, as follows: 47% through *Laminaria japonica* and *Ulva pertusa*, 31% through *Undaria pinnatifida* and 26% through *Eisenia bicyclis*, respectively.

From the results, it was assumed that the accumulation of radioactive cobalt by mollusca is affected by the species of seaweeds as food. A very high concentration of ingested radioactive cobalt in the midgut gland was seen on the autoradiograph of abalone samples.

1. INTRODUCTION

With the rapid increase in nuclear power generation in the last decade, more precise information on marine environmental contamination caused by the possible release of radionuclides with cooling water has been called for, in order to evaluate the radiation exposure to man and ensure the safety of nuclear facilities. Although several investigations have been carried out over a number of years on the radioactive contamination of marine organisms, knowledge of the accumulation of induced radionuclides by fishes, mollusca and seaweeds through the food chain in the sea is still scarce. Such information is very important; in Japan, for example, 10% of the total catch of fish (about 10 million tons per year) is caught in coastal areas where power reactors are most likely to be constructed. In this paper, special interest is placed on the critical pathway of radioactive cobalt, and the results of radioisotope tracer experiments under laboratory conditions are described.

Hiyama and Kahn [1] reported the uptake of ^{60}Co by several kinds of marine organisms and the influence of some environmental factors on this uptake. Shimizu et al. [2], van Weers [3] and Pentreath [4] studied the uptake of radioactive cobalt by the mussel, Mytilus edulis, which is considered a good indicator organism for radioactive contamination of the marine environment because of its world-wide distribution. Harrison [5] investigated the accumulation and loss of cobalt by marine clams, Mya arenaria, and calculated the rate constants of accumulation and loss, and the steady state concentration factors.

Reed [6] examined the uptake and excretion of ^{60}Co by black bullheads, Ictalurus melas, and noted that the biological half-lives varied according to the pathway of ^{60}Co intake, whether via water or via food. Kimura and Ichikawa [7] determined the absorption rate and retention rate of ingested ^{60}Co by the common goby, Acanthogobius flavimanus, as 4% of the dose administered by a single administration and 6% of the total dose administered by daily oral administration, respectively.

Lowman and Ting [8] investigated the uptake of cobalt by marine organisms and sediments using ionic and organic radiocobalt as tracers and indicated that there was a marked difference among the organisms in their responses to the chemical forms of cobalt in seawater.

Compared with these investigations carried out on fish, bivalves and other animals, only a few papers are concerned with the accumulation of cobalt by seaweeds. In view of this, several kinds of seaweed were used in the present work to observe uptake, distribution and loss of ^{60}Co and transfer of ^{60}Co in the food chain by feeding certain mollusca with contaminated seaweed.

2. MATERIALS AND METHODS

2.1. Tracer experiments

The seaweed was collected on the seashore of Ibaraki Prefecture, on the East coast of Mid-Japan, except Laminaria japonica, which was collected at Hokkaido, at the northern end of Japan, and precultured in aquariums under fluorescent illumination for a few days before the experiments.

The following species were selected:

Green algae: Ulva pertusa Kjellman

Brown algae: Laminaria japonica Areshoug, Eisenia bicyclis Setchell, Undaria pinnatifida Suringar, Hizikia fusiforme Okamura, Sargassum thunbergii O. Kuentze, Sargassum ringgoldianum Harvey

Red algae: Chondrus ocellatus Holmes, Ahnfeltia paradoxa Okamura.

Fifty litres of seawater was filtered with a membrane filter of 0.45 μm pore-size and filled into the aquarium equipped with a filter bed of quartz gravel. The tracer of radioactive cobalt ($^{60}CoCl_2$ in HCl, carrier free) was added to the seawater at a concentration of about 20 μCi/l and the seaweeds were introduced into the aquariums after the radioactivity of the seawater showed a steady value. The seawater was circulated through the filter bed and the temperature was maintained at 15 ± 1°C during the experiments. The lighting was by fluorescent illumination of 1000 lux with an intermittence of 12 hours per day.

A few individual plants of the seaweed taken up from the aquariums during the course of the experiment were rinsed briefly with non-radioactive seawater and divided into several parts, i.e. tip, centre and base, or blade

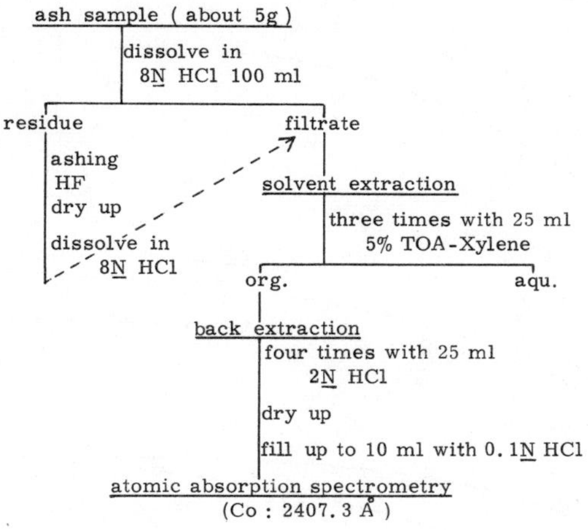

FIG.1. Scheme for analysis of cobalt in seaweeds.

and stem. Each part was weighed after moisture had been removed by paper towels and then radioactivities were measured by using a well-type scintillation detector for calculating concentration factors based on the radioactivity of the seawater taken simultaneously from the various aquariums.

Loss of radioactivity from the seaweeds was observed on samples transferred to running non-radioactive seawater after being contaminated with ^{60}Co in the uptake experiments.

2.2. Stable cobalt analysis

The concentration of stable cobalt in the seaweeds was determined by atomic absorption spectrometry. Before the analysis, the seaweeds were rinsed with filtered seawater and detached completely from adhering substances such as sand or other small organisms. Their weight was measured after moisture had been removed and the samples were ashed at 450°C in an electric furnace for 48 hours. The scheme of the chemical analysis was as shown in Fig.1. The concentration of stable cobalt in seawater was quoted from the results reported by Suzuki et al. [9] as approximately 0.14 μg/l in seawater collected from the same area as our samples were taken.

2.3. Microautoradiography

The part constituting the stem and base of blade was separated from Laminaria reared for 60 days in radioactive seawater and frozen rapidly in a mixture of acetone and dry ice (-78°C). The section of the frozen sample was prepared by using a microtome in 10 μm thickness and dried in a deep-freezer after being stretched on slide glasses. The surface of the samples was covered with the autoradiographic films (Fuji film, type ET-2F) and exposed for suitable periods. After development, the microautoradiographs were observed with a microscope.

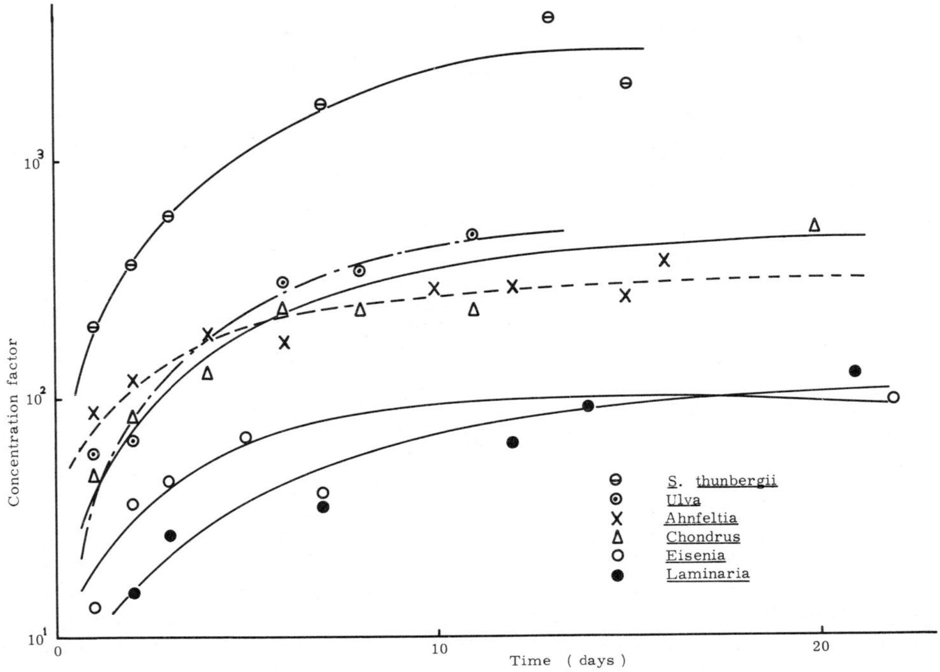

FIG.2. Uptake of ^{60}Co by several kinds of seaweed.

2.4. Feeding experiments

The transfer of ^{60}Co in the food chain was examined by feeding abalones, Haliotis discus, with three kinds of brown algae and one kind of green algae labelled with ^{60}Co. Three abalones were reared in running non-radioactive seawater and fed with the contaminated seaweeds in the amount of 2-3% of the body weight of the three abalones, ensuring complete ingestion within 1-2 hours.

From the results of observation of the whole-body radioactivity of the abalones measured for several days by using the "Armac" 4π-scintillation counter, it was assumed that digestion and excretion of the feeds were achieved within 48 hours after administration under conditions of about 15°C water temperature. Thus the absorption rate for ^{60}Co for the abalones was calculated as follows:

$$\text{Absorption rate (\%)} = \frac{\text{Radioactivity of abalone 48 h after ingestion}}{\text{Radioactivity of abalone just after ingestion}} \times 100$$

3. RESULTS AND DISCUSSION

3.1. Uptake of ^{60}Co by seaweeds

Figure 2 shows the uptake pattern of ^{60}Co on the parts of seaweeds where the concentration factor showed the highest values. Rapid uptake

TABLE I. CONCENTRATION FACTORS FOR RADIOACTIVE AND STABLE COBALT BY SEAWEEDS

Seaweed	Tracer experiment	Stable cobalt analysis	
	C.F.	Co content (μg/kg fresh)	C.F.
Green algae			
Ulva pertusa	439	48.0	340
Brown algae			
Laminaria japonica	138	9.4*	70
Eisenia bicyclis	34	28.5	200
Undaria pinnatifida	210	26.0	190
Hizikia fusiforme	127	15.0	120
Sargassum thunbergii	2574	264.9	1890
S. ringgoldianum	——	36.4	260
Red algae			
Chondrus ocellatus	833	120.7	860
Ahnfeltia paradoxa	290	127.6	910
Sea water		0.14* μg/ℓ	

* obtained by activation analysis.
C.F. = concentration factor

in the initial stage was followed by slow increase of the concentration factor and the uptake tended to reach an equilibrium state after about 10 days for every species of the seaweeds examined. Higher concentration factors were obtained in S. thunbergii than Laminaria and Eisenia, suggesting the contribution of surface adsorption in ^{60}Co uptake by seaweeds, since the latter two, which are relatively large brown algae, have smoother surfaces than the former.

The whole-body concentration factors for ^{60}Co for the seaweeds, calculated from the radioactivity ratio observed at the end of uptake experiments are shown in Table I, with the values obtained from the results of stable cobalt analyses. In comparing the two methods of calculation, the values of concentration factor showed relatively good agreement but a variation of about two orders of magnitude was observed among the different species of seaweed.

The concentration of stable cobalt in the seaweeds obtained in the present work, 9.4 μg/kg raw (0.11 μg/g dry) for Laminaria and 264.9 μg/kg raw (1.4 μg/g dry) for S. thunbergii, corresponds with the range of cobalt contents in most of the seaweeds reported by Fukai [10] as 0.1 to 1.5 μg/g dry.

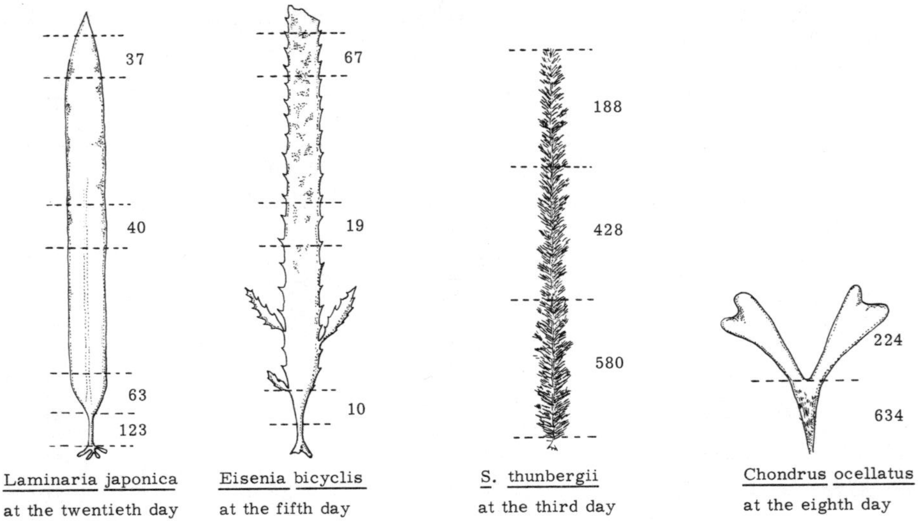

FIG.3. Concentration factors on the various parts of seaweeds.

A similar tendency was observed for the concentration of stable and radioactive cobalt on S. thunbergii, Chondrus and Ulva with the higher values but substantial comparison among the species was difficult because of insufficient data on green and red algae although these species showed higher concentration factors than brown algae, with the exception of S. thunbergii. Even in the same genus of sargasso, S. thunbergii showed a far higher value for the concentration factor compared with stable cobalt concentration in S. ringgoldianum, suggesting the peculiar species specificity of seaweeds for cobalt concentration.

3.2. Diversity of ^{60}Co uptake by the parts of seaweeds

Figure 3 shows the different values of concentration factor obtained in radioisotope tracer experiments according to the part of seaweed. In Laminaria, Chondrus and S. thunbergii the lower parts near the holdfast showed higher values than the upper parts near the tip, in contrast to Eisenia, where the value decreased with distance from the tip. The stem of Hizikia and Ahnfeltia gave higher concentration factors than the blade in the tracer experiments. On Hizikia higher contents of stable cobalt in the base than the tip were reported by Ishii et al. [11]. Bryan [12] also reported that the concentration of heavy metals (i.e. Cu, Zn, Mn, Fe and Pb) in Fucus vesiculosus collected in estuaries tended to increase with distance from the growing tips. The distribution pattern of ^{60}Co in Hizikia, S. thunbergii, Chondrus and Ahnfeltia agreed with the trend of stable cobalt or other heavy metals found in Hizikia and Fucus, but an exception to this trend is Laminaria, in the blade of which Yoshimura and Oishi [13] reported no characteristic distribution patterns for Fe, Al, Mn and Cu except for higher contents of Fe, Al and Mn in the apex. In the body of these seaweeds, growth is most active between blade and stem for Laminaria and Eisenia, whereas tip and blade represent the active growing parts in Hizikia, S. thunbergii, Chondrus and Ahnfeltia.

It is supposed, therefore, that cobalt is apt to be accumulated in higher concentrations in the older parts of seaweeds than the younger parts. In view of the great deal of surface adsorption that takes place with the accumulation of metals by seaweeds, the diverse distribution of cobalt in the various parts of seaweeds is considered due to the structural condition of the surfaces. Additional effects of adhering substances, such as microorganisms and inorganic or organic materials, which might provide the surface of seaweeds with a greater propensity for deposition of metals should also be taken into consideration in the actual marine environments.

3.3. Localization of ^{60}Co in seaweeds

The autoradiographs of Laminaria prepared in order to throw light on the localization of accumulated ^{60}Co in the seaweeds showed absorption of radioactivity in the inner layer of the seaweeds as well as adsorption on the surface, but the radioactivities measured on the samples divided into cortex part and inner layer showed a value about five times as high in the cortex part as in the inner part. It was therefore suggested that adsorption of cobalt on the surface of seaweeds plays an important role in the uptake of cobalt as one of the intake pathways at the initial stage, followed subsequently by accumulation mainly in the cortex parts.

3.4. Loss of ^{60}Co from seaweeds

Figure 4 shows the loss of radioactivity from seaweeds, and Table II gives the biological half-lives calculated from the excretion curves, for various parts of the seaweeds. Although the biological half-lives varied between 7.3 days for the tip and 79.6 days for the base of S. thunbergii at 15°C, the younger parts showed the trend of faster loss of ^{60}Co than the older parts, suggesting more firm fixation of cobalt on the older surface than the surface of growing younger parts, where the loss of cobalt was considered to be affected by metabolic activity of the seaweeds.

3.5. Transfer of ^{60}Co in food chain

The transfer of ^{60}Co in the food chain from contaminated seaweeds to mollusca was examined by feeding abalones, Haliotis discus, 25-30 g, with four kinds of seaweeds labelled with ^{60}Co. The absorption rates of ^{60}Co in the abalones, calculated by the equation previously given, are shown in Table III. There were some variations in the results: however, average values according to the different species of seaweed were obtained as follows: 47% for Laminaria and Ulva, 31% for Undaria, and 26% for Eisenia. The result showed that the accumulation of ^{60}Co by mollusca was also affected by the species specificity of seaweeds as food.

Kikuchi et al. [14] showed the food values of 20 kinds of seaweed as "Eisenia equivalent", i.e. growth in shell length of the young abalone fed with a certain seaweed relative to that with Eisenia, assuming the growth

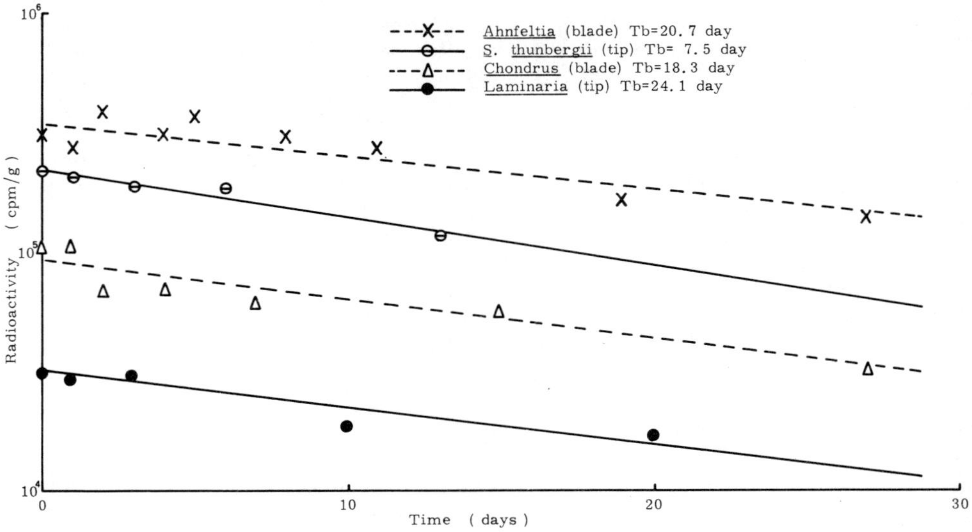

FIG.4. Loss of ^{60}Co from several kinds of seaweed.

TABLE II. BIOLOGICAL HALF-LIVES OF ^{60}Co BY SEAWEEDS

Seaweed	Period (days)	Temp. (°C)	Part	Biological half-life (day)
Brown algae				
Laminaria	35	15 ± 1	Tip	24.1
			Base	12.8
			Stem	36.7
S. thunbergii	13	15 ± 1	Tip	7.5
			Centre	12.1
			Base	79.6
Red algae				
Ahnfeltia	27	15 ± 1	Blade	20.7
			Stem	57.6
Chondrus	23	15 ± 1	Blade	18.3
			Stem	41.6

TABLE III. ABSORPTION RATE OF ^{60}Co IN ABALONES (*Haliotis discus*) TWO DAYS AFTER INGESTION OF LABELLED SEAWEEDS

Seaweed	No. of exp.	Total weight of three abalones	Quantity of food	Absorption rate
Brown algae				
Laminaria	1	72.7 g	1.9 g	49.5 %
	2	71.3	1.5	43.3
	3	71.2	1.7	44.8
	4	73.1	1.5	43.9
	5	70.0	2.0	52.0
	6	68.7	1.9	48.9
				Ave. 47.1 ± 3.3*
Undaria	1	90.6	2.1	34.1
	2	88.2	2.8	35.6
	3	71.7	2.2	15.5
	4	71.3	2.6	30.1
	5	73.4	2.9	32.9
	6	72.6	3.6	34.9
				Ave. 30.5 ± 7.6*
Eisenia	1	88.9	2.2	19.8
	2	90.4	1.6	25.7
	3	90.2	2.0	34.1
	4	89.3	2.5	22.2
				Ave. 25.5 ± 6.3*
Green algae				
Ulva	1	76.7	1.1	51.3
	2	74.4	1.4	41.6
	3	87.2	1.1	52.4
	4	81.8	1.3	41.5
				Ave. 46.7 ± 6.0*

* standard deviation.

with Eisenia as 1. The values for Undaria and Ulva are 1.02 and 0.58, respectively, showing the lower value for Ulva, which contributes more towards the ^{60}Co accumulation than the former. This fact suggests that the extent of contribution in the accumulation of ^{60}Co by abalones is less related to the food values of seaweeds as nutrition and more related to the physical or chemical states of the accumulated ^{60}Co on seaweeds or to the digestivity of the seaweeds for abalones.

On the rate of transfer, Reed [6] reported the absorption rate of ^{60}Co by the fish, Ictalurus melas, one day after feeding as 4.5% of the administered dose through a labelled worm. A similar result was also reported by Kimura et al. [7], namely that the common goby, Acanthogobius flavimanus, absorbed approximately 4% of administered ^{60}Co through a worm one day after feeding.

The absorption rates of ^{60}Co by several kinds of marine fish fed with labelled alfa-starch or crustacean Artemia, were estimated by Nakahara et al. [15] to be in the range of 3 to 16% of the administered dose.

Compared with these results, the relatively higher rates for transfer of ^{60}Co from seaweeds to mollusca are considered to be noteworthy, showing the important role of seaweeds in the food chain as a primary producer and radioactivity donor.

The distribution rates of ^{60}Co in the abalones fed with labelled seaweeds were about 9-11% in the foot, 89-91% in the viscera and 0.3-0.5% in the shell; the rates in fact were almost identical among the different samples used in the experiment. As compared therefore to the whole-body radioactivity, these percentages for the foot, viscera and shell were independent of the species of seaweed. The contribution of the midgut gland of the abalone to the high radioactivity in the viscera was proved by the autoradiograph of abalone samples and it was supposed that considerable amounts of radioactive cobalt were accumulated in abalones by biological processes.

REFERENCES

[1] HIYAMA, Y., KAHN, J.M., On the concentration factors of radioactive I, Co, Fe and Ru in marine organisms, Rec. Oceanogr. Works Japan 7 (1964) 79.

[2] SHIMIZU, M., KAJIHARA, T., SUYAMA, I., HIYAMA, Y., Uptake of ^{58}Co by Mussel, Mytilus edulis, J. Radiat. Res. 12 (1971) 17.

[3] Van WEERS, A.W., "Uptake and loss of ^{65}Zn and ^{60}Co by the mussel, Mytilus edulis", Radioactive Contamination of the Marine Environment (Proc. Symp. Seattle, 1972) IAEA, Vienna (1973) 385.

[4] PENTREATH, R.J., The accumulation from water of ^{65}Zn, ^{54}Mn, ^{60}Co and ^{59}Fe by the mussel, Mytilus edulis, J. Mar. Biol. Assoc. U.K. 53 (1973) 127.

[5] HARRISON, F.L., "Accumulation and loss of cobalt and caesium by the marine clam, Mya arenaris, under laboratory and field conditions", Radioactive Contamination of the Marine Environment (Proc. Symp. Seattle, 1972) IAEA, Vienna (1973) 453.

[6] REED, J.R., Uptake and excretion of ^{60}Co by black bullheads, Ictalurus melas (Rafinesque), Health Phy. 21 (1971) 835.

[7] KIMURA, K., ICHIKAWA, R., Accumulation and retention of ingested ^{60}Co by the common goby, Bull. Jap. Soc. Scient. Fisheries 38 (1972) 1097.

[8] LOWMAN, F.G., TING, R.Y., "The state of cobalt in seawater and its uptake by marine organisms and sediments", Radioactive Contamination of the Marine Environment (Proc. Symp. Seattle, 1972) IAEA, Vienna (1973) 369.

[9] SUZUKI, H., et al., Radioecological Studies on Marine Environment, National Institute of Radiological Sciences, NIRS-R-3 (1974) 50.

[10] FUKAI, R., Distribution of Cobalt in Marine Organisms, IAEA Radioactivity in the Sea, Publication No.23 (1968).
[11] ISHII, T., et al. (1974) National Institute of Radiological Sciences, Chiba-shi, Japan, unpublished data.
[12] BRYAN, G.W., The effects of heavy metals (other than mercury) on marine and estuarine organisms, Proc. R. Soc. (London), Ser. B 177 (1971) 389.
[13] YOSHIMURA, A., OISHI, K., Distribution of inorganic constituents of Kombu blade-I. On metallic ions, Bull. Jap. Soc. Scient. Fisheries 39 (1973) 317.
[14] KIKUCHI, S., SAKURAI, Y., SASAKI, M., ITO, T., Food values of certain marine algae for the growth of the young abalone, Haliotis discus Hannai, Bull. Tohoku Regional Fisheries Research laboratory No.27 (1967).
[15] NAKAHARA, M., et al., Radioecological Studies on Marine Environment, National Institute of Radiological Sciences, NIRS-R-3 (1974) 44.

DISCUSSION

G. STOTT: In the United Kingdom one type of seaweed is used to make a delicacy called laver bread. Another type is harvested for the production of alginic acid, and seaweed may also be dried and used in malt distilleries as a source of heat to swell the barley. Is seaweed used for similar purposes in Japan?

M. SAIKI: A popular Japanese food called "nori" is also made from laver (*Porphyra*). The "nori" is prepared in dried paper-like sheets, which are slightly toasted and eaten with soya-bean sauce and boiled rice. Some kinds of green algae are eaten in the form of a powder on boiled rice. Several species of brown algae are boiled with soya-bean sauce and sugar, and consumed somewhat in the same way as pickles. The gelatinous agar-agar is also prepared from red algae, and naturally we produce a considerable amount of alginic acid from seaweeds. The average consumption of seaweed by the Japanese is comparatively low, being about 4 g/day, though a higher consumption has, not unexpectedly, been observed in fishing villages on the coast, where the average intake has been recorded as 40 g (raw)/day for 3 months and 10 g (dry)/day for 9 months.

S. BONOTTO: According to published literature, Japan exports a large amount of *Porphyra*. I should like to know whether any contamination by ^{60}Co and other radionuclides has been detected in these marine algae.

M. SAIKI: Several years ago trace amounts of ^{60}Co were detected in sea mussels and brown algae, the result of contact with cooling water from a power reactor in Fukui Prefecture. The highest level was 1 pCi/g (wet), but this was in rare samples only. As far as laver (*Porphyra*) is concerned, no ^{60}Co has been detected, since *Porphyra* is usually produced on culture beds, which are always remote from reactor sites.

S. BONOTTO: You suggested that the uptake of ^{60}Co by marine algae is related to the metabolic activity of the plant. Do you know whether the photosynthetic capacity is greater in the apical region of the plant than at the base?

M. SAIKI: Growth is most active between blade and stem for *Laminaria* and *Eisenia*, whereas the apical region and blade were the most active growing parts in *S. thunbergii* and *Chondrus*. From this and the data shown in Fig.3, it is supposed that cobalt is preferentially accumulated in the older parts of seaweed, rather than in the younger parts. Adsorption of ^{60}Co in seaweed might be affected by the structural nature of the latter and also by substances adhering to the surfaces, such as micro-organisms. However, biological metabolic activity in seaweed may still be connected with the accumulation and loss of ^{60}Co, since other experimental work shows that the accumulation and loss of ^{60}Co in brown algae (Hizikia fusiforme) varies with exposure time to light (see "Radioecological studies on Marine Environment", NIRS-R-3, National Institute of Radiological Sciences, Japan (1974) 34).

J.D. MARSHALL: The experiments were carried out using a soluble form of cobalt (CoCl). Does this mean that the cobalt discharged to the environment is in a similar biologically available form?

M. SAIKI: Sometimes we detected trace amounts of ^{60}Co in molluscs and algae collected from the marine environment. We have never detected ^{60}Co in seawater coolant because of its extremely low concentration. Consequently we could get no information on the chemical form of ^{60}Co in the environment. It is likely that Co tends to complex easily in a marine environment. The biological activity of micro-organisms and bacteria in the formation of Co complexes may not be negligible. It should be noted that Co is a component of Vitamin B_{12}, enriched in sea mud.

V.A. KNIZHNIKOV: The algae you have studied contain alginic acid, which is capable of inhibiting the assimilation of many radionuclides from food, in particular ^{90}Sr. The assimilation of the latter radioisotope can in fact be reduced by a factor of 5-10 under the effect of alginic acid. In estimating the risk of contamination of algae by cobalt-60, have you considered this property of alginic acid to bind various radionuclides and micro-elements?

M. SAIKI: As you say, alginic acid plays an important role in the metabolism of Sr. In the radioactivity survey data on several kinds of algae compiled by Ueda et al. at the National Institute of Radiological Sciences (NIRS), it was suggested that the concentration of ^{90}Sr in seaweed varied with the concentration of alginic acid in seaweed. Another colleague of mine, Mr. Ishikawa of NIRS, is investigating the relation between the chemical constitution of polysaccharide in seaweed and the uptake of ^{106}Ru by the plant, but it seems rather difficult to get a clear correlation between these factors. It is known that the cell-wall of brown algae is enriched in alginic acid; however, reference to our Table I shows that a remarkably high Co concentration in brown algae was not observed. In the present context we have not sufficient data to discuss this question.

TRANSPORT AND ACCUMULATION OF RADIOCOBALT IN THE MARINE SEDIMENT OF URAZOKO BAY

Y. NAGAYA, K. NAKAMURA
National Institute of Radiological Sciences,
Chiba-shi, Japan

Abstract

TRANSPORT AND ACCUMULATION OF RADIOCOBALT IN THE MARINE SEDIMENT OF URAZOKO BAY.

Concentrations of ^{60}Co and other radionuclides in surface sediment samples were investigated in and around Urazoko Bay, Fukui Prefecture, Japan, where a nuclear power plant has been operating since 1969 and the radioactive waste effluent has been released into the sea. The main source of ^{90}Sr and ^{137}Cs was recognized to be radioactive fallout from nuclear explosions, whereas ^{60}Co was considered to originate from the nuclear power reactor. For evaluating radioactive pollution characteristics of the sea bottom, the behaviour of ^{60}Co was investigated using the ^{137}Cs concentration as an indicator of the radionuclide sorption capability or accumulation capability of the sediment. No seasonal variation was observed in the distribution pattern of the $^{60}Co/^{137}Cs$ ratio and the retention of the heavy initial discharge in 1969 is considered to dominate the radionuclide level in the sediment. The ^{60}Co contamination in the sediment, expressed as the $^{60}Co/^{137}Cs$ ratio, in this narrow and semi-closed bay was found to vary according to a simple exponential function of the distance from the discharge outlet. It was found that the contamination is spreading out gradually from Urazoko Bay to outer regions. Also the correlation between the amount of discharged radioactive waste and the variation of the ^{60}Co level in the sediment was investigated. The radionuclide extraction efficiency from the sediment for hydrochloric acid was examined and average rates of 70 to 80% for ^{137}Cs and ^{60}Co were obtained, but fluctuations in the efficiencies were sometimes large.

1. INTRODUCTION

Among the radionuclides in the waste effluent released from a nuclear power plant, ^{60}Co is known to be one of the most important in environmental pollution, because of its considerably high discharge rate and relatively long half-life. Several reports have presented information on the behaviour of radiocobalt in aquatic environment, especially in freshwater systems [1].

In the ecosystem of shallow coastal waters, the bottom sediment could be considered to be the most important indicator of radioactive pollution, because of its abundance, rather slow movement in the sea and large capability of accumulating radionuclides. Heft et al. [2] reported that the sediment of Humboldt Bay contributed remarkably to the retention of ^{60}Co, more than 95% of the total inventory, in the ecosystem.

Commercial nuclear power reactors are now being built and operated in Japan. However, only a few field studies on the radioactive contamination of coastal sediments around nuclear facilities have been reported.

Tsuruga nuclear power station (331 MW), Tsuruga, Fukui Prefecture, releases waste effluent into Urazoko Bay, a small, shallow and semi-closed inlet. As rates of dispersion and dilution of discharged radionuclides in such a bay are considered to be low in comparison with those in the open coast area, the contamination of the sediment in Urazoko Bay is expected to be fairly high. This report describes research on the assessment of the ^{60}Co pollution of the bottom sediment of this bay, taking into consideration the accumulation capability of the sediments.

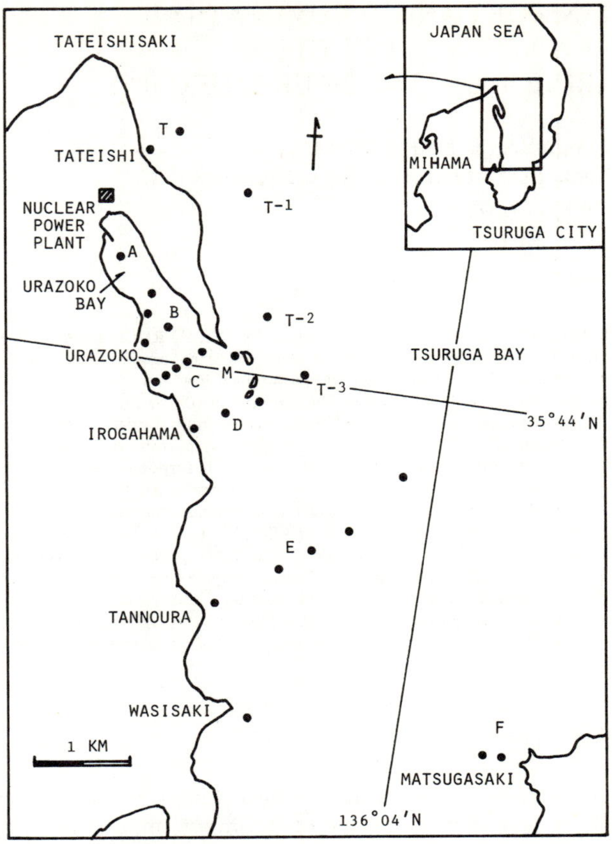

FIG. 1. Sampling stations.

2. MATERIALS AND METHODS

2.1. Collection of samples

From 1966 to 1974, 121 samples of sediment were collected from the 27 stations shown in Fig.1. For convenience, the stations were classified into seven areas (A, B, C, D, E, F and M), according to the south-eastward distances from the discharge outlet, and a reference area (T), where the contamination was expected to be lowest, was chosen in consideration of the geographical features.

Urazoko Bay is a small and narrow inlet, approximately 1 km wide by 2 km long, located along the west coast of Tsuruga Bay, and its maximum water depth is 18 to 20 m in the central zone of the inlet. The flow of terrestrial fresh water into Urazoko Bay is very poor, and around the mouth of the bay the invasion of sea water is observed throughout most of the year. Therefore the bay water shows physical and chemical characteristics rather similar to those of the sea water of the outer region [3]. The nuclear power station has its reactor cooling water inlet and outlet at the end of the bay.

All sediment samples were collected from the top 2 cm of the bottom surface, using the shaver type sediment collector [4,5].

Gravel was sifted out with a sieve of 2 mm mesh in sea water and the sediment was dried under an infra-red lamp.

2.2. Particle size distributions and radionuclide sorption capacities

The particle size distributions of typical sediment samples from each area were obtained by sifting and using a particle-size analyser (Shimadzu, type RS-11). For convenience, the sediment samples were classified into four categories according to their particle size distributions as follows: (1) that for which the fraction of particles having a diameter less than 0.075 mm is under 25% in weight (sand), (2) between 25 and 50% (silty sand), (3) between 50 and 75% (sandy silt) and (4) over 75% (silt). The total surface area of each particle size fraction of the sediment samples was calculated on the assumption that the sediment particles were spherical. The sediment samples collected were mostly silty sand or sandy silt.

The relative sorption capacities of sediment samples were investigated by a tracer experiment under the following experimental conditions: (1) the radionuclides used were ^{137}Cs and ^{60}Co in the form of chlorides, (2) the sea water used was 200 ml in volume, of pH 8.1 and at a temperature of 20°C, and (3) the sediment samples were presevered in sea water to prevent change of sediment property by drying out [6], and an equivalent amount to 200 mg dry weight of each sample was used in the experiment. The sediment samples were agitated in sea water containing the radionuclides to be sorbed, and after 5 hours the radioactivity remaining in the sea water was measured. The apparent distribution coefficient K is expressed as the ratio of the radioactivity per gramme of the sediment to that per millilitre of the sea water, and is taken as the indicator of the relative sorption capacity.

2.3. Radiochemical analyses

Before the power plant went into operation, concentrations of ^{90}Sr and ^{137}Cs in the sediment, originating from nuclear explosions, were determined as the background of artificial radioactivity in Urazoko Bay. After 200 g of dried sediment was treated with 6$\underline{N}$ hydrochloric acid, the extract was analysed radiochemically following the procedures described by Nagaya and Saiki [4].

For samples collected in 1969 and thereafter, ^{60}Co and ^{137}Cs were separated by extraction with hot 8$\underline{N}$ hydrochloric acid from 250 to 500 g of dried sediments. The ^{60}Co concentration was determined by the following method, which is essentially the same as that suggested by Shiozaki et al.[7]: namely, Fe, Mn, Ni, U, etc. were removed by ion-exchange processes as reported by Kraus and Moore [8], and Cu, Ru, etc. were separated by ion exchange – solvent extraction procedure [9].

Beta radioactivity of the nuclides was measured in the form of caesium chloroplatinate (^{137}Cs), cobalt anthranilate (^{60}Co) and yttrium oxalate (^{90}Y, the daughter nuclide of ^{90}Sr) using a low-background gas-flow counter (Tracerlab, Omni-Guard type).

2.4. Gamma-ray spectrometry

The identification of the radioisotopes of caesium and cobalt in the sediments and the extraction efficiencies of the radionuclides were determined by gamma-ray spectrometry. The gamma-ray spectra were obtained using a Ge(Li) detector (Horiba, type GU-C41) coupled with a 1024-channel pulse-height analyser (Northern Scientific, Model NS-710) and the radionuclide concentrations were calculated from the spectra using the method of Covell [10].

Because of the relatively low detection sensitivity in the determination of the radionuclide concentration, the gamma-ray spectromety is applicable only for the samples containing a considerably large amount of the radionuclide, i.e. several tens pCi/kg dry.

3. RESULTS AND DISCUSSIONS

3.1. Some physico-chemical properties of the sediment

3.1.1. Acid extraction efficiencies

The acid extraction efficiencies for ^{137}Cs and ^{60}Co were examined by comparing the gamma-ray spectra of the untreated sediment samples with those of the extraction residues. In Table I are shown the results for the samples collected in August 1973. The average extraction efficiency was about 70% for both ^{137}Cs and ^{60}Co, with the standard deviations less than 20%.

In Fig.2 the correlations between the analytical results obtained using gamma-ray spectrometry (ordinate) and those using radiochemical analysis (abscissa) are shown. The samples which show the statistically significant peaks of ^{137}Cs and ^{60}Co in their gamma-ray spectra were selected and compared. Linear relations were observed and confirmed by a least squares method calculation. The reciprocals of the gradients of the lines can be considered to be the average acid extraction efficiencies of the radionuclides, i.e. about 80%.

In spite of the occasional fluctuation, the acid extraction efficiencies of the artificial radionuclides are considered to be approximately constant, and in this report the concentration of the radionuclide in the sediment represents the amount of acid-extractable radionuclides.

3.1.2. Sorption capability of the sediment

In regard to physical and chemical properties, coastal sediments usually do not show uniform distribution even in very small areas, especially for complicated geographical features, such as bays and inlets. Since the physical and chemical properties vary, the radionuclide concentration also varies. Heft et al. [2] reported that "isolated high counting

TABLE I. ACID-EXTRACTION EFFICIENCIES OF THE RADIONUCLIDES FROM THE SEDIMENTS COLLECTED IN AUGUST, 1973

Sample No.	Area	Loss of weight* (%)	Extracted radionuclide (%)			
			^{137}Cs	^{60}Co	^{54}Mn	^{40}K
1	A	9.4	82	80	85	31
2	B	3.4	**	45	58	52
3	B	12.6	81	85	90	26
4	C	7.4	43	77	67	27
5	C	11.4	60	69	68	**
6	C	12.4	86	66	92	**
7	C	13.0	61	70	70	27
8	C	14.2	75	76	95	10
		Average	70 ± 16	71 ± 12	78 ± 14	29 ± 14

* Loss of weight by acid extraction.

** Below detection limit by gamma-ray spectrometry.

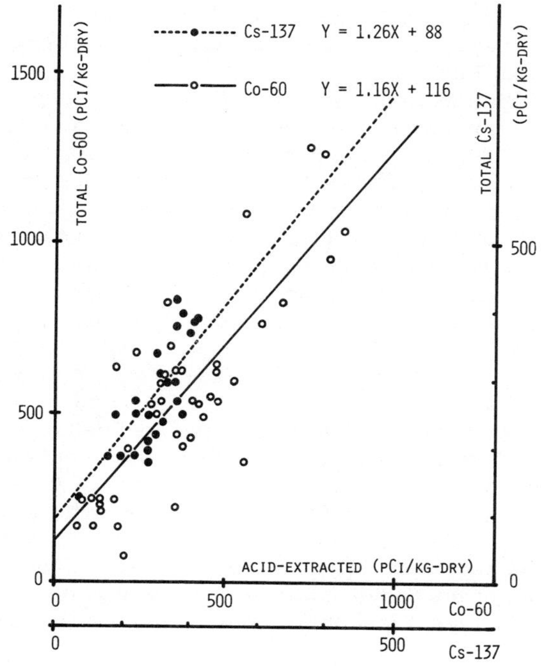

FIG. 2. Correlations of the amounts of total and acid-extracted radionuclides in the sediments.

TABLE II. THE APPARENT DISTRIBUTION COEFFICIENTS (K) OF RADIONUCLIDES FOR THE SEDIMENTS COLLECTED IN JULY, 1971

Area	Sediment	K-Cs*	K-Co*	K-Co/K-Cs
A	Silty sand	48 ± 2	169 ± 12	3.5 ± 0.3
C	Sandy silt	66 ± 6	231 ± 11	3.5 ± 0.4
D	Sand	32 ± 8	26 ± 5	0.8 ± 0.3
E	Sand	16 ± 2	36 ± 4	2.3 ± 0.4
M	Sand	5 ± 1	20 ± 1	4.0 ± 0.8
T	Sand	15 ± 1	16 ± 2	1.1 ± 0.2
			Average	2.5 ± 1.3

* Average of triplicate measurements.

samples" were occasionally found in the sediments of Humboldt Bay. Also in Urazoko Bay sediment, striking differences of ^{144}Ce contents were observed among sampling locations [11]. In view of this, it is concluded that the ranges and rates of radioactive contamination of sediments in an area are not indicated only by the results of radioactivity measurements, especially when the number of samples are rather small. Therefore auxiliary data, giving the radionuclide sorption capability of the sediment, are required for an evaluation of the radioactive contamination of the sediment of Urazoko Bay.

The apparent distribution coefficient, K, was often used as an index of the relative sorption capacity of the sediment. In Table II the apparent distribution coefficients of typical sediments from different areas are shown. Since the K values of the sediment samples vary greatly with variations in experimental conditions, they are rather unsuitable for direct use as the measure of relative sorption capacity or sorption capability. But in regard to the radionuclides, the ratio of the K values appears to be similar for all sediment samples. So the sorption capability for ^{137}Cs of the sediment in these areas corresponds roughly to that for ^{60}Co.

3.1.3. Particle size distributions and surface areas of the sediment

If the accumulation of radionuclides in the sediment were caused by sorption on the surface of suspended particles and the subsequent deposition of the particles on the sea bottom, the surface areas of the sediment particles should be an indicator of the radionuclide sorption capacity.

In Table III the radionuclide contents in the different particle size distributions of the sediment collected from area C in June 1974 are shown, and the correlation of the radionuclide contents with the surface areas or the weights of the sediment fractions can be expressed by the following linear function:

$$Y = a \cdot X + b$$

TABLE III. THE RADIONUCLIDE DISTRIBUTIONS IN THE DIFFERENT PARTICLE-SIZE FRACTIONS OF THE SEDIMENT COLLECTED FROM AREA C IN JUNE, 1974

Diameter (mm) range	mean	Weight (%)	Surface area* (%)	^{137}Cs (%)	^{60}Co (%)	^{54}Mn** (%)	^{40}K** (%)	^{238}U** (%)
2.000-0.500	1.250	3.2	0.2	0.3	1.6	2.2	2.0	5.1
0.500-0.250	0.375	8.5	1.4	3.9	3.2	3.2	8.5	10.6
0.250-0.125	0.188	23.0	10.1	16.2	10.8	19.6	23.1	19.4
0.125-0.075	0.100	16.9	10.4	15.5	10.5	12.5	20.1	17.5
Below 0.075	0.038	48.4	77.9	64.1	73.9	62.5	46.3	47.4
	Total	100.0	100.0	100.0	100.0	100.0	100.0	100.0

* Calculated using the mean diameter and the weight of the fraction.

** Determined by gamma-ray spectrometry.

TABLE IV. RELATION BETWEEN THE RADIONUCLIDE CONTENTS AND THE SURFACE AREA OR THE WEIGHT OF THE SEDIMENT FRACTION

X \ Y		Radionuclide content (%) ^{137}Cs	^{60}Co	^{54}Mn	^{40}K	^{238}U
Surface area (%)	a	0.76	0.93	0.70	0.45	0.46
	b	4.80	1.41	6.00	11.0	11.0
Weight (%)	a	1.40	1.64	1.39	0.96	0.92
	b	-8.00	-12.8	-7.75	0.79	1.59

where Y is the radionuclide content (%), X is the surface area (%) or the weight (%) of the sediment fractions, and a and b are constants. Calculated values of a and b are shown in Table IV. It is a matter of course that the contents of ^{40}K and ^{238}U are almost in direct proportion to the weights of the sediment fractions, because they are the natural radionuclides contained in the sediment particles. On the other hand, artificial radionuclides appear to be directly proportional to the surface areas, although substantial fluctuations are observed among the radionuclides. Therefore the surface areas of the sediment particles can sometimes be used as an indicator of the sorption capability of the sediment. Also, because of their similar correlation with the surface areas, ^{137}Cs content can represent the sorption capability of the sediment for ^{60}Co or ^{54}Mn.

3.2. Variations of the radionuclide concentrations

The ^{90}Sr and ^{137}Cs concentrations in sediment around the discharge outlet (area A) for samples collected from 1966 to 1971 are shown in Fig.3. Since no significant increase in the ^{90}Sr and ^{137}Cs levels was observed after the start of the power plant operation in 1969, it can be concluded that the amount of these radionuclides released from the power plant has been so small as to hardly contribute to the concentrations in the sediment, and almost all these nuclides in the bottom sediment of Urazoko Bay can be considered to be derived from radioactive fallout.

The levels of ^{137}Cs and ^{60}Co contained in the sediment collected from 1969 to 1974 are summarized in Table V. Since the existence of significant amounts of ^{134}Cs and ^{58}Co was not observed in the gamma ray spectra of the sediment samples, radiocaesium and radiocobalt obtained by chemical analysis were considered to be ^{137}Cs and ^{60}Co, respectively. The maximum concentration of ^{60}Co in the sediment of area A collected in June 1974 was 1 160 pCi/kg dry. Figure 4 shows the secular variations of the ^{137}Cs and ^{60}Co concentrations in the sediment collected from area A and the

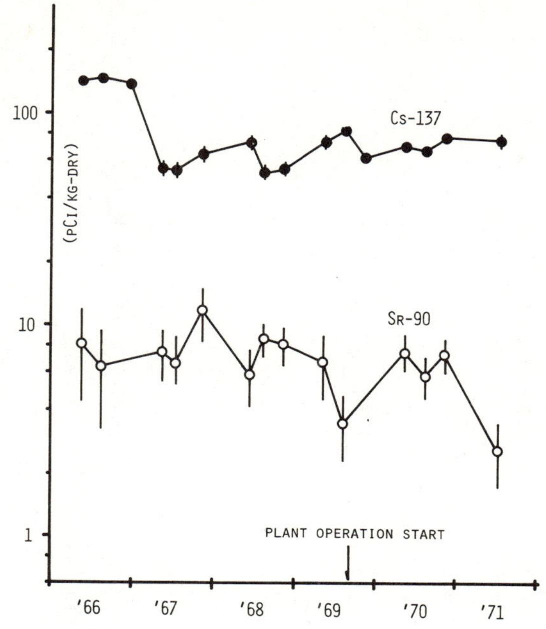

FIG. 3. Variations of ^{137}Cs and ^{90}Sr concentrations in the sediments collected in area A, during 1966-71.

discharged radioactivity over a period of 3 months expressed as the total gamma radioactivity, which includes about 30% ^{60}Co and a few % ^{137}Cs on the average. The content of ^{60}Co in the sediment increased abruptly just after the start of the operation of the power plant, in accordance with the high discharge rate, and it can be concluded that almost all the ^{60}Co in the sediment originated from the nuclear power reactor. The discharged radioactivity decreased markedly after a peak in early 1970, by almost two orders of magnitude; however, the ^{60}Co level in the sediment remained the same. This phenomenon indicates that the heavy initial load of ^{60}Co was retained in the sediment, because the processes of deposition and accumulation of the radionuclide are almost irreversible and desorption from the sediment is, at most, very slow.

The secular variation of the ^{60}Co concentration after late 1970 is rather similar to that of ^{137}Cs derived predominantly from radioactive fallout, suggesting that the pattern of variation is hardly caused by fluctuation of discharged radioactivity, but is dominated by differences in radionuclide sorption capability among the sediments collected for the different periods.

3.3. ^{137}Cs concentration as an indicator of sorption capability

As concluded in sections 3.1.2. and 3.1.3., the ^{137}Cs sorption capability or the surface area of the sediment can be used as the sorption capability

TABLE V. THE RANGES OF THE RADIONUCLIDE CONCENTRATIONS AND THE CONCENTRATION RATIOS OBSERVED DURING 1969 TO 1974

Area	^{137}Cs*		^{60}Co*		$^{60}Co/^{137}Cs$	
	Minimum	Maximum	Minimum	Maximum	Minimum	Maximum
A**	86 ± 2	169 ± 5	3 ± 2	1,157 ± 19	0.03 ± 0.01	8.0 ± 0.2
B	21 ± 2	206 ± 5	73 ± 5	724 ± 15	1.7 ± 0.1	5.5 ± 0.6
C	45 ± 1	490 ± 14	6 ± 1	609 ± 12	0.13 ± 0.01	4.5 ± 0.1
D	19 ± 2	184 ± 3	6 ± 1	197 ± 7	0.64 ± 0.02	3.8 ± 0.2
E	48 ± 1	152 ± 3	5 ± 1	85 ± 7	0.10 ± 0.02	0.70 ± 0.06
F	44 ± 9	128 ± 5	6 ± 1	13 ± 2	0.09 ± 0.02	0.27 ± 0.04
T	8.2 ± 0.6	11.9 ± 0.9	0.1 ± 0.7	14.0 ± 0.4	0.01 ± 0.04	1.28 ± 0.05
M	11.5 ± 0.9	16 ± 2	3 ± 1	142 ± 3	0.23 ± 0.07	9.8 ± 0.3
Others	4.6 ± 0.5	49 ± 2	0.2 ± 0.7	13 ± 2	0.02 ± 0.07	0.9 ± 0.2

* pCi/kg dry

** Including the data observed before the start of plant operation.

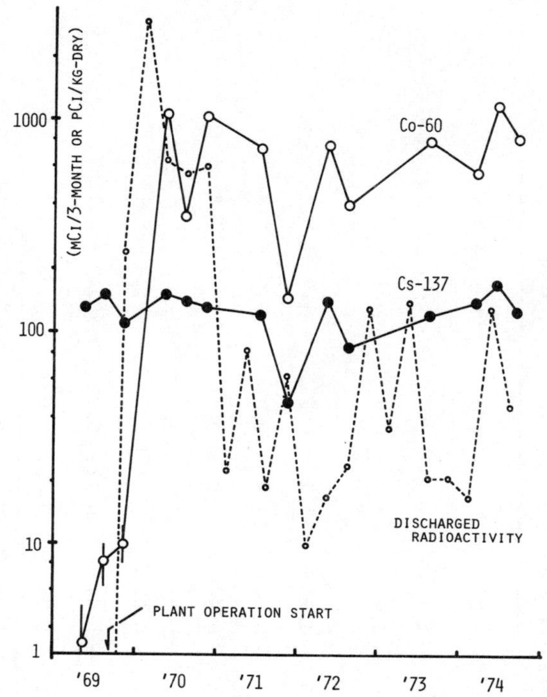

FIG. 4. Variations of the discharged radioactivity and the concentrations of ^{137}Cs and ^{60}Co in the sediments of area A, during 1969-1974.

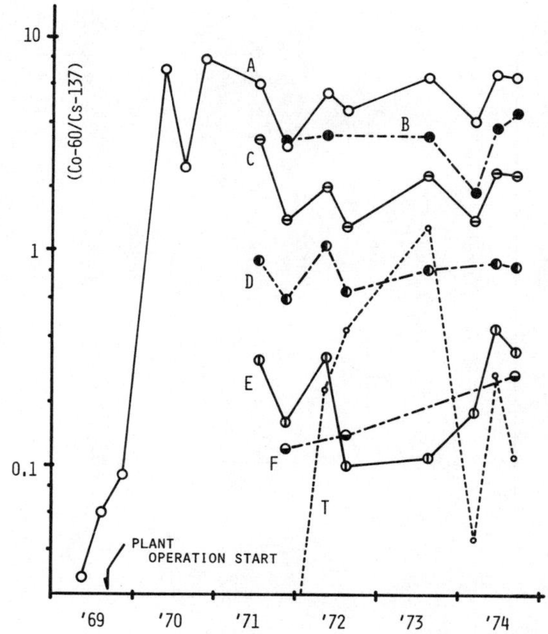

FIG. 5. Variations of $^{60}Co/^{137}Cs$ ratios in the sediments from seven areas.

TABLE VI. DISTRIBUTIONS OF THE RADIONUCLIDE CONCENTRATIONS AND THE CONCENTRATION RATIOS IN THE REGION AROUND AREA T IN 1974

	Date	Area T	Area T-1	Area T-2	Area T-3
	March	0.2 ± 0.7	0.2 ± 0.7	4.1 ± 1.0	1.5 ± 0.8
^{60}Co*	June	3.2 ± 0.9	3.3 ± 0.9	2.7 ± 0.9	4.9 ± 1.0
	Sept.	1.3 ± 0.4	3.0 ± 0.9	2.9 ± 0.9	4.1 ± 1.2
	March	9.1 ± 0.8	9.9 ± 0.9	5.3 ± 0.5	4.6 ± 0.5
^{137}Cs*	June	11.8 ± 0.9	10.5 ± 0.8	13.9 ± 0.0	9.9 ± 0.8
	Sept.	11.4 ± 0.9	10.9 ± 0.8	10.2 ± 0.8	11.9 ± 0.8
	March	0.02 ± 0.08	0.02 ± 0.07	0.77 ± 0.20	0.33 ± 0.18
$^{60}Co/^{137}Cs$	June	0.27 ± 0.08	0.31 ± 0.09	0.19 ± 0.07	0.49 ± 0.11
	Sept.	0.11 ± 0.04	0.28 ± 0.08	0.28 ± 0.09	0.34 ± 0.10
	Average	0.13 ± 0.04	0.20 ± 0.05	0.41 ± 0.08	0.39 ± 0.08

* pCi/kg dry

indicator of ^{60}Co for the sediment. But the determination of the apparent distribution coefficient and the measurement of the particle size distribution of the sediment are rather difficult and troublesome. On the other hand, it is clear that (1) since ^{137}Cs originates mainly from fallout and was distributed uniformly in sea water of the areas investigated [4,11], the ^{137}Cs concentration in the sediment samples can be considered to represent the ^{137}Cs sorption capability of the sediment in the areas, and (2) in general, the ^{137}Cs concentration in the sediment is considerably high and, therefore, radiochemical analysis can be performed without difficulty [4,11]. For these reasons, the ^{137}Cs content was used as an indicator of the ^{60}Co sorption capability of the sediment, and variations of the ^{60}Co contamination were investigated by comparing the ratio of the ^{60}Co and ^{137}Cs concentrations of each sample. In this connection, Pickering et al. [12] reported that the distribution patterns of ^{137}Cs and ^{60}Co released from nuclear facilities are similar in the bottom sediment of rivers.

3.4. Transport and accumulation of radiocobalt in the sediment

The secular variations of the $^{60}Co/^{137}Cs$ ratio are shown in Fig.5. In all the areas, with the exception of area T, similar variations were observed; however, after 1971 the fluctuations are not significant. Steep increase and abrupt decrease of the ratio observed in the sediment of area T may be caused by the gradual dispersion of the contamination to the outer regions of Urazoko Bay or by the inflow from the outer sea of pollutants, although the ^{60}Co concentration in the sediment is still low (14 pCi/kg-dry in August 1973) in comparison with that measured inside Urazoko Bay (790 pCi/kg dry at area A). To investigate the origin of radiocobalt in the sediment from area T, $^{60}Co/^{137}Cs$ ratios of the sediments around area T were calculated as shown in Table VI. The gradual decrease of the ratio from area T-3 to area T indicates clearly that the origin of ^{60}Co lies in Urazoko Bay.

The correlation between the ratio, observed from July 1971 to September 1974, and the distance from the discharge outlet are shown in Fig.6. Almost no secular variations were observed in the distribution pattern, suggesting that the transport and deposition rates of minute particles were practically invariable. In Urazoko Bay, a negligibly small inflow of terrestrial water and a regular movement of sea water throughout the year may establish uniform rates of dispersion. On the other hand, in the case of rivers, seasonal variations have been observed in the radioactivity of the bottom surface sediment, since the transport of suspension and suspended sediment particles is affected by the rate of inflowing water [13].

According to the data in Fig.6, correlation of ^{60}Co contamination with the distance from the discharge outlet can be expressed as a simple exponential function:

$$Y = c \cdot e^{-d \cdot X}$$

where Y is the $^{60}Co/^{137}Cs$ ratio of the sediment, X is the distance in kilometers and c and d are constants. The constants c and d seem to have different values in the distant region beyond area E. The

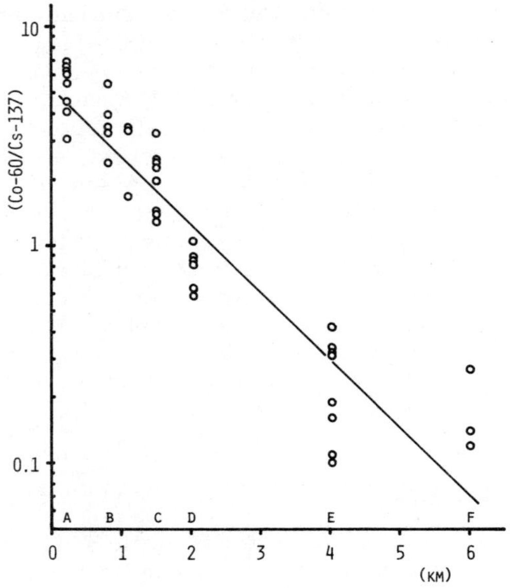

FIG. 6. The $^{60}Co/^{137}Cs$ ratios in the sediments versus the distance from the plant.

TABLE VII. CALCULATED VALUES OF c AND d

Date	c	d
July, 1971	7.5	0.81
Nov., 1971	3.4	0.63
May, 1972	6.1	0.87
Aug., 1972	3.2	0.64
Aug., 1973	8.7	1.09
Mar., 1974	4.5	0.80
June, 1974	7.0	0.75
Sept.,1974	5.9	0.59
Total	5.2	0.71

results of calculation of c and d are listed in Table VII. A similar distribution pattern for the contents of ^{144}Ce, ^{106}Ru, etc. was found in the coastal area around Windscale [14], while in the case of the discharge through the Columbia River, no definite correlation with the distance was observed in the coastal sea [15].

Since the constant d shows a significant fluctuation, the environmental conditions influencing the dispersion rate of ^{60}Co are considered to have been fairly stabilized and uniform in and around the bay. Besides, the constant c can be dominated by the radionuclide concentration. Since relatively small variation in the constant c is observed in Urazoko Bay, in spite of the remarkable decrease in the discharge rate, it is confirmed that the ^{60}Co exists in significant quantities in the sediment of the bay and the ^{60}Co level is found to be maintained nearly constant throughout the period of investigation.

Concerning the present situation of Urazoko Bay, it can be said that an apparent equilibrium has been established among the factors affecting the increase and decrease of ^{60}Co in the sediments, such as the physical decay of the deposited ^{60}Co, the desorption from sediment particles, the transport of particles and the sorption of freshly discharged radionuclides. Therefore as long as the release of radionuclides remains at the present rate, approximately the same level of ^{60}Co will be maintained in the sediment of Urazoko Bay, and even if the discharge rate is decreased, it might require a considerably long time for the lowering of the ^{60}Co level in the sediment.

For the evaluation of radioactive pollution of marine sediment around nuclear facilities, the radionuclide sorption capability of the sediment should be taken into consideration. In this report the concentration of fallout ^{137}Cs in the sediment was used as a measure of sorption capability and the transport and accumulation of ^{60}Co released from a nuclear power plant were successfully investigated, by comparing the ratio of ^{60}Co and ^{137}Cs concentrations in the sediment.

ACKNOWLEDGEMENTS

The authors wish to express their cordial thanks to Dr. M. Izawa and Dr. R. Ichikawa, National Institute of Radiological Sciences, for their useful suggestions. They are also grateful to the staff of Fukui Prefectural Fisheries Experimental Station and to the staff of Tsuruga Power Station, The Japan Atomic Power Company, for their aid in collecting samples and supplying useful information on the subject.

REFERENCES

[1] FUKAI, R., MURRAY, C.N., "Environmental behaviour of radiocobalt and radiosilver released from nuclear power stations into aquatic systems", Environmental Behaviour of Radionuclides Released in the Nuclear Industry (Proc. Symp. Aix-en-Provence, 1973), IAEA, Vienna, (1973) 217-247.

[2] HEFT, R.E., PHILLIPS, W.A., RALSON, H.R., STEELE, W.A., "Radionuclide transport studies in the Humboldt Bay marine sediment", Radioactive Contamination of the Marine Environment (Proc. Symp. Seattle, 1972), IAEA, Vienna, (1972) 593-614.

[3] FUKUI PREFECTURAL FISHERIES EXPERIMENTAL STATION, The Report on the Survey of the Environment of Fishing Ground prior to the Nuclear Reactor Construction (1967).

[4] NAGAYA, Y., SAIKI, M., Accumulation of radionuclides in coastal sediment of Japan (I). Fallout radionuclides in some coastal sediments in 1964-1965, J. Radiat. Res. 8 1 (1967) 37-43.
[5] SAIKI, M., KOYANAGI, T., NAGAYA, Y., UEDA, T., "Review of sampling and ashing methods", Reference Methods for Marine Radioactivity Studies, Technical Reports Series No. 118, IAEA, Vienna, (1970) 81-92.
[6] DUURSMA, E.K., BOSCH, C.J., Theoretical, experimental and field studies concerning diffusion of radioisotopes in sediments and suspended particles of the sea. Part B: Methods and experiments, Neth. J. Sea Res. 4 4 (1970) 395-469.
[7] SHIOZAKI, M., ODA, K., KIMURA, T., SETO, Y., "The artificial radioactivity in sea water", Researches in Hydrography and Oceanography (Commemoration Publ. of the Centenary of the Hydrographic Dept. of Japan), Tokyo (1972) 203-249.
[8] KRAUS, K.A., MOORE, G.E., Anion exchange studies VI. The divalent transition elements manganese to zinc in hydrochloric acid, J. Am. Chem. Soc., 75 (1953) 1460-1462.
[9] KORKISCH, J., AHLUWALIA, S.S., Separation of large amounts of iron (III) from cobalt, nickel and aluminium by combined ion exchange-solvent extraction, Anal. Chim. Acta 34 (1966) 308-313.
[10] COVELL, D.F., Determination of gamma-ray abundance directly from the total absorption peak, Anal. Chem. 31 (1959) 1785-1790.
[11] NAKAMURA, K., NAGAYA, Y., Accumulation of radionuclides in coastal sediment of Japan (II). Fallout radionuclides in some coastal sediments in 1966-1972, J. Radiat. Res. (1975) (to be published).
[12] PICKERING, R.J., CARRIGAN, P.H., Jr., TAMURA, T., ABEE, H.H., BEVERAGE, J.W., ANDREW, R.W., Jr., "Radioactivity in bottom sediment of the Clinch and Tennessee rivers", Disposal of Radioactive Waste into Seas, Oceans and Surface Waters (Proc. Symp. Vienna, 1966), IAEA, Vienna, (1966) 57-88.
[13] PERKINS, R.W., NELSON, J.L., HAUSHILD, W.L., Behaviour and transport of radionuclides in Columbia River between Hanford and Vancouver, Washington, Limnol. Oceanogr. 11 (1966) 235-248.
[14] JEFFRIES, D.F., Fission-product radionuclides in sediments from the North-East Irish Sea, Helgol. Wiss. Meersunters. 17 (1968) 280-290.
[15] FORSTER, W.O., "Radionuclide distribution in Columbia River and adjacent Pacific shelf sediments", Ch. 27, The Columbia River Estuary and Adjacent Ocean Water (PRUTER, A.T., ALVERSON, D.L., Eds.), Univ. Wash. Press, Seattle & London, (1972) 663-735.

DISCUSSION

K.C. PILLAI: Do you have any information on the clay and mineral content of the sediments? Can you explain why you were able to get only 72% extraction of ^{137}Cs from the sediments?

Y. NAGAYA: In our paper we only considered the particle size of the sediment. Of course, the mineral content can be determined. Intensive study is now going on in regard to the extractability of the radionuclides in the sediment.

J.A. HETHERINGTON: You have shown that the uptake of radionuclides by sediments depends on the grain-size of the latter, and you take this to mean that the degree of uptake is proportional to the surface area of the sediments. May I ask whether you have ever found that the activity per grain begins to increase again in the coarser sand fractions after having passed through a minimum in the fine sand fraction? I ask this because in the course of our work on coastal and estuarine sediments in the north-east Irish Sea we have noticed that the specific activity does pass through a minimum in the range 125-250 μm before increasing again in the range 350-400 μm. We do not have a complete explanation of this, but it is found that the C.E.C. shows a similar relationship with particle size, and it has been suggested that this might be due to the relative increase in oxidizable carbon.

Y. NAGAYA: We have not found any tendency for the sorption activity to increase in the case of coarser sediment fractions. Our Urazoka Bay sediment samples may of course have different properties from your samples, and this might account for the difference in sorption capability.

As regards the oxidation-reduction conditions we also found that, under anaerobic experimental conditions, the Urazoka Bay sediment shows a sorption capability for ^{95}Zr-^{95}Nb an order of magnitude higher than that under aerobic conditions. However, the difference is not so large in the case of ^{137}Cs or ^{60}Co.

Y. NISHIWAKI: The apparent distribution coefficients of radiocaesium for particle sediments given in this paper are about an order of magnitude smaller than those given in paper SM-198/16 by Pillai. May I ask Mr. Pillai for his views on the cause of this difference? In order to render these results comparable I think it is important to state the physico-chemical characteristics of the sediment particles and the conditions of the sea water at the locality where the studies were made.

K.C. PILLAI: I feel that the difference in the distribution coefficients of radiocaesium is due to the variation in sediment characteristics. In the case of sediments from the Bombay harbour area, the higher K_d factors are probably due to the clay mineral content of the sediment. This was also evident from the low extractability of ^{137}Cs from sediments. I agree with your comments.

J.A. HETHERINGTON: May I comment on Mr. Nishiwaki's observation that Pillai's value for the K_d of ^{137}Cs in sediment was an order of magnitude greater than the value reported by Nagaya. I think that in fact the difference between the two results can be reconciled if account is taken of the effect of the particle size of the sediments. Nagaya has already shown how this affects K_d and the fact that Pillai has indicated that his sediments were muddy, with a significant clay content, is probably the main reason why he finds a higher K_d for ^{137}Cs than Nagaya, whose sediments contain much more sand.

RADIOCAESIUM DYNAMICS IN A CONTAMINATED FLOODPLAIN ECOSYSTEM IN THE SOUTHEASTERN UNITED STATES

C.T. GARTEN Jr, J.B. GENTRY, J.E. PINDER III,
Rebecca R. SHARITZ, M.H. SMITH
Savannah River Ecology Laboratory, Drawer E,
Aiken, South Carolina,
United States of America

Abstract

RADIOCAESIUM DYNAMICS IN A CONTAMINATED FLOODPLAIN ECOSYSTEM IN THE SOUTHEASTERN UNITED STATES.

The fate of radiocaesium (primarily Cs-137) released into a small stream, Steel Creek, during the operation of nuclear production reactors has been studied. Releases from fuel element disassembly basins occurred concurrently with releases of heat-exchanger cooling water. Cessation of reactor operations reduced stream flow and temperature and permitted colonization of a radiocaesium-contaminated floodplain by plants and animals. Subsequent studies have shown (1) mean radiocaesium concentrations in the soil of approximately 100 pCi/g, (2) considerable spatial variability in sediment radiocaesium concentrations, (3) significant radiocaesium uptake by plants with soil-plant concentration factors usually greater than unity, and (4) appreciable radiocaesium concentrations in animals. Significant uptake by plants occurs because kaolinite, the principal clay mineral in the sediments, does not fix radiocaesium into unavailable forms. Radiocaesium uptake by plants exhibits considerable spatial heterogeneity with radiocaesium concentrations in animals being proportional to plant concentrations. Plant and soil concentrations are not correlated, and the factors causing spatial heterogeneity in plant uptake are currently being investigated. Variability in radiocaesium concentrations between individuals of animal species is less than that observed for plant species. This reduction in variation in concentrations probably occurs because animals are mobile and feed over a large area, thereby integrating widely varying concentrations in vegetation.

1. INTRODUCTION

Complex environmental problems have resulted from the geometrical increase in human numbers and increasing demands for energy to sustain the quality of life. Many resulting ecological or environmental problems are associated with the cycling of materials from industry into ecosystems. Understanding principles of mineral or elemental cycling in nature will enable us to better assess the future impact of industrial expansion and its related by-products, wastes, and effluents on ecosystems. Increased demands for nuclear power, in particular, necessitate a wider understanding of the fate of long-lived radionuclides in the environment.

Aquatic and terrestrial ecosystems on the U. S. Energy Research and Development Administration's Savannah River Plant (SRP), near Aiken, SC, offer unique opportunities to research problems associated with radionuclide cycling. Low-level releases of radionuclides have occurred from some of the existing nuclear production facilities at various times over the last twenty years. A focus for research by investigators at the University of Georgia's Savannah River Ecology Laboratory (SREL) has been a post-thermal reactor effluent stream, Steel Creek. Steel Creek is one of several streams on the SRP which has received cooling water from nuclear production facilities. Steel Creek is unique in that it has received more radioactive effluent than any other stream system on the SRP, yet the system is typical of small coastal plain streams in the southeastern USA.

Early in 1971, Steel Creek was chosen by SREL as the main site for studies of radionuclide dynamics in stream ecosystems. Among our objectives were: (1) estimate radionuclide concentrations in different components of the ecosystem, (2) identify major radionuclide pathways and (3) develop predictive relationships among trophic levels for radionuclide concentrations. The objective of this paper is to summarize information from past studies in Steel Creek and discuss the ramifications of our findings in view of a rapidly developing nuclear industry in the southeastern USA.

2. STEEL CREEK STUDY AREA

Steel Creek is a 20-km stream which drains about 90 km^2 and flows south toward the Savannah River Terraces subdivision of the Coastal Plain. Before entering the Savannah River, Steel Creek meanders through a cypress-hardwood swamp forest. The geologic formation underlying the area is the McBean, of Eocene origin, consisting predominantly of compact clayey sand and sandy clay. Unlike other unconsolidated sediments in the vicinity, clays of the McBean Formation have a relatively high ion-exchange capacity of up to 15.2 meg-100 g [1]. Steel Creek floodplain soils are chiefly of the Okenee series in the humic-gley soils group. These soils have fine sandy loam A horizons and sandy clay loam and sand B horizons and C horizons [2]. Bottom stream sediment minerals in this region are predominantly quartz in the sand and silt fractions and kaolinite in the clay fractions [3].

Prior to 1968, nuclear production reactors discharged large volumes of heat-exchanger cooling water into Steel Creek. From 1961 to 1970, purge water released from disassembly basins, which housed irradiated reactor fuel elements and a few defective experimental fuel assemblies, at the reactor facilities contained more than 20 detectable radionuclides. Most radionuclides had a half-life of days or months. Among the fission products released, radiocesium (Cs-137) was judged to be the most significant in a biological sense because of its long half-life (30 years). Over the 9-year period, approximately 260 curies of radiocesium were released into Steel Creek [4]. Annual releases varied from 5.2 to 53.5 curies.

Heated effluent from reactor facilities caused Steel Creek to exceed its original banks and contributed to a dissemination of radiocesium over the existing floodplain. When cooling water discharges ceased, the stream returned to normal flow and a large floodplain was exposed to colonization by vegetation and animals. Terrestrial plants and animals thus became exposed to radioactive wastes.

3. CORRELATIONS BETWEEN RADIOCESIUM CONCENTRATIONS IN ECOSYSTEM COMPONENTS

The ultimate fate of radiocesium in an ecosystem is a function of the way in which it is transferred through successive trophic

Table 1. Radiocesium concentrations (pCi/g dry wt) in selected components of the Steel Creek floodplain ecosystem. Time and place of sample collection is variable.

Component	Species or Type	n	$\bar{X}$ Conc. or Range of $\bar{X}$ Concentrations	Reference
Soil	Fine Clay	182	145.8-189.5	[7]
	Coarse Sand	114	13.1-23.5	[7]
Producers				
Herbaceous Leaves				
	Sagittaria latifolia	194	347.6-514.8	[6]
	" "	280	257.9-961.5	[25]
	Scirpus cyperinus	272	35.2-190.2	[25]
	" "	130	509.9	[13]
	Polygonum punctatum	138	265.4-323.4	[6]
	" "	294	115.8-635.5	[25]
	Andropogon spp.	102	494.4	[13]
	"	20	470.0-2300	[9]
Woody Leaves				
	Salix nigra	280	178.8-595.2	[25]
	" "	20	510-1600	[9]
	Myrica cerifera	59	134.3	[13]
	" "	18	540-1500	[9]
	Alnus serrulata	45	215.6	[13]
	" "	18	370-1000	[9]
Consumers				
Invertebrates				
	Coleoptera	50	440-990	[9]
	Orthoptera	50	400-1500	[9]
	Homoptera	50	150-600	[9]
	Araneae	50	260-970	[9]
Amphibians and Reptiles				
	Hyla cinerea	141	189.2	[26]
	**Natrix* spp.	13	122.2	[27]
Birds				
	*Herbivores	--**	32	[28]
	*Omnivores	--**	25	[28]
	*Carnivores	--**	12	[28]
Mammals				
	Sigmodon hispidus	91	607.9	Unpubl. Data
	Oryzomys palustris	47	214.2	" "

*pCi/gm wet weight; to approximate pCi/gm dry weight, multiply by 4.
**Sample sizes not reported.

Table 2. Correlation coefficients (r) between radiocesium concentrations in different plants and arthropod taxa from various sites within a contaminated stream community. With the exception of one value (NS) all r values are significant at the 0.05 level of probability.

	Plants				Arthropods		
	Andropogon	*Alnus*	*Myrica*	*Salix*	Araneae	Coleoptera	Homoptera
Plant Genera:							
Andropogon	1.00						
Alnus	0.85	1.00					
Myrica	0.70	0.74	1.00				
Salix	0.72	0.56^{NS}	0.70	1.00			
Arthropod taxa:							
Araneae	0.92	0.84	0.79	0.88	1.00		
Coleoptera	0.94	0.91	0.88	0.77	0.96	1.00	
Homoptera	0.86	0.65	0.79	0.94	0.93	0.88	1.00
Orthoptera	0.91	0.76	0.75	0.91	0.98	0.93	0.96

levels in the food chain. Radiocesium concentrations for plants and animals collected at various points along Steel Creek's floodplain from the entry point of reactor effluent to the Savannah River Swamp are presented in Table 1. The data are not comprehensive of all species for which radiocesium concentrations have been measured, but the animals and plants listed are representative of the dominant biota. Most concentrations are expressed on a gram oven dry weight basis. Radiocesium concentrations were determined by gamma spectrometry. Radiocesium concentrations in flora and fauna from the Steel Creek ecosystem are generally greater than those reported from a comparable old-field community on the drained White Oak Lake bed at Oak Ridge National Laboratory [5]. There are no consistent trends for increasing or decreasing radiocesium concentrations across major trophic levels in Steel Creek. There is an increase in concentrations from soils to producers; followed by a decrease in concentrations from producers to consumers.

Correlations among radiocesium concentrations in different components has been investigated by sampling vegetation and arthropods from ten different areas within Steel Creek. The degree of linear correlation between radiocesium levels in different ecosystem components is relatively high (Table 2). Correlations among plant genera range from 0.56 to 0.85 and are generally lower than correlations for either plant-arthropod relationships (range from 0.65 to 0.94) or arthropod-arthropod relationships (range from 0.88 to 0.98).

Correlations in radiocesium levels among the biological components of the system are higher than correlations between the soil and the biota. Radiocesium concentrations in *Sagittaria latifolia* and *Polygonum punctatum*, two herbaceous plant species common on the Steel Creek floodplain, were not correlated with total radiocesium concentrations in the soils on which they were growing [6]. Radiocesium uptake by plants in Steel Creek appears to be related to factors other than the concentration of the isotope in the soil. It has been estimated that approximately 85% of the variability in *Polygonum* radiocesium concentrations is caused by variation in external factors (Pinder and Sharitz, unpubl. ms.). Interrelationships among soil potassium, organic matter, and other edaphic factors may obscure simple correlations between plant and soil radiocesium concentrations. Brisbin *et al.* [7] reported a positive correlation ($r = 0.78$; $N = 52$) between the concentrations of radiocesium and available potassium in the soil, and high concentrations of potassium in the soil can reduce cesium uptake by plants [8]. Thus, the availability of potassium may decrease radiocesium uptake in highly contaminated Steel Creek soils.

These correlations indicate that radiocesium availability to the biota varies from area to area within Steel Creek. Radiocesium concentrations in consumers are highly correlated with plant radiocesium concentrations indicating that once radiocesium enters plants it moves freely to other components of the ecosystem with the concentrations in upper trophic levels being proportional to plant concentrations. The factors causing differential uptake by plants in different areas are as yet unknown. Plant concentrations do not merely reflect soil concentrations.

Table 3. Discrimination factors between precursor and acceptor compartments in five species of plants from Steel Creek.

Compartments		Discriminant Factors for Various Species				
Acceptor	Precursor	*Andropogon*	*Scirpus cyperinus*	*Alnus serrulata*	*Salix nigra*	*Myrica cerifera*
Fruit	Stem	2.44	2.22	2.27	--	--
Leaf	Stem	1.05	0.77	1.25	0.94	0.69
Stem	Root	0.38	0.31	--	--	--
Root	Soil	0.21	0.24	--	--	--

As previously mentioned, correlations between radiocesium concentrations in the animal components of the ecosystem are on the average higher than those for the plant components [9]. Movement by animals will cause them to feed from a wide range of radiocesium concentrations in vegetation. Similarities of functional roles among the arthropods studied (Table 2) may be responsible for the high correlations between concentrations in taxonomically distinct forms. Most of our understanding of isotope movement through ecosystems is derived from data on uptake and elimination by individuals and requires detailed knowledge of trophic relationships. The correlational approach does not require such information, although it too must ultimately depend upon biomass flow. Animal concentrations will be a function of their diet and the concentration of the isotope in the food. Even though food webs may be extremely complex and subject to seasonal alterations, the proportional flows of biomass must be relatively constant between components. Composition of the diet will vary over time for most animals, but, across species within a large taxonomic group (e.g., Coleoptera), the variability in mean dietary concentrations over time must be considerably reduced. Therefore, correlations between isotope concentrations in different taxa may be relatively invariant over time and space.

4. RADIOCESIUM AND POTASSIUM MOBILITY IN PLANTS

The relative mobility of two elements (e.g., A and B) between two compartments, precursor and acceptor, may be expressed by the discrimination factor,

$$D = \frac{\dfrac{\text{conc } A_a}{\text{conc } B_a}}{\dfrac{\text{conc } A_p}{\text{conc } B_p}} ,$$

where conc A_a and conc B_a express the elemental concentrations in the acceptor compartment and conc A_p and conc B_p are the concentrations in the precursor compartment. The discrimination factor is dimensionless and the concentrations may be expressed in any convenient units. Values of the discrimination factor greater than 1 indicate that element A is more mobile than B between the two compartments, whereas values less than 1 indicate that A is less mobile than B. The use of discrimination factors to assess radionuclide hazards has been questioned because discrimination factors vary over time and space [10, 11, 12]. However, for comparisons of elemental mobilities over short time intervals in a given ecosystem discrimination factors are still useful in describing the mobility of a radioactive element, such as Cs-137, relative to a stable element, such as K.

Discrimination factors (Table 3) were computed from ratios of pCi Cs-137/g to mg K/g for the relative mobility from the soil into two herbaceous (*Andropogon* sp., *Scirpus cyperinus*) and three woody (*Alnus serrulata*, *Salix nigra*, *Myrica cerifera*) plants. Discrimination factors were computed for transfers from soil to roots, roots to stems, stems to leaves, and stems to fruits. Stems are whole stems from herbs and twigs from woody plants. Fruits here refer to reproductive structures in general and include flowers inflorescences, seeds, and fruits. Cs-137 to K ratios were calculated from samples collected during autumn, 1971.

Cs-137 to K ratios in soils were expressed as picocuries of available Cs-137 to available potassium. Available Cs-137 was determined by leaching with 1 N HNO_3 [7]. Concentrations of available K in the soil were reported by Brisbin et al. [7]. The ratio of total Cs-137 to available K was 3204 pCi Cs-137/mg K. Cs-137 to K ratios in all plant organs were much less than that in the soil with the ratios in plant leaves ranging from 41 to 66 pCi Cs-137 to mg K.

Discrimination factors for Steel Creek indicate the following: (1) relative to potassium, radiocesium mobility from the soil to the roots is low; (2) radiocesium mobility from roots to stems is also less than that of ptoassium; (3) once radiocesium enters the stem it moves freely to leaves and may be concentrated in reproductive organs; and (4) among the species examined, discrimination factors for a given transfer are relatively invariant. The data suggest that the reduced Cs-137 to K ratios observed in plant leaves are the result of Cs-137 being less mobile between soils and roots than K. Discrimination factors computed using the soil as the precursor compartment and leaves as the acceptor compartment ranged from 0.051 for Myrica to 0.083 for Andropogon and indicated that the relative mobilities of radiocesium and potassium were similar in the Steel Creek plant species. Thus, plant species can show considerable variability in their mean radiocesium concentrations (Table 1 and [13]) but when the relative mobility of radiocesium and potassium are compared the species are found to behave in similar ways.

The soil to leaf discrimination factors were similar to those reported by Cline and Hungate [11] for bean plants growing in solutions of differing potassium concentrations. They were also similar to the discrimination factors reported by Cline [12] for beans grown in solution with low concentrations of potassium. Although the discrimination factors were similar, the Cs-137 to K ratios in the solutions were much greater than that in the Steel Creek soil. Cline [12] found a discrimination factor of 0.06 for new and 0.08 for old leaves when Cs-137 to K ratios in the solution were 8.5×10^4 pCi Cs/mg K. For Cs-137 to K ratios in solution that were comparable to the available Cs-137 to available K ratios in Steel Creek, Cline [12] found a discrimination factor approximately equal to 1. These results support Kornberg's [10] conclusion that radiocesium to potassium ratios vary too much to be used to extrapolate results from one area to another. However, the similarity in discrimination factors across species in Steel Creek suggests that these factors may be extrapolated from species to species within an ecosystem. The agreement in discrimination factors between Steel Creek and the solution culture experiments of Cline and Hungate [11] and Cline [12] are probably the result of wide variation in Cs-137 to K ratios in solution and do not necessarily represent similarities in radiocesium and potassium dynamics.

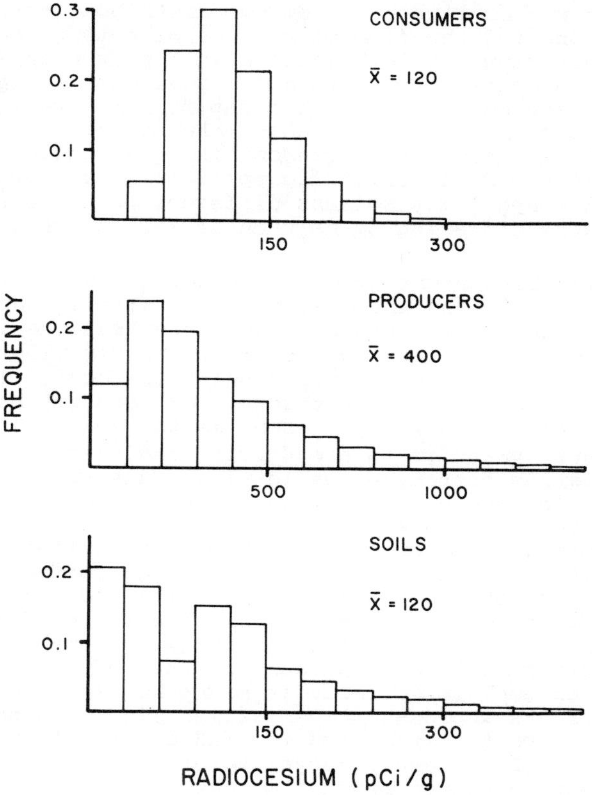

FIG. 1. Generalized frequency distributions of radiocaesium concentrations (pCi/g dry weight) in soil, producers and consumers from a contaminated southeastern coastal plain stream. $\overline{X}$ = mean value.

5. VARIABILITY IN CONCENTRATIONS ACROSS TROPHIC LEVELS

Figure 1 illustrates frequency distributions of radiocesium concentrations in three generalized ecosystem components: soils, producers and consumers. These distributions are idealized distributions compiled from numerous samples from each compartment. The distributions were created by assuming a log-normal distribution for radiocesium concentrations with distribution parameters typical of samples collected. Four data sets on radiocesium concentrations in soil, 11 sets on concentrations in producers, and 27 sets on concentrations in consumers were employed. The validity of the lognormal distribution as a general model for describing frequency distributions of radiocesium concentrations has been discussed by Pinder and Smith [14] and Schubert [15].

Soils show a bimodal distribution of radiocesium concentrations due to the predominance of either coarse or fine particle sediments [6, 7]. Therefore, the frequency distribution for the soil compartment was constructed from two distributions; one for the coarse and one for the fine particle sediments. The cumulative frequency distribution is bimodal with a peak near 20 pCi/g dry weight due to coarse sediments and a peak near 100 pCi/g dry weight due to fine sediments. As is evident from Figure 1, soil concentrations are highly variable. Recurrent flooding and stream meanderings have produced a random mixing of coarse and fine sediments on the Steel Creek floodplain. As a result, soil samples collected 1 m apart may possess different particle sizes and have concentrations differing by as much as 190% of their mean [7].

Radiocesium concentrations in producers are usually greater than those in soil and are also highly variable. However, the relative variability of radiocesium concentrations in producers is less than that of soils, and the frequency distributions of concentrations in producers are usually unimodal. It appears that producers are not only concentrating radiocesium, but they are also integrating the variability present in the soil. This integration occurs because plant roots are drawing radiocesium from soils of widely differing contamination levels. This reduction in variability in producers might not be expected if soil radiocesium concentrations were more uniform within small areas or if each plant's roots were drawing radiocesium from infinitely small volumes of soil. It is unlikely that physiological control of radiocesium concentrations contributes markedly to the reduced variability in producers since coefficients of variation for producer concentrations ranged from 40% to 160% with some species having frequency distributions comparable to that of soil. If the integrator concept is correct, it might be expected that large perennial plants would show less variability than small annuals due to (1) larger root systems drawing radiocesium from various sediments and (2) internal recycling of radiocesium. Limited data suggest that *Polygonum punctatum*, a small herbaceous annual, has higher variability in radiocesium concentrations than alder (*Alnus serrulata*), myrtle (*Myrica cerifera*), and willow (*Salix nigra*) trees.

A further reduction in relative variability occurs from producers to consumers. The coefficient of variation of radiocesium concentrations in producers is usually 80-120%, whereas the coefficient of variation in consumer populations is usually 20-40%. Consumers also exhibit variation in the coefficient of variation (10-140%) with some samples showing variability comparable to that observed in soils. Considerable variability was noted within relatively small taxonomic groups. Low radiocesium concentrations (< .01 ppb) and the wide range in observed coefficients of variation argue against reduced variation due to physiological control of radiocesium concentrations. Further integration of the variability in soil radiocesium concentrations by consumers is indicated.

Variability among herbivores, omnivores and carnivores was similar, indicating that the principal reduction in variability occurs in the producer to herbivore link of the food chain. Spatial variation in grazing, or grazing from a variety of species, means herbivores in different areas encounter similar average radiocesium concentrations in their diet. Thus, the variability in radiocesium concentrations between individual herbivores is less than among individual plants. Carnivorous species, whether they are generalists (i.e., feed on a variety of prey species) or specialists (i.e., feed on a single prey species), should show variability in radiocesium concentrations similar to that observed in herbivores. Generalists would encounter similar mean radiocesium concentrations in their diet, and specialists would encounter little variability in radiocesium concentrations between prey species. We hypothesize that sedentary herbivores which graze on a single plant will show greater variation between individuals than wide ranging herbivores which graze on a variety of species. Data to evaluate this prediction are not yet available.

The reduction in variability across trophic levels appears to be independent of radiocesium concentration factors. Variability in radiocesium concentrations decreased from soils to producers and again from producers to consumers. Concentration ratios between producers and soils are usually greater than 1, whereas consumer to producer concentration ratios are less than 1.

Leibscher and Smith [16] report greater variability in the concentrations of nonessential elements and micronutrients than macronutrients in humans. The concentrations of radiocesium and aluminum and several micronutrients are more variable than the concentrations of macronutrients (e.g., N, P, K, Ca) in *Sagittaria latifolia* and *Polygonum punctatum* (Sharitz, unpubl. data). Macronutrients may tend to have lower variability due to increased physiological control of concentrations. The concentrations of micronutrients and nonessential elements may experience less physiological control, and thus be greater affected by variations in concentration in the diet (or soil in the case of plants). If nonessential elements, such as radionuclides, show great variation in availability in soil, then the variability in their concentrations may decrease through the food chain.

6. IS STEEL CREEK REPRESENTATIVE OF THE SOUTHEASTERN USA?

It is difficult to generalize from a single stream system to an entire region, but in view of the rapid growth and expansion of the nuclear industry in the southeastern USA, it is desirable to consider if valid inferences concerning the fate of radionuclides in coastal plain ecosystems can be drawn from the Steel Creek experience. Perhaps the major characteristics which distinguish radionuclide dynamics in Steel Creek from studies in other areas of the southeastern USA are the availability of radiocesium to primary producers from soil and the subsequently large concentration ratios for the biota.

Plant-soil concentration factors in the southeastern USA are typically higher than values reported from other areas [17]. Dahlman et al. [17], using the data of Cummings et al. [18], calculated concentration factors ranging from 0.02 to 72.0 in oat plants growing on forty-three soils collected from across the southeast. In Steel Creek the concentration factors in leaves and roots of woody plants range from 3.8 to 7.6 [19] and concentration factors in herbaceous plants range from 0.6 in vegetation on soils of high radiocesium concentration to over 15 in vegetation on sediments with low radiocesium concentrations [6]. Thus, concentration factors in Steel Creek appear to be representative of the southeast in general and show almost as much variation from place to place and species to species within Steel Creek as is observed across the southeast.

The high concentration factors in Steel Creek are caused by the predominance of kaolinite as the basic clay mineral in the soil. Cummings et al. [20] reported that plant uptake of radiocesium increased as the ability of the clays in the soil to fix radiocesium into unavailable forms decreased. Kaolinite apparently does not fix radiocesium into unavailable forms [20]. Furthermore, Brisbin et al. [7], using either 1 N NaCl, 1 N HNO_3, or Triassic well water, were able to desorb more than 11% of the radiocesium from Steel Creek sediments. Illite, a clay mineral which fixes radiocesium into unavailable forms [17, 20, 21], is the principal clay mineral in the White Oak Lake bed at Oak Ridge, Tennessee. Using reagents similar to Brisbin et al. [7], Lominick and Gardiner [22] were unable to desorb more than 1% of the radiocesium from White Oak Lake soils. As a result of the fixing of radiocesium by illite, soil-plant concentration factors in the lake bed were low, ranging from 0.01 to 0.03. Dahlman et al. [17] have hypothesized that the higher soil-plant concentration factors observed in the southeastern USA are caused by the wide-spread occurrence of kaolinite in Coastal Plain soils. Thus, the presence of kaolinite in Steel Creek soils would explain the similarity between Steel Creek soil-plant concentration factors and those of Cummings et al. [18].

Therefore, it would seem appropriate to extrapolate the Steel Creek results to predict radiocesium availability and cycling in other southeastern ecosystems which contain kaolinite as the most abundant clay mineral. However, a word of caution is in order. The Steel Creek ecosystem has been subjected to stresses that are uncommon in most southeastern USA streams, i.e. increased stream flow and erosion along with thermal effluents. Our preliminary results indicate that these stresses have not significantly affected radiocesium cycling, but studies currently in progress on other SRP streams and thermal areas may yet demonstrate significant effects of these stresses on cycling processes.

7. SUMMARY

The early phases of our work involved quantification of radiocesium concentrations in abiotic and biotic components of the Steel Creek ecosystem. Emphasis was placed on the soil and dominant plant and animal species that occur in the creek and on the floodplain. Inventories of radiocesium could be calculated by multiplying the biomass of a component times its radiocesium concentration, but problems associated with estimating the biomass of biotic components and the temporal and spatial variability of radiocesium concentrations reduces the practical value of these inventories. Instead of developing our research in a systems analysis approach, we have chosen to strive for an understanding of radiocesium cycling through conceptual goals. Several important results have developed from our work.

First, the sediments in Steel Creek are the primary repository for radiocesium. However, unlike most areas where the radiocesium is fixed to clay minerals, the radiocesium in Steel Creek is not fixed by the kaolinite and is available for plant uptake. The plants are concentrating radiocesium up to 20 times soil concentrations and making the radiocesium available for entry into food chains. Thus, animals and plants from Steel Creek demonstrate high contamination levels. Radiocesium uptake by plants, and consequently the concentrations in animals, vary between areas in Steel Creek. The factors causing this variability are as yet unknown, and studies are currently in progress at SREL to identify these factors. Results of investigations by other researchers and preliminary SREL findings indicate that potassium availability and abundance of soil organic matter may be important in controlling radiocesium uptake by plants. Steel Creek is a successional ecosystem recovering from scouring of the stream course by heated effluents from reactor operations [23]. Plant succession is proceeding from herbaceous to shrubby species, and seedlings of the climax canopy species have begun to grow. Understanding the factors affecting radiocesium uptake by plants is important since as succession proceeds the organic matter content of the soil and the abundance of available nutrients will change. These changes may result in increased or decreased radiocesium uptake by plants. Such changes in radiocesium uptake need to be understood and predicted.

Second, the results of analyses of radiocesium to potassium discrimination factors indicate that differences between species in radiocesium concentrations, expressed as pCi/g, become less when concentrations are expressed as pCi/mg K. We do not imply that radiocesium is acting as an analog of potassium. The independence of radiocesium and potassium and the inability of cesium to substitute for potassium in plants have already been documented [10, 24]. However, radiocesium is not cycling through plants completely independent of other elements, and we are currently proceeding with studies to analyze the relationships between radiocesium cycling processes and the cycling of other elements.

Third, the reduction in variability of radiocesium concentrations from soils to producers to consumers we have termed the "integrator effect". This phenomenon has, to the best of our knowledge, not been reported for radiocesium or any other element. Whether this integrator effect occurs in other ecosystems and for other elements is unknown, but there seems to be little reason to doubt its existence elsewhere. Moreover, it cannot be determined at present whether this phenomenon is an important aspect of ecosystem cycling processes or just a curious result of food web dynamics. In either case, the reduction in the variability in contamination levels is an important point to consider in evaluating hazards to man from low-level contamination of food sources.

ACKNOWLEDGEMENTS

We thank the staff and students of the Savannah River Ecology Laboratory for their individual contributions to this research program and for allowing us to reference unpublished data. We also thank Mrs. E. M. James for assistance in preparation of the manuscript. This work was supported by contracts AT(38-1)-708, AT(38-1)-819 and AT(38-1)-310 between the U. S. Energy Research and Development Administration (formerly the U. S. Atomic Energy Commission) and the University of Georgia.

REFERENCES

[1] LANGLEY, T. M., MARTER, W. L., The Savannah River Plant, USAEC Rep. DP-1323 (1973).

[2] UNITED STATES DEPARTMENT OF AGRICULTURE, SOIL CONSERVATION SERVICE, National Cooperative Soil Survey, Fort Worth, Texas (1968).

[3] HAWKINS, R. H., Mineralogy and ion exchange characteristics of Savannah River Plant streambed sediments, USAEC Publ. DPST-71-332 (1971).

[4] MARTER, W. L., Radioactivity in the environs of Steel Creek, USAEC Rep. DPST-70-435 (1970).

[5] CROSSLEY, D. A., JR., "Comparative movement of ^{106}Ru, ^{60}Co and ^{137}Cs in arthropod food chains", Proc. 2nd Nat'l. Symp. on Radioecology (NELSON, D. H., EVANS, F. C., (Eds.)), Ann Arbor, Mich. (1969).

[6] SHARITZ, R. R., SCOTT, S. L., PINDER, J. E., III, WOODS, S. K., Uptake of radiocesium from contaminated floodplain sediments by herbaceous plants, Health Phys. 28 (1975) 23.

[7] BRISBIN, I. L., JR., BEYERS, R. J., DAPSON, R. W., GEIGER, R. A., GENTRY, J. B., GIBBONS, J. W., SMITH, M. H., WOODS, S. K., Patterns of radiocesium in the sediments of a stream channel contaminated by production reactor effluents, Health Phys. 27 (1974) 19.

[8] SQUIRE, H. M., MIDDLETON, L. J., Behaviour of Cs^{137} in soils and pastures a long term experiment, Rad. Bot. 6 (1966) 413.

[9] ANDERSON, G. E., GENTRY, J. B., SMITH, M. H., Relationships between levels of radiocesium in dominant plants and arthropods in a contaminated streambed community, Oikos 24 (1973) 165.

[10] KORNBERG, H. A., The use of element-pairs in radiation hazard assessment, Health Phys. 6 (1961) 46.

[11] CLINE, J. F., HUNGATE, F. P., Accumulation of potassium, $cesium^{137}$, and $rubidium^{86}$ in bean plants grown in nutrient solutions, Plant Physiol. 35 (1960) 826.

[12] CLINE, J. F., Effect of nutrient potassium on the uptake of caesium-137 and potassium and on discrimination factor, Nature (Lond.) 193 (1962) 1302.

[13] GARTEN, C. T., JR., BRIESE, L. A., GEIGER, R. A., SHARITZ, R. R., SMITH, M. H., "Radiocesium levels in vegetation colonizing a contaminated floodplain", Mineral Cycling in Southeastern Ecosystems (HOWELL, F. G., GENTRY, J. B., SMITH, M. H. (Eds.)), USAEC Symp. Series, CONF-740513 (in press).

[14] PINDER, J. E., III, SMITH, M. H., "Frequency distributions of radiocesium concentrations in soil and biota", Mineral Cycling in Southeastern Ecosystems (HOWELL, F. G., GENTRY, J. B., SMITH, M. H. (Eds.)), USAEC Symp. Series, CONF-740513 (in press).

[15] SCHUBERT, J., BRODSKY, A., TYLERS, S., The log-normal function of the distribution of strontium-90 and other fission products in humans. Health Phys. 13 (1967) 1187.

[16] LIEBSCHER, K., SMITH, H., Essential and nonessential trace elements, Arch. Environ. Health 17 (1968) 81.

[17] DAHLMAN, R. C., FRANCIS, C. W., TAMURA, T., "Radiocesium cycling in vegetation and soil", Mineral Cycling in Southeastern Ecosystems (HOWELL, F. G., GENTRY, J. B., SMITH, M. H. (Eds.)), USAEC Symp. Series, CONF-740513 (in press).

[18] CUMMINGS, S. L., JENKINS, J. H., FENDLEY, T. T., BANKERT, L., BEDROSIAN, P. H., PORTER, C. R., Cesium-137 in white-tailed deer as related to vegetation and soils of the southeastern United States, Radionuclides in Ecosystems (NELSON, D. J., Ed.) USAEC CONF-710501 (1971).

[19] GARTEN, C. T., JR., BRIESE, L. A., SHARITZ, R. R., GENTRY, J. B., "Season variation in radiocesium concentrations in three tree species", Mineral Cycling in Southeastern Ecosystems (HOWELL, F. G., GENTRY, J. B., SMITH, M. H. (Eds.)), USAEC Symp. Series, CONF-740513 (in press).

[20] CUMMINGS, S. L., BANKERT, L., GARRETT, A. R., JR., REGNIER, J. E., ^{137}Cs uptake by oat plants as related to the soil fixing capacity, Health Phys. 11 (1965) 1325.

[21] SAWHNEY, B. L., Sorption and fixation of microquantities of cesium by clay minerals: Effect of saturating cations, Soil Sci. Proc. 29 (1965) 25.

[22] LOMINICK, T. F., GARDINER, D. A., The occurrence and retention of radionuclides in the sediments of White Oak Lake, Health Phys. 11 (1965) 567.

[23] SHARITZ, R. R., IRWIN, J. E., CHRISTY, E. J., Vegetation of swamps receiving thermal effluents, Oikos 25 (1974) 7.

[24] NASON, A., McELROY, W. D., Modes of action of the essential mineral elements, Plant Physiol. 3 (1963) 451.

[25] BRIESE, L. A., GARTEN, C. T., JR., SHARITZ, R. R., "Distribution of radiocesium in vegetation along a contaminated stream", Mineral Cycling in Southeastern Ecosystems (HOWELL, F. G., GENTRY, J. B., SMITH, M. H. (Eds.)). USAEC Symp. Series, CONF-740513 (in press).

[26] DAPSON, R. W., KAPLAN, L., Biological half-life and distribution of radiocesium in a contaminated population of green treefrogs Hyla cinerea, Oikos 26 (1975) 39.

[27] STATON, M. A., BRISBIN, I. L., JR., GEIGER, R. A., Some aspects of radiocesium retention in naturally contaminated captive snakes, Herpetologica 30 (1974) 204.

[28] STRANEY, D. O., BEAMAN, B., BRISBIN, I. L., JR., SMITH, M. H., Radiocesium in birds of the Savannah River Plant, Health Phys. 28 (1975) 341.

DISCUSSION

Yu.V. SIVINTSEV: You showed that there is practically no correlation between the contents of ^{137}Cs in soil and plants. Does this mean that the main pathway for uptake of ^{137}Cs by the plants in the region you have considered is aquatic?

J.E. PINDER III: For several reasons we feel that the uptake of radiocaesium from water by plants is negligible. First, radiocaesium concentrations in the water are very low, less than 1 pCi/l. Second, plants growing partially submerged in water have radiocaesium concentrations similar to those in plants growing on non-flooded soils. Third, desorption studies have indicated that the radiocaesium in the soil is available for plant uptake. However, it is possible that some uptake from water does occur. It is also possible that partial submersion may result in leaching of radiocaesium from plant tissues.

Yu.V. SIVINTSEV: I note that many of the details given in your oral presentation are not mentioned in the text of your paper.

J.E. PINDER III: These details are drawn from publications referenced in our paper.

G. STOTT: That the animal species shows less variability in radiocaesium concentration may indeed be due, as stated in the paper, to the fact that animals are mobile. However, I should like to ask whether any measurements have been made on variability in the biological half-life of ^{137}Cs in the different animal species.

J.E. PINDER III: Biological half-lives have been determined for a number of species, but most of the information has not yet been published. The investigations in question show individual differences in biological half-lives.

K.C. PILLAI: You referred to the correlation between the organic matter content and radiocaesium content of soil. May I ask what range of values you obtained for organic matter in soils.

J.E. PINDER III: A mean organic matter concentration of 6.21% was reported by Brisbin et al. [7]. The same authors also report the correlation between organic matter and radiocaesium concentrations in soil.

S. BONOTTO: How many species of plants have you utilized for studying the uptake of radiocaesium?

J.E. PINDER III: We have determined radiocaesium concentrations in more than eleven species. The data from seven species have been published in references cited in the paper.

S. BONOTTO: Are the species you reported on in this work representative of the area you have investigated?

J.E. PINDER III: Yes, these species are the dominant plants in the area.

Y. NISHIWAKI: It is interesting to see that while the radiocaesium concentration in consumers and producers may be assumed to correspond to a log-normal distribution, the concentration in soils shows a bimodal distribution. You state that soils show a bimodal distribution of radiocaesium concentration owing to the predominance of either coarse-particle or fine-particle sediments. Does this mean that the size distribution of particle sediments is also bimodal? If so, what is the size range of the coarse and fine particle sediments in the locality you studied?

J.E. PINDER III: Soils tend to show bimodal distributions of particle sizes as a result of erosion and deposition caused by stream meanderings. The following particle-size distributions were determined by sieving oven-dried soils:

Particle size (mm)	Per cent composition by weight	
	Coarse	Fine
>2.0	2.6	2.1
1.0 - 2.0	4.9	4.5
0.5 - 1.0	16.4	12.0
0.25 - 0.5	46.7	35.5
0.1 - 0.25	25.0	29.7
<0.1	4.5	16.3

These results indicate that the principal difference between soil types is the absence of silt and clay-sized particles from the coarse soils. Most coarse soils contained less than 6% of the <0.1 mm particles, whereas most fine soils contained more than 10%.

UPTAKE OF COBALT-60 FROM SEA WATER AND FROM LABELLED FOOD BY THE COMMON SHRIMP *Crangon crangon* (L.)

A.W. van WEERS
Reactor Centrum Nederland,
Petten, N.H.,
The Netherlands

Abstract

UPTAKE OF COBALT-60 FROM SEA WATER AND FROM LABELLED FOOD BY THE COMMON SHRIMP *Crangon crangon* (L.).

The role of two different modes of uptake in the accumulation of ^{60}Co by the common shrimp (*Crangon crangon* (L.)) is the subject of the present study. The results show that accumulation of ^{60}Co from sea water is a slow process. The concentration factor for whole animals reached in one month was only about 13. Most of the activity accumulated from water appears to be associated with the exoskeleton. As a consequence, moulting has a pronounced effect on the uptake pattern of ^{60}Co and on the subsequent retention of the radionuclide by shrimps kept in non-radioactive sea water. After single feeding of shrimps with labelled mussel flesh, ^{60}Co is retained according to an exponential function with a short-lived and a long-lived component. The short-lived component has a mean biological half-life of 1.2 days and accounts for about 80% of the initial activity. About 20% of the initial activity is lost with a mean biological half-life of about 10 days. After repeated feeding of labelled mussel flesh the short-lived component in the ^{60}Co retention is virtually absent. Cobalt-60 taken up with food is localized mainly in the digestive gland and the concentration in the edible muscles from the abdomen is relatively small. It is concluded from the present study that direct uptake from sea water will play only a minor role in the accumulation of ^{60}Co in the internal organs of the shrimp. ^{60}Co released into the marine environment will be taken up by shrimps mainly from food. The results indicate a rapid turnover of ^{60}Co in shrimps.

INTRODUCTION

Cobalt-60 is one of the major long-lived neutron-activation products in the low-level liquid radioactive effluent of nuclear power plants.

Upon release into the marine environment this radionuclide may reach man by several routes. One of the potential pathways leading to radiation exposure is the use of shrimps for human consumption. Commercially important shrimp species like the common shrimp Crangon crangon(L.) have their habitat in estuarine and shallow coastal waters which are already being used for disposal of low-level liquid radioactive waste from nuclear power plants. The level of ^{60}Co to be expected in shrimps depends not only on release rates and physical dilution but also on the availability of the radionuclide for uptake and on the metabolism of cobalt in the organism. This paper presents the results of a laboratory study on the role of different modes of uptake in the accumulation of ^{60}Co by shrimps.

MATERIAL AND METHODS

Shrimps (Crangon crangon(L.)) weighing 0.6 to 2.2 g were collected at the North Sea shore near Petten and maintained in aquaria at 9-11 °C until needed. Freeze-dried mussels were given as food twice a week. Animals which had been acclimated in the aquaria for at least one week were used in the experiments. All uptake and retention experiments were carried out at a sea water temperature of 14-16 ^{o}C and were started after at least 3 days of acclimatation of the shrimps to the experimental conditions. The salinity of the filtered natural sea water used in the experiments ranged from 30 to 34 $^{o}/_{oo}$. The shrimps were kept individually in perforated polyethylene holders. For the experiments on the uptake of ^{60}Co from sea water 8 to 10 shrimps were placed in sea water containing 3 or 15 μCi ^{60}Co per liter. ^{60}Co was used in the cloride form (code CKS-1, The Radiochemical Centre, Amersham). The specific activity was > 100 μCi/μg. The uptake media were renewed several times during the experiments and 10 ml samples were regularly taken for measurement of the radionuclide concentration. The variation of the ^{60}Co concentration did not exceed $\pm$ 5 % of the average concentration. In the experiments on the uptake and retention of ^{60}Co from food the digestive glands of mussels, which had taken up ^{60}Co from sea water for 3 to 5 days, were used as labelled food.

During the retention experiments the shrimps were kept in continuously refreshed sea water and were fed freeze-dried musselflesh once in every two or three days.

Measurements of the radionuclide concentration in shrimps were made by whole-body γ-counting of living animals utilizing a 3"x3" well-crystal coupled to a single-channel analyser. The time course of uptake and retention could thus be followed for each individual shrimp. The same equipment was used for all other ^{60}Co measurements including the regular counting of a ^{60}Co standard sample.

RESULTS

^{60}Co uptake from sea water

The accumulation of ^{60}Co from sea water has been expressed as the increase of the concentration factor for the whole animal with time. The concentration factor is defined as the ratio between the ^{60}Co concentration in shrimps (cpm/g fresh weight) and the average concentration in the uptake medium (cpm/ml). Figure 1 shows the results for a group of eight shrimps kept in ^{60}Co-spiked sea water for 41 days. The mean concentration factor inceases slowly to a value of about 13 at day 31. A steady state concentration is still not reached at that time. Between day 31 and the end of the uptake experiment 6 shrimps moulted. As is shown for two of these animals most of the ^{60}Co taken up is lost at moulting. Moulting usually occurs

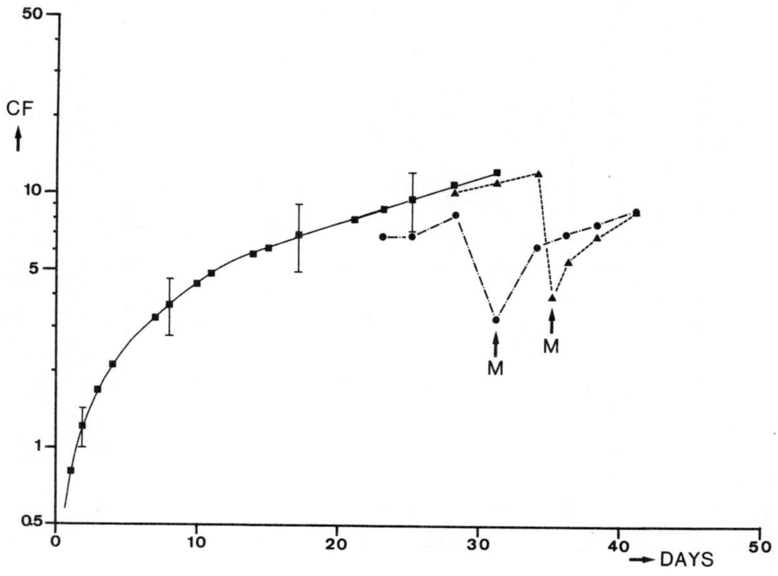

FIG. 1. Uptake of ^{60}Co from sea water by shrimps. The symbol ■ and vertical bars represent the mean ± 1 standard deviation of 8 or 6 individuals; ▲ and ●, data for two individual shrimps. CF: concentration factor, ↑M: moulted.

overnight, consequently the exact time-lapse between moulting and detection of moulting is not known. Because of the temporary increased ^{60}Co uptake following moulting the loss as a result of moulting is higher than the measured loss. The latter ranged from 41 to 71 percent. Animals that moulted towards the end of a second uptake experiment of 21 days duration lost at moulting 60 to 83 percent of their activity. Cast exoskeletons continued to accumulate ^{60}Co from sea water.

Shrimps which had accumulated ^{60}Co from sea water were subsequently placed in non-radioactive sea water and the loss of ^{60}Co by each individual shrimp was followed for about one month. As is shown in Figure 2 the rate of loss until moulting varied for individual shrimps. Relatively high initial rates of loss were observed for animals which had moulted two days before the end of the preceding uptake (shrimp 38-6). At the first moult all shrimps lost most of their retained activity. In Table I the data on ^{60}Co loss at the first moult are summarized.

All shrimps except one that moulted after 7 days retention, lost between 86 and 97 % of their retained activity. Only a few percent of the initial activity remained after moulting. No shrimp moulted shortly after the end of the uptake period. As a consequence, the fraction initially associated

TABLE I. LOSS OF COBALT-60 BY SHRIMPS AT FIRST MOULT DURING RETENTION AFTER ^{60}Co-UPTAKE FROM SEA WATER

shrimps, no.	duration of uptake period (Days)	duration of retention until moulting (Days)	percentage of initial activity present before moulting	percentage of initial activity present after moulting	percentage of activity lost at moulting
35-7		7	78	19.5	75
35-9		20	62	6.5	90
35-1	41	24	55	4.0	93
35-8		24	54	4.2	92
35-10		24	60	4.0	93
38-4		11	71	1.9	97
38-5		18	55	2.2	96
38-7	21	18	59	4.4	93
38-8		21	56	3.3	94
38-1 a)		29	28	3.2.	89
38-6 a)		29	27	3.8	86

a) Moulted two days before the end of the uptake period.

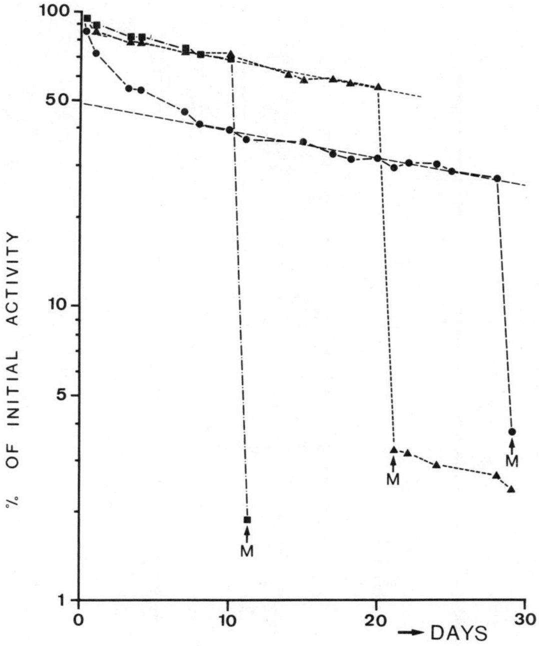

FIG.2. Loss of ^{60}Co by three shrimps after 21 days' uptake of ^{60}Co from sea water. The shrimp with the higher initial rate of loss (●: shrimp 38-6) had moulted two days before the end of the uptake period. ↑M : moulted.

with the exoskeleton could not be measured directly. However, from the extrapolation to time zero of the linear parts of the retention curves, it is concluded that, with two exceptions, 70 to 90 % of the initial activity was bound by the exoskeleton. The corresponding values for the two shrimps that had moulted two days before the end of the preceding uptake period were 49 % and 53 % respectively.

^{60}Co uptake and retention from food

Shrimps were given a single feeding of ^{60}Co labelled musselflesh and the subsequent retention of ^{60}Co was measured for each individual shrimp. The retained activities during the loss period were expressed as percentages of the initial activity measured within one hour after labelling. The retention of ^{60}Co after single uptake from food appeared to be described

TABLE II. PARAMETERS FOR RETENTION OF ^{60}Co BY SHRIMPS AFTER SINGLE UPTAKE FROM LABELLED FOOD (temp.: 14-16°C)

number of shrimps	biological half life of short-lived component A (Days)		biological half life of long-lived component B (Days)		ratio A_o/B_o	
	range	mean (s.d.)	range	mean (s.d.)	range	mean (s.d.)
42	0.5-2.6	1.2 (0.5)	4.9-17.3	9.8 (2.7)	1.1-9.8	3.8 (2.2)

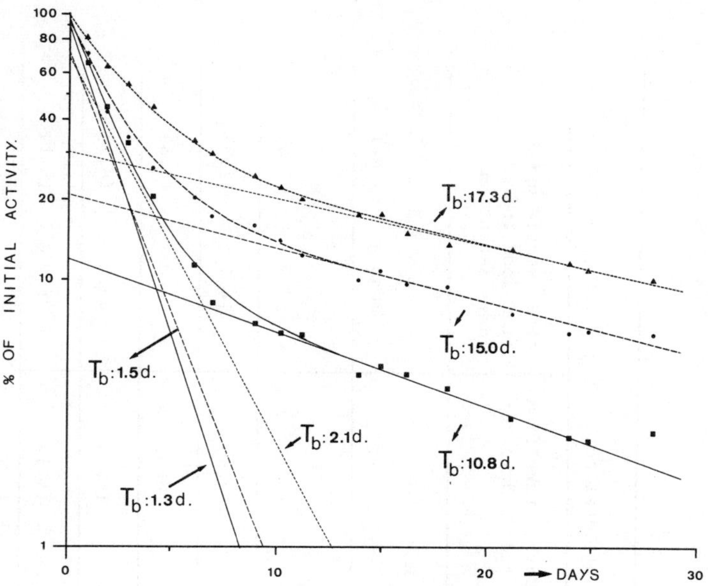

FIG. 3. Loss of ^{60}Co by three shrimps after single uptake of ^{60}Co with labelled food. For each shrimp the computed retention curve is shown together with the regression of the short-lived component and the long-lived component. T_b: biological half-life.

by an exponential model with a short-lived and a long-lived component, according to:

$$Y_t = A_o . e^{-\lambda_A t} + B_o . e^{-\lambda_B t}$$

where Y_t = fraction of initial activity retained at time t

A_o = fraction of initial activity present in the short-lived component A at t = 0

B_o = fraction of initial activity present in the long-lived component B at t = 0

λ_A and λ_B = rate constants for biological elimination of ^{60}Co from component A and B respectively.

Measured retention data for 42 individual shrimps were fitted to this model using a method of least squares. In Figure 3 the results for 3 animals are shown. The retention parameters for all shrimps are summarized in Table II. About 80 % of the initial activity was lost with a mean biological half-life of 1.2 days. The remaining 20 % was excreted with a mean biological half life of 9.8 days. No effect of moulting on the loss of ^{60}Co could be

TABLE III. DISTRIBUTION OF COBALT-60 IN SHRIMPS DETERMINED BY DISSECTION OF FREEZE-DRIED ANIMALS

parts of the body	percentages of total activity					
	after 18 days uptake from sea water [a]		after 18 days uptake from sea water and 39 days retention [b]		after single uptake from food and 12.5 days retention	
	n = 2		n = 3		n = 5	
			mean	(s.d.)	mean	(s.d.)
Appendages of cephalothorax and abdomen	46.4	36.7	36.0	(3.4)	7.7	(1.4)
Dorsal exoskeleton	32.5	40.5	42.6	(12.4)	14.7	(2.8)
Muscles of abdomen	0.5	1.6	1.6	(0.6)	4.0	(1.6)
Digestive gland	1.6	3.7	3.3	(1.5)	43.2	(8.7)
Remaining parts	19.0	17.6	20.2	(4.1)	29.1	(6.1)

[a] No moulting occurred shortly before the end of the uptake period.

[b] No moulting occurred during the retention period.

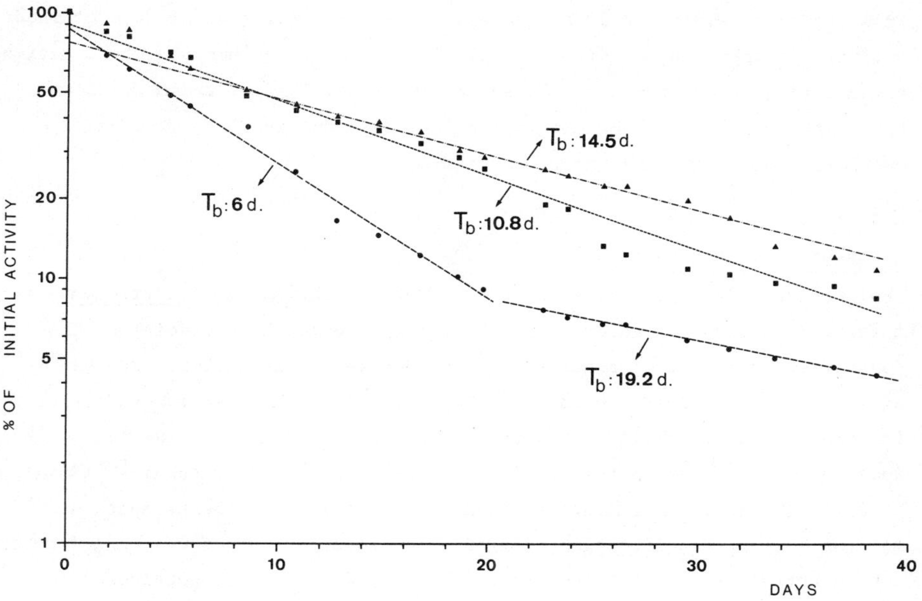

FIG. 4. Loss of ^{60}Co by three shrimps after repeated uptake of ^{60}Co with labelled food. T_b: biological half-life.

observed early in the retention period. However when moulting occurred after most of the activity had been lost from the animal the cast exoskeleton could contain up to 20 percent of the retained activity.

Repeated feeding

Repeated feeding of labelled musselflesh once in every three days resulted in a slow increase of the ^{60}Co concentration in the shrimps. Before each feeding most of the activity from the previous uptake was already excreted. The ^{60}Co retention by shrimps, which were given labelled food 20 times in 60 days, differed from retention after single uptake only with respect to the virtual absence of the short-lived component (Figure 4).

Distribution of ^{60}Co in the shrimp

A number of shrimps was freeze-dried, dissected and the distribution of ^{60}Co over several parts of the body was determined. The result given in Table III shows that after ^{60}Co uptake from sea water the internal organs of the shrimp contain only a few percent of the total activity (Table III column 2). In the absence of moulting the distribution does not change during retention (Table III, column 3). In contrast, after single uptake from food followed by 12.5 days retention, a large fraction of the retained activity is found in the digestive gland (Table III, column 4). The fraction present in muscles from the abdomen is low, regardless of the mode of uptake. The activity of remaining parts includes contributions from gills, muscles, stomach and the ventral exoskeleton.

DISCUSSION AND CONCLUSIONS

The results of the present study show that the shrimp *Crangon crangon* accumulates ^{60}Co from sea water mainly by adsorption to the exoskeleton. The concentration factor of about 13 for non-moulted whole shrimps, reached after one month uptake, is low when compared with published concentration factors of stable cobalt in Crustacea. The latter range from 144 to 590 for whole animals [1]. From the stable cobalt data reported by Fukai [2], Shimizu et al. [1] calculated a concentration factor of 232 for the carapace and 72 for the soft parts of *Crangon crangon*. They used a value of 0.5 μg/l for the concentration of cobalt in sea water. Fukai and Murray [3] concluded from recent data on the cobalt concentration in sea water that the approximate concentration factor of cobalt in soft parts of Decapod Crustacea is 800.
An equilibrium concentration of ^{60}Co in shrimp was not attained in the present study. However, the low concentration factor for whole animals and the large fraction of ^{60}Co being bound by the exoskeleton indicate that in soft parts a concentration factor of ^{60}Co, comparable to the value for stable cobalt, will not be reached by direct uptake of ^{60}Co from sea water. This most probably explains the low concentration factors observed by Hiyama and Khan [4] and by Shimizu et al [1] in experiments on the uptake of ^{60}Co by the related shrimp *Leander pacificus*. Consequently, ^{60}Co released into the marine environment will be taken up by shrimps mainly through the food chain. Intervening trophic levels are accordingly involved in the transfer of ^{60}Co to man through consumption of shrimps.
The long-lived fraction of ^{60}Co taken up from food will determine the long-term accumulation of ^{60}Co in shrimps. Assuming that the observed mean biological half life of about 10 days applies also under natural conditions, 90 % of the equilibrium concentration will be reached after about one month (3 half lives) of constant uptake of ^{60}Co from food. The turnover rate of ^{60}Co in shrimps might be influenced by environmental and physiological conditions such as the composition of the food, the rate of food intake and the reproductive phase of the shrimp. It seems however unlikely that under field conditions the influence of the uptake of ^{60}Co from food on the accumulation of ^{60}Co by shrimps will be essentially smaller.

Uptake from food resulted in a relatively high concentration of ^{60}Co in the digestive gland. The concentration in abdominal muscles, which constitute most of the edible portion of shrimps, is much lower. A similar distribution of ^{60}Co in the crab Carcinus maenas was reported by Amiard-Triquet and Amiard [5]. This indicates that the human radiation exposure resulting from consumption of shrimps tends to be overestimated when concentration factors for whole animals or whole soft parts are used in the dose estmate.

ACKNOWLEDGEMENT

The author wishes to thank Mrs A.H.A.Disco-De Neyn and Mr E.E.A.Noordeloos for their assistance.

REFERENCES

[1] SHIMIZU, M., KAJIHARA, T., HIYAMA, Y., Rec. Oceanogr. Wks Japan 10 2 (1970) 137.

[2] FUKAI, R., Distribution of Cobalt in Marine Organisms, Radioactivity in the Sea, Publ. No. 23, IAEA, Vienna (1968) 19 pp.

[3] FUKAI, R., MURRAY, C.N., in Environmental Behaviour of Radionuclides Released in the Nuclear Industry (Proc. Symp. Aix-en-Provence, 1973), IAEA, Vienna (1974) 217.

[4] HIYAMA, Y., KHAN, J.M., Rec. Oceanogr. Wks Japan 7 2 (1964) 79.

[5] AMIARD-TRIQUET, C., AMIARD, J.C., Rev. Int. Oceanogr. Méd. 33 (1974) 49.

DISCUSSION

V.V. PISAREV: How many moults does the shrimp undergo before reaching an age at which it is suitable for consumption?

A.W. van WEERS: According to R. Meixner (Ber. Dtsch. Wiss. Komm. Meeresforschung 20 2 (1969) 93) it takes about 25 moults from metamorphosis to the reproductive phase, when the shirmps are also of marketable size.

S. BONOTTO: Could you comment a little more on the procedure used in the whole-body γ-counting of live animals? Also, do you have special reasons for measuring the activity of single shrimps?

A.W. van WEERS: I use a 3 in.× 3 in. NaI scintillation crystal with a 2-in.-deep well. The animals are placed in a perforated tube and counted upside down for, usually, not longer than one minute. The reason for measuring individual shrimps is that in this way you can register feeding and moulting and observe the effect thereof on radionuclide accumulation or retention.

A.R. BERG: In experiments with labelled food, the radioactivity of the alimentary bolus introduces a systematic error since it is ingested, but is not absorbed, by the tissues, and will be rapidly eliminated via the faeces (egestion). In relation to this comment, do you not think that the short biological half-life component of retention might be related to egestion and not to true excretion?

A.W. van WEERS: The short-lived component, with a half-life of about one day, can be explained by both rapid excretion and loss of non-absorbed food with faeces. The retention data do not permit discrimination between the different processes since the initial loss can only be described as loss from a single component. The measurements were started within one hour of giving the labelled food, so the loss with faeces is included in the retention data.

A.R. BERG: What would be the percentage distribution of ^{60}Co activity among the various organs if the radioactivity of the alimentary bolus were subtracted, and how would it then compare with the distribution after accumulation from water?

A.W. van WEERS: The distribution data for uptake from food were obtained by dissection 12.5 days after a single uptake of labelled food. During that period the animals received non-labelled food once every two or three days, which certainly cleared the stomach and the gut of labelled material, so there seems to be no reason why one should allow for the activity remaining in these organs.

A.A. FRAIZIER: You conclude that ^{60}Co exposure determined on the basis of shrimp consumption tends to be overestimated if one takes the concentration for the entire shrimp organism. Is it not possible that in the process of culinary preparation (which frequently involves the whole animal) some of the radiocobalt is transferred to parts which are normally eaten?

A.W. van WEERS: I have no data directly related to your question, but ^{60}Co released from the non-edible parts could be adsorbed on edible parts. To prepare shrimps for consumption, fresh whole animals are boiled for a short time. It would take a very peculiar mechanism to cause any marked change in radionuclide distribution during preparation.

S. FOWLER: I find the high ^{60}Co content in moults interesting. We have looked at the trace-metal content of crustacean moults and compared the results obtained with whole-body and/or muscle-tissue values; for most elements we find that moult tissue contains much higher levels than either the edible tissue or the animal as a whole. In the case of cobalt we find that in crustaceans more than 50% of the content resides in metabolically inert moult tissue that constitutes only about 10% of the animal's weight. The general tendency towards association of elements or radionuclides with moults is important in the context of monitoring. If only a few whole crustaceans are used to make a sample, wide variation between samples can be expected since the organisms which make up the sample will be in different intermoult stages. If crustaceans are to be used for monitoring purposes, large numbers of individuals are necessary to yield a value which is truly representative of the overall levels contained in these organisms. It would probably be best to use only the edible tissues of crustaceans for monitoring.

A.W. van WEERS: The high ^{60}Co content of the moult indicates that some non-metabolic adsorption process will also be responsible for the relatively high trace-metal content of moults observed for a number of crustaceans. The distribution of the total trace-element content over edible and non-edible parts may be size-dependent and also dependent on the moulting stage. With regard to ^{60}Co monitoring, I agree that large samples of edible parts have to be analysed. Stable-cobalt analyses of edible and non-edible parts will indicate what differences might arise in the levels of radiocobalt.

Gwyneth D. HOWELLS: The shrimp is an omnivorous feeder — do you have any information on the uptake or availability of ^{60}Co incorporated in a mixed diet or in other individual food items?

A.W. van WEERS: The availability of ^{60}Co for uptake may indeed depend on the diet of the shrimp, which is known to include a variety of organisms, but I did not study this. I only compared the ^{60}Co retention from the diet of digestive gland tissue, labelled through ^{60}Co uptake by living mussels, with the retention from digestive gland tissue labelled by drying a few millilitres of a ^{60}Co tracer solution on freeze-dried material. I found no significant difference in ^{60}Co retention from these two differently labelled materials.

L.F. FARGES: In your experiments you used only small numbers of shrimps, e.g. 8-10 in some experiments and only 3 in others. Does this not cause doubts regarding the interpretation of your measurements?

A.W. van WEERS: In my opinion one can get valuable information from experiments carried out with a limited number of animals, particularly when the radioisotope kinetics are checked by making repeated measurements on individuals without sacrificing them. Even with a relatively small number of individuals in the sample, the standard deviations of measured and computed parameters will give information on the magnitude of the individual variation. The average age of the individuals seems more important than sample size, since food uptake rate, turnover of stable elements and radionuclides, and surface-to-volume ratio can be strongly age-dependent.

ГИГИЕНИЧЕСКИЕ КРИТЕРИИ К ОЦЕНКЕ ЗАГРЯЗНЕНИЯ РАДИОНУКЛИДАМИ ПРИБРЕЖНЫХ МОРСКИХ ВОД

Д.И.ГУСЕВ
Институт биофизики
Министерства здравоохранения СССР,
Москва,
Союз Советских Социалистических Республик

Доклад представлен В.А. Книжниковым

Abstract–Аннотация

HEALTH CRITERIA FOR ESTIMATING ACCEPTABLE CONTAMINATION OF COASTAL WATERS BY RADIONUCLIDES.

The paper deals, in the context of nuclear power development, with working limits on radionuclide concentrations in seawater, standardized for coastal waters in areas where fishing takes place and edible seaweeds and bottom invertebrates are collected. The author discusses permissible levels of seawater contamination in the neighbourhood of beaches periodically washed by radioactive water. Standard coefficients are given expressing the radiation dose to critical organs of man (mrem/μCi) in the uptake of radionuclides by the digestive tract. These can be used to estimate individual and collective doses to the population.

ГИГИЕНИЧЕСКИЕ КРИТЕРИИ К ОЦЕНКЕ ЗАГРЯЗНЕНИЯ РАДИОНУКЛИДАМИ ПРИБРЕЖНЫХ МОРСКИХ ВОД.

В докладе рассмотрены в аспекте развития ядерной энергетики вопросы о рабочих пределах концентраций радионуклидов в морской воде, нормированных для прибрежных морских вод применительно к зонам рыбного промысла, добычи пищевых водорослей и донных беспозвоночных. Рассмотрен вопрос о допустимых уровнях загрязнения морской воды в зоне пляжей, периодически заливаемых радиоактивной водой. Даны нормированные коэффициенты, характеризующие дозы облучения критических органов человека (мбэр/мкКи) при поступлении радионуклидов в пищеварительный тракт, которые могут быть использованы для оценки индивидуальных и коллективных доз облучения населения.

Несмотря на уменьшение в последние годы интенсивности глобальных выпадений, загрязнение радионуклидами морей, особенно прибрежных вод, продолжается. Наряду со стратосферными выпадениями продуктов ядерных взрывов в прибрежные зоны морей радионуклиды попадают со сбросами жидких радиоактивных отходов (Ирландское и Северное моря, залив Твэрен Балтики) и стоками рек (Колумбия, Рона, Дунай и др.). В последние годы осуществляется широкая программа строительства атомных электростанций (АЭС) на побережье морей и океанов (США, Япония, Великобритания, Италия, СССР, ГДР и др.). В печати обсуждаются различные аспекты строительства плавучих АЭС на специальных платформах в прибрежных водах около крупных городов.

Вместе с малоактивными жидкими стоками и газоаэрозольными выбросами от АЭС в водоем неизбежно попадает некоторое количество радионуклидов. Опыт эксплуатации АЭС показал, что в водоемы-охладители с одной атомной электростанции ежегодно попадает до десятка кюри смеси кобальта-60, марганца-54, цинка-65, цезия-137, йода-131, рутения-106, доли кюри стронция-90, а также сотни кюри трития [1].

При аварийных ситуациях не исключается возможность попадания бо́льшего количества активности, в том числе более широкого спектра осколочных радионуклидов.

Наконец, нельзя не учитывать возрастающее количество судов с ядерными энергетическими установками и перспективы изменения структуры источников электроэнергии в ближайшие десятилетия.

Все это свидетельствует о необходимости усиления внимания гигиенистов и других специалистов к вопросу санитарной охраны прибрежных морских вод от радиоактивного загрязнения. Мы считаем наиболее важной задачей сегодняшнего дня в охране прибрежных морских вод разработку надежных, всесторонне обоснованных нормативов рабочих пределов концентраций радионуклидов в морской воде с учетом многопрофильного использования морей населением. Рабочие пределы концентраций, а не ПДК должны быть основой для инженеров-проектантов атомных энергетических комплексов на побережье морей.

Вполне очевидно, что названные выше источники не могут существенно повысить степень радиоактивности морской среды мирового океана в целом. Вместе с этим они могут являться причиной образования локальных очагов повышенной радиоактивности морской воды, донных отложений, литоральной зоны, пляжей, а также флоры и фауны, включая промысловые морские организмы. При этом следует учитывать, что прибрежная зона морей наиболее продуктивна, здесь сосредоточены основные места нереста и добычи ценных пород рыб.

Наши многолетние исследования были направлены на разработку предельно допустимых концентраций и допустимых рабочих пределов загрязнений морской воды радионуклидами с учетом сложных закономерностей поступления последних по биологической цепочке: морская вода-гидробионты (рыба, донные беспозвоночные, пищевые водоросли)-человек.

Для этого были сформулированы требования к ПДК, определены на основе собственных и всех доступных литературных материалов предельные коэффициенты накопления радиоактивных и стабильных изотопов ряда химических элементов в съедобных частях морских рыб, пищевых водорослях, мягких тканях моллюсков, ракообразных и иглокожих, обобщены данные о потреблении приморским населением рыбы и морепродуктов и обоснована парциальная допустимая доза облучения критических органов человека, относимая на загрязнение радионуклидами морской среды. Специальный комплекс исследований был выполнен для нормирования уровней загрязнения морской воды в зоне расположения пляжей, периодически заливаемых радиоактивной водой.

В основе нормирования концентраций радионуклидов в прибрежных морских водах необходимо ориентироваться на рекомендации Международной комиссии по радиологической защите и Международного агентства по атомной энергии о критических группах населения и критических путях воздействия радиационного фактора с учетом возможного вклада всех других источников облучения, кроме природного радиоактивного фона и медицинских процедур.

В связи с многосторонним использованием населением морской среды и продуктов моря, а также заботой о воспроизводстве биологических ресурсов моря требования к допустимым концентрациям радионуклидов в морской воде целесообразно подразделить на радиационно-гигиенические и радиобиологические.

Гигиенические требования сводятся к тому, чтобы неограниченное использование населением продуктов морского промысла, использование мест купания и пляжей, а также воздействие загрязнения морской среды на население в процессе профессиональной деятельности не сопровождались облучением критических групп населения выше парциальных допустимых доз, выделяемых на данный источник ионизирующего облучения. В качестве верхнего предела такой дозы считаем целесообразным принять 10% предела дозы, регламентированного для отдельных лиц из населения "Основными нормами безопасности при защите от излучения", МАГАТЭ,1968 [2], совпадающего с таковыми по НРБ СССР, 1969 [3]. Следовательно, за счет загрязнения морской среды предел дозы у критических групп населения на все тело, гонады и красный костный мозг принят нами равным 50 мбэр/год; для костной ткани, щитовидной железы, кожных покровов всего тела – 300 мбэр/год и для всех других отдельных критических органов – 150 мбэр/год. Эти величины, как показал анализ, являлись фактическими предельными дозовыми нагрузками у приморского населения (исключая зоны преднамеренного загрязнения сбросами отходов и зоны местных выпадений продуктов взрыва) в периоды интенсивных выпадений в связи с ядерными испытаниями. Можно полагать, что они вполне обеспечат также потребности бурно развивающейся ядерной энергетики. В рассматриваемом аспекте нельзя согласиться с подходами к нормированию загрязнения морепродуктов и морской воды некоторых зарубежных авторов [4-6], которые на этот источник облучения расходовали полную дозу допустимого облучения критических групп населения. В настоящее время на население воздействует целый комплекс источников ионизирующих излучений, в том числе некоторые из них глобального характера, не поддающиеся регулированию в национальных масштабах. Поэтому распределение предела дозы по источникам облучения является насущной необходимостью.

Радиобиологическое или экологическое требование к допустимым концентрациям радионуклидов в морской воде сводится к тому, чтобы регламентированные верхние пределы концентраций исключали поражающее действие на наиболее ранимые звенья водных биоценозов, в том числе и на генетическом уровне. По мнению специалистов, таким звеном являются эмбриональные стадии развития рыб и других представителей морской фауны [7].

Промежуточным критерием для расчета рабочих пределов концентраций радионуклидов в морской воде является допустимый уровень радиоактивности рыбы и морепродуктов. Он зависит от дозы облучения критических органов, выделяемой на эту часть рациона, и от количества потребляемых продуктов. Среднемировое потребление продуктов моря в расчете на одного человека небольшое. В СССР вылов рыбы составляет около 20 кг/год на душу населения. Однако у нас существуют большие группы приморского населения и коллективы с нормированным рационом, которые потребляют около 80 кг рыбы в год. В некоторых районах приморское население ежегодно расходует в пищу до 26 кг водорослей и мяса донных беспозвоночных. Указанные количества продуктов моря близки к тем значениям, которые рекомендованы группой экспертов МАГАТЭ,1960 [8] применительно к рациону населения, у которого вся потребность в белках удовлетворяется за счет морской пищи. Следовательно, они могут быть приняты в качестве показателя потребления продуктов моря критическими группами населения.

ТАБЛИЦА 1. КОЭФФИЦИЕНТЫ ПРЕДЕЛЬНОГО НАКОПЛЕНИЯ РАДИОАКТИВНЫХ И СТАБИЛЬНЫХ ИЗОТОПОВ В СЪЕДОБНОЙ ЧАСТИ ПРОМЫСЛОВЫХ МОРСКИХ ОРГАНИЗМОВ ДЛЯ НОРМИРОВАНИЯ ЗАГРЯЗНЕНИЯ МОРСКОЙ ВОДЫ

№№ пп	Химический элемент	Рыбы		Мясо моллюсков	Мясо ракообразных	Мясо иглокожих	Водоросли
		Мышцы, мягкие ткани	Кости				
1	Кальций	2	200	10	10	10	10
2	Хром	2000	-	500	400	-	1000
3	Марганец	130		10000	7500	200	1200
4	Железо	2000		10000	2500		10000
5	Кобальт	80		200	600	240	430
6	Цинк	3000		10000	5000	1400	1000
7	Стронций	2	200	10	10	1	50
8	Иттрий	10		18	80		800
9	Цирконий	150		33	100		1000
10	Ниобий	100		140	100		2000
11	Молибден	20		100	100		15
12	Рутений	100		10	10	10	2000
13	Кадмий	5		10000	1	100	1500
14	Сурьма	140		100	70	70	200
15	Йод	15		70	50	60	2000
16	Цезий	25	3	50	25	10	70
17	Церий	25	100	100	100	15	700
18	Полоний	15	50	25	25		1000
19	Плутоний	25		25	45		2000

При таком подходе, когда на загрязнение морской среды выделяется 10% предела дозы облучения и принимается потребление морепродуктов на уровне величин, характерных для критической группы, допустимые концентрации радионуклидов в съедобной части рыбы составляют величину, равную 1 СДК радионуклида в питьевой воде, а в мясе беспозвоночных и в пищевых водорослях – по 5 СДК, в питьевой воде – по НРБ – 69 или "Основным нормам безопасности при защите от излучения", 1968 г.

Рыба и морепродукты с уровнями радиоактивности выше этих показателей могут использоваться в пищу только при условии пропорционального уменьшения их количества в рационе по сравнению с вышеуказанным потреблением критическими группами или при условии увеличения выделяемой парциальной дозы допустимого облучения за счет других источников.

Мерой концентрирующей способности тканей водных организмов в отношении радионуклидов в гигиене и радиоэкологии принят предельный коэффициент накопления (КН) или фактор концентрирования. В итоге

ТАБЛИЦА II. РАБОЧИЕ ПРЕДЕЛЫ КОНЦЕНТРАЦИЙ РАДИОНУКЛИДОВ В МОРСКОЙ ВОДЕ (Ки/л) ПРИ ДЛИТЕЛЬНОМ ЗАГРЯЗНЕНИИ РАЙОНОВ С РАЗЛИЧНЫМ ПРОМЫСЛОВЫМ ИСПОЛЬЗОВАНИЕМ

№№ пп	Радионуклид	Рыбопромысловая зона	Районы добычи			Зона смешанного промысла
			моллюсков	ракообразных	пищевых водорослей	
1	Хром-51	$7 \cdot 10^{-10}$	$1 \cdot 10^{-8}$	$1 \cdot 10^{-8}$	$7 \cdot 10^{-9}$	$7 \cdot 10^{-10}$
2	Марганец-54	$1 \cdot 10^{-9}$	$6 \cdot 10^{-11}$	$8 \cdot 10^{-11}$	$5 \cdot 10^{-10}$	$1 \cdot 10^{-10}$
3	Железо-55	$4 \cdot 10^{-10}$	$4 \cdot 10^{-10}$	$2 \cdot 10^{-9}$	$4 \cdot 10^{-10}$	$4 \cdot 10^{-10}$
4	Железо-59	$3 \cdot 10^{-11}$	$3 \cdot 10^{-11}$	$1 \cdot 10^{-10}$	$3 \cdot 10^{-11}$	$3 \cdot 10^{-11}$
5	Кобальт-57	$5 \cdot 10^{-9}$	$1 \cdot 10^{-8}$	$3 \cdot 10^{-9}$	$4 \cdot 10^{-9}$	$3 \cdot 10^{-9}$
6	Кобальт-58	$1 \cdot 10^{-9}$	$2 \cdot 10^{-9}$	$7 \cdot 10^{-10}$	$1 \cdot 10^{-9}$	$7 \cdot 10^{-10}$
7	Кобальт-60	$4 \cdot 10^{-10}$	$9 \cdot 10^{-10}$	$3 \cdot 10^{-10}$	$4 \cdot 10^{-10}$	$3 \cdot 10^{-10}$
8	Цинк-65	$3 \cdot 10^{-11}$	$5 \cdot 10^{-11}$	$1 \cdot 10^{-10}$	$5 \cdot 10^{-10}$	$3 \cdot 10^{-11}$
9	Стронций-89	$1 \cdot 10^{-9}$	$6 \cdot 10^{-9}$	$6 \cdot 10^{-9}$	$1 \cdot 10^{-9}$	$1 \cdot 10^{-9}$
10	Стронций-90	$4 \cdot 10^{-11}$	$2 \cdot 10^{-10}$	$2 \cdot 10^{-10}$	$4 \cdot 10^{-11}$	$4 \cdot 10^{-11}$
11	Иттрий-90	$2 \cdot 10^{-9}$	$5 \cdot 10^{-9}$	$1 \cdot 10^{-9}$	$1 \cdot 10^{-10}$	$1 \cdot 10^{-10}$
12	Цирконий-95	$4 \cdot 10^{-10}$	$1 \cdot 10^{-8}$	$3 \cdot 10^{-9}$	$3 \cdot 10^{-10}$	$3 \cdot 10^{-10}$
13	Ниобий-95	$1 \cdot 10^{-9}$	$4 \cdot 10^{-9}$	$5 \cdot 10^{-9}$	$3 \cdot 10^{-10}$	$3 \cdot 10^{-10}$
14	Молибден-99	$2 \cdot 10^{-9}$	$2 \cdot 10^{-9}$	$2 \cdot 10^{-9}$	$1 \cdot 10^{-8}$	$2 \cdot 10^{-9}$
15	Рутений-103	$1 \cdot 10^{-9}$	$5 \cdot 10^{-8}$	$5 \cdot 10^{-8}$	$3 \cdot 10^{-10}$	$3 \cdot 10^{-10}$
16	Рутений-106	$1 \cdot 10^{-10}$	$6 \cdot 10^{-9}$	$6 \cdot 10^{-9}$	$3 \cdot 10^{-11}$	$3 \cdot 10^{-11}$
17	Кадмий-109	$4 \cdot 10^{-8}$	$1 \cdot 10^{-10}$	$1 \cdot 10^{-8}$	$6 \cdot 10^{-10}$	$1 \cdot 10^{-10}$
18	Кадмий-115 м	$5 \cdot 10^{-9}$	$1 \cdot 10^{-11}$	$1 \cdot 10^{-9}$	$8 \cdot 10^{-11}$	$1 \cdot 10^{-10}$
19	Кадмий-115	$7 \cdot 10^{-9}$	$2 \cdot 10^{-11}$	$2 \cdot 10^{-9}$	$1 \cdot 10^{-10}$	$2 \cdot 10^{-10}$
20	Сурьма-124	$2 \cdot 10^{-10}$	$1 \cdot 10^{-9}$	$2 \cdot 10^{-9}$	$5 \cdot 10^{-10}$	$2 \cdot 10^{-10}$
21	Сурьма-125	$7 \cdot 10^{-10}$	$5 \cdot 10^{-9}$	$7 \cdot 10^{-9}$	$2 \cdot 10^{-9}$	$7 \cdot 10^{-10}$
22	Йод-131	$1 \cdot 10^{-10}$	$1 \cdot 10^{-10}$	$2 \cdot 10^{-10}$	$1 \cdot 10^{-11}$	$1 \cdot 10^{-11}$
23	Цезий-134	$3 \cdot 10^{-10}$	$1 \cdot 10^{-9}$	$2 \cdot 10^{-9}$	$6 \cdot 10^{-10}$	$3 \cdot 10^{-10}$
24	Цезий-137	$6 \cdot 10^{-10}$	$1 \cdot 10^{-9}$	$3 \cdot 10^{-9}$	$1 \cdot 10^{-9}$	$6 \cdot 10^{-10}$
25	Церий-141	$3 \cdot 10^{-9}$	$4 \cdot 10^{-9}$	$4 \cdot 10^{-9}$	$6 \cdot 10^{-10}$	$6 \cdot 10^{-10}$
26	Церий-144	$4 \cdot 10^{-10}$	$6 \cdot 10^{-10}$	$6 \cdot 10^{-10}$	$1 \cdot 10^{-10}$	$1 \cdot 10^{-10}$
27	Полоний-210	$5 \cdot 10^{-11}$	$1 \cdot 10^{-10}$	$1 \cdot 10^{-10}$	$4 \cdot 10^{-11}$	$4 \cdot 10^{-11}$
28	Плутоний-239	$2 \cdot 10^{-10}$	$1 \cdot 10^{-9}$	$5 \cdot 10^{-10}$	$1 \cdot 10^{-11}$	$1 \cdot 10^{-11}$

изучения этого вопроса установлены величины предельных КН стабильных и радиоактивных изотопов в съедобных частях морских рыб, пищевых водорослях и мясе беспозвоночных (табл. I). Они могут быть приняты для нормирования в качестве верхнего предела концентрирующей способности морских промысловых организмов в отношении радионуклидов с большими и средними периодами полураспада для случаев длительного загрязнения морской воды.

ТАБЛИЦА III. РАБОЧИЕ ПРЕДЕЛЫ КОНЦЕНТРАЦИЙ РАДИОНУКЛИДОВ В МОРСКОЙ ВОДЕ И ПЕСКАХ ПЛЯЖЕЙ ПО ДОЗЕ ВНЕШНЕГО ОБЛУЧЕНИЯ ОТ ПЛЯЖА

№№ пп	Радионуклид	Рабочие пределы загрязнения песка пляжа (Ки/кг)	Предельный коэффициент накопления в песке	Рабочие пределы концентраций в морской воде (Ки/л)
1	Хром-51	$2 \cdot 10^{-6}$	1000	$2 \cdot 10^{-9}$
2	Марганец-54	$4 \cdot 10^{-8}$	400	$1 \cdot 10^{-10}$
3	Железо-59	$3 \cdot 10^{-8}$	1000	$3 \cdot 10^{-11}$
4	Кобальт-60	$1 \cdot 10^{-8}$	1000	$1 \cdot 10^{-11}$
5	Цинк-65	$6 \cdot 10^{-8}$	60	$1 \cdot 10^{-9}$
6	Цирконий-95	$5 \cdot 10^{-8}$	150	$3 \cdot 10^{-10}$
7	Ниобий-95	$5 \cdot 10^{-8}$	150	$3 \cdot 10^{-10}$
8	Рутений-103	$3 \cdot 10^{-7}$	600	$5 \cdot 10^{-10}$
9	Кадмий-109	$1 \cdot 10^{-5}$	100	$1 \cdot 10^{-7}$
10	Сурьма-124	$3 \cdot 10^{-8}$	300	$1 \cdot 10^{-10}$
11	Цезий-134	$3 \cdot 10^{-8}$	300	$1 \cdot 10^{-10}$
12	Цезий-137	$7 \cdot 10^{-8}$	300	$2 \cdot 10^{-10}$
13	Церий-141	$2 \cdot 10^{-6}$	1400	$1 \cdot 10^{-9}$
14	Церий-144	$1 \cdot 10^{-5}$	1400	$7 \cdot 10^{-9}$

В связи со значительным отличием величин КН в разных группах промысловых организмов (табл.I) санитарная значимость загрязнения в одних и тех же концентрациях районов с различным промысловым использованием будет неодинакова. Поэтому нормирование морской среды целесообразно проводить дифференцированно, применительно к зонам различного морского промысла. В этих целях ниже рассмотрен вопрос о рабочих пределах концентраций радионуклидов в воде рыбопромысловой зоны, зоны добычи морских водорослей, зоны промысла беспозвоночных (моллюсков, ракообразных) и для районов смешанного промысла. Для последних в качестве рабочего предела рекомендуется принимать наименьший показатель пределов концентраций, установленных применительно к каждому отдельному виду промысла.

В табл.II представлены результаты таких расчетов для 28 радионуклидов, гигиенически наиболее значимых при попадании в прибрежные морские воды с учетом реальных источников их образования, степени радиотоксичности и закономерностей миграции к человеку по водным биологическим цепочкам.

В связи с размещением АЭС на побережье морей и рек возникает необходимость оценки возможных доз облучения населения от пляжей, которые могут периодически заливаться водой с наличием радионуклидов. Это обусловлено тем, что зона акватории, прилегающая к АЭС, является общедоступной для населения без каких-либо ограничений. Немаловажное значение в гигиеническом отношении имеет решение и обратной задачи, т.е. определение допустимых рабочих пределов концентраций радионуклидов в морской воде и песках пляжа, исходя из заданной дозы облучения. Для решения этих вопросов нами были выполнены специальные исследования. Было установлено, например, что в условиях Балтики при санаторном и неорганизованном отдыхе население находится на пляжах в среднем не более 200 часов в год. Если на этот источник облучения выделить дозу 5 мбэр/год или 1% от годового предела дозы для отдельных лиц из населения, то мощность дозы гамма-излучения на пляже без учета естественного фона должна быть не выше 25 мкР/ч. Соответствующие этой мощности дозы величины рабочих пределов концентраций бета-гамма-излучателей в песке пляжа приведены в табл.III, колонка 3. Расчет был произведен по формуле, рекомендованной для протяженного источника с полубесконечным пространством.

Для расчетов допустимой концентрации радионуклидов в морской воде на основе допустимого загрязнения песков пляжей была выполнена серия опытов по установлению предельных коэффициентов накопления радионуклидов в мелкозернистых песках пляжей Балтийского моря с преимущественным содержанием фракций менее 1 мм. Полученные данные указаны в табл.III, колонка 4. В итоге этой работы были рассчитаны рабочие пределы концентраций радионуклидов в морской воде, при которых облучение населения от пляжей составит не более 5 мбэр/год (табл.III).

Для облегчения оценки доз облучения критических органов человека при пероральном поступлении различных радионуклидов, например, с морепродуктами, был произведен расчет годовых доз облучения в мбэр на 1 мкКи годового поступления. При этом поступление радионуклида может приниматься как однократным, так и фракционированным. Расчет производился для растворимых и нерастворимых соединений радионуклидов, поступающих в пищеварительный тракт. Полученные при этом данные представлены в табл.IV. Эти сведения могут быть использованы как для оценки индивидуальных, так и коллективных доз облучения населения, обусловленных, например, потреблением радиоактивных морепродуктов.

После определения суммарного количества поступивших населению радионуклидов с общим количеством потребленного продукта, пользуясь данными табл.IV, легко рассчитать коллективную дозу облучения населения за счет рассматриваемого источника. Если целенаправленно будет организован дозиметрический контроль вылавливаемой рыбы и морепродуктов в зонах, подверженных влиянию источника загрязнения водоема, то таким путем представляется возможным подойти к оценкам обобщенного показателя — социальной стоимости (в гигиеническом значении этого термина) того или иного объекта, являющегося источником загрязнения морской среды. Например, для АЭС можно оценить дозой облучения населения и соответственно вероятным числом лейкемий, частотой возникновения злокачественных новообразований последствия сброса отходов в водоем (если таковой производится) на единицу активности сброса, на единицу выработанной электроэнергии или единицу электрической мощности реактора. Решением обратной задачи представляется возможным с вероятностных позиций подойти к нормированию сбросов радиоактивных отходов в прибрежные зоны морей.

ТАБЛИЦА IV. НОРМИРОВАННЫЕ ПОКАЗАТЕЛИ ДОЗ ОБЛУЧЕНИЯ КРИТИЧЕСКИХ ОРГАНОВ ЧЕЛОВЕКА (мбэр/мкКи) ПРИ ПОСТУПЛЕНИИ РАДИОНУКЛИДОВ ЧЕРЕЗ ОРГАНЫ ПИЩЕВАРЕНИЯ

№№ пп	Радионуклид	Растворимые соединения		Нерастворимые соединения	
		Критический орган	мбэр/мкКи	Критический орган	мбэр/мкКи
1	Тритий (HTO, T_2O)	Ткани тела	0,2	-	
2	Углерод-14	Жировая ткань	2,3	-	
3	Натрий-22	Весь организм	16,9	ЖК(НТК)	62,5
4	Натрий-24	ЖК(НТК)	10,0	ЖК(НТК)	68,2
5	Фосфор-32	Костная ткань	200,0	ЖК(НТК)	83,3
6	Сера-35	Гонады	10,0	ЖК(НТК)	6,8
7	Кальций-45	Костная ткань	411,1	ЖК(НТК)	15,0
8	Хром-51	ЖК(НТК)	1,25	ЖК(НТК)	1,15
9	Марганец-52	ЖК(НТК)	57,7	ЖК(НТК)	62,5
10	Марганец-54	ЖК(НТК)	15,0	ЖК(НТК)	15,6
11	Марганец-56	ЖК(НТК)	15,6	ЖК(НТК)	18,75
12	Железо-55	Селезенка	2,4	ЖК(НТК)	0,83
13	Железо-59	ЖК(НТК)	31,9	ЖК(НТК)	35,7
14	Кобальт-57	ЖК(НТК)	3,5	ЖК(НТК)	5,0
15	Кобальт-58	ЖК(НТК)	15,0	ЖК(НТК)	20,8
16	Кобальт-60	ЖК(НТК)	38,5	ЖК(НТК)	53,6
17	Медь-64	ЖК(НТК)	5,8	ЖК(НТК)	18,75
18	Цинк-65	Весь организм	6,32	ЖК(НТК)	10,7
	- " -	Простата	15,62	-	
	- " -	Печень	15,0	-	
19	Рубидий-86	Весь организм	9,26	ЖК(НТК)	79,0
	- " -	Поджелудочн.железа	27,8	-	
20	Стронций-85	Весь организм	6,6	ЖК(НТК)	10,7
21	Стронций-89	Костная ткань	312,5	ЖК(НТК)	68,2
22	Стронций-90	Костная ткань	9375,0	ЖК(НТК)	53,6
23	Иттрий-90	ЖК(НТК)	93,7	ЖК(НТК)	93,7
24	Иттрий-91	ЖК(НТК)	71,4	ЖК(НТК)	71,4
25	Цирконий-95	ЖК(НТК)	30,0	ЖК(НТК)	30,0
26	Ниобий-95	ЖК(НТК)	19,5	ЖК(НТК)	19,5
27	Молибден-99	Почки	10,7	ЖК(НТК)	48,4
	- " -	ЖК(НТК)	79,0	-	
28	Рутений-103	ЖК(НТК)	22,7	ЖК(НТК)	23,4
29	Рутений-106	ЖК(НТК)	156,3	ЖК(НТК)	156,3
30	Серебро-110 м	ЖК(НТК)	62,5	ЖК(НТК)	62,5
31	Кадмий-109	ЖК(НТК)	10,7	ЖК(НТК)	10,7
32	Кадмий-115 м	ЖК(НТК)	75,0	ЖК(НТК)	75,0
33	Кадмий-115	ЖК(НТК)	55,5	ЖК(НТК)	51,7
34	Сурьма-122	ЖК(НТК)	65,2	ЖК(НТК)	65,2

№№ пп	Радионуклид	Растворимые соединения		Нерастворимые соединения	
		Критический орган	мбэр/мкКи	Критический орган	мбэр/мкКи
35	Сурьма-124	ЖК (НТК)	83,3	ЖК (НТК)	83,3
36	Сурьма-125	ЖК (НТК)	18,75	ЖК (НТК)	19,0
37	Йод-126	Щитовидная железа	2142,8	ЖК (НТК)	27,7
38	Йод-129	- " - - " -	100,0	ЖК (НТК)	8,8
39	Йод-131	- " - - " -	1875,0	ЖК (НТК)	29,2
40	Цезий-134	Весь организм	73,7	ЖК (НТК)	47,0
41	Цезий-137	- " - - " -	41,7	ЖК (НТК)	42,8
	- " -	Печень	107,1	-	
	- " -	Селезенка	83,3	-	
	- " -	Мышечная ткань	78,9	-	
42	Церий-141	ЖК (НТК)	21,4	ЖК (НТК)	21,4
43	Церий-144	ЖК (НТК)	156,25	ЖК (НТК)	156,25
44	Празеодим-143	ЖК (НТК)	38,5	ЖК (НТК)	38,5
45	Прометий-147	ЖК (НТК)	8,3	ЖК (НТК)	8,3
46	Вольфрам-181	ЖК (НТК)	5,2	ЖК (НТК)	5,8
47	Вольфрам-185	ЖК (НТК)	15,6	ЖК (НТК)	17,0
48	Ртуть-203	Почки	107,1	ЖК (НТК)	17,0
49	Свинец-210	Почки	1364,0	ЖК (НТК)	10,7
	- " -	Весь организм	5200,0	-	
50	Полоний-210	Селезенка	2589,0	ЖК (НТК)	65,2
51	Плутоний-238	Костная ткань	750,0	ЖК (НТК)	68,0
52	Плутоний-239	- " - - " -	833,0	ЖК (НТК)	65,2
53	Уран-235	Почки	500,0	ЖК (НТК)	68,0

Представленные в этом сообщении рабочие пределы концентраций радионуклидов в морской воде рекомендуются для практического использования при решении санитарных вопросов защиты окружающей среды. При обсуждении материалов желательно получить мнение радиоэкологов в аспекте ожидаемых биологических последствий загрязнения морской среды на уровне предложенных рабочих пределов. Имеющиеся в СССР материалы по этому вопросу свидетельствуют о приемлемости нормативов и по радиобиологическим критериям безопасности воспроизводства промысловых морских организмов.

ЛИТЕРАТУРА

[1] ЧОПОРНЯК, А.Б., АЭС и Окружающая Среда (обзор), М., 1974 г.
[2] Основные Нормы Безопасности при Защите от Излучения, Серия изданий по безопасности №9, Издание 1967 г., Вена, МАГАТЭ, 1968.
[3] Нормы Радиационной Безопасности (НРБ-69), М., Атомиздат, 1972 г.

[4] ATEN, A.H.W. Jr., "Permissible concentrations of radionuclides in sea water", Health Physics, Pergamon Press, 6 (1961) 114-25.
[5] DUNSTER, H.S., Disposal of Radioactive Liquid Wastes into Coastal Waters, Progress in Nuclear Energy, Series XII, vol.I Pergamon Press, London 1 (1959) 566-582.
[6] Radioactive Wastes Disposal into Atlantic and Gulf Coastal Waters, NAS-NRC, Publ. № 655, Washington, 1959.
[7] ПОЛИКАРПОВ, Г.Г., Радиоэкология Морских Организмов, М., Атомиздат, 1964.
[8] Удаление Радиоактивных Отходов в Моря, Доклад специальной группы экспертов (февраль 1960 г.), МАГАТЭ, Вена, 1961 г.

DISCUSSION

J.A. HETHERINGTON: My first question concerns the dose limits which you are using for setting derived working limits. In relation to pathways causing irradiation via internal contamination you appear to be using a limit which is equivalent to 10% of the one recommended by the ICRP, whereas in the case of irradiation via external exposure, following occupancy of contaminated beaches, you use a limit which is equivalent to only 1% of the ICRP value. Why do you use different limits for internal and external exposure?

V.A. KNIZHNIKOV: I believe one reason why the dose adopted by the author for marine products is ten times greater than that adopted for the beaches is that marine products are of greater significance for mankind than beaches. A second reason, I think, is that the correlation between the values of 10% for marine products and 1% for beaches is suitable for estimating working levels.

J.A. HETHERINGTON: You say you do not think it is right, when assessing the radiation exposure pathways associated with one particular discharge, to take into account other discharges which may be contributing to the same pathways. Why do you think it is not necessary to consider all contributors to a given exposure pathway?

V.A. KNIZHNIKOV: For the radiation dose to critical population groups the author takes only a fraction of the dose recommended by the ICRP and the IAEA for individual members of the population because of the fact that the critical groups can be exposed to additional irradiation from other sources, for instance fall-out or migration of natural radioactive elements from mineral fertilizers through the food chains.

R. BITTEL: Are your results obtained from in situ measurements or from experiments? In this connection I should like to draw your attention to the joint research undertaken by the Commissariat à l'Energie Atomique and the Institut National de la Recherche Médicale on the transfer of elements and radionuclides in marine chains under controlled conditions.

V.A. KNIZHNIKOV: Thank you for this useful information. In reply to your question the results were, as far as I know, obtained in situ from the seas around the north and north-east shores of our country.

B.G. BLAYLOCK: Did I understand correctly that at concentrations of radionuclides in the seawater there would be no effects on the most sensitive link in the aquatic ecosystem, the embryonic fishes, and no genetic effect on the aquatic organisms?

V.A. KNIZHNIKOV: During the embryogenic stage of fish, relatively low doses of irradiation can, actually, cause damage to the developing eggs, resulting in abnormalities. However, for the fish population or the fish reproductive capacity to be adversely affected, the levels of radionuclide contamination of the water would have to be thousands of times higher than those permitted from the viewpoint of human radiological safety.

S. BONOTTO: You have reported a concentration factor of 10 000 for algae. For what element and what algae?

V.A. KNIZHNIKOV: The maximum concentration factor of 10 000 is for iron. I do not know what algae this relates to.

METHOD OF MONITORING AND EVALUATION OF EFFECTS ON AND DOSES TO MAN AND AQUATIC BIOTA

(Sessions VIII and IX)

Chairmen

J.C. ANCELLIN (France)

P.O. AGNEDAL (Sweden)

DOSE ESTIMATION AND PREDICTION OF RADIATION EFFECTS ON AQUATIC BIOTA RESULTING FROM RADIOACTIVE RELEASES FROM THE NUCLEAR FUEL CYCLE*

B.G. BLAYLOCK, J.P. WITHERSPOON
Environmental Sciences Division,
Oak Ridge National Laboratory,
Oak Ridge, Tenn.,
United States of America

Abstract

DOSE ESTIMATION AND PREDICTION OF RADIATION EFFECTS ON AQUATIC BIOTA RESULTING FROM RADIOACTIVE RELEASES FROM THE NUCLEAR FUEL CYCLE.

Aquatic organisms are exposed to radionuclides released to the environment during various steps of the nuclear fuel cycle. Routine releases from these processes are limited in compliance with technical specifications and requirements of federal regulations. These regulations reflect I.C.R.P. recommendations, which are designed to provide an environment considered safe for man. It is generally accepted that aquatic organisms will not receive damaging external radiation doses in such environments; however, because of possible bioaccumulation of radionuclides there is concern that aquatic organisms might be adversely affected by internal doses. The objectives of this paper are: (1) to estimate the radiation dose received by aquatic biota from the different processes and determine the major dose-contributing radionuclides, and (2) to assess the impact of estimated doses on aquatic biota.

Dose estimates are made by using radionuclide concentration measured in the liquid effluents of representative facilities. Where measurements of concentrations are not available, predicted radioactive releases to the aquatic environment are used for dose calculations. Although radioactive releases from reactors used to generate electrical energy have received the most attention, and are the best documented, this evaluation indicates the potential for a greater radiation dose to aquatic biota from the nuclear fuel supply facilities (i.e. mining and milling).

The effects of chronic low-level radiation on aquatic organisms are discussed from somatic and genetic viewpoints. On the basis of the body of radiobiological evidence accumulated up to the present time, no significant deleterious effects are predicted for populations of aquatic organisms exposed to the estimated dose rates resulting from routine releases from conversion, enrichment, fabrication, reactors and reprocessing facilities. At the doses estimated for milling and mining operations it would be difficult to detect radiation effects on aquatic populations; however, the significance of such radiation exposures to aquatic populations cannot be fully evaluated without further research on effects of chronic low-level radiation.

Introduction

Aquatic organisms are exposed to many different radionuclides released to the environment from various facilities of the nuclear fuel cycle. Routine releases from these processes are restricted in compliance with the International Commission on Radiological Protection (I.C.R.P.), which recommends limits for radiation doses for members of the general public [1]. These ICRP recommendations are designed to provide assurances that man-made environmental radioactivity does not exceed levels considered safe for man. It is generally accepted that aquatic organisms will not receive damaging external radiation doses in such environments; however, because of possible bioaccumulation of radionuclides there is concern that aquatic organisms might be adversely affected by internal doses.

* Research supported by the Energy Research and Development Administration under contract with the Union Carbide Corporation.

This paper will address the effect on aquatic biota of radioactive releases from the different processes in the enriched uranium dioxide fuel cycle. Specific objectives are: (1) to estimate the radiation doses received by aquatic biota from the different process effluents and determine the major dose contributing radionuclides, and (2) to assess the impact of estimated doses on aquatic biota.

The different processes of the nuclear fuel cycle considered here are: uranium mining, uranium milling, conversion facilities, uranium enrichment, fuel fabrication, light water reactors and fuel reprocessing.

Model Facilities

In predicting radioactive releases to the aquatic environment, model facilities were assumed for the different processes in the nuclear fuel cycle. The models represent the methods most commonly used in the United States and the size of the facility is representative of the average facility. When possible, actual measurements of radionuclide concentrations in the liquid effluents were used as the source term for calculating the radiation dose rate to aquatic biota. The measurements were scaled to represent the liquid effluents from a model facility.

Dose Calculations

Aquatic biota living in the vicinity of nuclear facilities can be exposed to low levels of radioactivity released in the liquid effluents. The organisms can receive a radiation dose from internal emitters as a result of radionuclides assimilated from food and absorbed from water. External exposure can result from immersion in water that contains the radioactivity; in addition, some types of biota can receive an external exposure from radionuclides accumulated in sediments

The BIORAD [2] computer code was used to calculate internal and external radiation doses to aquatic biota. For dose calculations it was assumed that the radionuclide concentrations in water remain constant and that the biota reach a steady-state concentration.

The effective absorbed energy used in the calculation of internal dose from each photon-emitting radionuclide is a function of the size of the organism. A conservative approach has been used because all BIORAD calculations of internal dose to aquatic plants, invertebrates and fish have utilized effective absorbed energies for 30-cm diameter tissues. In almost all cases the effective diameter of aquatic biota is less than 30 cm; however, effective absorbed energies corresponding to this large diameter were chosen because they maximize the dose from gamma photons and x-rays. This conservative allocation is a safeguard against the possibility that the bioaccumulation factor or the radiosensitivity has been underestimated for a certain type of organism.

Since aquatic organisms may concentrate radionuclides in their tissues to higher levels than the concentration of radionuclides in the ambient water, the dose from internal emitters will generally contribute the major part of the dose to the organisms. The bioaccumulation factors [3-5] used in the dose calculations for aquatic plants, invertebrates and fish represent the higher values found in the literature except for a few radionuclides for which existing data support smaller values than the highest found. In actual exposure situations, the accumulation of radionuclides is expected to be small because of dietary dilutions by noncontaminated foods or because of consumption of aquatic plants with bioaccumulation factors lower than those assumed.

TABLE 1

Estimated Internal Dose to Aquatic Biota in mrads/yr from Radioactive Effluents from a Model Uranium Mining Operation

Radionuclides	Microcurie/ml	Aquatic Plants	Invertebrates	Fish
U-234	5.3E-09[a]	4.9E+03	4.9E+02	4.9E+01
U-235	5.3E-09	4.6E+03	4.6E+02	4.6E+01
U-238	5.3E-09	4.3E+03	4.3E+02	4.3E+01
RA-226	7.4E-09	3.8E+04	3.8E+03	7.6E+02
TH-230	1.3E-09	1.8E+03	5.8E+02	3.5E+01
PB-210	3.3E-10	1.2E+01	6.2E_00	1.9E+01
PO-210	4.7E-09	9.7E+03	9.7E+04	2.4E+02
TOT DOSE		6.3E+04	1.0E+05	1.2E+03

[a]Read as 5.3×10^{-9}.

Uranium Mining

About half of the uranium ore in the United States is produced by open pit (strip) mining [6]. The model mine complex occupies about 3000 acres and produces about 1600 MT ore per day [7].[1] This amount is equivalent to about 5.3 times the annual fuel requirement for a model light water reactor (1000 MWe). The liquid effluents from a mining operation consist of runoff from mining water which is pumped from the mine to keep it dry while the ore is extracted.

The drainage water contains relatively large amounts of suspended solids and is passed to a settling basin. Part of the water becomes run-off, but the bulk recycles through natural seepage and evaporation. The run-off contains natural uranium, radium, thorium and other radionuclides. When it is economically feasible, the uranium is recovered from the drainage water before it is discharged.

The assumption was made that the model mine pumped 1500 gpm of drainage water into a settling basin and that ten percent (150 gpm) was discharged as run off. Since most uranium mines in the United States are located in the arid southwest, the receiving habitat was assumed to be a small stream. The concentrations of radionuclides in Table 1 were obtained by diluting measured concentrations of radionuclides in mining water [7] in a 5 cfs (140 ℓ/s) stream.

Internal dose rate calculations for aquatic plants, invertebrates, and fish showed that Ra-226 and Po-210 were the greatest dose-contributing radionuclides. Aquatic invertebrates received the greatest dose, 97 rads/yr from Po-210. This was due to the bioaccumulation factor of 2×10^4 which was

[1] In this paper MT is used as an abbreviation for metric tons or tonnes; gpm for gallons per minute; cfs for cubic feet per second.

used in the dose calculations. The water immersion dose to the biota exposed to radionuclides in the mining water was 0.017 mrad/yr for the gamma dose and 0.21 mrad/yr for the beta plus gamma dose.

Uranium Milling

Uranium mills extract U_3O_8 from ores which contain from four to six lbs. of uranium per ton. Extraction is accomplished by leaching crushed and ground ore with either an acid or alkaline solution. The acid leach process, which is the most commonly used, accounts for 80% of the annual production of U_3O_8 in the United States. After the ore is leached the resultant solution containing uranium goes through a solvent-extraction process in which the uranium is purified and concentrated. The residual (tailings) which represents the crushed ore minus most of the uranium is released as a slurry into a tailings pond.

The model uranium mill is located near a uranium mine and processes 600 000 MT of ore/yr. IT releases 2 500 MT/day of waste liquid into a tailing retention pond system. The retention system will permit the evaporation of most of the waste liquid; however, it is expected that there will be some seepage from the retention pond. Assuming a seepage rate of 300 ℓ/m (80 gpm), the estimated concentrations of radionuclides in a 5 cfs (140 ℓ/s) steam are given in Table 2 [8].

The resulting internal dose calculations to aquatic biota show that ^{230}Th and ^{226}Ra are the greatest dose contributors. Thorium-230 is released in the highest concentration in the seepage water and has a bioaccumulation of 1.5×10^3 in aquatic plants. As a result, aquatic plants receive the highest total dose, 1200 rads/yr.

Conversion Facilities - Uranium Hexafluoride Production

The next step in the nuclear fuel cycle is the conversion of the solid U_3O_8 from mills into the volatile uranium hexafluoride (UF_6) for subsequent isotopic enrichment. Two processes are used for UF_6 production: the dry hydrofluor method and the wet-solvent-extraction methods. Both processes are used to produce almost equal portions of total UF_6 in the United States. The hydrofluor method releases radioactivity primarily in the gaseous and solid states, while the solvent extraction method releases more of the radioactive waste in liquid effluents.

Since the wet-solvent-extraction method releases more radioactivity to the aquatic environment, this method was selected for use in the model plant. The model plant processes about 5000 MTU and is capable of supplying the annual fuel requirements of 27.5 model light water reactors [6]. A UF_6 plant uses large quantities of water for chemical processes and cooling systems. Therefore, it must be located near a reliable water source, such as a river or lake. The assumption was made that the model plant is located on a 1300 cfs (36.8×10^3 ℓ/s) river.

Irradiated materials are not handled by conversion facilities, and all radionuclides present in the liquid effluents occur, to some extent, in nature. The concentrations of radionuclides measured in the liquid effluent from a conversion plant [9] were used to obtain the radioactive source terms in Table 3. The assumption was made that 1125 gals/min (31.8×10^3 ℓ/s) were discharged into a 1300 cfs stream.

TABLE 2

Estimated Internal Dose to Aquatic Biota in mrads/yr from Radioactive Effluents from a Model Uranium Milling Operation

Radionuclides	Microcurie/ml	Aquatic Plants	Invertebrates	Fish
U-234	9.6E-09[a]	8.8E+03	8.8E+02	8.8E+01
U-235	9.6E-09	8.3E+03	8.3E+02	8.3E+01
U-238	9,6E-09	7.7E+03	7.7E+02	7.7E+01
TH-230	7.9E-07	1.1E+06	3.5E+05	2.1E+04
RA-225	1.3E-09	6.7E+04	6.7E+03	1.3E+03
TOT DOSE		1.2E+06	3.5E+05	2.2E+04

[a]Read as 9.6×10^{-9}.

TABLE 3

Estimated Internal Dose to Aquatic Biota in mrads/yr from Radioactive Effluents from a Model Conversion Facility

Radionuclides	Microcurie/ml	Aquatic Plants	Invertebrates	Fish
U-234	5.4E-10[a]	4.9E+02	4.9E+01	4.9E+00
U-235	5.4E-10	4.6E+02	4.6E+01	4.6E+00
U-238	5.4E-10	4.3E+02	4.3E+01	4.3E+00
RA-226	9.6E-13	4.9E+00	4.9E-01	9.9E-02
TH-230	3.8E-13	5.1E-01	1.7E-01	1.0E-02
TOT DOSE		1.4E+03	1.4E+02	1.4E+01

[a]Read as 5.4×10^{-10}.

The total internal dose rate calculated for aquatic plants was 1.40 rads/yr and 0.14 rads/yr for aquatic invertebrates. The major dose-contributing radionuclides were uranium-234, 235 and 238.

Isotopic (Uranium) Enrichment

The concentration of U-235 is about 0.7% in the UF_6 produced in the conversion plants. The U-235 must be enriched to 2 to 4 percent before fuel can be fabricated. Enrichment is accomplished in a process called gaseous diffusion which involves some 1700 process stages.

TABLE 4

Estimated Internal Dose to Aquatic Biota in mrads/yr from Radioactive Effluents from a Model Uranium Enrichment Facility

Radionuclide	Concentration (μCi/ml)	Algae	Invertebrates	Fish
Tc-99	1.6E-8[a]	1.1E0[b]	1.4E-1	4.2E-1
U-234	1.2E-9	1.1E3	1.1E2	1.1E1
U-235	5.6E-11	4.8E1	4.8E0	4.8E-1
U-236	1.7E-11	1.5E1	1.5E0	1.5E-1
U-238	7.1E-10	5.7E2	5.7E1	5.7E0
Total		1.7E3	1.7E2	1.8E1

[a]Read as 1.6×10^{-8}.

[b]Read as 1.1.

The model plant produces 10 500 metric tons of separative work units per year, enough to furnish the annual fuel for 90 model light-water reactors of 1000 MW [6]. The plant must be located in an area where large quantities of cooling water are available. Radionuclides are released in liquid wastes from process cleanup operations and from auxiliary production facilities. The radionuclide concentrations in Table 4 represent the measured concentrations in the liquid effluents from the Oak Ridge Gaseous Diffusion Plant after discharge into a 1300 cfs stream.

Uranium-234 was the major dose-contributing radionuclide with aquatic plants receiving the highest dose, 1.7 rads/yr.

Fuel Fabrication

Fuel fabrication plants convert UF_6 to UO_2 by a wet process in which ammonium hydroxide is reacted with steamed UF_6 to produce a slurry, which is then dried and converted to UO_2 powder. This powder is then compacted into pellets for placement in cladding to form fuel rods.

The model fuel fabrication plant produces 900 MT of fuel per year. This annual production will furnish the yearly requirements of 26 light-water reactors. The plant required about 425 000 gal/day of water which is used for cooling process equipment and diluting liquid process wastes prior to release It is assumed that liquid effluents are released to a freshwater river which has a minimum flow of 1300 cfs. Table 5 gives the concentrations of radionuclides in the river and the radiation dose to aquatic biota [10]. Uranium 234 and 238 are important contributors to dose.

TABLE 5

Estimated Internal Dose to Aquatic Biota in mrads/yr from Radioactive Effluents from a Model Fuel Fabrication Facility

Radionuclide	Microcurie/ml	mrad/yr			
		Plants	Invertebrates	Fish	Waterfowl
U-234	9.53E-8[a]	1.03E+0	1.24E+2	2.07E+1	2.07E-1
U-235	3.35E-9	3.37E-2	4.04E+0	6.73E-1	6.69E-2
U-236	4.85E-9	5.00E-2	6.01E+0	1.00E+0	1.00E-2
U-238	1.85E-8	1.75E-1	2.09E+1	3.49E+0	3.49E-2
Th-231	3.35E-9	3.98E-1	1.33E-1	7.95E-3	3.72E-6
Th-234 & Pa-234	1.85E-8	1.09E+1	3.69E-1	2.22E-1	3.92E-3
Total		1.28E+1	1.58E+2	2.61E+1	2.62E-1

[a]Read as 1.6×10^{-8}.

Light Water Reactors

Fuel elements are used in reactors to furnish heat for steam electric power plants for the generation of electricity. Over 50 nuclear power plants are currently in operation in the United States, and approximately 60 more are under construction. The model light-water reactor produces a 1000 MW of electricity and requires 35 metric tons of UO_2 fuel per year to operate.

During the operation of nuclear power reactors, radionuclides are formed by fission of the nuclear fuel and by neutron activation of structural materials, corrosion products, and impurities in reactor coolant water. A small fraction of these radioactive materials enter the plant's waste system and are channeled into various effluent streams.

The concentration of radionuclides used in calculating doses to aquatic biota living in the vicinity of a nuclear power station are given in Table 6 for pressurized water reactors (PWR's) and in Table 7 for boiling water reactors (BWR's). The radionuclide concentration was obtained by taking the highest average annual concentration of each radionuclide measured in the effluents of ten pressurized water reactors and nine boiling water reactors [11]. The specific radionuclides in effluents vary from plant to plant because of fuel performance, reactor power production, and different designs in reactor and waste treatment systems. Therefore all of the radionuclides listed in Table 6 and Table 7 would not be found in the effluents of all reactors. Kaye and Rohwer [12] listed 89 radionuclides which were most often considered in environmental impact statements for light-water reactors; however, only 28 radionuclides were identified in the effluents of pressurized water reactors and 23 in the effluents of boiling water reactors [11].

TABLE 6

Estimated Internal Dose to Aquatic Biota in mrads/yr from Radioactive Effluents from a Pressurized Water Reactor Fresh Water and Saline Water Habitats

		Fresh Water			Saline Water		
Radio-nuclides	Micro-curie/ml	Aquatic Plants	Inverte-brates	Fish	Algae	Mollusks/Crustaceans	Fish
H-3	7.6E-06[a]	1.4E+00	1.4E+00	1.4E+00	1.4E+00	1.4E-00	1.4E+00
C-14	1.1E-10	4.9E-01	9.7E-01	4.9E-01	1.1E-04	1.1E-04	1.1E-04
NA-24	2.7E-09	1.3E+01	2.3E+00	2.7E+00	1.3E-01	1.3E-01	1.3E-01
MN-54	8.4E-12	8.0E-01	3.2E+00	8.0E-03	8.0E-01	4.0E+00	2.4E-01
CR-51	3.5E-09	6.5E-02	3.2E-02	6.5E-02	1.6E+00	1.6E+00	1.6E-01
FE-59	2.8E-10	4.2E+00	1.4E+01	4.2E-01	2.5E+01	8.5E+01	4.2E+00
CU-64	1.7E-11	7.7E-02	7.7E-02	1.5E-02	7.7E-02	3.9E-01	7.7E-02
CO-57	4.9E-12	1.6E-03	1.6E-03	1.6E-04	8.2E-04	8.2E-02	8.2E-04
CO-58	4.0E-09	9.0E+00	9.0E+00	9.0E-01	4.5E+00	4.5E+00	4.5E+00
CO-60	2.5E-09	1.4E+01	1.4E+00	7.7E-01	7.7E-01	3.8E+01	3.8E+00
ZN-65	1.3E-10	7.7E-01	7.7E+00	7.7E-01	7.7E-01	3.8E+01	3.8E+00
SR-89	4.2E-11	2.1E-01	4.3E-02	2.1E-03	8.5E-03	4.3E-04	4.3E-04
SR-90	2.6E-12	2.7E-02	5.3E-03	2.7E-04	1.1E-03	5.6E-05	5.3E-05
NB-95	1.6E-10	1.2E+00	1.5E-01	4.6E+01	1.5E-01	3.1E-01	1.5E-01
ZR-95	4.0E-11	8.2E-01	5.5E-03	2.7E-03	8.2E-01	8.2E-02	2.5E-02
MO-99	1.2E-10	1.2E+00	1.2E-02	1.2E-02	1.2E-01	1.2E-01	1.2E-02
RU-106	3.9E-10	2.1E+00	3.1E+00	1.0E-01	1.0E+01	1.0E+00	3.1E-02
AG-100M	4.6E-13	2.9E-03	1.1E-02	3.3E-05	1.4E-02	7.2E-02	1.4E-02
SB-124	1.1E-11	5.2E-01	3.4E-03	3.4E-04	3.4E+00	3.4E-01	3.4E-01
I-131	3.9E-09	1.3E+00	1.6E-01	4.9E-01	3.2E+02	3.2E+00	6.5E-01
I-132	5.7E-11	7.2E-02	9.0E-03	2.7E-02	1.8E+01	1.8E-01	3.6E-02
I-133	7.2E-09	4.5E+00	5.6E-01	1.7E+00	1.1E+03	1.1E+01	2.2E+00
I-134	1.5E-08	1.7E+01	2.1E+00	6.3E+00	4.2E+03	4.2E+01	8.4E+00
I-135	5.0E-09	4.8E+00	6.0E-01	1.8E+00	1.2E+03	1.2E+01	2.4E+00
CS-134	5.3E-14	8.7E-05	1.1E-04	4.8E-04	1.1E-05	5.4E-05	3.3E-05
CS-137	1.3E-08	1.1E+01	1.4E+01	5.7E+01	1.4E+00	7.2E+00	4.3E+00
LA-140	4.2E-12	7.5E-01	1.5E-01	3.7E-03	4.5E-02	1.5E-02	4.5E-03
CE-144	2.9E-10	3.6E+01	7.1E+00	1.8E-01	2.1E+00	7.1E-01	2.1E-01
W-187	1.0E-12	1.5E-02	1.3E-04	1.5E-02	1.3E-03	1.3E-03	1.3E-04
TOT DOSE		1.2E+02	7.9E+01	1.2E+02	6.9E+03	1.4E+03	3.9E+01

[a] Read as 7.6×10^{-6}.

Dose calculations were based on the concentrations of radionuclides in the effluents. Although the concentrations of radionuclides will decrease as the discharged effluents mix with the receiving body of water, thereby reducing the radiation dose to the biota, the maximum concentrations to which the aquatic biota might be exposed were used in the dose calculations.

Nuclear reactors are located on fresh water streams, estuaries and open sea coasts; thus radioactive liquid effluents are released into saline as well as freshwater aquatic habitats. Consequently, dose rates were calculated for aquatic biota exposed to radioactivity in saline and freshwater for PWR's and BWR's, Table 6 and 7. Dose rates to aquatic biota in saline water were greater for algae and mollusks/crustaceans than for aquatic plants and invertebrates in fresh water as a result of higher bioaccumulation factors;

TABLE 7

Estimated Internal Dose to Aquatic Biota in mrads/yr from Radioactive Effluents from a Boiling Water Reactor in Fresh Water and Saline Water Habitats

		Fresh Water			Saline Water		
Radio-nuclides	Micro-curie/ml	Aquatic Plants	Inverte-brates	Fish	Algae	Mollusks/ Crustaceans	Fish
H-3	3.6E-07[a]	6.8E-02	6.8E-02	6.8E-02	6.8E-02	6.8E-02	6.8E-02
NA-24	1.2E-09	6.0E+00	1.0E+00	1.2E+00	6.0E-02	6.0E-02	6.0E-02
MN-54	1.1E-08	1.0E+03	4.0E+03	1.0E+01	1.0E+03	5.0E+03	3.0E+02
CR-51	2.8E-08	5.1E-01	2.6E-01	5.1E-01	1.8E+01	1.3E+01	1.3E+00
FE-59	1.8E-10	2.7E+00	8.6E+00	2.7E-01	1.6E+01	5.4E+01	2.7E+00
CU-64	2.2E-10	1.0E+00	1.0E+00	2.1E-01	1.0E+00	5.2E+00	1.0E+00
CO-58	2.2E-09	5.0E+00	5.0E+00	4.0E-01	2.5E+00	2.5E+02	2.5E+00
CO-60	8.2E-08	4.6E+02	4.6E+02	4.6E+01	2.3E+02	2.3E+04	2.3E+02
ZN-65	1.1E-09	6.4E+00	6.4E+01	6.4E+00	6.4E+00	3.2E+02	3.2E+01
SR-89	2.1E-09	1.1E+01	2.2E+00	1.1E-01	4.3E-01	2.2E-02	2.2E-02
SR-90	3.5E-10	3.6E+00	7.2E-01	3.6E-02	1.4E-01	7.2E-03	7.2E-03
Y-90	3.5E-10	2.9E+01	5.8E+00	1.5E-01	1.7E+00	5.8E-01	1.7E-01
MO-99	1.8E-09	1.8E+01	1.8E-01	1.8E-01	1.8E+00	1.8E+00	1.8E-01
SB-124	2.6E-10	1.2E+01	7.7E-02	7.7E-03	7.7E+01	7.7E+00	7.7E+00
I-131	1.8E-07	6.0E+01	7.5E+00	2.3E+01	1.5E+03	1.5E+01	3.0E+00
I-132	7.2E-12	9.2E-03	1.1E-03	3.4E-03	2.3E+00	2.3E-02	4.6E-03
I-133	1.4E-09	9.4E-01	1.2E-01	3.5E-01	2.4E+02	2.4E+00	4.7E-01
I-134	9.9E-13	1.1E-03	1.4E-04	4.2E-04	2.8E-01	2.8E-03	5.6E-04
CS-134	8.8E-09	1.4E+01	1.8E+01	7.2E+01	1.8E+00	9.1E+00	5.4E+00
CS-137	1.6E-08	1.4E+01	1.8E+01	7.1E+01	1.8E+00	8.9E+00	5.3E+00
LA-140	1.1E-09	2.0E+02	3.9E+01	9.9E-01	1.2E+01	3.9E+00	1.2E+00
W-187	5.4E-12	8.3E-02	6.9E-04	8.3E-02	6.9E-03	6.9E-03	6.9E-04
NP-239	1.5E-09	8.3E+00	3.3E+00	8.3E-02	4.0E-01	6.7E-01	6.7E-01
TOT DOSE		1.9E+03	4.6E+03	2.3E+02	3.1E+03	2.9E+04	5.9E+02

[a]Read as 3.6×10^{-7}.

TABLE 8

Estimated Internal Dose to Aquatic Biota in mrads/yr from Radioactive Effluents from a Model Fuel Reprocessing Plant

Radionuclide	Concentration (μCi/ml)	Algae	Invertebrates	Fish
H-3	1.7E-07[a]	3.2E-02	3.2E-02	3.2E-02
I-129	1.2E-10	8.2E-03	1.0E-03	3.1E-03
I-131	3.0E-10	9.9E-02	1.2E-03	3.7E-02
RU-103	1.8E-10	3.0E+00	4.4E-01	1.5E-02
RU-106	9.8E-10	5.1E+01	7.7E+00	2.5E-01
CS-134	5.2E-10	5.3E+00	1.1E+00	4.2E+00
CS-137	2.6E-10	1.4E+00	2.8E-01	1.3E+00
TE-127m	1.4E-11	8.2E-02	8.2E-03	8.3E-04
TE-129m	5.7E-12	2.3E-02	1.2E-02	1.2E-03
SR-89	3.9E-11	2.0E-01	4.0E-02	2.0E-03
SR-90	3.7E-11	3.8E-01	7.7E-02	3.8E-03
Y-90	3.7E-11	3.1E+00	6.2E-01	1.6E-02
Y-91	7.2E-11	4.0E+00	7.9E-01	2.0E-02
ZN-95	1.2E-10	2.5E+00	1.7E-02	8.4E-03
NB-95	2.2E-10	1.7E+00	2.1E-01	6.4E+00
AG-110m	1.2E-12	7.5E-03	2.9E-02	6.7E-05
SB-125	3.9E-12	3.7E-02	2.4E-04	2.4E-05
CE-141	2.2E-11	3.4E-01	8.6E-02	2.2E-03
CE-144	3.7E-11	3.6E+00	9.0E-01	2.2E-02
PM-142	4.8E-11	3.1E-01	6.2E-02	1.5E-03
EU-154	3.3E-12	4.1E-01	8.1E-02	2.0E-03
EU-155	3.1E-12	4.6E-03	9.2E-04	2.3E-05
U-234	3.7E-14	1.7E-06	2.0E-04	3.4E-05
U-235	8.2E-17	3.5E-08	4.wE-06	7.1E-07
U-236	1.4E-15	6.1E-07	7.4E-05	1.2E-05
U-237	1.5E-15	6.2E-07	7.4E-05	1.2E-05
PU-238	3.4E-12	1.3E+00	3.7E-01	1.3E-02
PU-239	4.0E-13	1.4E-01	3.9E-02	1.4E-03
PU-240	5.7E-13	2.0E-01	5.6E-02	2.0E-03
PU-241	1.2E-10	1.8E+00	5.3E-01	1.8E-02
AM-241	7.7E-15	4.1E-03	8.2E-04	2.0E-05
AM-243	8.7E-15	4.4E-02	8.3E-03	2.2E-04
CM-242	8.2E-12	6.1E+01	1.2E+01	3.1E-01
CM-244	1.2E-12	6.7E+00	1.3E+00	3.3E-02
TOT DOSE		1.5E+02	2.7E+01	1.2E+01

[a]Read as 1.7×10^{-7}.

however, dose rates were greater for fish in fresh water. The isotopes of iodine are more important dose contributors in saline water than in fresh water because of the difference in bioaccumulation. Cobalt-60 and ^{54}Mn, which have relatively high bioaccumulation factors, are important dose contributors in both fresh and saline water. In comparing dose-contributing radionuclides from BWR's and PWR's, the isotopes of cesium (^{134}Cs and ^{137}Cs) are of greater importance in the effluents of PWR's as a result of decreased releases of other radionuclides. Conversely, the release of tritium in the liquid effluents of PWR's is about an order of magnitude higher; however, because of bioaccumulation of 1.0 and a low dose rate factor, tritium is not an important dose contributor.

Fuel Reprocessing

Spent fuel from a light-water reactor still contains usable fissonable material. Fuel reprocessing recovers this unused material by separating it from the radioactive product in spent portions. This reclaimed fuel is then cycled back to an enrichment plant and incorporated into new feed material for fuel fabrication.

In fuel reprocessing a mechanical shear and nitric acid leach system is used to segment and dissolve fuel elements. The unused portion of fuel is then purified and recovered by a solvent extraction method. The purified uranium product is converted to UF_6 and shipped to isotopic enrichment facilities. The residue from these processes is highly radioactive. Most of the fission products and transuranic elements built up in the fuel elements, which power light-water reactors, are now waste products of the reprocessing facility.

One commercial reprocessing plant has been in operation in the United States for about six years. This plant is currently shut down for extensive modification. An additional plant is under construction and is nearing completion [7].

The model plant will process 1500 metric tons of fuel per year and has no radioactive liquid effluent. However, it was assumed that radioactive materials from the gaseous effluents would be deposited in a fresh water river at the same rates and amounts as on a similar area of land [13]. To evaluate aquatic pathways leading to a potential radiation dose to aquatic biota, it was assumed that a segment of a river 1 mile long by 0.1 mile wide by 3 m deep is located 0.5 miles from a model plant in the direction of the prevailing wind. All radionuclides remain in the water at a steady state with no further dilution by volume flow or settling out.

The concentration of radionuclides in the water and the radiation dose to aquatic invertebrates and fish are given in Table 8. In general, dose to algae and invertebrate is due primarily to radionuclides of Cm, Ru, Cs and Y. The dose to fish is heavily influenced by radionuclides of Cs, Cm and Nb. Algae receive the greatest dose because of bioaccumulation factors.

Effects of Radioactive Effluents on Aquatic Biota

Potential exposure to radiation from radionuclide concentrations which may result from releases from nuclear facilities was estimated in the previous sections. This section will attempt to assess the effects of radiation dose on aquatic biota from these releases. There is a lack of radiation effects information on aquatic biota; also in most cases the radiation dose in experimental work is orders of magnitude higher than that experienced by natural populations exposed to effluents from nuclear facilities. Thus extrapolation from controlled laboratory experiments will be necessary.

The effects of radiation on aquatic biota have been reviewed by Templeton et al. [14] and Auerbach et al. [15]. More recently a comprehensive review has been completed by Ophel et al. [16]. In general these reviews indicate that radiation effects would not be detected at the dose rates estimated for aquatic biota living in the effluents released from most nuclear facilities. With the exception of milling and mining operations most dose rates are estimated to be less than 30 rads per year.

In estimating doses from milling and mining operations a conservative approach was taken (i.e. the upper limits for bioaccumulation factors and radionuclide concentrations were used in dose estimations). Thus, the doses are upper limits or overestimates. The highest dose rates were estimated for aquatic biota receiving effluents from milling operations. Aquatic plants were estimated as receiving the highest doses, 1200 rads/yr, with aquatic invertebrates receiving 360 rads/yr and fish 22 rads/yr, Table 2. The next highest annual dose was received by aquatic invertebrates (100 rads/yr) from mining water run-off Habitats in which such doses occur should be in the restricted zone of the facility and considered part of the radioactive waste disposal system. Nevertheless, one can examine the effects of chronic irradiation on aquatic biota receiving equivalent dose rates.

The highest estimated radiation dose to plants (1200 rad/yr) would be expected to produce little or no measurable effects. Lower plants such as algae have a relatively high degree of radioresistance compared to higher plants. Chronic dose rates in the range of 3 to 4 rads/day have been shown to have effects on radiosensitive higher plants such as conifers but, at most, only slight growth-inhibiting effects [17].

Most of the studies which have been made on natural populations exposed to chronic radiation higher than background levels have been conducted at the Oak Ridge National Laboratory (ORNL). Studies have been conducted on populations of fish, snails, and insects that inhabit White Oak Lake. This lake, located in the ORNL reservation, has served as a final settling basin for low-level radioactive effluent since 1943 [18]. The natural populations which live in this environment have been exposed to chronic radiation for many generations. Gambusia affinis, the mosquito fish, which received an estimated dose of 11 rads/day, had a significantly greater number of dead embryos and abnormal embryos than did control populations [19].

In related laboratory experiments the effects of chronic irradiation at dose rates of 1.3, 2.5 and 5.4 rads/h were investigated for a period of 47 days [20]. No excessive mortality occurred at any of the doses. After 47 days the surviving fish were sacrificed for pathological studies. No histological effects were detected in the hemopoietic organs; however, atrophy of the testis was detected after 18 days at all dose rates including dose rates of 1.3 rad/h. Based on the results of these experiments it would be exceedingly difficult to detect these types of results on fish that were exposed to chronic irradiation at a dose rate of 22 rads/yr (100 mrads/day), the highest estimated dose rates for fish from any of the nuclear facilities.

Field and laboratory studies were conducted on the snail, Physa heterostroph from White Oak Lake [21,22]. The dose rate received by these snails at the time of the study was calculated as 0.65 rad/day; however, the population had been exposed to higher dose rates in the past. A larger number of eggs per capsule was observed in snails from White Oak Lake, but a decrease in the number of capsules per snail occurred. A series of laboratory experiments were conducted to determine the effects of 1, 10 and 25 rads/hour on survival size and reproduction of P. heterostropha. A dose rate of 1 rad/h (8760 rads/yr) produced no significant effects while a dose rate of 10 rads/h affected all parameters tested.

This section is not intended to be a review of the literature on the effects of radiation on aquatic biota, but the previous examples are representative. Based on these studies and on the previously mentioned reviews [14,15,16], it would be exceedingly difficult to detect somatic or reproductive effects on aquatic populations receiving a dose of 1 rad/day or less.

The problem of genetic effects of chronic low-level radiation cannot be dismissed. If it is accepted that the mutation rate in organisms is linear in relation to radiation dose and that there is no threshold value for the production of mutations, an increase over the background levels of radiation would increase the mutation rates. Very few studies have been carried out on the long-term effect of chronic low-level radiation on natural aquatic populations. A cytogenetic study was started in 1960 on the Chironomus (midge) larvae which live in White Oak Lake [23,24]. The frequency and kind of chromosome aberrations observed in the White Oak Lake population were compared with control populations. The calculated dose received by Chironomus larvae in the White Oak Lake population was 230 rads/yr. More chromosome aberrations were being produced in the irradiated populations than in the control populations. However, these aberrations did not persist or become established in the population and were eliminated by selection or genetic drift. Similar effects would be expected in aquatic populations exposed to mining and milling effluents, but the significance of these effects on aquatic populations is difficult to assess.

The effects of an increased mutation rate on aquatic organisms has recently been addressed by Templeton et al. [25]. The argument was put forth that any prediction of the effects of an increased mutation rate on fish and other aquatic organisms resulting from an increase in the levels of environmental radiation must be made within the perspectives of the reproductive rate of the species and the value of one individual to the population. The same criteria cannot be used to assess and evaluate the consequences of an increased mutation rate for aquatic populations as are used for human populations. For humans, a great value is placed on the individual members and many individuals with relatively low adaptive values are maintained in the population. On the contrary, for aquatic organisms whose reproductive rates are generally very high and on which the selective pressures are strong, the value of one or even thousands of individual organisms to the population is rather insignificant insofar as the long-term structure and fate of the population are concerned. In such populations, often much less than one percent of the viable zygotes are normally expected to mature to adulthood and to reproduce, i.e., to comprise the effective gene-pool. Even if we make the most conservative assumption that all induced mutations are harmful to the population, we would predict that, even so, no significant deleterious effects are likely to be produced on populations of aquatic organisms at the dose rates estimated for all the nuclear facilities related to the nuclear fuel cycle with the exception of milling and mining operations.

Conclusion

The potential for greater radiation doses to aquatic biota seems to be associated with the nuclear fuel supply facilities (i.e. mining and milling). The higher doses are due to naturally occurring radionuclides, ^{226}Ra, ^{210}Po and ^{230}Th. This evaluation of dose to aquatic organisms supports the increased attention now being given to environmental problems related to the final disposition of radioactive wastes such as mill tailings, which contain uranium and its daughter radionuclides. Because of the number, the importance, and the production of fission radionuclides, radioactive releases from nuclear power plants have received the most public attention. The nuclear reactors are also the best monitored facilities; however, on a comparative basis, the radiation dose estimated for aquatic biota in the effluent of a nuclear power reactor is less than for most of the other facilities.

The dose rate estimates are conservative and are probably overestimates of the doses received by aquatic biota from radioactive releases from the different facilities of the nuclear fuel cycle.

In conclusion we would predict that somatic or genetic effects produced on aquatic biota at the dose rates estimated for conversion, enrichment, fuel fabrication, nuclear power reactors, and reprocessing facilities would not significantly affect the exposed aquatic populations. Dose rate estimates for aquatic biota from milling and mining operations are much higher than estimates from other facilities in the nuclear fuel cycle. Additional information from long-term studies on aquatic populations exposed to chronic low-level irradiation would be necessary to fully evaluate the effects of such dose estimates. However, even at these higher dose rates (aquatic plants, 3.3 rads/day, invertebrates 1 rad/day and fish 0.06 rad/day) based on the information now available, radiation effects on aquatic populations probably would not be detected and the populations would continue without obvious detrimental effects.

References

[1] I.C.R.P., Recommendations of the International Commission on Radiological Protection (Adopted 17 September 1965), Publication 9, Pergamon Press, New York (1966).

[2] MCKAY, L. R., A Methodology for Calculating Radiation Doses from Radioactivity Released to the Environment. ORNL-4992, (In Press) 1975.

[3] CHAPMAN, W. H., FISHER, H. L., and PRATT, M. W., Concentration Factors of Chemical Elements in Edible Aquatic Organisms, University of California, Lawrence Radiation Laboratory, UCRL-50564 (1968).

[4] REICHLE, D. E., DUNAWAY, P. B. and NELSON, D. J., Turnover and concentration of radionuclides in food chains, Nucl. Safety 11 1 (Jan-Feb 1970).

[5] POLIKARPOV, G. F., Radioecology of Aquatic Organisms, Reinhold, New York (1967).

[6] WITHERSPOON, J. P. Environmental Considerations in the Nuclear Fuel Cycle, In Proceedings of the Symposium on Ecology, Energy and Economics, Carroll Business Bulletin Vol. 13, 4 (March 1974).

[7] Environmental Survey of the Uranium Fuel Cycle, U.S. Atomic Energy Commission, Fuels and Materials, Directorate of Licensing, Wash-1248, April 1974.

[8] Environmental Analysis of the Uranium Fuel Cycle, Part I - Fuel Supply, U. S. Environmental Protection Agency, Washington, D. C., EPA-520/9-73-003-B, October 1974.

[9] Draft Environmental Statement, Related to the Kerr-McGee Nuclear Corporation, Sequoyah Uranium Hexafluoride Plant, Fuels and Materials Directorate of Licensing, U. S. Atomic Energy Commission, Docket No. 40-8027, April 1974.

[10] PECHIN, W. H., BLANCO, R. E., DAHLMAN, R. C., FINNY, B. C., LINDAUR, R. B., and WITHERSPOON, J. P. Correlation of Radioactive Waste Treatment Costs and the Environmental Impact of Waste Effluents in the Nuclear Cycle for Use in Establishing "As Low as Practical" Guide - Fabrication of Light-Water Reactor Fuel from Enriched Uranium Dioxide, ORNL-TM-4902 (May 1975) 208 pp.

[11] Report on Releases of Radioactivity in Effluents and Solid Waste from Nuclear Power Plants for 1972, Directorate of Regulatory Operations, USAEC, August, 1973.

[12] KAYE, W.V. and ROHWER, P. S., "Radiological assessment of nuclear power stations". In Advances in Radiation Biology. (LETT, J. T. and ADLER, H., Eds.). Academic Press, New York. Vol. 5 (1975) 47-81.

[13] FINNY, B. C., BLANCO, R. E., DAHLMAN, R. C., KITTS, F. G., and WITHERSPOON, J. P., Correlation of Radioactive Waste Treatment Costs and the Environmental Impact of Waste Effluents in the Nuclear Fuel Cycle for Use in Establishing "As Low as Practicable" Guide - Nuclear Fuel Reprocessing, ORNL-4901 (In Press) 1975.

[14] TEMPLETON, W. J., NAKATANI, R. E., and HELD, E., "Radiation effects," in Radioactivity in the Marine Environment, National Academy of Sciences, Washington, D. C. (1971).

[15] AUERBACH, S. I., NELSON, D. J., KAYE, S. V., REICHLE, D. E. and COUTANT, C. C., "Ecological considerations in reactor power plant siting," in Environmental Aspects of Nuclear Power Stations, IAEA-SM-146/53, Vienna (1971) 804-820.

[16] OPHEL, I. L., HOPPENHEIT, M., and ICHIKANA, R., Effects of Ionizing Radiation on Aquatic Organisms. In Proceedings of a Panel on the Effects of Ionizing Radiation on Aquatic Organisms and Ecosystems, International Atomic Energy Agency, Vienna, 18-22 April, 1974.

[17] SPARROW, A. H., Comparison of the tolerance of higher plant species to acute and chronic exposures of ionizing radiation, Japan, J. Genet. 40: 12-37, 1965.

[18] LEE, P. K. and AUERBACH, S. I., "Determination and evaluation of the radiation field above White Oak Lake," (1960) USAEC Rpt. ORNL-2755.

[19] BLAYLOCK, B. G. The fecundity of a Gambusia affinis affinis population exposed to chronic environmental radiation," Radiation Research 37 1 (1959) 108-117.

[20] COSGROVE, G. E. and BLAYLOCK, B. G., Acute and Chronic Irradiation Effects in Mosquito Fish at 15°C or 25°C, in Third Internaional Symposium on Radioecology, AEC Conference 710501, May 10-12, 1971.

[21] COOLEY, J. L., Effects of chronic irradiation and temperature on populations of the aquatic snail Physa heterostropha, Ph.D. Thesis, University of Tennessee, Knoxville, Tennessee (August 1970).

[22] COOLEY, J. L., Effects of chronic environmental radiation on a natural population of the aquatic snail Physa heterostropha. Radiation Research 54 1 (1973) 130-140.

[23] BLAYLOCK, B. G., Chromosomal aberrations in a natural population of Chironomus tentans exposed to chronic low-level radiation. Evolution 19 3 (1965) 421-429.

[24] BLAYLOCK, B. G., Chromosomal polymorphism in irradiated natural populations of Chironomus. Genetics 53 1 (1966) 131-136.

[25] TEMPLETON, W. L., BLAYLOCK, B. G., HOLDEN, M. J., WOODHEAD, D. S., MUKERJEE, R. N., RAVERA, O., BERNHARD, M., and S'ZTANYIK, L., Effects of Ionizing Radiation on Aquatic Populations and Ecosystems. In Proceedings of a Panel on the Effects of Ionizing Radiation on Aquatic Organisms and Ecosystems, International Atomic Energy Agency, Vienna, 18-22 April, 1974.

DISCUSSION

Yu.V. SIVINTSEV: Why are there no BWR ^{135}I data in Table VII, and where did you get the values for the magnitude and isotopic composition of the discharges, which greatly affect the dose calculations? I should also like to know what value you assumed for the residence time of the wastes in calculating the magnitude of radionuclide discharges into the aquatic environment.

B.G. BLAYLOCK: The radionuclide concentrations were obtained using the highest average annual concentration for each radionuclide measured in the effluents from nine boiling-water reactors. The particular radionuclides in the effluents vary from plant to plant. Iodine-135 was not detected in the effluents of any of the nine boiling-water reactors, so it was not considered in dose calculations. Kaye and Rohwer [12] give a list of 89 radionuclides which featured most often in environmental impact statements for light-water reactors; however, only 23 radionuclides were measured in the effluents of boiling-water reactors and only these 23 were used for dose calculations. For dose calculations it was assumed that the radionuclide concentrations in water remain constant and the biota reached a steady-state concentration.

P. BEAU: Is it realistic to assume that reprocessing plants release no liquid effluents? Is this hypothesis not likely to modify the relative importance of the various stages of the fuel cycle?

B.G. BLAYLOCK: The United States has no commercial reprocessing plant in operation at present. One commercial plant operated for about six years, but it is currently shut down for extensive modifications. An additional plant is under construction and nearing completion. Specifications for the model reprocessing plant assume no liquid radioactive effluents. These assumptions may not be realistic on a cost benefit basis; however, we are currently assuming that radioactive liquid effluents will not be released. Under these conditions the relative importance of the various stages of the fuel cycle will not be altered.

To obtain source terms for the aquatic environment it was assumed that radioactive materials from the gaseous effluents would be deposited in a freshwater river at the same rate and in the same amounts as on a similar area of land.

Suzanne VIGNES: I understand that you observed no effects of chronic irradiation on the fecundity of invertebrates in the lake. Is it correct that with the larvae of Gambusia affinis you even found increased fecundity for single doses below 100 rad?

B.G. BLAYLOCK: We did not examine the effects of chronic irradiation on fecundity in Chironomus in White Oak Lake, but the number of embryos per brood was examined in the mosquito fish, Gambusia affinis. The results of this study have been reported in Radiation Research (Ref.[19] of the paper). I will try to summarize the study.

Females of the ovoviviparous species Gambusia were dissected shortly before they were to give birth to their young, and the numbers of viable embryos, dead embryos, and abnormal embryos were recorded. Although there was a significantly greater number of dead embryos and abnormal embryos in the irradiated population, there was also a significantly greater number of viable embryos relative to the control group. It was suggested that the increased fecundity of the females in the Gambusia population in White Oak Lake may be an adjustment to the increased mortality of the embryos and could be attributed to radiation. In this respect radiation would be analogous to an environmental factor that increased mortality rates.

Similar results have been obtained by Donaldson at the University of Washington* in long-term studies on the effect of chronic low-level gamma radiation on the Chinook salmon. The proportion of salmon irradiated in the embryonic stage with 0.5 rad/day that returned to spawn was greater than in the case of the control fish and they produced a greater number of eggs.

In another study conducted on the snail Physa heterostropha, in White Oak Lake, Cooley (Ref.[22] of the paper) found an increased number of eggs per capsule but a decrease in the number of egg capsules per snail. The radiation dose received by these snails at the time of the study was calculated as 0.65 rad/day; however, the population had previously been exposed to higher doses. Data from the field study supported by laboratory experiments led Cooley to hypothesize that the reduced frequency of capsule production is compensated for by the increased number of eggs per capsule.

J.A. HETHERINGTON: With regard to Mr Beau's question, it would be unfortunate should the impression be given that Mr. Blaylock's main conclusions would need to be altered if the ground rules were changed to permit the discharge of liquid radioactive wastes from the reprocessing plant. My colleague, Mr. Woodhead, has considered the consequences — for the natural marine resources — of discharges of radionuclides from the Windscale plant into the Irish Sea. Theoretical calculations, laboratory studies and environmental measurements have been combined to show that the discharge of the major fission products results in negligible hazard to the major marine resources. He is now extending his work to the important transuranic nuclides and this will include ^{242}Cm, which is present in Windscale effluents at a concentration equivalent to a few per cent of that of ^{239}Pu.

---

* DONALDSON, L.R., and BONHAM, K., Effects of low-level chronic irradiation of Chinook and Coho salmon eggs and alevins, Trans. Amer. Fish Soc. 93 (1964) 333-341.

CONVENTIONAL AND RADIOACTIVE POLLUTANTS OF INLAND WATERS IN AUSTRIA

L.J. OTTENDORFER
Federal Institute for Water Control,
Vienna, Austria

Abstract

CONVENTIONAL AND RADIOACTIVE POLLUTANTS OF INLAND WATERS IN AUSTRIA.

The hydrological and orographic situation of Austria is described. Water quality and the organization of its control are discussed. The results of these investigations are converted into water quality maps issued in regular intervals. The importance of pollution abatement in lakes is mentioned. The Austrian Danube is taken as a typical example of a conventionally polluted river. Complications arising from the construction of hydroelectric plants are discussed and the different sedimentation and the risks of deoxygenation are considered.

The radiological control of inland waters was started in 1961 after the nuclear test series. The data collected from 16 sampling points at monthly intervals serve now as the guideline for the issue of licences for nuclear power plants and the limits for radiological releases. The radiological data so far collected by the Federal Ministry of Health and Environmental Protection cover atmospheric and aqueous contamination. A Federal Law regulates the maximum permissible impacts on the population and the environment. All relevant data are issued regularly by the above mentioned Ministry. The influence of already existing and planned nuclear power stations in the neighbouring countries (upstream of Austria: Switzerland, Federal Republic of Germany, Czechoslovakia) is considered. Radiological control in connection with the first Austrian nuclear plant (capacity 700 MW) and the following ones is discussed. The initiatives of several international organizations to install a multinational control system are mentioned. The system would be planned for all riparian countries of the Danube catchment area.

1. INTRODUCTION

The hydrological situation of the territory of Austria is quite favourable in comparison with other European countries. The annual precipitation amounts to about 1200 mm/a, making a total of 100×10^9 m^3. A further 35×10^9 m^3 of surface water flows from Bavaria into Austria through the rivers Inn and Danube. The surface runoff of 54×10^9 m^3 flowing towards Czechoslovakia, Hungary and Yugoslavia, corresponds to a precipitation of 654 mm/a. The main part of the Austrian water resources stems from snow and ice deposits in the alpine regions which regenerate every year.

The orographic structure defines three main catchment areas (Fig.1):

(I) Rivers north of the main alpine region, feeding to the Austrian Danube; some of these rivers cross the territory of Bavaria

(II) Rivers south of this system, running towards the Mur and Drau; a number of small rivers traverse the Austro-Hungarian border. All these rivers finally flow into the Danube on Hungarian or Yugoslavian territory

(III) The province of Vorarlberg, the most western part of Austria, delivers its surface runoff to the Alpine Rhine and the Lake of Constance.

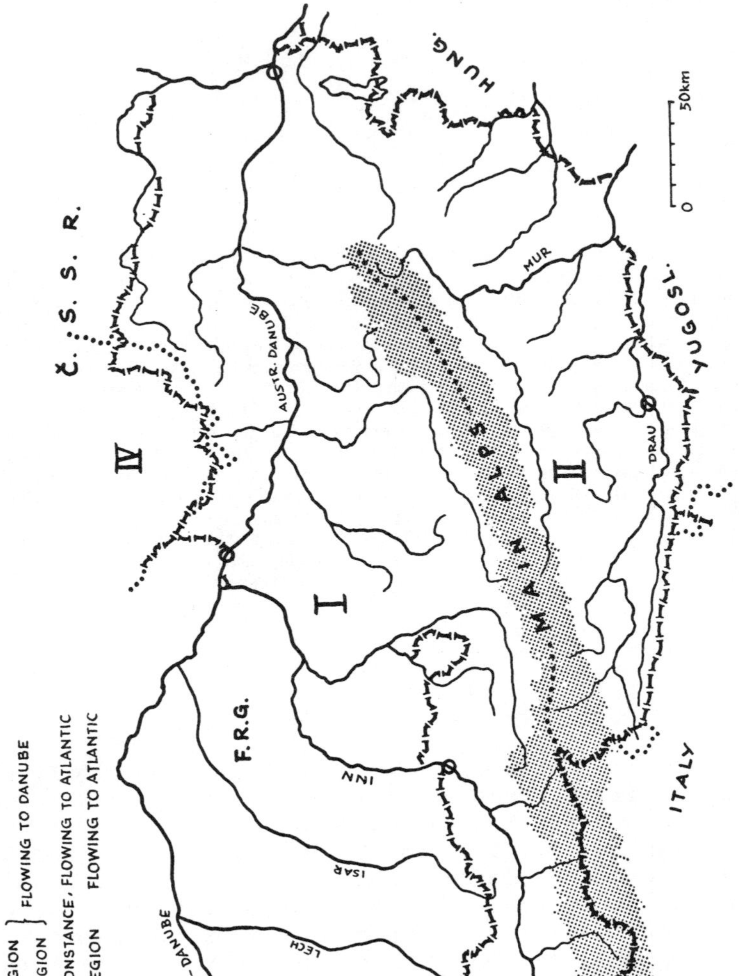

FIG. 1. Main catchment areas in Austria.

A negligible part of Austrian territory belongs (IV) to the catchment area of the River Moldau and is therefore, via the river Elbe, connected with the North Atlantic.

2. CONVENTIONAL POLLUTION

The control of water quality and pollution is, as in any other industrialized country, of crucial importance for environmental policy, health and economy. Immediately after World War II, when trade and industry were at a complete standstill, it was essential to start production of needed goods as quickly as possible. Clearly considerations of water quality and environmental protection had to be postponed. The difficult situation in the World Market in the 1950s again caused hesitation and delays in the construction of industrial and municipal treatment facilities.

Work connected with the water quality of surface waters in Austria is regularly controlled by the competent authorities of each of the nine Austrian provinces. All major Austrian rivers are periodically surveyed with the aim of issuing water quality maps for each of the nine provinces at about five-yearly intervals and keeping the general water quality map of Austria up to date. The classification of river quality is based on the well known saprobic system of Kolkwitz-Marsson-Liebmann. This method, using a limited number of indicator species, is supplemented by bacteriological and of course physico-chemical investigations. The Austrian Danube and some rivers in the border areas are inspected twice a year.

A water quality map, issued by the Federal Ministry of Agriculture and Forestry shows the development of water quality of all Austrian rivers. More detailed information can be taken from special maps of each of the nine provinces which are issued at five-yearly intervals (scale 1:200 000; small scale size DIN A4).

A special problem arises in connection with the numerous lakes in Austria which are of great importance for their recreational value and of course to the tourist industry. The enormous increase in mass tourism, with all its negative effects like insufficient sanitary equipment, concentration of housing facilities and hotels without proper sewage treatment, has made it necessary to provide sewage collectors along the lake shores and to construct central sewage treatment plants. In some cases even the usual biological treatment is insufficient. Advanced treatment is necessary in order to remove the main nutrients nitrogen and phosphorus. The selection of the most suitable sites for sewage collectors, pumping stations and treatment plants has to be considered with greatest care, using systems analysis and mathematical modelling wherever possible. Experience gained by international organizations such as WHO, IIASA and IAEA in these fields have been of great value in solving these problems.

The Austrian Danube may be considered as a characteristic example of surface water pollution. Along the Austrian Danube — and to some extent the River Inn — and the more important tributaries feeding these two rivers, one finds a large variety of communities and industrial sites such as steel mills, chemical works, pulp and paper mills, tanneries, and sugar and food factories. The effect of combined effluents from all these industries is nevertheless not critical since the comparatively high flow rate of the Danube has so far safeguarded rapid biodegradation. The developments of recent decades, however, have added another complication through the construction of hydro-electric power plants.

The slope of the Austrian Danube of 0.43‰, resulting from a height difference of about 140 m for about 300 km length, shows that even this large river has to be regarded as typically alpine in character. The flow rate at average discharge ranges between 2 to 4 m/s. The construction of hydroplants results first of all in a reduction of flow rate which again allows minerals and other suspended solids to form sediment. These sediments accumulate a wide range of micropollutants including not only persistent organic compounds but also practically all inorganic constituents.

It is well known that various radionuclides accumulate in these deposits. All the rivers coming from the main alpine system, consisting of ancient rock, carry comparatively high amounts of radionuclides. It may be noted that the natural radioactivity is used for therapeutical and balneological purposes at Gastein and a few other places.

The main aim of water quality preservation is to maintain water quality class II wherever possible or at least class II-III. Under these conditions it can be expected that self-purification and biological and physical re-oxygenation help to keep the required quality. The critical point is reached when excessive loads result in heavy oxygen consumption. When faced with significant temperature increase from meteorological conditions or by releases of thermally loaded effluents from power stations and industrial sites this self-purification capacity may no longer be capable of compensating for the oxygen demand. The resulting conversion to the anaerobic state implies in most cases a complete loss of the usual aquatic biocoenosis and fishlife. Even after the restoration of aerobic conditions it takes the water body several weeks to recover. It is therefore absolutely necessary to take all possible measures to prevent this critical deoxygenation. The change from the aerobic to the anaerobic state of a water body is accompanied not only by a number of significant biological but also of chemical processes.

The release of hydrogen sulphide by anaerobic decomposition of organic compounds leads to the precipitation of ferrous sulphide and a number of other metal sulphides which are finally found in the sediment. Many radionuclides will also be co-precipitated. This fact has to be kept in mind when the constituents of sediments have to be assessed. Fortunately there are only a few rivers in Austria where sedimentation under anaerobic conditions may occur.

Most of the conventional pollution of the Austrian Danube originates from cities like Linz with its concentration of various industrial sites like steel mills, pulp and paper mills and chemical works and from some of the major tributaries which also receive and transfer a substantial pollution load from industrial and sanitary wastes. A large amount of pollution comes from the capital Vienna, delivering the sanitary wastes of about 2 million people and a large amount of industrial and trade effluents. A central treatment plant, at present under construction, will go into operation by 1977 or 1978. A noticeable improvement in water quality is expected to occur at the point of the frontier crossing with Czechoslovakia.

In late autumn some additional pollution is produced by the wastes of several sugar mills. This usually starts in the first week of October, ending about the New Year. Some of these plants feed their wastes into the Danube or into rivers near the Austro-Hungarian or the Austro-Czechoslovakian border. Some are equipped with modern treatment facilities including closed circuits, chemical flocculation and sedimentation. Currently under study is the possible effect of these untreated or treated wastes on the alkali metal content of the receiving water bodies (^{40}K, $^{134,137}Cs$) and perhaps some other radionuclides of interest.

3. CONTROL OF RADIOACTIVE CONTAMINATION

As a consequence of the large-scale nuclear tests in the atmosphere in 1961 and thereafter, the monitoring of Austrian surface water bodies and the atmosphere for radioactive contamination has been established. The Federal Institute for Water Quality is responsible for the regular monitoring of the radioactivity of water bodies including surface water and groundwater. Samples from 16 selected points are analysed at monthly intervals.

This monitoring programme was of great value when the construction of nuclear power plants in Austria was considered. The figures so far obtained helped considerably in the determination of basic values and the calculation of maximum permissible loads for the environment of these stations. These limits have been laid down in the radiological and the water-use licences for the first power plant situated at Zwentendorf/Danube, about 40 km west and upstream of Vienna. Decisions on radiological data are based on the ICRP recommendations.

The Federal Institute for Water Quality was instructed to start an extensive examination of the radiological situation in the vicinity of Zwentendorf at the beginning of 1973. The monitoring programme includes a total of 296 samples per year, including total $\alpha + \beta$, ^{40}K, net $\alpha + \beta$ and total α-activity. Determination of tritium[1] and other specific radionuclides of interest will be included as soon as possible when the necessary equipment is available. Samples include groundwater, Danube water, rainwater, seston, water weeds (mainly Fontinalis), fish (mainly Chondrostoma nasus and Abramis brama), grass and soil (for further details see Ref.[5]).

Besides the radiological control, the thermal effects of the cooling water on the Danube have been carefully studied. The plant, with a capacity of about 700 MW, will release a maximum of 35 m^3/s of cooling water which is to be taken from the Danube and returned with a temperature rise of 13 degC. This will result in a temperature rise in the Danube of 0.7 degC under minimum flow conditions. The thermal pollution risks are considered to be almost as important as the radiological impacts. Both aspects have to be taken into account not only with Austrian stations but also for the effects of nuclear power stations in the upstream neighbouring countries (Switzerland, Federal Republic of Germany, Czechoslovakia). Special investigations have been carried out for several years in connection with the planned Swiss plant at Rüthi near the Austrian border. Basic values of groundwater and surface waters and also possible atmospheric contamination have been determined.

---

[1] These determinations are carried out at present by the Bundesversuchs- und Forschungsanstalt Arsenal, Geotechnisches Institut, Vienna.

The radiological load of the River Inn and the Austrian Danube derives from three different sources:

Natural radioactivity (base rock from the Alps)
Contamination by atmospheric fallout
Possible contamination by radionuclides released from nuclear power plants.

Two of the 16 sampling points mentioned above were chosen for the control of the River Inn at Kirchbichl before this river enters German territory and for the control of the Danube at Engelhartszell, some distance downstream of Passau where the Inn and the Danube form the Austrian Danube. Any noticeable influence from Bavarian nuclear power plants can eventually be measured in these samples. These values gain increased importance in view of the Austrian nuclear power stations being planned near these two rivers.

The results of all radiological investigations carried out by different university and government institutions are regularly published by the Federal Ministry of Health and Environmental Protection [1]. The results of radiological determinations in surface waters for the years 1960-1970 carried out by the radiological department of the Federal Institute for Water Quality have been published separately [2-4].

These reports contain the results of various radiological investigations, such as:

General environmental radioactivity
Radioactivity in foodstuffs
Atmospheric precipitation
Surface water, groundwater, rainwater tanks
Radiation exposure to man and biota.

Some typical results of radiological surface water measurements for the years 1971-1974 are given in Figs 2-4. The remaining $\alpha + \beta$ activity (^{40}K excluded) ranges round about 2 pCi/l, with some peaks up to 7-10 pCi/l.

It can be seen that all levels of radiation exposure to man and biota are far below the limits determined in the Austrian Federal Radiation Protection Law (Strahlenschutzgesetz 1969; Strahlenschutzverordnung 1972).

The monitoring of the conventional and radiological pollution of water bodies in Austria is not only of basic interest for the country itself. Since Austria is situated in the central part of Europe, all its rivers cross several European countries before they reach the Black Sea – those belonging to the Danube catchment area – or the North Sea – those flowing towards the Rhine and the Lake of Constance, or towards the Moldau.

It is not surprising that a number of international organizations expressed interest in multinational surveying and monitoring programmes in which the Austrian rivers – particularly the Danube – play an important role.

For the scientific part of these investigations the Joint Working Group for Danube Research, a sub-society of the Societas Internationalis Limnologiae, must be mentioned. The activities of this Working Group started in 1956. Scientists of all the Danube countries cooperate in 15 specialized groups, one of these dealing with radiological problems.

Programmes for the surveying and monitoring of the Danube have been suggested by WHO, IAEA, and ECE. A Convention for the Protection of

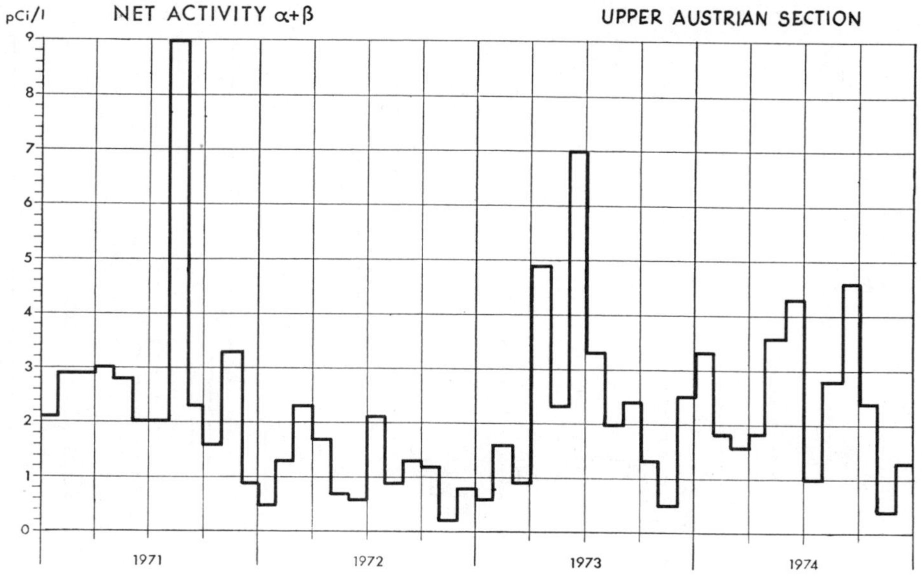

FIG. 2. Radiological measurements: Inn/Kirchbichl.

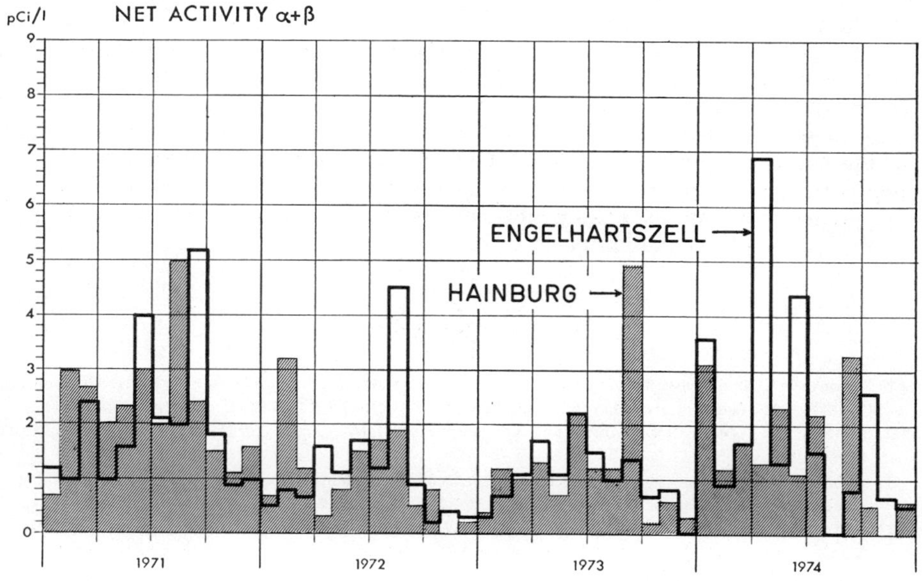

FIG. 3. Radiological measurements: Donau/Engelhartszell and Hainburg.

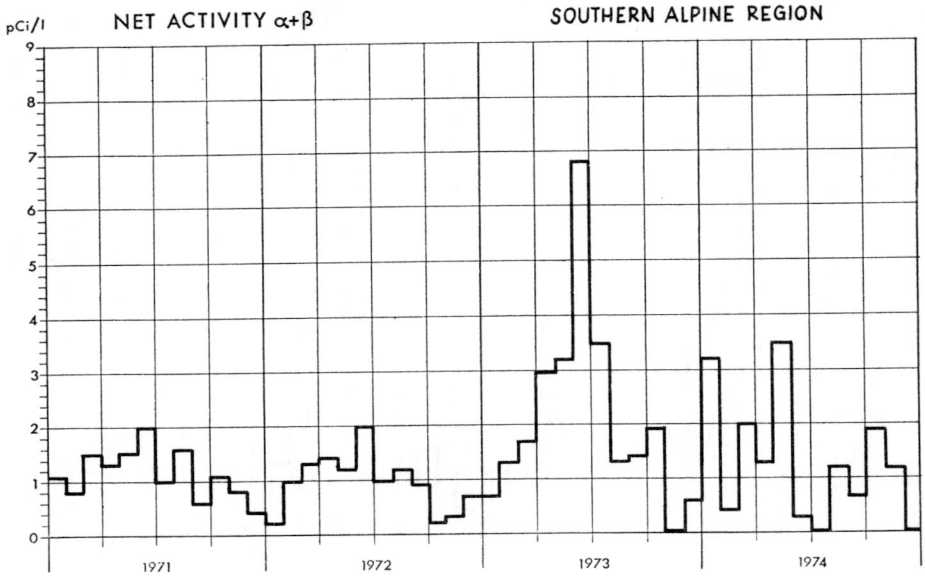

FIG. 4. Radiological measurements: Drau/Neubrücke.

International Water Courses has been prepared by the European Council and also by the European Communities. A similar initiative came from the Romanian Government.

It is obvious that an international river of the size and importance of the Danube attracts the interest of a large number of authorities and international organizations dealing with environmental problems. This may in some cases lead to duplication of effort, but in view of the importance of the numerous problems connected with environmental and radiation protection, it is certainly better to have several institutions with all their resources working on these problems than to feel that no organization really cares about the future.

REFERENCES

[1] Radioaktivitätsmessungen in Österreich, 1.-10. Jahresbericht, Bundesministerium für soziale Verwaltung (1961-1970); Bundesministerium für Gesundheit und Umweltschutz 1971-1975 (under preparation).

[2] FRANTZ, A., Die Radioaktivität in der österreichischen Donau, Archiv für Hydrobiologie, Suppl. XXX (1967) 340-363.

[3] FRANTZ, A., Die österreichischen Kernreaktoren und ihre bisherige Auswirkung auf Grund- und Oberflächenwasser, Wasser und Abwasser, Wien (1968) 121-140.

[4] FRANTZ, A., Studie über die Radioaktivität im Wasser unter besonderer Berücksichtigung österreichischer Verhältnisse. Gefahren und Möglichkeiten zur Aufrechterhaltung der Wasserversorgung, Wasser und Abwasser, Wien (1970) 9-42.

[5] FRANTZ, A., WANDERER, E., Umweltbedingungen für das erste österreichische Kernkraftwerk, Wasser und Abwasser, Wien (1974).

DISCUSSION

S. BONOTTO: Could you tell us what plants and animals are used in Austria for radioactivity surveillance?

L.J. OTTENDORFER: The following species have been used in Austria for control of radioactive contamination:

At present

Plants	Animals
Fontinalis	Fish (flesh, skeleton, skin, scales):
Cladophora	Barbus barbus
Grass	Chondrostoma nasus
Clover	Abramis brama
Turnip leaves	Salmotrutta fario
	Gobio gobio

On occasion

Plants	Animals
Cinclodotus	Hirudineae
Sphaerotilus natans	Gastropoda
	Fish: Leuciscus cephalus
	Leuciscus idus
	Leuciscus leuciscus
	Salvelinus alpinus
	Chondrostoma genei
	Sardinius erythrophth.
	Rutilus rutilus
	Tinca vulgaris

ASSESSMENT OF DOSES IN THE ENVIRONMENT FOR LIQUID RELEASES FROM NUCLEAR POWER REACTORS

J. KASTNER, J.S. BLAND
Division of Technical Review,
Office of Nuclear Reactor Regulation,
U.S. Nuclear Regulatory Commission,
Washington, D.C.,
United Stated of America

Presented by D.N. Edgington

Abstract

ASSESSMENT OF DOSES IN THE ENVIRONMENT FOR LIQUID RELEASES FROM NUCLEAR POWER REACTORS.

In preparing an Environmental Impact Statement for a commercial power reactor, it is a requirement that the potential doses in the environment from the liquid releases be assessed. The fundamental equation for the calculations of this assessment is $R_{ipr} = C_{ip}U_pD_{ipr}$ where R_{ipr} is the dose rate to organ r from nuclide i via pathway p, C_{ip} is the concentration of nuclide i in the media of pathway p, U_p is the usage associated with pathway p, and D_{ipr} is the dose factor of nuclide i, pathway p, and organ r. A comparison of the projected potential doses, as calculated in their Environmental Impact Statements, and the potential doses, based on the actual 1973 radioactive releases from operating plants, shows that the majority of the LWRs have kept their off-site doses well within 'Appendix I' design objectives.

1. INTRODUCTION

The preparation of an Environmental Impact Statement prior to the licensing of a commercial power reactor requires the assessment of the radiological impact of the proposed facility. Detailed evaluations are presented of the impact in the public domain of liquid releases to the environment, including pertinent exposure pathways and probable exposures of individuals and population groups. The method of assessment is the same for any facility; only the details are different.

The prime determination is the radiation exposure to man which arises from the ingestion and external presence of various radionuclides. Both individual and population doses are evaluated and predicted in the impact statement.

In the development of the radiological impact statement we start with the facility itself, work through the environmental population factors operating between the points of release and the points of public exposure, and then consider the potential radiation doses to the public.

An example will be given of the evaluation of dose for the liquid releases from boiling and pressurized water reactors. The pathways considered include those of drinking water, dissolved noble gases, swimming, investion of fish, crustacea, etc. Bioaccumulation factors for fresh and salt water are applied in the calculation of dose to man via ingestion of aquatic foods.

TABLE I. CALCULATED POTENTIAL DOSES

Reactor type	Principal ingestion	Predicted releases		Measured releases	
		Principal organ	Dose (mrem/a)	Dose (mrem/a)	Major nuclide
PWR	Fish	Body	2.0	1.3	Cs-137 98%
PWR	Fish	Body	0.044	0.0022	Mn-54 95%
PWR	Water	Thyroid	0.25	0.00012	H-3 97%
PWR	Water	Thyroid	5.6	0.48	I-131 94%
PWR	Seafood	Thyroid	10	0.0013	I-131 98%
BWR	Seafood	GI-LLI	5.8	0.3	Co-58 74%
PWR	Fish	Body	0.77	0.41	Cs-134 54% Cs-137 43%
BWR[a]	Seafood	Thyroid	70	0.4	I-131 99%
BWR[a]	Seafood	Thyroid	1.2	4.6	I-131 90%
PWR	Fish	Body	0.12	0.086	Cs-134 46% Cs-137 53%

[a] Radwaste system to be augmented.

To test our predictions we have evaluated the doses based on measured liquid releases from about forty plants. It is demonstrated that the majority of light water reactors keep off-site doses well within their design criteria (Table I).

2. FUNDAMENTAL EQUATION FOR CALCULATION OF RADIATION DOSE

Equation (1) is the fundamental equation for the calculation of radiation dose[1] to man via liquid effluent pathways.

$$R_{ipr} = C_{ip} U_p D_{ipr} \qquad (1)$$

where

R_{ipr} = dose rate to organ r from nuclide i via pathway p (mrem/a)

C_{ip} = concentration of nuclide i in the media of pathway p (pCi/kg)

U_p = usage, the exposure rate or intake rate associated with pathway p (h/a or kg/a, [or l/a], as appropriate)

D_{ipr} = dose factor: a value specific to a given radionuclide, pathway, and organ which can be used to calculate radiation doses from the intake of a radionuclide (mrem/pCi intake) or from exposure to a given concentration of a radionuclide in water (mrem/h per pCi/l)

[1] The term 'dose' used here and throughout this paper means the 'equilibrium dose', which is equal to the 'dose commitment' for most radionuclides. For cases where the effective half-life is long and an equilibrium condition dose does not occur, the annual dose commitment is the dose to the individual received over his lifetime as a consequence of the intake during the first year.

The three factors comprising Eq. (1) are discussed in the following subsections, almost all of which were taken directly from WASH-1258 [1]. (An updated version of the portion of WASH-1258 describing models and computer programs is contained in BNWL-1754 [2]).

Concentrations in environmental media, C_{ip}

The concentrations C_{ip} in environmental media of interest at the points where calculations should be made, need not be evaluated explicitly; rather, C_{ip} is estimated from the mixing ratio M_p, the bioaccumulation factor B_{ip}, the radionuclide release rate Q_i, and other terms presented in the pathway equations that appear later in this discussion.

Usages, U_p

The second term of Eq. (1), the usage term U_p, is, as indicated above, expressed as a consumption rate (kg/a or l/a) or as an exposure time (h/a), as appropriate for the pathway under consideration.

The NCR staff encourages the use of site-specific data[2] whenever possible, provided that such data are sufficiently documented.

Dose factors, D_{ipr}

Equations for calculating internal dose factors are derived from those given by the ICRP [3] for body burden and MPC. Effective adsorbed energies for the radionuclides are calculated from the ICRP model. Where data are lacking, metabolic parameters for the Standard Man are used for other ages as well.

The dose factors for external exposure were derived on the assumption that the contaminated medium is large enough to be considered an 'infinite volume' relative to the range of the emitted radiations. Under this assumption, the energy emitted per gram of medium is equivalent to the energy absorbed per gram of medium. One then merely converts MeV per disintegration per gram to rem and corrects for the differences in energy absorption between air or water and tissue, and for physical geometry of each specific exposure situation.

The dose from submersion in water is an external dose to either the skin only, or both the skin and total body, depending upon the penetrating power of the radiation emitted by the waterborne radionuclides. Only beta and gamma radiation that could penetrate 7 mg/cm^2 of tissue is considered in calculating skin dose. Gamma radiation dose at 5-cm depth in tissue is used for calculating external dose to the total body (and for internal organs). These dose factors have units of mrem/h per pCi/l water.

Material deposited from sedimentation in an aquatic system or from irrigation water onto the ground represents a fairly large, nearly uniform,

---

[2] In selecting usage values not only the present land and water uses should be considered, but also changes in land and water uses made possible by such activities as chemical pollution abatement. Radioactive material released into waterways may include long-lived radionuclides that have potential for build-up in sediments and biota, causing their persistence for many years — perhaps beyond the lifetime of the nuclear power station. It is important that long-term use of a waterway not be restricted by the discharge of radioactive material for LWR stations.

thin sheet of contamination. The factors for converting surface contamination in pCi/m^2 to gamma dose rate at one metre above a uniformly contaminated plane have been described by Soldat and others [4-6]. Dose factors for exposure to soil sediment have units of mrem/h per pCi/m^2 surface.

A set of dose factors for 45 radionuclides was originally calculated for the Year 2000 model [5]. These have since been recalculated using recent decay scheme information [7] and the list was expanded to include 115 nuclides. The revised list is given in BNWL-1754 [2] and contains several radionuclides for which the daughter is not listed separately (e.g. Ru-Rh-106, Cs-137-Ba-137m, Ce-Pr-144). In those instances, the daughter's decay energy has been included in the parent for the purpose of calculating the dose factor.

3. EQUATIONS FOR LIQUID PATHWAYS

This section develops the set of equations required for the liquid pathway model. The principal difference between pathways is the manner in which the radionuclide concentrations are calculated. The doses from the five pathways should be added to determine the total dose. To meet the 'as low as practicable' design objective annual dose of 5 mrem, the total dose R_r to any organ r associated with these liquid pathways p should not exceed 5 mrem/a:

$$R_r = \sum_{p=1}^{5} R_{pr} \leq 5 \text{ mrem/a}$$

(a) Potable water

The dose rate from ingestion of water is calculated from the following equation:

$$R_{pr} = 1119 \sum_{i=1}^{115} \frac{Q_i M_p}{F} \exp(-\lambda_i t_p) \, U_p D_{ipr} \tag{2}$$

Symbols for this equation are defined at the end of Section 3.

The summation process adds the dose contribution from each of the 115 nuclides that appear in the dose factor library, yielding the total dose for the pathway-organ combination selected.

The term Q_i/F in Eq.(2) defines the concentration of nuclide i in the effluent at the point of discharge. The expression $(Q_i M_p/F)\exp(-\lambda_i t_p)$ yields the concentration at the time the water is consumed. This latter concentration is the term C_{ip} in Eq.(1). The transit time t_p is set equal to 12 hours to allow for radionuclide travel from the point of release to the water plant intake (assumed to be located immediately downstream), as well as to allow for travel time through the water plant and its associated distribution system. No credit is given for radionuclide removal by water treatment plants except in instances where sufficient documentation is available.

It should be noted that, depending on the hydrological dispersion model employed, the mixing ratio M_p or dilution factor may not be explicitly defined. In those instances (e.g. build-up of activity in a cooling pond), the relative concentration in the mixed stream (compared to the effluent concentration) may be supplied as a function of radiological decay constant, with any potential effluent recycling taken into account.

(b) Aquatic foods

The concentrations of radionuclides in aquatic foods are directly related to the concentrations of the nuclides in water. Equilibrium ratios between the two concentrations, called bioaccumulation factors in this report, can be found in the literature [8, 9]. The inclusion of the bioaccumulation factor B_{ip} in Eq.(2) yields Eq.(3), which is suitable for calculation of internal dose from consumption of aquatic foods.

$$R_{pr} = 1119 \frac{U_p M_p}{F} \sum_{i=1}^{115} Q_i B_{ip} D_{ipr} \exp(-\lambda_i t_p) \tag{3}$$

The transit time t_p is set equal to 24 hours to allow for radionuclide travel through the food chain, as well as to allow for food preparation time.

(c) Shoreline deposits

The calculation of dose rate from exposure to contaminated shoreline involves two additional steps. The accumulation of radionuclides in the sediment must be estimated. The width of the contaminated shoreline must be taken into account.

The calculation of sediment load, transport, and concentrations of radionuclides associated with suspended and deposited materials is complex. One method of addressing this problem was used in the Year 2000 Study [5]. A simplified scheme for obtaining an order-of-magnitude estimate of the concentration of shoreline sediments was required for the dose calculation scheme presented herein. Eqs (4) and (5) may be used for estimating such contamination levels.

$$C_{is} = K_c \frac{C_{iw} [1 - \exp(-\lambda_i t)]}{\lambda_i} \tag{4}$$

where

C_{is} = concentration of nuclide i in sediment (pCi/kg)

C_{iw} = concentration of the nuclide in water adjacent to the sediment (pCi/l)

λ_i = decay constant of the nuclide[3]

K_c = an assumed constant in units of litres per kilogram per day

t = length of time the sediment is exposed to the contaminated water, nominally taken to be the operating lifetime of the facility.

In the original evaluation of the equation, λ_i was chosen to be the radiological decay constant, but the true value should include an 'environmental' removal constant. The relationship was tested and the value of K_c was derived for several radionuclides by utilizing data on water and sediment samples collected over a period of several years in the Columbia

[3] If the presence of a radionuclide in water and sediment is controlled primarily by radioactive equilibrium with its parent nuclide, then the water concentration and half-life of the parent should be used in Eqs (4) and (5).

River between Richland, Washington, and the river mouth and in Tillamook Bay, Oregon, 75 km south of the river mouth [10, 11]. Since the primary use of the equation is to facilitate estimates of the exposure rate from gamma emitters one metre above the sediment, an effective surface contamination was estimated. This surface contamination was taken to be all the nuclides contained within the top 2.5 cm (1 in.) of sediment.[4] The dose contribution from the radionuclides at depths below 2.5 cm was ignored.

The resulting equation is as follows:

$$S_i = 100 T_i C_{iw} W \left[1 - \exp(-\lambda_i t)\right] \quad (5)$$

where

S_i = 'effective' surface contamination (pCi/m^2) to be used in subsequent calculations

T_i = radiological half-life of nuclide i (days)

W = shore width factor

Shore width factors were derived from data given in Fig. 3.1(5) of Ref. [12].

The combination of Eqs (4) and (5) into the general Eq. (1) leads to the following equation for calculating the radiation dose from exposure to shoreline sediments:

$$R_{pr} = U_p \sum_{i=1}^{115} S_i D_{ipr} = 100\, U_p W \sum_{i=1}^{115} C_{iw}\, T_i\, D_{ipr} \left[1 - \exp(-\lambda_i t)\right] \quad (6)$$

$$= 111\,900 \frac{U_p M_p W}{F} \sum_{i=1}^{115} Q_1\, T_i\, D_{ipr} \exp(-\lambda_i t_p) \left[1 - \exp(-\lambda_i t)\right] \quad (7)$$

(d) Swimming and boating

The equation for calculating the external dose to the skin and total body from swimming (water immersion) or boating (water surface) is as follows:

$$R_{pr} = 1119 \frac{U_p M_p}{F K_p} \sum_{i=1}^{115} Q_1 D_{ipr} \exp(-\lambda_i t_p) \quad (8)$$

where K_p is 1 for swimming and 2 for boating (4π geometry for swimming and 2π geometry for boating).

Definition of symbols used in Section 3

R_{pr} = total dose rate to organ r from all the nuclides i in pathway p (mrem/a)

Q_i = release rate of nuclide i (Ci/a)

F = flow rate of the liquid effluent (ft^3/s)

t_p = transit time required for nuclides to reach the point of exposure. For internal dose, t_p is the total time elapsed between release of the nuclides and ingestion of food or water (h)

[4] With a mass of 40 kg/m^2 of surface.

U_p = usage, the exposure rate or intake rate associated with pathway p (h/a or kg/a as appropriate)

M_p = mixing ratio (reciprocal of dilution factor) at the point of exposure (or the point of withdrawal of drinking water or point of harvest of aquatic food) (unitless)

λ_i = radioactive decay constant of nuclide i (h^{-1})

1.1×10^3 = a constant that converts from (Ci/a)/(ft^3/s) to pCi/l (multiplied by an additional factor of 100 for shoreline sediments)

B_{ip} = bioaccumulation factor for nuclide in pathway p (pCi/kg per pCi/l)

D_{ipr} = dose factor: a number specific to a given radionuclide, pathway, and organ which can be used to calculate radiation doses from the intake of a radionuclide (mrem/pCi intake) or from exposure to a given concentration of a radionuclide in water (mrem/h per pCi/l)

W = shoreline width factor, dimensionless

T_i = radioactive half-life of nuclide (days)

t = period of time in which sediment is exposed to the contaminated water, nominally taken to be the operating lifetime of the facility (h)

K_p = geometry factor, dimensionless, for swimming and boating

n = number of radionuclides in dose factor library (at present, n = 115)

The equations given yield the dose rate to various organs of an adult individual from the exposure pathways mentioned. Appendix I of 10 CFR Part 50 requires [13] that the annual dose or dose commitment to any organ of an individual from the sum of the exposure pathways from liquid effluents should not exceed 5 mrem.

4. APPLICATION TO OPERATING POWER PLANTS

Liquid releases

An example of the application of these methods to operating power plants can be given by Dresden 1.

The radioactive discharge to water from Dresden 1, at the point where someone could drink the water, gave a concentration of 2.5×10^{-8} μCi/ml with a nuclide distribution of both fission and activation products.

Dresden also released 1.5×10^{-8} $\mu Ci/cm^3$ of ^3H. Now, for the mix of fission and activation products, not including ^3H, one can calculate a dose factor of 2×10^7 (mrem/a)/($\mu Ci/cm^3$).

This means, if one should drink this water at the normal rate of 1200 cm^3/d for a year, the whole body dose will be $2.5 \times 10^{-8} \times 2 \times 10^7$ = 0.5 mrem/a. Similarly, for tritium the dose factor is about 10^5 (mrem/a)/($\mu Ci/cm^3$); thus, $1.5 \times 10^{-8} \times 1 \times 10^5 = 1.5 \times 10^{-3}$ mrem/a.

The authors have also considered the impact of noble gases dissolved in liquid effluents. Xenon-133 is the predominant isotope found, and the maximum annual average concentration measured in reactor effluents has been about 4×10^{-8} pCi/cm^3. For this source the immersion dose rate, rad/h, is approximately given by 2 EC, where E = energy (in MeV/dis) = 0.65, and C = concentration (in μCi/g). Thus, one expects about 5×10^{-5} mrem/h.

For 100 hours of swimming, the dose is 5×10^{-3} mrem and the dose to an aquatic organism living in the water for a year is less than 0.5 mrad. Hence we conclude that the contribution of this source to radiological impact is small indeed.

Actually the average volume of water discharged from the plant and the river at that point should be considered in determining the effective dilution and, thus, the correct concentration to be used in the dose calculations. A figure of man·rems can be derived by multiplying the dose value, including effect of dilution, by the number of people drinking untreated river water at various points.

Another aspect should be considered in the case of activity discharged to water, and that is bioaccumulation in fish. Bioaccumulation factors in rivers have been observed to be as high as 1000x for caesium, with comparable values for cobalt, iron, zinc, and manganese. For the Dresden mix, fresh fish would provide a whole body dose factor of about 10^8 (mrem/a)/(μCi/cm^3) for nuclides other than ^{3}H. The national daily intake of fresh fish is assumed to be 20 g/day.

In some other areas of the country one would have to consider molluscs and crustacea which have different bioaccumulation factors. For fish it might be better to consider the total catch and estimate the man·rem resulting from it.

Estimates also have been made of the dose rates to individuals and total dose rates to population for cities using rivers for drinking water supply — on the assumption that 5 Ci/a are released of fission and activation products and 1000 Ci of ^{3}H. For example, for Minneapolis-St. Paul, with Mississippi flow rate of 5000 ft^3/s and 10^6 population, the dose rate estimate would be 0.1 mrem/a to the individual and therefore about 100 man·rem/a. Natural background = $0.1 \times 10^6 = 10^5$ man·rem/a.

Analysis of experience in the prediction of impact

Although the details of the impact (both doses and exposure pathways) are functions of the site, the maximum envelope of predicted impacts is not only insignificant when compared with natural background radiation but is generally well below the guidelines of 'As Low As Practicable' (ALAP) objectives of proposed Appendix I to 10 CFR 50 (Ref.[13]), the dose criteria against which judgement of the effectiveness of radwaste systems is made; this is presented in Ref. [14], which is the result of the public hearing connected with Ref. [1]. It can be said that, for the most part, normal radioactive releases are not critical factors in siting. On the other hand, determination of critical, site-dependent exposure pathways does provide the bases for effective environmental monitoring.

Table I compares some predicted doses to man with those calculated on the basis of measured liquid releases. These selected reactors were the only ones to release significant quantities of radioactive effluents during 1973.

In conclusion, it can be stated that the majority of LWRs have kept their off-site doses well within Appendix I design objectives. In several cases the radionuclide distribution was different from that predicted but in no case was there an actual significant impact on man or biota in the environment.

REFERENCES

[1] USAEC, Final Environmental Statement Concerning Proposed Rule Making Action: Numerical Guides for Design Objectives and Limiting Conditions for Operation to Meet the Criterion 'As Low As Practicable' for Radioactive Material in Light-Water-Cooled Power Reactor Effluents, USAEC Report No. WASH-1258 (July, 1973).

[2] SOLDAT, J.K., ROBINSON, N.M., BAKER, D.A., Models and Computer Codes for Evaluating Environmental Radiation Doses, Pacific Northwest Laboratories, USAEC Report No. BNWL-1754 (in press).

[3] ICRP, Report of ICRP Committee II on Permissible Dose for Internal Radiation, International Commission on Radiological Protection, ICRP Publication 2, Pergamon Press (1959).

[4] SOLDAT, J.K., Modeling of Environmental Pathways and Radiation Doses from Nuclear Facilities, Pacific Northwest Laboratories, USAEC Report No. BNWL-SA-3939 (1971).

[5] FLETCHER, J.F., DOTSON, W.L., (compilers), HERMES — A Digital Computer Code for Estimating Regional Radiological Effects From The Nuclear Power Industry, Hanford Engineering Development Laboratory, USAEC Report No. HEDL-TME-71-168 (1971).

[6] SOLDAT, J.K., "Conversion of survey meter readings to concentration ($\mu Ci/m^2$)", Item 04.3.4 in: Emergency Radiological Plans and Procedures, (K.R. HEID, Ed.), Hanford Laboratories, USAEC Report No. HW-70935 (1962).

[7] LEDERER, C.M., HOLLANDER, J.M., PEARLMAN, I., Table of Isotopes, 6th Edn, John Wiley and Sons Inc. (1967).

[8] THOMPSON, S.E., BURTON, C.A., QUINN, D.J., NG, Y.C., Concentration Factors of Chemical Elements in Edible Aquatic Organisms, Lawrence Radiation Laboratory, USAEC Report No. UCRL-50564, Rev. 1 (Oct. 1972).

[9] FREKE, A.M., A model for the approximate calculation of safe rates of discharge of radioactive wastes into marine environments, Health Phys. 13 (1967) 743.

[10] NELSON, J.L., "Distribution of sediments and associated radionuclides in the Columbia River below Hanford", p. 3.80 in: Hanford Radiological Sciences Research and Development Annual Report for 1964 (PEARCE, D.W., GREEN, J.K., Eds.), Pacific Northwest Laboratories, USAEC Report No. BNWL-36 (1965).

[11] TOOMBS, G.L., CULTER, P.B. (compilers), Comprehensive Final Report for the Lower Columbia River Environmental Survey in Oregon, June 5, 1961 - July 31, 1967, Oregon State Board of Health, Div. of Sanitation and Engineering (1968).

[12] Handbook of Radiological Protection, Part I: Data, prepared by a panel of the Radioactivity Advisory Committee (DUMSTER, H.J., Chairman), Dept. of Employment, Dept. of Health and Social Security, Min. of Health andSocial Services, Northern Ireland, SNB 11 360079 8, Her Majesty's Stationery Office, London (1971).

[13] Code of Federal Regulations CFR 10, Energy, Part 50, Appendix I, published by: Office of Federal Register, National Archives + Records Service, General Services Administration, USA (1974 revised).

[14] Concluding Statement of Position of the Regulatory Staff. Public Rulemaking Hearing on: Numerical Guides for Design Objectives and Limiting Conditions for Operation to Meet the Criteria 'As Low As Practicable' for Radioactive Material in Light-Water-Cooled Nuclear Power Reactors, Docket No. RM-50-2, February 20, 1974.

DISCUSSION

The replies by Mr. Kastner were obtained after the meeting.

P. BEAU: What is the relative importance of the different irradiation pathways in most cases? Are the boating and swimming pathways not generally negligible?

J. KASTNER: The relative importance of the pathways depends mainly on two factors: (1) the radionuclides released, and (2) the aquatic bio-accumulation factors of these radionuclides.

Because of the bioaccumulation factors, the ingestion of aquatic foods is usually the critical pathway. An exception to this, depending on the mixture and quantity of radionuclides released, is when drinking water (fresh water sites only) is the critical pathway.

The major nuclides that are important in this pathway are ^{3}H and ^{131}I. The unity aquatic bioaccumulation factor for H and the relatively low bioaccumulation for I, in addition to the larger individual consumption rate of water, sometimes makes drinking water the critical pathway. Generally, the swimming and boating pathways are insignificant.

F.O. HOFFMAN: The paper quotes a value of 20 g/d for the national average fish consumption. I assume that this value is used for calculating collective doses. Is it also used for calculating the dose to the individual or is some other value used? If another value is used, could you say what it is and how it was derived?

J. KASTNER: The 20 g/d national average fish consumption is used only when site specific data are not available. In most instances the population doses would be evaluated utilizing the actual fish harvest in the receiving water body. When such information is not available, then the 20 g/d average consumption is used with an appropriate population (site-specific).

In the case of a hypothetical maximum exposed individual, a consumption rate of 50 g/d is used. This consumption rate was derived by BNWL on the basis of a regional population survey (Ref. [5] of the paper).

J.A. HETHERINGTON: In connection with the question about the scale of the doses which may be received as a consequence of swimming and boating in water receiving nuclear waste, it may be useful to provide some information from an actual situation. If one considers the ^{137}Cs, for example, being discharged to the north-east Irish Sea, one can show that the annual whole-body dose to people who bathe in the sea regularly throughout the year would be of the order of a few microrems. This may be compared with the dose received by people eating fish which have reconcentrated ^{137}Cs; for a consumption rate of 250 g/day this dose would be of the order of 100 millirem/year. Similar calculations for other nuclides which contribute to the external dose to bathers give similar results, and it may be concluded that for all circumstances the doses from radioactivity retained in the water will be of minor importance compared with those pathways which involve a reconcentration of activity. The only possible exception is tritium. This nuclide is not reconcentrated by either biota or sediments and the activity in the water phase may have a radiological significance comparable with that reaching the food chains. Even in this case, however, it is found that the important exposure pathway is the inhalation of tritiated water vapour by the inhabitants of coastal communities and not the immersion dose to bathers.

IAEA-SM-198/51

BIOCHEMICAL AND PHYSIOLOGICAL CHANGES IN EGYPTIAN NILE FISH SUBJECTED TO VARYING LEVELS OF GAMMA IRRADIATION

H.M. ROUSHDY, H.S. EL-KASHEF
National Centre for Radiation Technology,
Atomic Energy Authority

A.E. IMAM
Barrage Institute of Inland Water and Fish Culture,
Cairo, Egypt

Abstract

BIOCHEMICAL AND PHYSIOLOGICAL CHANGES IN EGYPTIAN NILE FISH SUBJECTED TO VARYING LEVELS OF GAMMA IRRADIATION.

Radiation is nowadays to be considered as a new parameter in the ecology of water masses. Aquatic organisms, perhaps more than any other group of organisms, are directly exposed to radiation hazard and may be subjected to continuous low-level exposure from bottom material and from internal sources accumulated within their own bodies, originating partly from radionuclides released from nuclear facilities into the aquatic environment. In recent years, a large number of papers have been published on the uptake, concentration and release of radioactive material by aquatic organisms. However, radiation experiments on fish, a major source of food for human consumption,are still very rare and mostly restricted to studies on the effect of irradiation on eggs and larvae.

Since the study of the radiation effect on living aquatic organisms, particularly fish, is important in connection with the problems of preserving water resources for the benefit of mankind, the work presented here has been done to ascertain the effect of varying dose-levels of gamma irradiation on two common Egyptian Nile fish species, the catfish *Clarias lazera* and the *Tilapia nilotica*. Investigations carried out on *Clarias lazera* involved blood and muscle analyses as well as growth rate measurements. The results obtained showed impaired haematological levels, changes in weight of muscle proteins and, chiefly, retardation in growth rate. Investigations carried out on *Tilapia nilotica* revealed changes in the activity of certain digestive enzyme systems, glucose level in blood and concentration of the glycogen store in liver and muscles. In discussing the results obtained the authors have taken the relevant literature into consideration.

INTRODUCTION

In the water environment there is a natural background radioactivity originating from dissolved radionuclides that are added to the water from both the atmosphere and the land. In addition there is an artificial radioactivity that can be attributed to radioactive material produced by nuclear explosions and to radioactive waste of all kinds discharged into the water masses from land-based installations or by nuclear-powered ships. The problem of devising safe means of disposing of radioactive wastes has long concerned the attention of many authors but no completely satisfactory solution has yet been found.

A large number of publications exist on the uptake, concentration, retention and release of radioactive material by aquatic organisms, aiming at evaluation of the possible hazards resulting from the uptake of radioactivity by human beings, through the food chain, from aquatic food products. Because fish are particularly important as a source of food for human consumption, the physiology of the uptake, retention and excretion of radionuclides has been studied more closely in fish than in any other aquatic organisms.

The following radioisotopes have been recorded in fish contaminated by artificial radioactivity induced into the environment by nuclear explosions: ^{89}Sr, ^{90}Sr, ^{90}Y, ^{95}Zr, ^{95}Nb, ^{106}Ru, ^{113}Cd, ^{115}Cd, ^{137}Cs, ^{140}Ba, ^{140}La, ^{144}Ce, ^{54}Mn, ^{55}Fe, ^{59}Fe and ^{65}Zn [1]. In biological studies of the nuclear test sites on the Bikini and Eniwetok atolls [2], it was noticed that among herbivorous, omnivorous and carnivorous fish, the first group always had the greatest amount of radioactivity and the minimum concentrations were found among the last group.

Aquatic organisms, perhaps more than any other group of organisms, are directly exposed to radiation hazard and may be subjected to long-continued exposure: from radionuclides contained in the water environment, from cosmic rays, from bottom material and finally from internal sources accumulated within their own body.

Great attention has been given in recent years to the possible damaging effects of low concentration levels of added radioactivity in the water environment on the development and survival of fish eggs and larvae [3-6]. On the other hand, the effect on fish of low-level radiation has received very little attention and irradiation experiments on fish are still exceedingly rare owing, in part, to the difficulties encountered in experimentation.

Investigation of the radiation effects on living aquatic organisms, particularly fish, is extremely important in connection with the problems of preserving water resources for the benefit of mankind. As a result of irradiation, the fish population of contaminated streams may suffer deleterious effects, manifested by significant biological changes, e.g. shortened life-span, slower growth rate and possibly decreased fertility of the breeding stock [7].

The effects of ultrasonic waves can in many ways be considered comparable. A number of studies have been carried out to determine the effect of ultrasonic irradiation on living aquatic organisms, their various organs, structures and tissues. The findings aroused so much interest that published work on these lines has more than doubled over the past ten years. In addition to the harmful effects, exposure to ultrasound has been shown to be capable of accelerating certain processes, e.g. growth rate in *Leistes reticulatus* and apparent precocity in ovarian development [8].

In a review of literature on radiation effects on marine animals [9], it was found that, in general, the lower and more primitive forms were more resistant to ionizing radiation than the more evolved species.

The Nile is known to originate from rocky areas, forcing its way along thousands of kilometres of various types of land formations. Through the possible erosion of rocks and sediments containing radioactive isotopes, there is always the possibility of water contamination by different radionuclides. Furthermore, naturally occurring radioactive deposits of dark sand, rich in radioactive thorium, are known to be abundant in the northern parts of Egypt around the Nile estuaries near Damietta and Rosetta. The Nile water, being rich in clay content and phytoplankton population, is expected to be able to fix considerable quantities of radionuclides which may come to be introduced into the Nile stream.

Attempts have been made to study the effect of whole-body gamma irradiation on different Nile fish species, and some of the data have already been published [10-14]. Nevertheless, the authors feel that the information so far obtained is somewhat incomplete and fragmentary. Moreover, the syndrome involved in the production of radiation damage in fish still awaits further investigation.

The aim of the present paper is to survey the work carried out so far by the authors to determine the effect of whole-body gamma irradiation of two species of fresh water fish common in Egyptian inland waters.

MATERIAL AND METHODS

In this work the carnivorous catfish Clarias lazera, Cuvier & Valenciennes, and the herbivorous fish Tilapia nilotica were chosen as representative examples of the common species of Egyptian Nile fish. The fish were collected from the experimental ponds of the Barrage Institute of Inland Water and Fish Culture and reared in large glass aquaria maintaining their oxygen demands and food requirements.

Clarias lazera of two different ages were placed under experimentation; the juveniles, 12-14 cm in average length and less than one year old, and the adults, 29-34 cm in average length and about 2 years old. Fish received whole-body irradiation at the exposure levels of 100, 200 and 400 R at the respective dose rates of 5, 10 and 20 R/min using a 230-Ci cobalt-60 'Agromat' Gamma Irradiation Facility.

Tilapia nilotica on the other hand, 15 cm in average length, were irradiated at the exposure levels of 500 and 1000 R at a dose rate of 106 R/s using a 2500-Ci cobalt-60 indoor facility type Gammacell-220.

After varying intervals of post-irradiation time, the reared fish were taken for experimentation.

Investigations carried out on Clarias lazera involved growth-rate measurements besides blood and muscle analyses. Measurements included periodical length and weight determinations, and the weight/length ratio was calculated for comparative evaluation. Blood analyses included erythrocytic count and evaluation of haemoglobin concentration, haematocrit value and creatinine concentration. Muscle analyses on the other hand involved determinations of total protein value and water content.

Blood was collected by severing the caudal peduncle. Potassium oxalate was used as anticoagulant. Haemoglobin concentration was evaluated by using the cyanomethaemoglobin spectrophotometric method after Larsen [15]. Haematocrit values were determined by the Wintrobe method [16]. Creatinine concentration in blood was evaluated by using the Jaffé reaction and protein-free filtrate prepared according to Folin and Wu [17] and noting the modification made by Haden [18].

For muscle analyses, the fish was rapidly dissected and the muscles were carefully removed and investigated for protein value and water content. For muscle protein determination, the Kjeldahl method was used as described by Kirk [19].

Work carried out on Tilapia nilotica included investigations of the sucroclastic enzyme activity in intestinal and hepatopancreatic tissue homogenates, the glucose level in blood as well as the stored glycogen and free reducing sugars in liver and muscles.

For determining the hydrolytic activity of sucroclastic enzymes, tissue homogenates containing the enzymes were allowed to act on their specific substrates for 30 min at 37°C in a total reactive volume of 2.25 ml and at optimal pH conditions. The increase in the free reducing sugars, taken as a measure of enzyme activity, was determined by the method of Nelson [20] and Somogyi [21]. The developed colour was read on a Beckman spectrophotometer model DU 2400.

For the estimation of blood sugar concentration, blood was collected via the opercular lid through heart puncturing. The method of Nelson and Somogyi was used with slight modification. Glycogen store in liver was determined by using the method of Kemp and Kitis [22]. Free reducing sugars in liver and muscles were determined in deproteinized tissue extracts by the method of Mendle and Mayers [23]. The values for sugar content were calculated as mg glucose per 100 mg tissue.

RESULTS

The data obtained are, for convenience, collectively presented in graphs. Comments on the results will be restricted mainly to whether they are significantly different from the normal average values or not.

(a) Haematological levels

As shown in Fig. 1, whole-body gamma irradiation of *Clarias lazera* causes a significant drop in erythrocytic count, being more intense after the higher radiation dose level. Adult males were on the average slightly more affected than the females. Restoration of average control values takes place towards the fourth post-experimentation week except for the higher radiation level of 400 R, when the condition of prolonged anaemia persists until the end of the 10-week experimental period.

The patterns of the change in haemoglobin concentration as shown in Fig. 2, and of the change in haematocrit values as shown in Fig. 3, seem to parallel, in general, the trend described above with the erythrocytic count.

(b) Creatinine level in blood

The experiments showed no significant change in blood creatinine level resulting from irradiation apart from an initial transitory rise with the higher radiation dose levels (Fig. 4).

(c) Protein muscles

As shown in Fig. 5, whole-body irradiation of *Clarias lazera* causes a slight increase in the percentage protein dry weight of muscle. The data in the figure are taken from adults, but other experiments showed that the increase in the percentage protein of muscle is more pronounced in juveniles irradiated at the lower levels of 100 and 200 R; the higher dose level of 400 R on the other hand caused significant decrease in this percentage for both adults and juveniles.

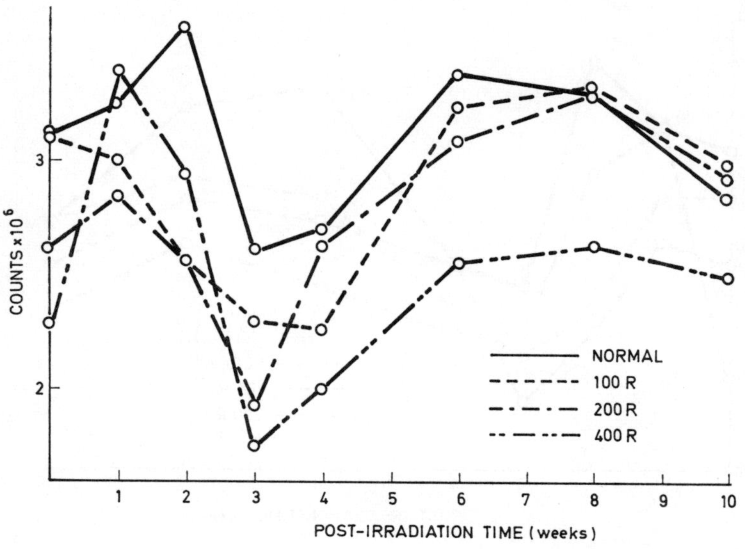

FIG. 1. Clarias lazera: changes in erythrocytic counts after different radiation exposures.

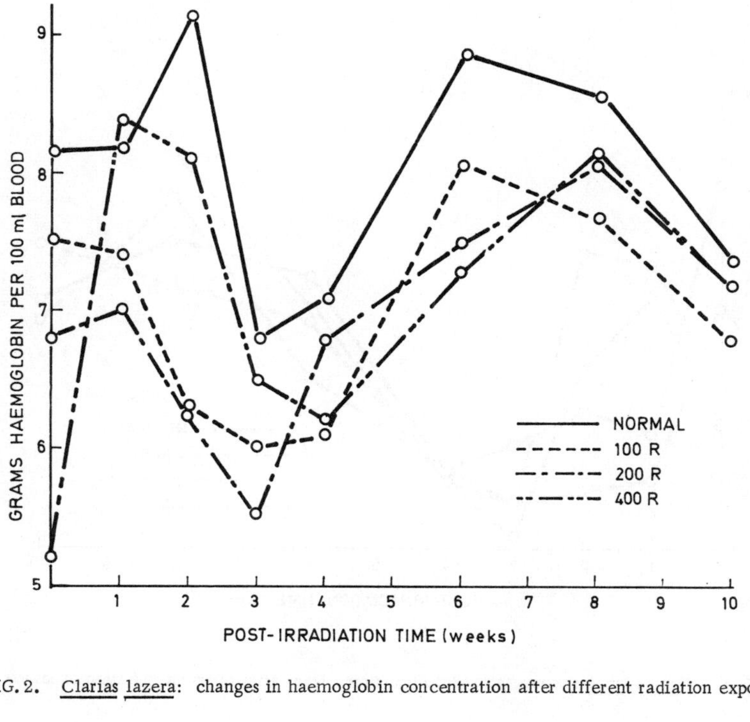

FIG. 2. Clarias lazera: changes in haemoglobin concentration after different radiation exposures.

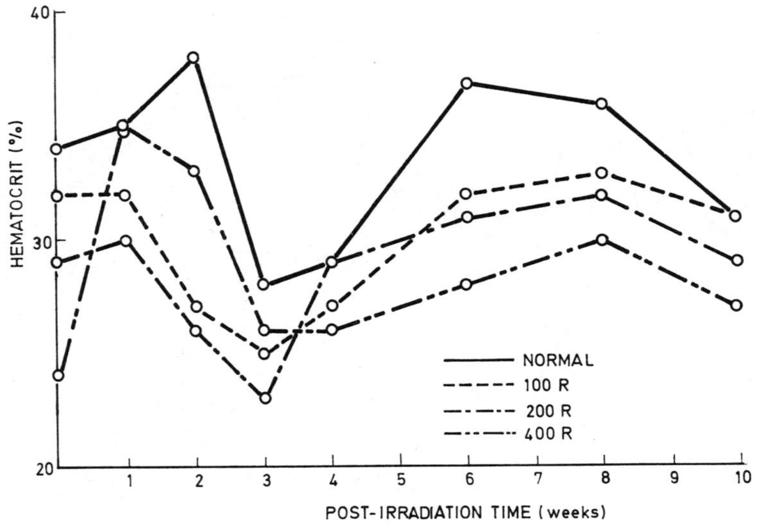

FIG. 3. Clarias lazera: changes in haematocrit values after different radiation exposures.

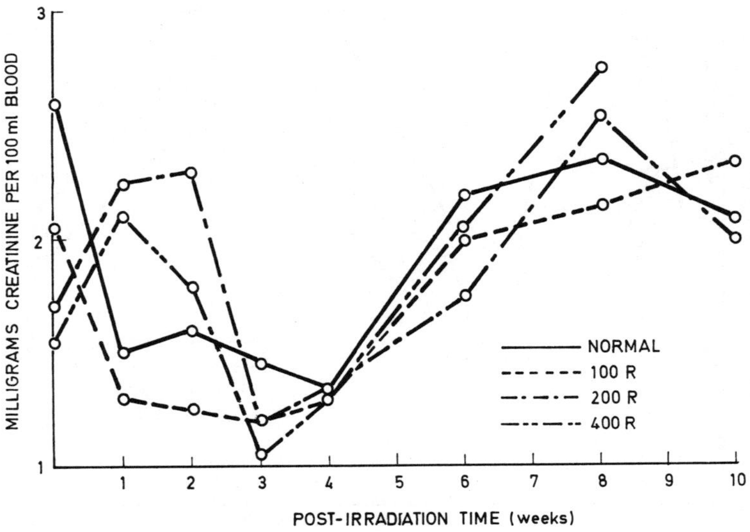

FIG. 4. Clarias lazera: changes in blood creatinine level after different radiation exposures.

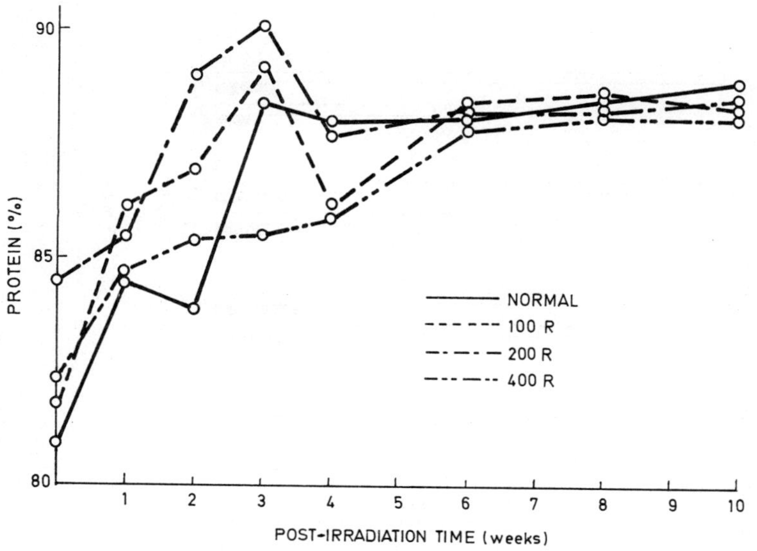

FIG. 5. Clarias lazera: changes in percentage protein dry weight after different radiation exposures.

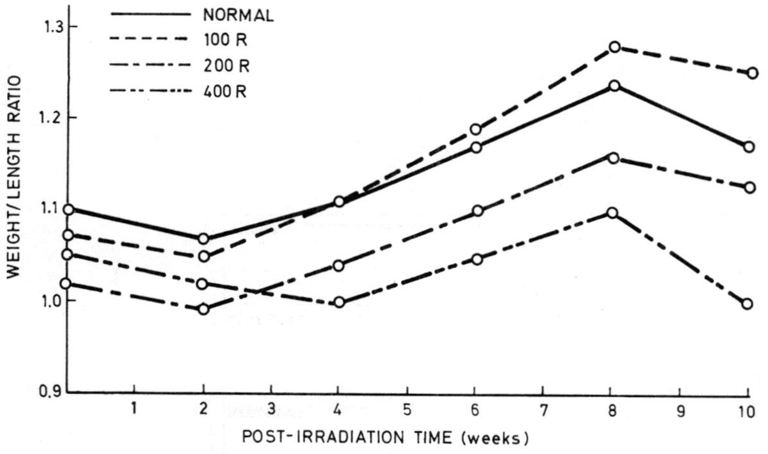

FIG. 6. Clarias lazera juveniles: changes in weight/length ratio after different radiation exposures.

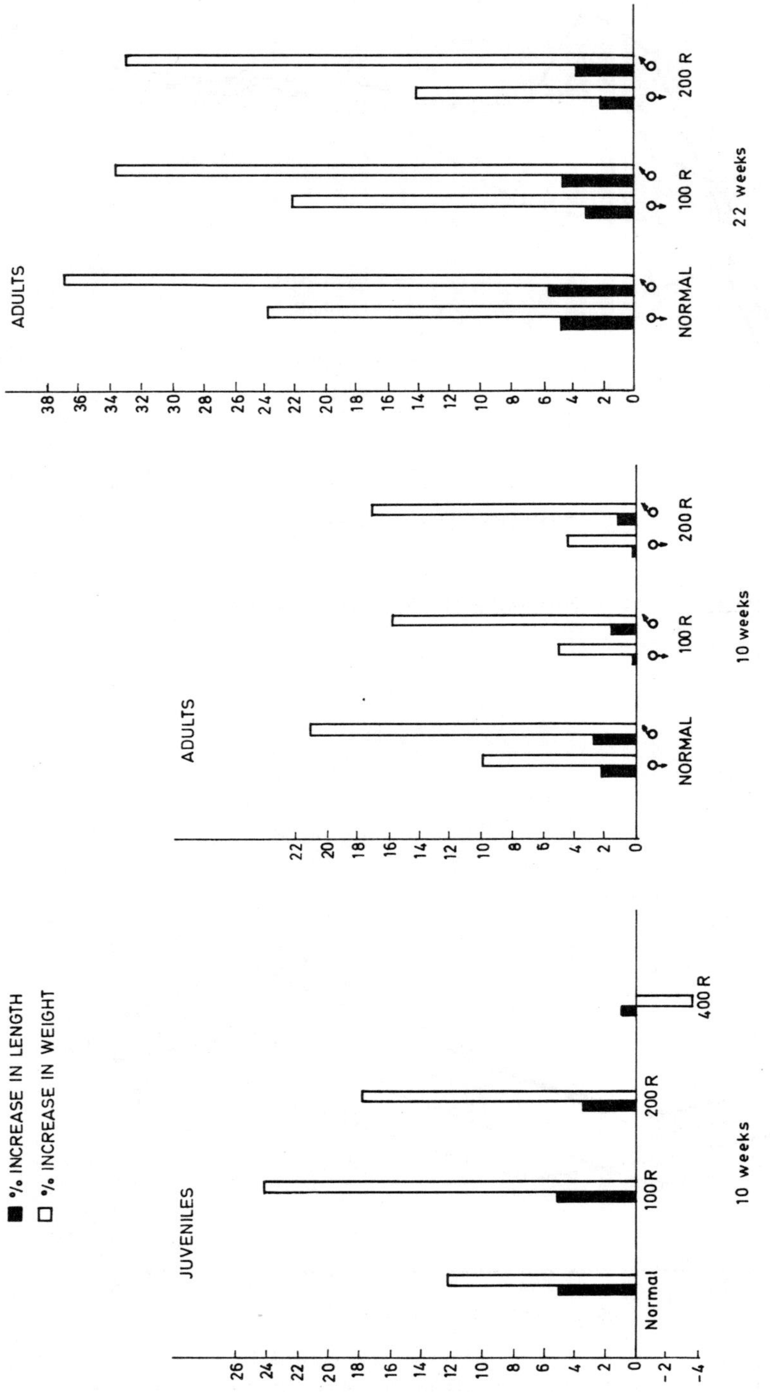

FIG. 7. Clarias lazera adults and juveniles, males and females: increase in growth rate after different radiation exposures.

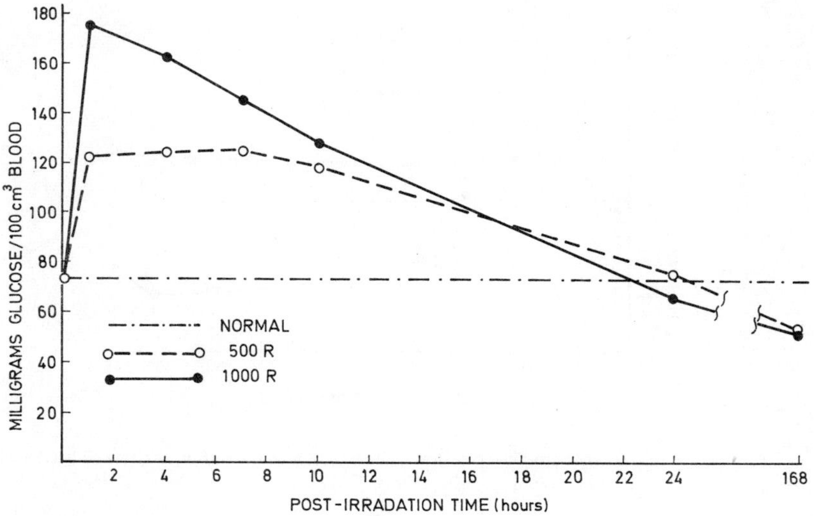

FIG. 8. Tilapia nilotica: changes in blood glucose level after different radiation exposures.

(d) Rate of growth

Whole-body gamma irradiation of Clarias lazera at the low exposure levels of 100, 200 and 400 R causes, in general, significant retardation in rate of growth of both juvenile and adult males and females. Retardation in the rate of increase in length of growing fish is especially marked in adult females. Parallel results are recorded for the percentage increase in weight. In juvenile fish, as shown in Figs 6 and 7, radiation levels of 100 and 200 R seem to accelerate the rate of increase in weight while the higher level of 400 R results in significant retardation. In juveniles, the weight/length ratio increases only after the radiation level of 100 R.

(e) Blood glucose level

As shown in Fig. 8, whole-body gamma irradiation of Tilapia nilotica results in the marked initial hyperglycemia being more intensive after the higher radiation level of 1000 R than after the lower level of 500 R. Later, the blood glucose level progressively decreases and a case of hypoglycaemia seems to dominate.

(f) Glycogen store in liver

As shown in Fig. 9, whole-body irradiation of Tilapia nilotica causes an initial transitory rise in glycogen store concentration in liver. After the lower radiation level of 500 R, the glycogen store shows a prolonged higher level than normal. The higher radiation level of 1000 R, on the other hand, results in a measurable decrease in glycogen store concentration prevailing for the rest of the 40-day experimental period.

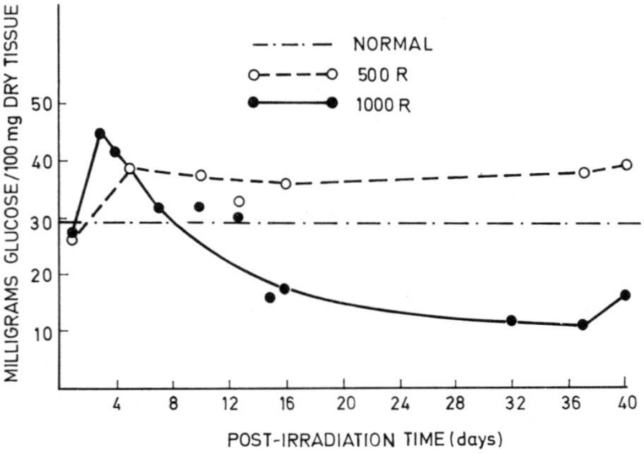

FIG. 9. Tilapia nilotica: changes in glucogen store concentration in liver after different radiation exposures.

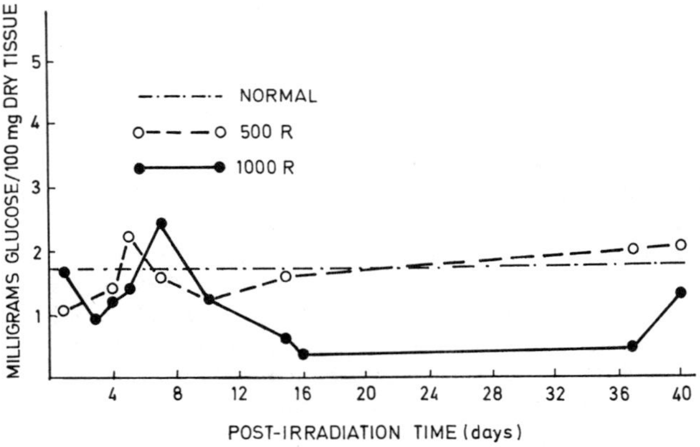

FIG. 10. Tilapia nilotica: changes in glycogen store concentration in muscle after different radiation exposures.

(g) Glycogen store in muscle

Figure 10 shows that whole-body irradiation of Tilapia nilotica causes, in general, no significant change in the glycogen store in muscle after radiation at 500 R. The higher level of 1000 R, on the other hand, results in prolonged significant decrease in glycogen concentration in muscle lasting till the end of the experimental period.

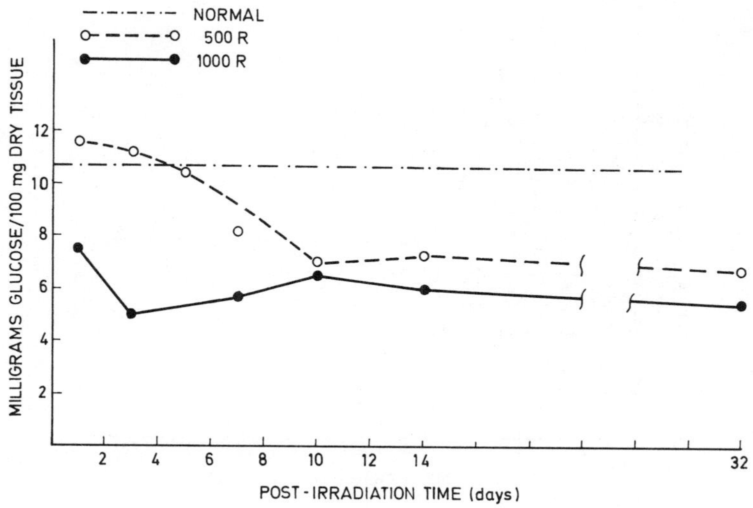

FIG. 11. Tilapia nilotica: changes in free reducing sugars in liver after different radiation exposures.

(h) Free reducing sugars in liver

Whole-body irradiation of Tilapia nilotica results in a noticeable drop in the concentration of free reducing sugars in hepatopancreatic tissue. The drop is more marked after 1000 R, lasting till the end of the experimental period (Fig. 11).

(i) Hydrolytic enzyme activities

The digestive action of intestinal and hepatopancreatic homogenates has been investigated on different carbohydrate substrates, lactose, maltose, starch, glycogen, lichenin and cellulose, at different pH. The results showed that amylase was the most active sucroclastic enzyme in Tilapia nilotica.

As shown in Figs 12 and 13, whole-body gamma irradiation of Tilapia nilotica at 500 and 1000 R results in marked deactivation of the hydrolytic activity of amylase in both the intestinal (Fig. 12) and pancreatic (Fig. 13) tissue homogenates. The inhibition is stronger after the higher radiation level and especially pronounced with pancreatic amylase.

DISCUSSION

It is a well known fact that any type of radiation, external or internal, given to animals causes alterations which may influence the course of tissue response. Radiobiological studies have established that certain animal

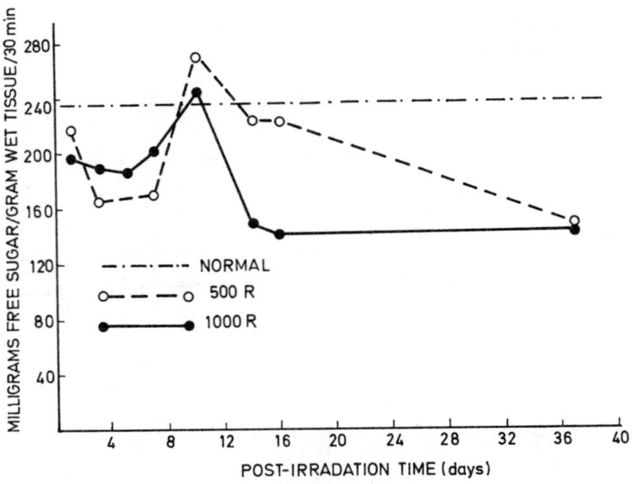

FIG. 12. Tilapia nilotica: changes in hydrolytic activity of intestinal amylase after different radiation exposures.

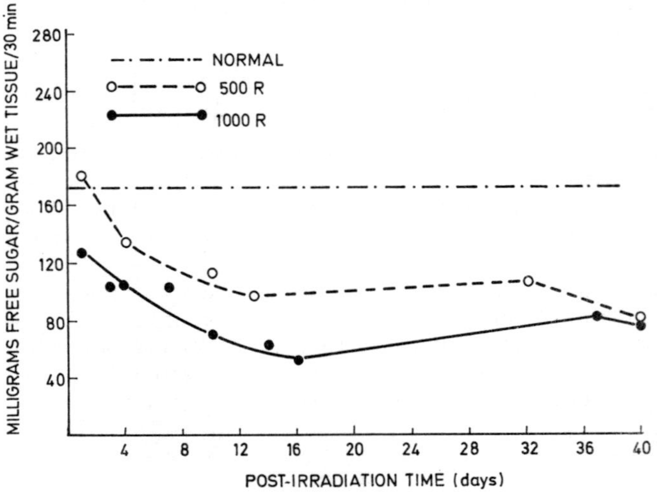

FIG. 13. Tilapia nilotica: changes in hydrolytic activity of pancreatic amylase after different radiation exposures.

cells, namely haematopoietic and reproductive stem cells, are comparatively very radiosensitive. However, most of these studies have been carried out on mammalian animals.

In the present work, attention has been drawn to the biochemical effects of whole-body gamma irradiation in fish and its influence on the rate of growth of Clarias lazera. As blood is known to be one of the most radiosensitive tissues, haematological studies have been carried out. The study of blood and blood-forming organs seems necessary to explain the post-irradiation changes. The radiation-induced change in the haematological levels of Clarias sp. seems likely to result from inhibition of formation of new blood cells as well as from direct destruction.

Unlike the excretion of urea, the creatinine output is practically independent of the protein level of the blood. Creatinine excretion is therefore considered to be an index of the magnitude of the metabolism of the tissues and especially the muscles. Owing to the practical difficulty of collecting urine samples from fish, only blood creatinine level has been determined in this work.

In the course of the work slight but significant acceleration in growth rate was traced after whole-body gamma irradiation of juvenile Clarias lazera at 100 R. This acceleration was mainly due to increase in weight rather than increase in length. Irradiation at the same dose level decreases the growth rate in adults, both male and female.

These results may be partly explained on the basis of the different susceptibility to radiation effects according to age. A similar increase in the growth rate in fish, as well as in many other animal species, has been reported in the literature [8]. It has also been reported by Loeb [24] that, according to the intensity of the irradiation, either (a) growth processes are at first stimulated and then undergo depression, or (b) an initial period of retardation takes place before the period of accelerated growth. Some authors, however, have claimed that there was no evidence in favour of stimulation of growth by any amount of radiation [25].

Whether or not the exposure levels used in the present study are injurious to Clarias lazera needs still further investigation. Before arriving at a definite conclusion, investigations should be carried out on a large number of specimens.

The rapid changes produced by whole-body gamma irradiation in the blood glucose level of Tilapia nilotica appear to be of great interest and permit conclusions about the activity status of the autonomic nervous system.

While the initially induced hyperglycaemia may be considered as the result of change in function of the autonomic nervous systems, possibly due to increased adrenalin secretion, the subsequent hypoglycaemia can be understood as being due to retarded absorption of carbohydrates from the intestine, leading to the absence of materials for glycogenesis in circulation [26].

The decreased level of free reducing sugars in the hepatopancreas, on the other hand, can be attributed to an increased tendency for the free sugars to escape into the blood circulation or to be utilized in the biosynthesis of glycogen. This may be explained as a direct effect of radiation on the processes of glycogenesis, glycogenolysis and glyconeogenesis.

The effect of ionizing radiation on enzyme systems has been the subject of repeated studies by many investigators, but results are still contradictory. With few exceptions, ionizing radiation seems to leave oxidative enzymes unaffected, while the majority of hydrolytic enzymes may show increased or decreased activity [27].

In the present work generally, the radiation dose level of 1000 R had a stronger inhibitory effect on the amylolytic activity than 500 R. It can accordingly be assumed that such a response follows the general rule of dose-effect relationship.

The changes induced by whole-body gamma irradiation on enzyme amylase activity seem to be correlated to complex factors: e.g. alterations in the properties of the enzyme itself, functional radiation damage developed in the site of synthesis of the enzyme and/or its place of action, and changes in enzyme formation, pancreatic secretion and intestinal absorption. Some of these changes may be indirectly related to toxic substances produced by radiation e.g. water products, which may either affect the enzyme activity itself or influence the rate of its synthesis [28], or alter the enzyme properties [29].

Through recent developments in nuclear energy, the availability of high-intensity sources of artificially radioactive materials has resulted in the possibility of exposure of a large portion of the population to radiation hazard. Besides having a natural radioactivity of its own, the aquatic environment is subjected to artificial radiocontamination which stems from radioactive material produced by nuclear test explosions, fallout and radioactive waste of all kinds discharged into the water. Radioactive wastes can be attributed to the medical, agricultural and industrial utilization of radioisotopes besides the operation of land-based power stations and the increasing use of nuclear-powered ships.

Technical progress in the radiation equipment field has greatly increased the intensity and energy of available radiations. The nuclear reactor is playing an increasingly important role as a source of power. The radiological impacts of releases from nuclear facilities into the aquatic environment seem to underlie many problems of major importance which one would expect to stimulate more comprehensive investigations.

REFERENCES

[1] MORI, T., SAIKI, M., in: Research on the Effect and Influence of Nuclear Bomb Test Explosions, National Academy of Science, Japan 2 (1956) 1205.
[2] DONALDSON, L.R., USAEC Report AECU-3412 (1954).
[3] SALBERG, A.N., J. Exp. Zool. 78 (1938) 441.
[4] RUGH, R., CLUGSTON, H., Biol. Bull. 108 (1955) 318.
[5] POLIKARPOV, G.G., IVANOV, V.N., Probl. Ichthy. 3 (1961) 20.
[6] POLIKARPOV, G.G., IVANOV, V.N., Sevast. Biol. St. Rep. 13 (1961).
[7] KRUMHOLZ, L.A., Bull. Am. Mus. Nat. His. 110 (1956).
[8] BOWES, D.W., Diss. Cath. Univ. Am. Press (1960).
[9] DONALDSON, L.R., FASTER, R.F., Nat. Acad. Sci., Nat. Res. Council. Rep. No. 96 (1957).
[10] EL-KASHEF, H.S., ROUSHDY, H.M., M. Sc. Thesis Fac. Sci. Cairo Univ. (1968).
[11] IMAM, A.E., ROUSHDY, H.M., M.Sc. Thesis Fac. Sci. Cairo Univ. (1969).
[12] ROUSHDY, H.M., EL-KASHEF, H.S., Isotope Radiat. Res. 2 (1969) 27.
[13] ROUSHDY, H.M., EL-KASHEF, H.S., Isotope Radiat. Res. 3 (1970) 49.
[14] ROUSHDY, H.M., IMAM, A.E., Z. Naturforsch. B 26 (1971) 962.
[15] LARSEN, H.N., Prog. Fish Cult. 26 (1964) 11.
[16] WINTROBE, M.M., Folia Haematol. 51 (1934) 32.
[17] FOLIN, O., WU, H., J. Biol. Chem. 38 (1919) 81.
[18] HADEN, R.L., J. Biol. Chem. 56 (1923) 469.

[19] KIRK, P.L., Adv. Protein Chem. 3 (1947) 139.
[20] NELSON, N., J. Biol. Chem. 153 (1944) 375.
[21] SOMOGYI, M., J. Biol. Chem. 160 (1945) 62.
[22] KEMP, KITIS, Biochem. J. 56 (1954) 646.
[23] MENDLE, A.K., MAYERS, D.K., Biochem. J. 56 (1954) 639.
[24] LOEB, L., Am. J. Roentgenol. Radium Ther. 9 (1922) 497.
[25] BLOOM, W., Histopathology and Irradiation of External and Internal Sources, McGraw-Hill (1948).
[26] ELLINGER, F., Medical Radiation Biology, Charles Thomas (1957).
[27] ROTH, J.S., EICHEL, H.J., WASE, A., ALPER, C., BOYD, M.J., Archs Biochem. Biophys. 44 (1953) 95.
[28] GRAY, L.H., Brit. J. Radiol. 26 (1953) 609.
[29] COELHO, R., Energ. Nucl. (Paris) 9 (1967) 189.

DISCUSSION

A.R. BERG: I am impressed by the number of parameters which have been followed in your interesting study. I have three questions about the figures. First, in Figs 1-5, why do the curves for the various treatments and for the control not all start from the same point?

H.S. EL-KASHEF: As can be expected, laboratory investigations could not be carried out immediately after radiation exposure, since the radiation facility used is not located at the same laboratory building. The lapse of time before the first investigations and the variation in exposure levels were the reasons for the curves for the various treatments and the control not starting from the same point.

A.R. BERG: On the left of Fig. 7, how can one interpret the fact that growth in length is equal for the control and for the 100 R treatment, while growth in weight is higher for the 100 R treatment than for the control?

H.S. EL-KASHEF: Growth in weight does not necessarily parallel an increase in length. Growth in weight could be attributed to increased protein synthesis, fat deposition, increased water content, etc. The decrease in length on the other hand could be related to actual growth factors including cell mitotic activity, which is known to be seriously inhibited by radiation exposure.

A.R. BERG: In Figs 8-13, why do you not indicate the time-dependent variation for the control?

H.S. EL-KASHEF: Average values have been taken into consideration for the normal control fish for the sake of comparison. Variation in time for the control value was found to be within the normal range.

Suzanne VIGNES: Young tissues are generally regarded as more radiosensitive than adult tissues, but your experience seems to indicate the contrary. Could the increased growth observed in young fish irradiated at 100 and 200 R not be due to stimulation of the recovery processes in the stem cells, which would result in accelerated cell division, and would not this mechanism be dominant in comparison with the mechanism of increased cellular metabolism?

H.S. EL-KASHEF: The results presented confirm the well known concept that young living tissues are more radiosensitive than adult ones. This has been confirmed by our findings, showing that the acceleration of growth rate induced by lower levels of radiation exposure and loss of weight caused by higher exposure levels were only recorded in juvenile fish, and not in adult. Acceleration of growth rate following low-level radiation exposure could be explained by increased rate of metabolism and, as might be expected, protein synthesis in muscles. In addition, changes in water balance due to radiation exposure are apparently a contributory factor underlying changes in body weight. Loss of weight effected by a higher level of radiation exposure suggests significant inhibition of anabolic activity in parallel with increased catabolism.

S. BONOTTO: In Fig. 8 of your paper, did you treat the control fish (non-irradiated) in the same way as the irradiated fish (apart from the irradiation)? Irradiation in the small chamber of the gamma cell could cause stress effects in Tilapia nilotica, with a consequent possible change in the blood glucose level.

H.S. EL-KASHEF: Changes in glucose level in Tilapia nilotica possibly caused by the irradiation process in the small chamber of the gamma cell seem to be of very little significance in our work. Fish placed in small water containers were gently introduced into the gamma cell for a few seconds. Control fish were held and transported in the same way but naturally not in the gamma cell. The initially recorded hyperglycaemia lasting for about 20 hours is attributed to radiation effects and proved to be dose-dependent.

S. BONOTTO: Do you know whether in Tilapia nilotica the blood glucose levels shows a circadian rhythm?

H.S. EL-KASHEF: To my knowledge, nothing has been mentioned in the literature suggesting such a rhythm. The blood glucose level estimated in the normal control showed, during the whole experimentation period, only reasonable fluctuations within the normal biological range.

АНАЛИЗ КРИТЕРИЕВ, ОПРЕДЕЛЯЮЩИХ РАЗМЕРЫ ДОПУСТИМОГО РАДИОАКТИВНОГО СБРОСА АЭС В РЕЧНЫЕ СИСТЕМЫ

В.В.ПИСАРЕВ, И.С.ЦЫБИЗОВ
Институт прикладной геофизики,
Москва,
Союз Советских Социалистических Республик

Abstract–Аннотация

AN ANALYSIS OF CRITERIA FOR DETERMINING THE EXTENT OF PERMISSIBLE RADIOACTIVE WASTE DISPOSAL FROM NUCLEAR POWER STATIONS INTO RIVER SYSTEMS.

The economic advantages of siting nuclear power stations close to rivers are given by the fact that use can then be made of a very cheap direct-flow cooling system and by the possibility of disposal of radioactive wastes. However, if advantage is to be taken of these opportunities, waste disposal from the nuclear power station must be strictly regulated (subject to standards) in order to ensure the radiation safety of the population using the river and the territory adjacent to it in one way or another. A similar problem arises if a nuclear power station is regarded as a potential radiation risk in view of the possible accidental entry of radioactivity into the river waters. The maximum permissible disposal of liquid radioactive waste into river systems must be determined on the basis of the specific physicogeographical conditions, account being taken of the use made of the river and of the river valley and also of social and economic factors affecting the population groups critically involved. Estimates of the maximum permissible discharge of liquid radioactive wastes into rivers in specific cases have to be made in terms of quantitative characteristics or criteria suitable for practical use. This paper considers three groups of such criteria: hydrological, ecological and radiological. The hydrological criteria are used for measuring the mixing and dilution of the radioactive wastes in river flows and for determining the nature and scale of interaction between contaminated river waters and the tidal parts of the river valleys which are submerged during floods and high tides. The ecological criteria establish quantitative relations between the content of radionuclides in river waters and that in the parts of the environment which are of interest to us; in addition, they are a prerequisite for forecasting and analysing radiation conditions. As radiological criteria it is proposed to use the maximum river-water concentrations of nuclides, these being determined by taking into account the various ways in which radioisotopes affect the human organism. The values for maximum concentrations are an obvious and convenient instrument for practical use in that they express the dose limits established for irradiation of individual members of the population and, at the same time, take into account the distribution characteristics of radioisotopes in the various parts of the environment. As a result, they can be used, if due attention is paid to the hydrological factors governing the diffusion of radioactive impurities in river flows, to determine permissible release of liquid radioactive waste from nuclear power stations into river-bed flows.

АНАЛИЗ КРИТЕРИЕВ, ОПРЕДЕЛЯЮЩИХ РАЗМЕРЫ ДОПУСТИМОГО РАДИОАКТИВНОГО СБРОСА АЭС В РЕЧНЫЕ СИСТЕМЫ.

Возможности использования наиболее дешевой прямоточной системы охлаждения в сочетании с возможностью удаления радиоактивных отходов предопределяет экономическую целесообразность размещения АЭС вблизи рек. Реализация этой возможности должна опираться на строгую регламентацию (нормирование) сбросов АЭС с тем, чтобы обеспечить радиационную безопасность населения, так или иначе использующего реку и прилегающую к ней территорию. Аналогичная задача возникает и при оценке АЭС как потенциального источника радиационной опасности в случае аварийного поступления радиоактивности в речные воды. Размеры допустимого удаления жидких радиоактивных отходов в речные системы должны устанавливаться с учетом конкретных физико-географических условий, специфики использования реки и ее речной долины, а также с учетом социально-экономических особенностей критических групп населения. Для того, чтобы оценить величину допустимого сброса жидких радиоактивных отходов в реки в той или иной конкретной ситуации, необходимо располагать удобными для практического использования количественными характеристиками или критериями. В настоящем докладе рассматриваются три группы таких критериев: гидрологическая, экологическая и радиологическая. Гидрологические критерии определяют величины смешения и разбавления радиоактивных отходов в русловых потоках, контролируют характер и масштабы взаимодействия загрязненных речных вод с пойменными участками речных долин, затопляемых в периоды половодий и паводков. Экологические критерии устанавливают количественные связи между содержанием радионуклидов в речных водах и их содержанием в интересующих нас объектах внешней среды, создают предпосылки для прогноза и анализа радиационной обстановки. В качестве радиологических критериев предлагается использовать величины предельных концентраций радионуклидов в речных водах, установленных с учетом различных путей воздействия радиоактивных изотопов на организм человека. Величины предельных концентраций являются наглядным и удобным для практического использования выражением установленных пределов доз облучения отдельных лиц из населения и вместе с тем учитывают специфику распределения радиоактивных изотопов в различных объектах внешней среды. Это и позволяет использовать их совместно с гидрологическими факторами, контролирующими рассеяние радиоактивной примеси в русловых потоках, для определения величины допустимого сброса жидких радиоактивных отходов АЭС в речные системы.

1. ВВЕДЕНИЕ

Использование атомной энергии в мирных целях и прежде всего широкое развитие атомной электроэнергетики вызывает необходимость уделять значительное внимание проблеме радиационной безопасности населения и охраны окружающей среды.

Одной из составных частей этой проблемы является разработка общих принципов обеспечения радиационной безопасности при проектировании и эксплуатации АЭС. При современной технологии нормальная работа энергетических реакторов практически неизбежно сопровождается образованием радиоактивных отходов, часть которых удаляется в поверхностные воды.

Возможность использования наиболее дешевой прямоточной системы охлаждения в сочетании с удобством удаления малоактивных отходов предопределяет экономическую целесообразность размещения АЭС вблизи рек. Реализация такой возможности должна опираться на строгую регламентацию (нормирование) сбросов АЭС с тем, чтобы обеспечить радиационную безопасность населения, так или иначе использующего речные воды и территорию речных долин.

Под нормированием жидких слабоактивных отходов, удаляемых в речные системы атомными электростанциями, исследовательскими центрами и другими ядерными объектами, обычно понимается установление таких ограничений, налагаемых на выпускаемые отходы, которые с учетом рассеяния радиоактивных продуктов в речных потоках гарантирует населению, проживающему вблизи реки, полную радиационную безопасность.

Требования, предъявляемые к сбросу жидких отходов АЭС, сводятся к установлению предельных уровней концентраций лимитирующих радионуклидов и допустимого размера водного сброса.

Размеры допустимого выпуска радиоактивных продуктов в речные системы должны устанавливаться с учетом конкретных физико-географических условий, специфики использования реки и прилегающей к ней местности, а также с учетом социально-экономических особенностей критических групп населения.

Для того, чтобы оценить величину допустимого сброса жидких радиоактивных продуктов в реки в той или иной конкретной ситуации, необходимо располагать удобными для практического использования количественными характеристиками или критериями. Установление таких критериев и дает возможность определить для каждого объекта свои дифференцированные нормы годового сброса радиоактивных отходов.

Ниже рассматриваются три группы таких критериев: гидрологическая, экологическая и радиологическая.

2. ГИДРОЛОГИЧЕСКИЕ КРИТЕРИИ

Гидрологические критерии определяют величины смешения и разбавления радиоактивных отходов, сбрасываемых в русловые потоки.

Следует иметь в виду, что возможности рассеяния радиоактивных отходов в речных потоках по сравнению с разбавлением радиоактивных газов и аэрозолей, выбрасываемых в атмосферу, относительно невелики и, в конечном счете, ограничены величиной отношения расхода реки и объема выпускаемых отходов.

В соответствии с принятыми в Советском Союзе правилами охраны поверхностных вод от загрязнения сточными водами при определении величины возможного разбавления загрязнений для незарегулированных рек принимается минимальный среднемесячный расход воды в реке 95% обеспеченности [1].

Особое место при оценке допустимого сброса радиоактивных отходов занимает определение возможного масштаба и времени взаимодействия загрязненных речных вод с пойменными участками речных долин, затопляемых в периоды половодья и паводков.

Наконец, при оценке допустимого сброса радиоактивных продуктов весьма существенно учесть и такие факторы, как геометрия выпуска, наличие термической стратификации, струйности и т.п.

3. ЭКОЛОГИЧЕСКИЕ КРИТЕРИИ

Миграция радиоактивных изотопов, поступающих в поверхностные водотоки, сводится к их рассеянию, переносу и накоплению. Наиболее ярко выраженными накопительными (кумулятивными) свойствами в пресноводных экосистемах обладают водные организмы и донные отложения. Масштабы загрязнения окружающей среды заметно увеличиваются в результате весенних разливов рек, а также в результате осуществления различных ирригационных мероприятий, которые приводят к накоплению радиоактивных материалов в почвах речных долин.

Значение почвы как источника радиоактивного загрязнения объектов окружающей среды определяется ее особым положением в сложной экологической наземной системе. В результате обменных процессов вместе со стабильными химическими элементами радионуклиды поступают в растительность, а затем с кормами в сельскохозяйственных животных и, наконец, с пищевыми продуктами в организм человека.

Наибольшее внимание исследователей в настоящее время привлекают уже не столько сами пути миграции радиоактивных продуктов из воды или почвы в продукты питания, они достаточно хорошо изучены, а те количественные характеристики, которые определяют переход радионуклидов от "звена" к "звену" в пищевых цепочках человека.

ТАБЛИЦА 1. ПАРАМЕТРЫ, ХАРАКТЕРИЗУЮЩИЕ ПЕРЕХОД НЕКОТОРЫХ РАДИОНУКЛИДОВ В ПОЧВУ, РАСТИТЕЛЬНОСТЬ И ПРОДУКТЫ ПИТАНИЯ [2-9]

Изотоп	Коэффициент накопления K_1 рыба/вода	Коэффициент распределения K_2 дерново-луговая почва/вода	Коэффициенты перехода K_3			
			K_{31} трава/почва	K_{32} молоко/рацион коров	K_{33} мясо/рацион коров	K_{34} картофель/почва
^{24}Na	100	–	-	-	-	-
^{32}P	2000	-	-	-	-	-
^{60}Co	400	1000	$7 \cdot 10^{-2}$	-	-	-
^{65}Zn	1000	1450	$3 \cdot 10^{-1}$ [a]	$1 \cdot 10^{-2}$ [a]	$4 \cdot 10^{-4}$ [a]	$1 \cdot 10^{0}$
^{90}Sr	30	1400	1,74	$1{,}6 \cdot 10^{-3}$	$3 \cdot 10^{-3}$	$2 \cdot 10^{-2}$
^{137}Cs	1000	1000	$1{,}9 \cdot 10^{-1}$	$8 \cdot 10^{-3}$	$8 \cdot 10^{-1}$	$3 \cdot 10^{-2}$
^{144}Ce	30	250	$3{,}0 \cdot 10^{-2}$	$2 \cdot 10^{-4}$	-	$1 \cdot 10^{-3}$

[a] Коэффициенты K_{31}, K_{32}, K_{33} рассчитаны на основании данных, приведенных в работах [2-5].

В настоящем докладе мы намеренно отвлекаемся от разнообразных и многочисленных природных процессов, определяющих специфику распределения радионуклидов в различных объектах внешней среды, а рассматриваем только их количественное выражение, которое с известными приближениями может быть представлено в виде постоянных безразмерных коэффициентов или параметров.

К подобным параметрам следует отнести, во-первых, коэффициенты накопления K_1, отражающие процесс накопления радиоактивных продуктов главным образом в организмах водных экосистем, например, в пресноводной рыбе.

Во-вторых, к таким параметрам относятся коэффициенты распределения K_2, характеризующие поступление радионуклидов из речных вод в пойменные отложения в период весеннего половодья или в результате орошения. Коэффициент распределения представляет собой отношение удельной активности почвы к удельной активности воды с учетом соотношения их твердой и жидкой фазы, которое обычно равно 20 [2].

В-третьих, в эту же группу параметров входят коэффициенты перехода K_3, отражающие дискриминацию радионуклидов при их движении последовательно по звеньям различных пищевых цепей. Коэффициенты перехода определяются обычно соотношениями типа: удельная активность травы/удельная активность почвы, молоко/трава, мясо/трава и т.д.

Совокупность перечисленных параметров, величины которых контролируются экологическими закономерностями, и составляет группу экологических критериев.

Основное значение экологических критериев состоит в том, что они позволяют по известному содержанию радионуклидов в речных водах заранее оценить вероятные уровни радиоактивного загрязнения интересующих нас объектов внешней среды и тем самым создают необходимые предпосылки для прогноза и анализа радиационной обстановки.

Усиленное изучение закономерностей поведения и распределения в окружающей среде таких долгоживущих радионуклидов, как ^{90}Sr и ^{137}Cs, предопределило, к сожалению, недостаточное внимание к изучению поведения во внешней среде целого ряда других радиоактивных изотопов. Некоторое представление о возможных величинах параметров K_1, K_2, K_3 для нескольких радионуклидов дают данные, приведенные в табл.I.

Величины коэффициентов K_1, K_2, K_3 зависят от целого ряда факторов, обусловленных как свойствами самих радионуклидов и их химических соединений, так и свойствами окружающей среды, включая и населяющих ее организмов. Перечисленные обстоятельства и являются одной из основных причин, предопределяющих необходимость дифференцированного подхода при установлении величины допустимого сброса жидких радиоактивных отходов, удаляемых в поверхностные водотоки.

4. РАДИОЛОГИЧЕСКИЕ КРИТЕРИИ

Специфика облучения населения, проживающего в непосредственной близости от рек, в которые производится удаление жидких радиоактивных отходов, состоит в том, что получаемая им доза внутреннего и внешнего облучения формируется под влиянием комбинированного воздействия целого ряда факторов.

Внутреннее облучение может быть связано с использованием для питья речных вод без какой-либо предварительной очистки, потреблением загрязненной рыбы, мяса и молока сельскохозяйственных животных в результате их содержания на пойменных лугах, потреблением загрязненных овощей, выращиваемых на затопляемых или поливных землях.

Внешнее облучение может быть обусловлено полями излучений, образующимися над водной поверхностью рек и загрязненными пойменными участками.

Существуют и некоторые другие пути воздействия радиации на человека, вкладом которых в интегральную дозу можно пренебречь. Это облучение от донных отложений рек, загрязненных сельскохозяйственных орудий и орудий рыбной ловли, а также внутреннее облучение за счет вдыхания загрязненной пыли при сельскохозяйственных работах [10].

Дозы облучения от атмосферных выпадений радиоактивных веществ в настоящем сообщении не рассматриваются.

Каждому из путей облучения соответствует своя составляющая часть дозы. Сумма составляющих всех возможных путей облучения определяет интегральную или полную дозу облучения населения.

Для рассматриваемого контингента, состоящего главным образом из сельских жителей, который отнесен к критической группе отдельных лиц из населения, облучение регламентируется пределом доз, рекомендованных МКРЗ и принятых советским законодательством [11].

Практическая реализация этих рекомендаций для нормирования жидких радиоактивных отходов означает установление таких величин содержания радиоактивных изотопов в речных водах, которые с учетом местных условий не будут создавать дозы, превышающей установленный годовой предел.

Учитывая, что изотопы, присутствующие в речных водах, различаются по своим физико-химическим свойствам, биологическому действию и т.п., целесообразно вначале оценить величины предельных концентраций для отдельных радионуклидов. Это позволяет одновременно установить составляющие интегральной дозы и оценить критические пути их воздействия на человека.

Наиболее удобно такие оценки выполнить для некоторой гипотетической, достаточно приближенной к реальным условиям модели. Схема подобной модели, отражающая возможные пути облучения населения, проживающего вблизи равнинных рек, характерных для большей части территории Советского Союза, приведена на рис.1.

Величина предельной концентрации для отдельного радионуклида $ПК_i$, Ки/л, может быть получена из совместного решения следующей системы линейных уравнений:

$$D_1 = A_1 \cdot ПК_i$$

$$D_2 = A_2 \cdot ПК_i$$

$$D_n = A_n \cdot ПК_i$$

$$D = D_1 + D_2 + \ldots + D_n$$

где D_1, D_2,...D_n – составляющие интегральной дозы, бэр/год.

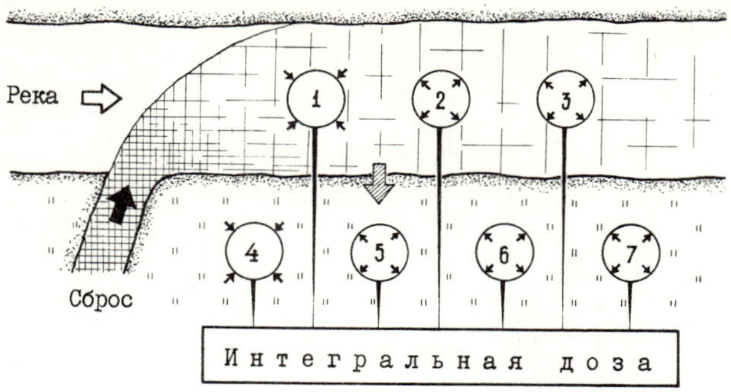

Рис.1. Схема формирования интегральной дозы облучения критической группы населения. 1 – водная поверхность; 2 – вода для питья; 3 – рыба; 4 – пойма реки; 5 – молоко; 6 – мясо; 7 – сельскохозяйственные продукты.

$A_1, A_2, \ldots A_n$ – постоянные коэффициенты, учитывающие распределение i-го радионуклида в объектах внешней среды, годовое поступление радионуклида в организм человека и особенности его биологического действия, бэр·л/Ки·год.

D – установленный годовой предел дозы, бэр/год.

В общем виде величина $ПК_i$ определяется соотношением

$$ПК_i = \frac{D}{A_1 + A_2 + \ldots + A_n} \text{ Ки/л}$$

В качестве исходных материалов для расчета использованы рекомендованные и принятые в СССР нормы облучения отдельных лиц из населения, основывающиеся на пределах годового поступления радионуклидов в организм и пределах доз на критические органы человека [11].

Условия загрязнения рассматриваемых здесь объектов окружающей среды учтены с помощью параметров, приведенных в табл.1.

Поскольку в Советском Союзе снабжение продуктами питания осуществляется централизованно, для рассматриваемой группы населения наиболее вероятно систематическое потребление только таких продуктов местного производства как рыба, молоко, мясо и некоторые виды овощей. Общее годовое потребление продуктов принято на основании рекомендованного рациона питания для населения СССР [12].

Продолжительность внешнего облучения жителей при проведении различных работ на реке и нахождении их на загрязненных участках местности принята равной трем месяцам из расчета 12 часов ежедневно.

Величины $ПК_i$ и составляющие интегральной дозы для некоторых изотопов, присутствующих в жидких сбросах ядерных установок, приведены в табл.II.

Данные, приведенные в таблице, получены исходя из семи наиболее вероятных путей облучения, показанных на рис.1. В тех случаях, когда в рассматриваемой модели один или несколько путей облучения

ТАБЛИЦА II. СОСТАВЛЯЮЩИЕ ИНТЕГРАЛЬНОЙ ДОЗЫ И ВЕЛИЧИНЫ ПРЕДЕЛЬНЫХ КОНЦЕНТРАЦИЙ ($ПК_i$) НЕКОТОРЫХ РАДИОАКТИВНЫХ ИЗОТОПОВ

Изотоп	$T_{1/2}$	Составляющие интегральной дозы[a]							$ПК_i^b$
		Внешнее облучение		Внутреннее облучение					
		водная поверхность	пойменные земли	питьевая вода	рыба	молоко	мясо (говядина)	картофель	
^{24}Na	14,9 часов	2,75	-	3,95	93,3	-	-	-	$1,1 \cdot 10^{-9}$
^{32}P	14,5 дней	-	-	0,60	99,4	-	-	-	$1,1 \cdot 10^{-10}$
^{56}Mn	2,6 часа	1,00	-	3,00	96,0	-	-	-	$3,1 \cdot 10^{-9}$
^{60}Co	5,25 лет	0,49	63,9	1,27	34,3	-	-	0,04	$4,4 \cdot 10^{-10}$
^{65}Zn	245 дней	0,15	18,1	0,66	61,5	4,12	0,17	15,3	$6,2 \cdot 10^{-10}$
^{90}Sr	28,4 лет	-	-	2,92	7,93	85,4	c	3,74	$1,2 \cdot 10^{-11}$
^{137}Cs	30 лет	0,06	3,62	0,75	71,8	21,6	c	2,16	$1,1 \cdot 10^{-10}$
^{144}Ce	284,5 дней	c	0,27	24,2	73,0	0,58	-	1,94	$2,9 \cdot 10^{-9}$

[a] В % от установленного годового предела дозы.
[b] Предельные концентрации изотопов, соответствующие установленному годовому пределу дозы.
[c] Величины концентраций менее $5 \cdot 10^{-3}$ %.

не реализуется, предельная концентрация может быть соответственно увеличена, т.е. ПК радиоактивных изотопов может изменяться в зависимости от местных условий.

При нахождении в речных водах смеси радиоактивных изотопов их соответствующие концентрации C_1, C_2, C_n должны быть ограничены исходя из следующего соотношения

$$\frac{C_1}{ПК_1} + \frac{C_2}{ПК_2} + \ldots + \frac{C_n}{ПК_n} \leqslant 1$$

Анализируя составляющие интегральной дозы, прежде всего следует отметить, что вклад в интегральную дозу, получаемую населением в результате потребления речной воды для питья, за исключением ^{144}Ce, не превышает 4 %.

Основным критическим путем облучения населения является потребление загрязненных продуктов питания. Для таких радионуклидов как ^{24}Na, ^{32}P, ^{56}Mn, ^{65}Zn, ^{137}Cs и ^{144}Ce основная часть установленного предела дозы формируется за счет потребления загрязненной рыбы. Для ^{90}Sr критическим путем облучения является потребление загрязненного молока.

С точки зрения внешнего облучения, наибольшую опасность представляет ^{60}Co, который в основном и обуславливает облучение населения при его нахождении на загрязненной пойме.

Рассчитанные с учетом различных путей поступления величины предельных концентраций радиоактивных изотопов в поверхностных водотоках являются наглядным и удобным для практического использования выражением установленных пределов доз облучения отдельных лиц из населения.

Значение предельных концентраций состоит в том, что они, наряду с требованиями радиационной безопасности, учитывают также специфику распределения радиоактивных изотопов в различных объектах внешней среды. Это обстоятельство позволяет рассматривать предельные концентрации в качестве радиологических критериев, которые совместно с гидрологическими факторами, контролирующими рассеяние радиоактивной примеси в русловых потоках, и определяют величину допустимого сброса жидких радиоактивных отходов АЭС в речные системы.

ЛИТЕРАТУРА

[1] Правила Охраны Поверхностных Вод от Загрязнения Сточными Водами, СССР, 1961 г.

[2] ТИМОФЕЕВ-РЕСОВСКИЙ, Н.В., ТИТЛЯНОВА, А.А. и др., "Поведение радиоактивных изотопов в системе почва-раствор", гл.II, Радиоактивность Почв и Методы ее Определения, АН СССР, изд-во "Наука", 1966 г.

[3] ЭЙЗЕНБАД, М., Радиоактивность Внешней Среды, Атомиздат, М., 1967 г.

[4] БАРХУДАРОВ, Р.М., ЗАЙЦЕВА, А.Ф., МАРЕЙ, А.Н., "Основные принципы санитарного нормирования радиоактивных веществ в воде поверхностных водоемов", Environmental Behaviour of Radionuclides Released in the Nuclear Industry (Proc.Symp. Aix-en-Provence, 1973) IAEA, Vienna (1973) 625.

[5] Радиобиология и Радиоэкология Сельскохозяйственных животных, М., Атомиздат, 1973 г.

[6] ФЕДОРОВ, Е.А., ПРИСТЕР, Б.С. и др., "Биологическое действие и поведение радиоактивных продуктов деления в сельскохозяйственных цепочках", 4th Int.Conf.Peaceful Uses At.Energy (Proc.Conf. Geneva 1971) 11, UN and IAEA, Vienna (1972) 663.

[7] ИЛЬИН, Д.Н., МОСКАЛЕВ, Ю.И., "О распределении,выведении и коэффициенте накопления стронция-90, цезия-137 и фосфора-32 у рыб", В кн. Распределение, Биологическое Действие и Миграция Радионуклидов, М., Медгиз, 1961 г.

[8] HARLEY, R.S., Uptake of radionuclides by water algae and fish, Health Physics 10 4 (1964).

[9] ЛЕБЕДЕВА, Г.Д., "Накопление церия-144 годовиками-карпами и элодеей", Мед. Радиология, №4, 1959 г.

[10] Frittelli, L., "Application of system analysis methodology to the determination of the limiting radiological capacity of the area surrounding a nuclear facility", Environmental Surveillance around Nuclear Installations (Proc.Symp. Warsaw, 1973) 1, IAEA, Vienna (1974) 425.

[11] Нормы Радиационной Безопасности (НРБ-69), М., Атомиздат, 1972 г.

[12] Концепция Биологического Риска Воздействия Ионизирующего Излучения, М., Атомиздат, 1973 г.

DISCUSSION

H.P. JAMMET: How do you deal with the cumulative effect of irradiations from different installations?

V.V. PISAREV: If several nuclear power stations are situated in the basin of the same river, the overall discharge of radionuclides is determined taking account of the contributions of the various installations.

P. BEAU: Are the limits imposed on nuclear power stations just concentration limits for river water or are there any other limits to be observed? For example, a limit on the concentration in the effluent, a limit on the annual amount of activity discharged, etc.

V.V. PISAREV: In the USSR the limit prescribed for the discharge of radioactive substances takes account of both air and water releases, and the dose to individual members of the population may not exceed 500 mrem/a. The doses received by the population are then insignificant.

G. BOERI: In the Soviet Union, is it permissible for individual members of the population to receive the full dose limit of 500 mrem/a via exposure pathways related to liquid effluents alone?

V.V. PISAREV: This limit of 500 mrem/a for the category "individual members of the population" takes into account both aquatic and air releases. In the paper it is assumed that the installation makes no releases into the air.

Suzanne VIGNES: You spoke of a concentration factor of 2000 for ^{32}P, but the literature usually quotes a factor of 100 000. Can you comment on this?

V.V. PISAREV: On the basis of our analysis of the relevant studies (see Table I), we believe that our concentration factor of 2000 for fish is more realistic.

СИСТЕМА РАДИАЦИОННОГО КОНТРОЛЯ, МЕТОДЫ И АППАРАТУРА ДЛЯ ИЗМЕРЕНИЙ ВОДЫ НИЗКОЙ УДЕЛЬНОЙ АКТИВНОСТИ

Ю.В. СИВИНЦЕВ
Институт атомной энергии им.И.В.Курчатова,
Москва,
Союз Советских Социалистических Республик

Abstract–Аннотация

A SYSTEM OF RADIATION MONITORING, AND METHODS AND EQUIPMENT FOR MEASURING WATER OF LOW SPECIFIC ACTIVITY.

The author considers criteria for the radiation protection of the population and the environment. He describes the role of procedures for monitoring waters of low specific activity in the framework of a system for ensuring the radiation safety of the population living near a nuclear power station. The main technical characteristics (background, efficiency, sensitivity) of the laboratory equipment for gamma spectrometric analysis of water samples of low specific activity are discussed.

СИСТЕМА РАДИАЦИОННОГО КОНТРОЛЯ, МЕТОДЫ И АППАРАТУРА ДЛЯ ИЗМЕРЕНИЙ ВОДЫ НИЗКОЙ УДЕЛЬНОЙ АКТИВНОСТИ.

Рассмотрены критерии радиационной защиты населения и окружающей среды. Описаны место и роль системы контроля вод низкой удельной активности в системе обеспечения радиационной безопасности населения, проживающего вблизи АЭС. Охарактеризованы основные физические параметры (фон, эффективность, чувствительность) лабораторной установки для гамма-спектрометрического анализа проб воды низкой удельной активности.

Введение

При решении проблемы радиационной безопасности по мере развития масштабов и мощности ядерных установок возникали новые задачи. На I этапе была поставлена и успешно разрешена задача обеспечения радиационной безопасности персонала ядерных установок (так наз. категории А облучаемых лиц). На II этапе главное внимание привлекла проблема радиационной защиты населения, проживающего вблизи крупных ядерных установок (категория Б - отдельные лица из населения). В настоящее время для ядерной энергетики одной из основных стала задача обеспечения радиационной безопасности всей человеческой популяции (категория В - население в целом). В ближайшие годы несомненно начнется изучение проблемы радиационной защиты биосферы, которое потребует координации усилий в международном масштабе. Полагая, что такой же подход необходим и для категории В, рассмотрим вопросы обеспечения радиационной безопасности лиц категории Б, возникающие при удалении жидких радиоактивных отходов с крупных ядерных установок.

I. Критерии радиационной защиты населения и окружающей среды

В соответствии с подходом, сложившимся в научной литературе и на практике, под радиационной безопасностью мы понимаем обеспечение условий, при которых дозы облучения определенной группы людей не превышают соответствующих дозовых пределов, устанавливаемых компетентными органами. Как известно, рекомендации в этой области вырабатывает Международная комиссия по радиологической защите (МКРЗ), а нормативные документы - Национальные комиссии по радиологической защите (НКРЗ).

В 40-50-ые годы рекомендации МКРЗ [1] и соответствующие отечественные нормативные документы [2] регламентировали не только допустимые дозы внешнего и внутреннего облучения человека, но и предельно допустимые мощности дозы внешнего облучения и предельно допустимые концентрации (ПДК) радионуклидов в воздухе и в воде. Как отдельные ученые [3], так и органы надзора [4] рассматривали последние в качестве основного критерия радиационной опасности. В итоге такого подхода в указанный период в практике проектирования и эксплуатации систем радиационной безопасности на ядерных установках фактически доминировал принцип ограничения мощности дозы, т.е. радиационного поля, воздействующего на человека, а не дозы, т.е. результата взаимодействия поля и облучаемой среды.

Регламентация мощности дозы внешнего облучения в сочетании с ограничением продолжительности работы в радиационно опасной зоне (так наз. "защита временем") и разработкой методов экранирующей защиты и дистанционного обращения с радиоактивными материалами ("защита ослаблением и расстоянием") оказалась эффективным способом обеспечения радиационной безопасности персонала ядерных установок.

Регламентация ПДК подавляющего большинства радионуклидов (за исключением радиоактивных изотопов благородных газов - аргона, криптона, ксенона - и короткоживущих изотопов углерода, азота и кислорода) базировалась на особой опасности внутреннего облучения человека. В частности, учитывалась практическая невозможность применения методов защиты временем, расстоянием и ослаблением. Вследствие этого при расчетах ПДК в воздухе и воде исходили из крайних предположений. Так, принималось, что рассматриваемый радионуклид постоянно присутствует, например, в воздухе, который в течение всей жизни вдыхает человек, проживающий вблизи ядерной установки. Для получения численных значений ПДК использовали биофизические сведения о допустимой величине дозы внутреннего облучения и коммуникационных константах в живом организме, т.е. скоростях отложения и выведения тех или иных элементов для различных органов и тканей. В отношении дозы внутреннего облучения предполагалось, что она обусловлена инкорпорированным радионуклидом, достигшим равновесной активности q_0 (мккюри) в критическом органе.

Вследствие такого нереалистического, консервативного подхода численные значения ПДК оказались исключительно низкими и зачастую лежащими за пределами чувствительности соответствующей спектрометрической и радиометрической аппаратуры. В целях получения достоверных данных об уровнях радиоактивных загрязнений объектов внешней среды, например речной воды, службам внешней дозиметрии приходилось использовать трудоемкую процедуру предварительного концентрирования радионуклидов из больших объемов. Помимо отмеченного практического неудобства упомянутые методы анализа не давали возможности предупреждать случаи повышенного загрязнения внешней среды и приводили бы к регистрации таких событий *post factum*.

Исследования больших контингентов из категорий А и В с помощью спектрометров излучений человека (СИЧ) показали, что как число случаев измеримого внутреннего загрязнения, так и активности инкорпорированных радионуклидов весьма малы. Так, по нашим данным, полученным с помощью установки СИЧ-IM [5,6], даже на испытательном реакторе РФТ доля персонала с измеримым (с чувствительностью порядка $10^{-3}\ q_0$ для осколочных изотопов) внутренним загрязнением не превышает 3-5%. В 90% таких случаев уровни внутреннего загрязнения составляют менее $10^{-2}\ q_0$.

Большое количество аналогичных данных позволило сделать вывод, что регламентация ПДК обеспечила не только достаточную, но даже избыточно высокую защиту от внутреннего облучения для лиц категорий А и Б. В значительной степени это было обусловлено отмеченным выше обстоятельством применения ПДК в качестве количественного критерия для мгновенных концентраций радиоактивных веществ в воздухе и воде. В действительности, как это ясно следует из описанного выше подхода МКРЗ и НКРЗ, получаемые численные значения являются среднегодовыми допустимыми концентрациями (СДК) радионуклидов в воздухе и воде.

Учитывая накопленный опыт по обеспечению радиационной безопасности населения вблизи крупных ядерных установок, НКРЗ при Министерстве Здравоохранения СССР при разработке действующих с 1969 г. в нашей стране "Норм радиационной безопасности (НРБ-69)" четко охарактеризовала упомянутые критерии как СДК и сохранила их только для целей оперативного контроля в наблюдаемой зоне. При этом специально отмечено, что для "оценки дозы облучения отдельных лиц из населения определяющим является годовое поступление радиоактивных изотопов в организм, а не их среднесуточная концентрация в воде и воздухе" [7].

МКРЗ пошла еще дальше и вообще упразднила понятие ПДК, подчеркнув необходимость исключительного применения величин дозовых пределов при оценках радиационной опасности [8]. На основе накопленного опыта можно с уверенностью утверждать, что как для персонала ядерных установок, так и для населения, проживающего в их окрестности, система дозовых пределов, разработанная МКРЗ, является достаточной.

Задача охраны внешней среды вблизи крупных ядерных установок, таких как ядерный реактор, атомная электростанция (АЭС) или радиохимический завод, требует выработки соответствующих количественных критериев для защиты внешней среды от радиоактивных отходов (РАО), в том числе жидких РАО. В настоящее время при весьма ограниченном количестве исходной информации о влиянии РАО на отдельные экосистемы или регионы (и тем более на биосферу в целом) в качестве такого временного количественного критерия целесообразно применять величину предельно-допустимой мощности выброса (ПДВ) радиоактивных веществ с ядерной установки во внешнюю среду. Численное значение ПДВ находят по соотношению

$$R_0 = \sum_i R_{0i} = \sum_i C_{0i} \cdot S$$

где C_{0i} – СДК i-ого радионуклида (кюри/литр);

S – диспергирующая способность реки или прибрежного района моря для жидких РАО или приземного слоя атмосферы для газообразных РАО (литр/сек).

Современная стратегия радиационной защиты базируется на убеждении, что если обеспечена радиационная безопасность человека, то тем самым достигается необходимая радиационная защита и любых цепей, звеньев и ячеек в биосфере. До тех пор пока в биосфере не будут обнаружены живые организмы, более чувствительные к действию ионизирующих излучений, чем человек, при расчетах ПДВ следует использовать величины СДК, рекомендованные НКРЗ для категории Б облучаемых лиц.

Идея и первые методы измерения мощности выброса радиоактивных веществ, удаляемых с ядерных реакторов, были предложены нами и развиты совместно с А.И.Докучаевым, В.Т.Корнеевым и Л.М.Лузановой в Институте атомной энергии им. И.В.Курчатова в начале 50-х годов [9]. Исследования по обоснованию ПДВ с 1957 г. ведутся группой сотрудников Института Прикладной Геофизики под руководством Е.Н.Теверовского [10].

В нашей стране НКРЗ одобрила метод расчета и численные значения ПДВ газообразных РАО. Ведется работа по подготовке аналогичных количественных критериев для жидких РАО [11]. В заключение данного раздела отметим, что концепция ПДВ позволяет существенно упростить систему радиационного контроля за РАО на ядерных установках в режиме их рутинной эксплуатации, а также ограничить суммарное количество радиоактивных веществ, удаляемых с ядерной установки во внешнюю среду.

II. Место и роль системы контроля вод низкой удельной активности

Современные комплексы радиационного контроля на крупных ядерных установках, например АЭС, включают в свой состав три системы - технологического контроля, биологической и внешней дозиметрии. Четкую границу между этими системами провести трудно, поскольку часто один и тот же детектор ионизирующих излучений может давать информацию всех трех типов. В самом общем виде можно лишь указать, что основное назначение системы биологической дозиметрии - регистрация уровней облучения персонала (категории А), внешней дозиметрии - получение аналогичных данных для лиц категории Б.

Методы и аппаратура для измерений воды низкой удельной активности играют доминирующую роль при контроле жидких РАО в системе внешней дозиметрии.

На этапах выбора площадки, строительства и предпусковых работ такие методы и аппаратура применяются для экспериментального определения диспергирующей способности водного бассейна, вблизи которого располагается ядерная установка, суточных и сезонных колебаний этого важного гидрологического параметра, а также в целях радиометрической съемки и картографирования фоновых характеристик для объектов внешней среды - воды, дисперсной фазы, илов, гидробионтов, почвы и растительности. В период пусковых и наладочных работ и опытной эксплуатации перечисленные задачи дополняются экспериментальным исследованием мощности выброса и радиоизотопного состава жидких РАО, а также степени изменчивости этих характеристик при различных режимах эксплуатации ядерной установки. Этот период максимального накопления информации в системе внешней дозиметрии должен завершиться получением данных о так наз. критическом пути и о величине допустимой мощности выброса жидких РАО.

Опираясь на упомянутые обобщающие характеристики, на стадии рутинной эксплуатации ядерной установки обширную систему радиометрических наблюдений за объектами внешней среды можно свернуть и довести радиационный контроль до разумного минимума.

Анализ возможных путей оптимизации системы радиационного контроля для жидких радиоактивных отходов (РАО) привел нас к заключению, что в целях обеспечения радиационной безопасности необходимо и достаточно осуществлять регистрацию мощности выброса, т.е. скорости удаления жидких РАО, в точке их сброса.

В виду относительно медленных изменений диспергирующей способности рек (или прибрежных районов моря) приемлемым является определение мощности выброса R за достаточно большой интервал времени, например за сутки (R , кюри/сутки). В общем случае

$$R \text{ (кюри/сутки)} = \sum_i R_i \text{ (кюри/сутки)}$$

где R_i - мощность выброса i -ого радионуклида, из числа присутствующих в сбросе в значительном количестве.

Помимо сокращения объема контроля определение мощности выброса удобно и для оценки относительной радиационной опасности сбросов. Упоминавшиеся выше последние рекомендации МКРЗ [7] и НКРЗ [6] наряду с величинами допустимых доз облучения содержат численные значения предела годового поступления P_i радионуклида с водой и продуктами питания. Сопоставляя значения мощности выброса R_i (кюри/сутки) и P_i (мккюри/год), легко получить безразмерные величины, характеризующие радиационную опасность RH удаляемых радионуклидов в условных единицах радиационной опасности (е. РОП):

$$RH = \sum_i (RH)_i = 3 \cdot 10^{-9} \sum_i R_i/P_i \ [\text{е. РОП}]$$

Знание абсолютной активности и компонентного состава радионуклидов в составе жидких РАО позволяет: I) провести прямое сопоставление с величиной ПДВ ; 2) выделить главные источники радиационной опасности среди различных компонентов сброса ; 3) не контролировать те радионуклиды, вклад которых в величину $\sum_i (RH)_i$ меньше заданной величины (например, I%).

Опыт эксплуатации свидетельствует, что изотопный состав жидких РАО отличается достаточным постоянством особенно в период рутинной эксплуатации ядерных реакторов. Это позволяет осуществлять непрерывную регистрацию мощности сброса методом измерения интенсивности γ -излучения от соответствующего трубопровода. Кроме первичной градуировки показаний такого интенсиметра необходимость анализа изотопного состава проб сбросных вод возникает только в тех случаях, когда отмечено существенное возрастание мощности выброса. Согласно современным представлениям МКРЗ и НКРЗ при уровнях облучения,менее I/3 соответствующего дозового предела, радиационный контроль, проводимый в целях радиационной безопасности, можно не осуществлять. Поэтому в качестве рабочего уровня, превышение которого требует проведение компонентного анализа сбросных вод, целесообразно выбрать значение 0,3 ПДВ.

Этот метод имеет особые преимущества, если система удаления жидких РАО с ядерного реактора (или др. крупной ядерной установки) спроектирована и эксплуатируется на основе схемы, исключающей загрязнение объектов внешней среды при возрастании мощности выброса свыше ПДВ. Главным принципом такой схемы является применение минимум 2-х емкостей для сбора протечек радиоактивных вод или дебалансных вод в системе замкнутых циркуляционных контуров. В каждый момент времени одна из этих емкостей используется для сбора, другая - для удаления жидких РАО после предварительного компонентного анализа представительной пробы и под непрерывным радиационным контролем мощности выброса. При получении сигнала о превышении рабочего уровня переключением соответствующей запорной аппаратуры сброс жидких РАО во внешнюю среду прекращается. При непрерывном удалении больших объемов жидких РАО с ядерной установки контроль мощности выброса позволяет своевременно переключить отходы повышенной удельной активности для сбора в буферном хранилище.

Ш. Лабораторная установка для гамма-спектрометрического анализа вод низкой удельной активности

При контроле жидких РАО, удаляемых с ядерных реакторов, применяется метод лабораторного анализа представительных проб с помощью γ-спектрометрической аппаратуры [8]. В подавляющем большинстве случаев этот метод позволяет относительно быстро получить достоверную информацию о компонентном составе и активности радионуклидов даже при низкой удельной активности пробы воды и избежать трудоемкой процедуры радиохимического анализа смеси изотопов.

Нами разработана и в течение ряда лет успешно эксплуатируется на нескольких ядерных реакторах лабораторная установка (экспериментальный стенд) для γ-спектрометрического анализа проб воды низкой удельной активности. Необходимая чувствительность (порядка 10^{-9} кюри в пробе объемом 3,75 л) достигнута за счет комплекса мер по обеспечению высокой эффективности регистрации γ-излучения пробы и снижения фона, обусловленного естественной радиоактивностью детектора и экранирующей защиты.

В состав стенда входят стальной низкофоновый домик, высокочувствительный низкофоновый сцинтилляционный детектор (сцинтиблок), источник стабилизированного высокого напряжения, многоканальный амплитудный анализатор и вспомогательная аппаратура.

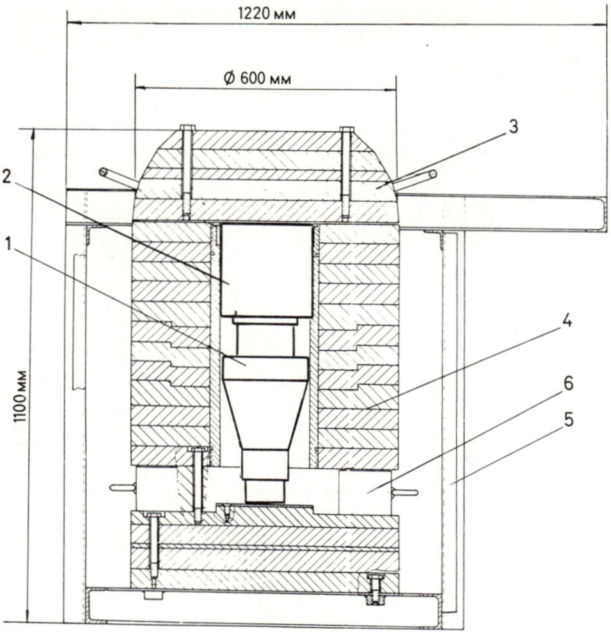

Рис.1. Конструкция стенда для измерений проб воды низкой удельной активности: 1 –сцинтиблок, 2 – контейнер с пробой воды, 3 – откатывающаяся крышка, 4 – стальная защита, 5 – рама, 6 – съемная деталь.

Стальной низкофоновый домин (СНФ), толщина защитных стенок которого составляет не менее 20 см для любого направления, имеет колодец глубиной 620 мм и диаметром 204 мм (рис. I). Домик изготовлен из обычной стали 3, которая, как показали специальные исследования, выполненные в ИАЭ [I2], отличается малой собственной радиоактивностью и кроме того имеет относительно невысокую стоимость. Толщина защиты выбрана для практически полного поглощения мягкой компоненты космического излучения, в результате чего фоновая скорость счета примененного сцинтилляционного детектора снижается примерно в 40 раз. Защита из свинца, хотя она и сократила бы размеры домика, не используется из-за наличия в этом материале природных радиоактивных примесей.

В колодце домика под крышкой (3), откатывающейся по верхней части рамы (5), размещены сцинтиблок (I) и медный стакан (2) для пробы воды. Размеры стакана Ø I90xI60 мм, объем

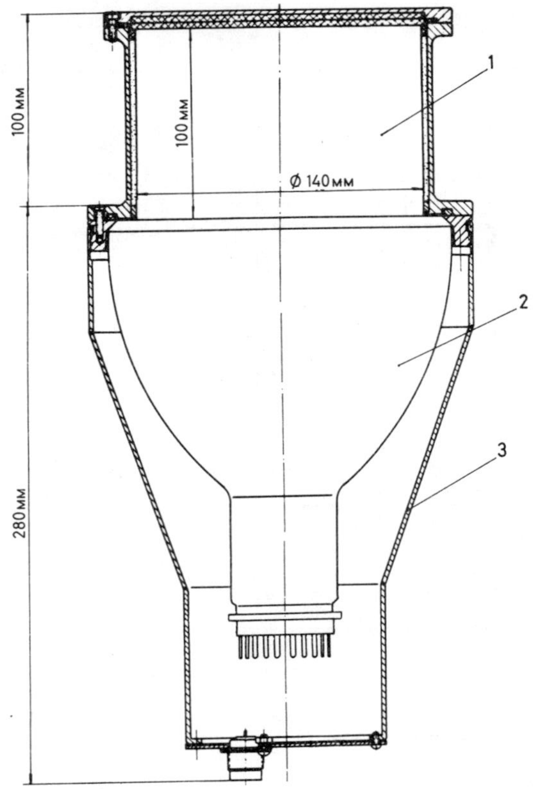

Рис.2. Конструкция сцинтиблока: 1 – кристалл NaJ(Tl), 2 – ФЭУ-49, 3 – кожух.

4,5 литра. Сцинтиблок опирается на "юбочку" – конструктивную деталь, препятствующую изгибу соединительного кабеля. Для введения кабеля в колодец предусмотрено окно, закрываемое съемной деталью (6).

Высокочувствительный низкофоновый сцинтиблок состоит из монокристалла NaJ(Tl) диаметром 140 мм и высотой 100 мм и фотоэлектронного умножителя ФЭУ-49 с диаметром фотокатода 160 мм (рис. 2). Кристалл (1) приклеен силиконовым или эпоксидным клеем к фотокатоду ФЭУ (2), специально подобранного к данному кристаллу для получения наилучшей разрешающей способности. Место склейки герметизировано из-за большой гигроскопичности кристалла. Для улучшения светособирания кристалл окружен слоем тефлоновой крошки или порошкообразной окиси магния со всех сторон за исключением торца, обращенного к ФЭУ.

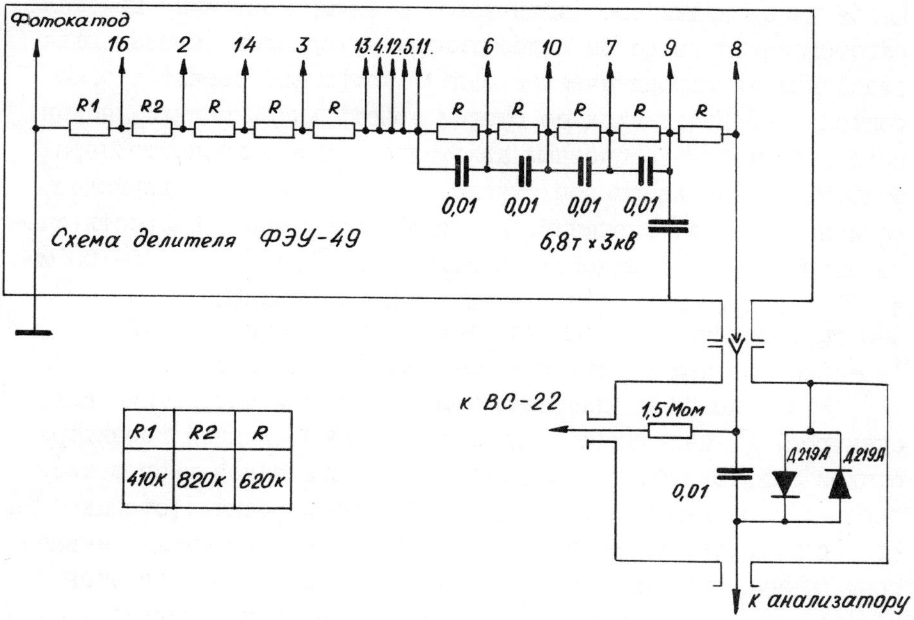

Рис.3. Схема включения сцинтиблока.

Кристалл и ФЭУ заключены в светонепроницаемый кожух (3), выполняющий также функции антимагнитной защиты. Конструкции домика СНФ и сцинтиблоков разработаны в ИАЭ под руководством И.В.Наумова.

Высокое напряжение (около I200–I600 в) от высоковольтного выпрямителя ВС-22 или батарейного источника питания БИП для ФЭУ и сигнал с него подают по одному и тому же кабелю РК-I. Делитель напряжения на ФЭУ собран по схеме, представленной на рис. 3.

С целью получения максимальной чувствительности предприняты специальные меры по снижению содержания природных радиоактивных элементов в монокристалле, фотоумножителе и конструктивных элементах сцинтиблока. В этой связи особое внимание уделено уменьшению концентрации калия (в том числе его естественно-радиоактивного изотопа K^{40} с периодом полураспада $I,3 \cdot I0^{9}$ лет и энергией γ-квантов I,46 Мэв) в исходном йодистом натрии, стекле колбы фотоумножителя, аппаратуре, применяемой для выращивания монокристаллов и варки стекла, и др. В разработанном сцинтиблоке применены ФЭУ-49, монтируемые в колбе из бескалийного (натрового) стекла, контейнер кристалла изготовляется из особо малоактивных материалов (нержавеющая сталь или электролитическая медь), устранено стекло в окне контейнера. На стендах со старой конструкцией сцинтилляционного детектора (со счетным кристаллом) для окна контейнера было применено кварцевое стекло, практически не содержащее радиоактивных загрязнений. В результате фоновая скорость счета сцинтиблоков с монокристаллами $NaJ(Tl)$ ⌀ 140x100 мм и ФЭУ-49, примененных на таких стендах, понижена до 4-6 тыс. имп/час.кг сцинтиллятора в диапазоне энергий 120-3000 кэв при измерениях под защитой 20 см стали.

Сцинтиблок предназначен для измерений энергетических спектров γ-излучения анализируемых проб. Важной характеристикой такого детектора является энергетическое разрешение, определяемое как ширина пика полного поглощения (фотопика) на его полувысоте, отнесенная к энергии регистрируемых γ-квантов. Обычно эту величину определяют с помощью γ-квантов Cs^{137} (E_γ = 660 кэв) и выражают в процентах. Энергетическое разрешение сцинтиблоков с монокристаллами $NaJ(Tl)$ ⌀ 140x100 мм и ФЭУ-49, разработанных в ИАЭ и использованных на описываемых стендах, составляет ~10%.

Импульс с выхода сцинтиблока через переходную коробку с высокочастотными разъемами подается на вход многоканального амплитудного анализатора. Схема соединений в переходной коробке представлена на том же рис. 3. На большинстве стендов использован многоканальный амплитудный анализатор АИ-128-1 ("Тензор-1"). При наличии сложной смеси γ-излучателей оказалось удобным применять систему АРИС - специализированное устройство для автоматической расшифровки измеренных спектров [13].

Чувствительность стенда ЭНАСТ характеризуют величиной минимально детектируемой активности (МДА). По определению МДА-активность, которую можно измерить на данной установке за полное время измерения T так, чтобы относительная погрешность δ измерения скорости счета источника (пробы воды) не превышала заданного значения. Выражение для определения МДА имеет следующий вид:

$$A_{min} = \frac{1 + 2\delta\sqrt{n_\varphi T}}{2{,}22 \cdot \delta^2 \cdot F T}$$

где $n_ф$ – скорость счета фона, имп/мин ;
T – полное время измерения, мин ;
F – эффективность регистрации γ -квантов данного излучателя.

При определении МДА по этой формуле следует учитывать, что такая оценка справедлива только в том случае, если заданная относительная погрешность δ не превышает 25% [I4] .

Рассчитанная с помощью этой формулы МДА Na^{24} для E_γ = 2,76 Мэв, которую можно измерить на стенде с использованием детектора с кристаллом $NaJ(Tl)$ размерами Ø I40xI00 мм, равна $I,3.I0^{-I0}$ кюри для T = 60 мин (30 мин измерения фона + 30 мин измерения эффекта с фоном), δ = 25% при фотоэффективности регистрации γ -квантов с энергиями 2,76 Мэв равной 0,67%.

Для определения фотоэффективности регистрации γ -квантов различных энергий была проведена градуировка стенда. В медный стакан наливали по 3,75 л растворов различных изотопов, после чего стакан устанавливали на детектор в рабочем положении и регистрировали аппаратурный спектр. По результатам измерений определяли фотоэффективность регистрации F - отношение числа импульсов, зарегистрированных в фотопике, к полному количеству квантов, испускаемых выбранным γ -излучателем.

Для градуировки были использованы следующие радиоактивные изотопы: Hg^{203} (E_γ = 0,279 Мэв) ; J^{I3I} (0,364 Мэв) ; Cs^{I37}(0,662 Мэв) ; Co^{60} (I,I7 и I,38 Мэв) ; K^{40} (I,46Мэв) и Na^{24} (I,38 и 2,76 Мэв). Абсолютную активность вводимых источников определяли с помощью 4π-счетчика β -излучения. Кривая фотоэффективности сцинтиблока для γ -излучающих радиоактивных веществ, растворенных в 3,75 л воды, построена на рис. 4. В качестве иллюстрации на рис. 5 представлены аппаратурные спектры, которые зарегистрированы при измерениях проб воды, содержащих Co^{60} и Na^{24}. Градуировочные работы проведены сотрудником ИАЭ В.А.Канарейкиным.

При многочисленных измерениях особое внимание уделяется тщательному сохранению радиационной чистоты внутренней поверхности колодца СНФ домика и отсутствию нефиксированных долгоживущих γ -загрязнений на пробах, вносимых внутрь колодца. В случае измерения проб, содержащих долгоживущие β-, γ- излучатели, контейнер с пробой (воды и т.п.) размещают под за-

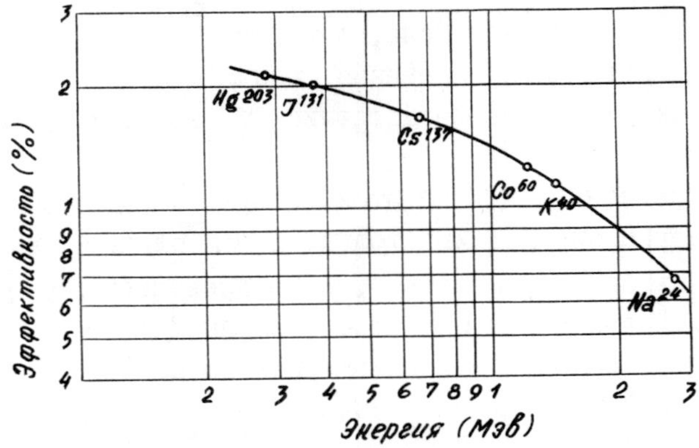

Рис.4. Фотоэффективность сцинтиблока для проб воды, содержащих γ-излучатели разных энергий.

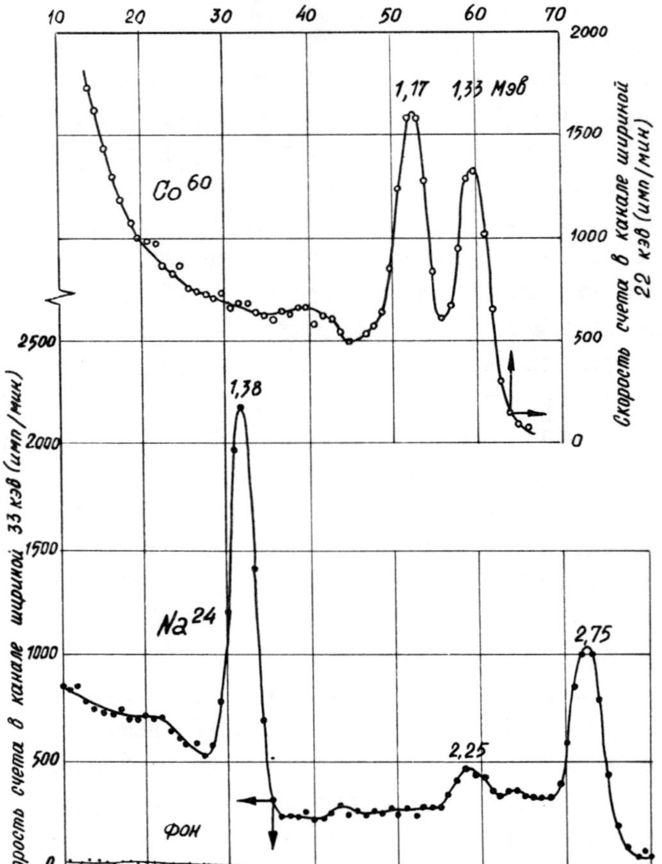

Рис.5. Аппаратурные спектры для проб воды, содержащих ^{60}Co ($2\cdot10^{-7}$ Ки) и ^{24}Na ($3{,}6\cdot10^{-7}$ Ки).

щитой только после погружения его в пластикатовый мешок или другую негигроскопическую и герметичную оболочку. Для проб воды, существенно различающихся по активности, соответственно закрепляют отдельные канистры и медные стаканы, снабдив их необходимой маркировкой для низкой и высокой активности. Фон в домике перед измерением пробы регистрируют, разместив на сцинтиблоке соответствующий медный стакан, заполненный чистой водой.

ЛИТЕРАТУРА

[1] Recommendations of the International Commission on Radiological Protection (ICRP), Publ. 2, Oxford, 1959. Пер. с англ., Радиационная Защита (Рекомендации МКРЗ, вторая публикация), М., Госатомиздат, 1961.

[2] Меры Защиты Работающих с Радиоактивными Веществами, М., Медгиз, 1958.

[3] ГУСЕВ, Н.Г., О Предельно Допустимых Уровнях Ионизирующих Излучений, М., Медгиз, 1961.

[4] МАРЕЙ, А.Н., ТАРАСЕНКО, Н.Ю., "Задачи санитарного контроля за состоянием внешней среды", В Сборнике радиохимических и дозиметрических методик, М., Медгиз, 1959.

[5] СИВИНЦЕВ, Ю.В., АРУТИНОВ, О.М., КАНАРЕЙКИН, В.А., ПАНОВ, М.А., "Спектрометр излучений человека", Атомная Энергия 18 2 (1965) 141.

[6] СИВИНЦЕВ, Ю.В., КАНАРЕЙКИН, В.А., "Модернизация установки СИЧ-1 и ее градуировка", Доклад на Всесоюзном Симпозиуме по Изучению Радиоактивности Человека, Обнинск, 1969.

[7] Нормы Радиационной Безопасности (НРБ-69), М., Атомиздат, 1970 (II изд. – 1972).

[8] Recommendations of the ICRP, Publ. 9, Oxford, 1966.

[9] СИВИНЦЕВ, Ю.В., Радиационная Безопасность на Ядерных Реакторах, М., Атомиздат, 1967.

[10] ТЕВЕРОВСКИЙ, Е.Н., РУЖЕНЦОВА, И.Н., "Предельно допустимые выбросы радиоактивных веществ в атмосферу", Атомная Энергия 37 3 (1974) 239.

[11] ПИСАРЕВ, В.В. и ЦЫБИЗОВ, И.С., "Анализ критериев, определяющих размеры допустимого радиоактивного сброса АЭС в речные системы", Труды настоящего симпозиума.

[12] СИВИНЦЕВ, Ю.В., СИТНИКОВ, Г.А., ЮЗГИН, В.С., "Исследование защиты спектрометра излучений человека", Доклад на Всесоюзном Симпозиуме по Измерению Радиоактивности Человека, Обнинск, 1969.

[13] ВЫСОТСКИЙ, Е.Д., СИВИНЦЕВ, Ю.В., "Специализированное полуавтоматическое устройство для регистрации и расшифровки спектров", Ibid, 1969.

[14] ДЕМЕНТЬЕВ, Б.А., Измерение Малых Активностей Радиоактивных Препаратов, М., Атомиздат, 1967.

DISCUSSION

K.E. WHITE: I should like to take up your remark that, contrary to United Kingdom practice, you use pre-atomic age steel for low-background shielding. We in the United Kingdom are well aware of the advantage of pre-war steel for shielding and use it widely, particularly for whole-body counting rooms. We have plenty of old guns and battleships to furnish material for this purpose.

Yu.V. SIVINTSEV: I may mention, in this connection, that we made a comparative study of the residual background count-rate of large scintillation detectors contained inside shielding of modern steel and of "pre-atomic" steel. This study was performed using the steel chamber of the SICh-1M spectrometer at the I.V. Kurchatov Institute of Atomic Energy, which is mentioned in the paper, and a similar steel chamber manufactured by the British firm "Nuclear Enterprises" for the Institute of Medical Radiology at Obninsk, near Moscow.

Alternate measurements of the background of "potassium-free" scintillation detectors (of Soviet and British manufacture) in these two chambers showed that the intrinsic radioactivity of the modern steel is just as low as that of the pre-1945 steel. However, the construction of low-background installations calls for radiation control of the materials employed — we encountered cases where the background was increased through radioactive contamination of the steel by both natural (U, Th) and artificial (^{60}Co) sources. The results of these studies are described in detail in Ref.[12] of the paper.

K.E. WHITE: You quoted background rates of 4 to 6 counts/h for the large sodium-iodide crystal. Were these figures for one energy channel of your multi-channel analyser?

Yu.V. SIVINTSEV: The background count mentioned was 4 - 6 counts/h per gram of scintillator, which corresponds to an overall count-rate of about 10 counts/s in the energy range 120-3000 keV.

K.E. WHITE: Was your total gamma counting of liquid effluent carried out using a flow-through cell around the crystal, or do you take individual 4-litre samples?

Yu.V. SIVINTSEV: Individual samples were measured in copper vessels with a capacity of 4 litres.

J.O.H. SNIHS: When you derive the permissible releases you make conservative assumptions as regards correlations between releases and radiation doses to man. In your calculations do you consider the dose commitment for, say, a year of practice, and if so how conservative are you in your assumptions regarding the biological half-life of the radionuclides released in the water? Do you use the physical half-life or an assumed or known biological half-life?

Yu.V. SIVINTSEV: Our approach to the calculation of maximum permissible release is conservative mainly in that account is taken of concentration in the "water — mud — fish" food chain. The paper by V.V. Pisarev showed that, if the concentration factor in this food chain is taken into account, the numerical value of the mean annual permissible concentration of radionuclides in the liquid effluents must be considerably reduced. In addition, a basic assumption made is that the liquid radioactive wastes are discharged continuously.

Under actual conditions at Soviet nuclear power stations liquid radioactive wastes are not discharged into the aquatic environment, water being used solely for cooling the condensers of the non-radioactive, secondary circuit. Control studies on the water and aquatic organisms in the river Don in the neighbourhood of the Novovoronezh nuclear power station, and on the river Neva during testing of the first nuclear-powered icebreaker "Lenin", showed that no radioactive contamination (apart from that due to global fallout) could be detected in these aquatic ecosystems. The results of these studies have appeared in the following publications:

1. SIVINTSEV, Yu.V., et al., Studies on radiation purity in respect of atmospheric air and the water of the river Neva in the region where mooring tests were carried out with the nuclear-powered icebreaker "Lenin" (in Russian), Atomnaya Energiya 10 3 (1961) 253.

2. SIVINTSEV, Yu.V., STEFANOVICH, A.N., Three years service confirms radiation safety of icebreaker "Lenin", Bull. of Int. Assoc. of Navigation Congresses 2 8 (1963) 5.

3. PETROS'YANTS, A.M., Radiation conditions in the neighbourhood of the Novovoronezh Nuclear Power Station (in Russian), Atomnaya Energiya 21 6 (1966) 492.

4. PETROS'YANTS, A.M., Radiation conditions in the neighbourhood of the I.V. Kurchatov Nuclear Power Station at Beloyarsk (in Russian), Atomnaya Energiya 23 1 (1967) 38.

5. OVCHINNIKOV, F.Ya, et al., Five years operational experience with the Novovoronezh Nuclear Power Station (in Russian), Atomnaya Energiya 27 4 (1969) 274.

J.D. MARSHALL: Could you say how the data from the gamma-spectrometer are processed? How do you derive the activity for each individual isotope from a spectrum for several isotopes?

Yu.V. SIVINTSEV: The spectrum obtained with the equipment described is analysed with the aid of calibration spectra, and the corresponding weighting factors are obtained. Details of the method and equipment used for automatic analysis of the measured spectrum are given in Ref. [13] of the paper.

Josette M. MATUTANO: You mentioned a sensitivity limit of 10^{-9}Ci. Is this the volumetric activity, i.e. 10^{-9} Ci/m^3, or the total activity of your sample?

Yu.V. SIVINTSEV: The value given is the total activity of the sample. The minimum detectable activity is 10^{-9}Ci in a sample of volume 3.75 litres.

Josette M. MATUTANO: Is your sample measured in liquid form or is it previously concentrated?

Yu.V. SIVINTSEV: The main advantage of using this method is that one can obtain information on the activity and composition of the water sample without having to subject it to radiochemical treatment. The measurement is performed directly with a water sample taken from the liquid radioactive waste discharge system.

M.J. OWERS: Your measurements on low-activity water are gamma measurements. I should be interested to know how you carry out low-level beta and alpha measurements, in particular the former.

Yu.V. SIVINTSEV: To analyse the beta-activity of low specific activity waste waters, use is made of high-sensitivity radiometers with detectors consisting of plastic surface-activated scintillators. This method was developed in the USSR and details have been published by the authors. Radiometers with a sensitivity of 10^{-9}-10^{-10}Ci/l have been produced in this way for beta-emitters, with a maximum beta spectrum limit higher than 0.3 MeV. An important advantage of these detectors is that their measurements are independent of the colour and turbidity of the samples. For measurements on water of extremely low specific activity, and for the measurement of ^{3}H and ^{14}C, recourse must be had to radiochemical concentration, the processed samples being introduced in the form of a gel into a liquid scintillator. The following three references give a full description of the procedures in question:

1. ARSAEV, M.I., et al., A surface-activated plastic scintillation detector for measuring the specific beta-activity of aquatic environments (in Russian), Yadernoe Priborostroenie, No. 13, Atomizdat, Moscow (1970) 14.

2. ARTEMENKOVA, L.M., et al., Methods and equipment for monitoring of open reservoirs, drinking water, and waste water discharged from nuclear power plants (in Russian), Radioisotopnye pribory i apparatura (Proc. Comecon Symp., Leipzig) 3 (1970) 14.

3. BACHURIN, A.V., et al. A portable device for the detection of low specific beta-activities in aqueous solutions (in Russian), Yadernoe Priborostroenie, No. 14, Atomizdat, Moscow (1971) 111.

G.BOERI: Do you base your discharge limits on maximum permissible concentrations or on maximum yearly amounts of radioactivity?

Yu.V. SIVINTSEV: As can be seen from the relationship given in our paper, the discharge limit, i.e. the maximum amount of radioactive substances which may be discharged into the atmospheric or aquatic environment per unit time, is based on the mean annual permissible concentrations.

G. BOERI: Do the discharge limits take account of the environmental capacity of specific sites and of all possible exposure pathways, or do you have national or regional discharge limits?

Yu.V. SIVINTSEV: The discharge limits are worked out both for the particular site and for the entire area (e.g. River Don, Lake Bajkal, etc.).

FACTORS DETERMINING THE ENVIRONMENTAL CAPACITY FOR RECEIVING RELEASES FROM NUCLEAR FACILITIES IN THE AQUATIC ENVIRONMENT

(Session X)

Chairman

L.J. OTTENDORFER (Austria)

RADIOLOGICAL EFFECTS OF A NUCLEAR POWER PLANT ON A RIVER SYSTEM, AS DEMONSTRATED BY THE GUNDREMMINGEN BWR ON THE DANUBE

H. HERRMANN, M. RUF, K. HÜBEL, W. LÜNSMANN
Bayerische Biologische Versuchsanstalt,
Munich, Federal Republic of Germany

Abstract

RADIOLOGICAL EFFECTS OF A NUCLEAR POWER PLANT ON A RIVER SYSTEM, AS DEMONSTRATED BY THE GUNDREMMINGEN BWR ON THE DANUBE.

The Gundremmingen Nuclear Power Plant (boiling water reactor, 237 MW(e)), on the Upper Danube, has been in operation since 1967. Radiological data have been collected throughout its period of operation. The behaviour of the radioactive waste products in the ecological system of the Upper Danube (water, sediments, suspended solids, water-plants, fish) has been analysed in connection with environmental contamination, the uptake capacity of the Danube, and the possible pathways to man. As a result of the investigations, it seems possible to build further nuclear power plants on the Danube, if their rates of release of radioactivity are similar to those at Gundremmingen.

Most of the world's electrical energy at the present time is produced by thermal power stations, which mainly use fossil fuels. The limiting factor in the production of electrical energy is the limited reserves of these fossil fuels and — if we consider the German part of the river Danube — the almost complete exploitation of the existing water power by hydro-electric stations. So nuclear plants become an additional primary energy source for satisfying the rapid increase in electrical energy demand. The nuclear power plant at Gundremmingen on the Danube, with a capacity of 237 MW(e), has been in operation since 1967.

Since then the Bayerische Biologische Versuchsanstalt in Munich, supported by various Ministries, has been collecting and interpreting the radiological data related to this plant.

The following are analysed: water samples upstream and downstream (up to 10 km) of the discharge point of the reactor; sediment samples at various distances from the plant, also upstream and downstream (up to 150 km) of the discharge point; samples of seston (suspended solids), fish (flesh and bones) and water plants. In addition, the reactor effluent before and after mixing with the cooling water, and this mixture after finally joining the Danube, are analysed. The water and the sediment are analysed from samples taken throughout each month, whereas the other samples are taken monthly at random.

Table I shows the artificial radionuclides in the Danube water at Gundremmingen after complete mixing with the reactor waste water and the cooling water. This mixture of radionuclides results from both fallout activity and from the reactor discharges. The values for the Danube water above the discharge point of the power plant show that about half this activity (in the surface water plus reactor effluent plus cooling water), with the exception of the tritium, results from fallout.

TABLE I. ARTIFICIAL RADIONUCLIDES IN DANUBE WATER AT GUNDREMMINGEN AFTER MIXING WITH REACTOR EFFLUENT AND COOLING WATER
Annual averages in picocuries per litre (pCi/l): - indicates not detected

Nuclide \ Year	1967	1968	1969	1970	1971	1972	1973
H-3	900	600	900	800	800	800	320
Co-58	0.07	0.04	0.04	0.03	0.04	0.07	0.04
Co-60	0.02	0.02	0.02	0.005	0.02	0.03	0.02
Sr-89	0.48	0.30	0.74	0.45	0.72	0.24	0.68
Sr-90	0.34	0.22	0.43	0.25	0.31	0.26	0.17
Zr-Nb-95	-	0.01	0.11	0.30	0.29	0.07	0.06
Ru-103	0.05	0.02	-	-	-	-	-
Ru-106	0.07	0.06	0.36	0.60	0.37	0.07	0.05
Sb-124	0.06	-	0.01	0.09	0.10	0.02	0.05
Sb-125	-	-	-	0.04	0.03	0.01	0.02
I-131	0.16	0.10	0.16	0.23	0.50	0.11	0.11
Cs-134	-	-	0.01	0.003	-	-	-
Cs-137	0.15	0.13	0.09	0.10	0.11	0.09	0.06
Ba-La-140	0.10	0.04	0.02	0.12	0.10	0.03	0.02
Ce-141/Ce-144	0.01	0.01	0.22	-	0.21	0.19	0.08
Totals (without H-3)	1.5	0.9	2.3	1.7	2.8	1.2	1.3

Table II shows the activity of the reactor effluent (mixed with the cooling water) before it reaches the Danube water. With the aid of the Ge(Li) detector 25 nuclides were determined; those with more than 1% of the whole activity – tritium excluded – are shown in the table. The greatest part of the total activity comes from tritium, namely 99%. If we only consider the other nuclides (i.e. all except tritium) we have the following:

Sr-89:	46%	Co-58:	2%	Co-60:	2%
I-131:	20%	Cs-134:	3%	Zr-95:	1%
Cs-137:	4%	Ba-140:	3%	Nb-95:	1%
Ba-133:	4%	La-140:	3%	Ce-141:	1%
Sr-90:	3%	Ba-131:	2%	Ce-144:	1%

Thus, if we exclude tritium, it can be said that half the activity of the fission and activation products is due to strontium-89, while the other nuclides have a very small share of the total. The radionuclides that represent less than 1% of the total are not listed.

Table III gives a summary of the activity levels in the Danube caused by the Gundremmingen plant from 1967 to 1973. There are considerable variations from year to year. The activity of fission and activation products (tritium excluded) ranges between 1.2 and 3.8 Ci, the tritium figures varying between 18 and 134 Ci.

The discharge of radioactive effluents from nuclear plants into surface waters causes accumulation processes to take place in living things (fish, water plants) and inanimate matter (sediments and seston) in the water. In hydrobiological material radioactive concentrations can occur that are a thousand times as high as the activity levels of the water. (Table IV shows figures for sediments and seston from 1967 to 1973.)

The concentration factor is defined as the ratio between the activity of the hydrobiological material and the activity of the surrounding water. Table V shows these concentration factors (annual averages for 1973) for sediments, seston, fish flesh, fish bone and water plants. Because of the varying discharges of activity from the nuclear plant, the varying fallout rates and also the often widely varying biotopes, a state of equilibrium will never be attained.

At the present state of technology it is still true that a small amount of radioactivity reaches the atmosphere and the receiving water. It is now essential to know what is the receiving capacity of surface waters (here the Danube) for radioactive waste products, which are the most important radionuclides and what are the critical pathways for these radioactive waste products to man. Only when these parameters are known can one make a radioactivity evaluation based on the environment controls round the reactor.

The possible critical pathways to man are as follows. First is the atmosphere, and then comes drinking water. In the Upper Danube (Bavarian Danube) no drinking water is directly drawn off, but at several points there are connections between Danube water and groundwater. Other pathways are via food, but in these regions of Bavaria precipitation is fairly high, so it is normally not necessary to use Danube water for irrigating fields and meadows, whereas in other Danube countries this use of the Danube waters is far more common and the possibility of this pathway to man is greater. Another pathway is the direct contact of man with the Danube water, e.g. by boating, swimming or working in or on the river.

TABLE II. ACTIVITY OF RADIONUCLIDES (Ci) RELEASED INTO RECEIVING WATER AT GUNDREMMINGEN: ANNUAL TOTAL

Nuclide \ Year	1967	1968	1969	1970	1971	1972	1973
H-3	26.5	21.4	17.8	30.5	44.72	82.634	133.798
Co-58	0.31	0.21	0.17	0.21	0.055	0.028	0.032
Co-60	0.09	0.11	0.08	0.035	0.028	0.032	0.025
Sr-89	1.33	0.69	1.57	1.77	2.175	1.307	0.544
Sr-90	0.09	0.02	0.085	0.21	0.211	0.110	0.037
I-131	0.71	0.53	0.68	0.28	1.045	0.247	0.239
Cs-134	-	-	0.04	0.02	0.052	0.043	0.031
Cs-137	0.57	0.69	0.21	0.09	0.085	0.068	0.044
Ba-131							0.028
Ba-133							0.049
Ba-La-140	0.44	0.21	0.17	0.07	0.068	0.031	0.062
Totals (without H-3)	3.5	2.4	3.0	2.7	3.8	2.0	1.2

TABLE III. REVIEW OF ACTIVITY RELEASED AT THE NUCLEAR POWER STATION AT GUNDREMMINGEN

Item / Year	1967	1968	1969	1970	1971	1972	1973
Annual quantity of discharge of the Danube at Gundremmingen (m^3)	4.42×10^9	5.34×10^9	4.39×10^9	7.08×10^9	3.46×10^9	3.20×10^9	4.42×10^9
Waste quantities released into the Danube (m^3)	2.13×10^4	1.50×10^4	1.07×10^4	8.85×10^3	9.81×10^3	9.27×10^3	7.62×10^3
Activity of fission and activation products (without H-3) (Ci)	3.5	2.4	3.0	2.7	3.8	2.0	1.2
H-3 activity (Ci)	26	21.4	17.8	30.5	44.7	82.6	133.8
Increase in Danube water activity, annual average, without H-3 (pCi/l)	0.8	0.4	0.7	0.4	1.1	0.6	0.3
Increase in Danube water activity due to H-3, annual average (pCi/l)	6	4	4	4	13	25.8	30.3
Fall-out concentration in the Danube water above Gundremmingen, annual average (pCi/l)	0.7	0.5	1.6	1.3	1.7	1.7	1.2
H-3 concentration in the Danube water above Gundremmingen, annual average (pCi/l)	900	600	900	800	800	800	530

TABLE IV. CONCENTRATION OF RADIONUCLIDES IN SUSPENDED SOLIDS AND SEDIMENTS OF THE DANUBE
Activity in pCi/g fresh weight (annual average values): - indicates not detected

Nuclide	Sediments							Suspended solids (seston)						
	1967	1968	1969	1970	1971	1972	1973	1967	1968	1969	1970	1971	1972	1973
K-40	2.67	3.67	1.85	2.88	2.41	1.98	2.96	2.13	3.09	2.19	2.22	2.52	1.82	2.23
Co-58	0.38	0.30	0.21	0.08	0.39	0.48	0.12	0.50	0.72	0.10	0.07	0.90	0.19	0.34
Co-60	0.05	0.10	0.11	0.07	0.12	0.32	0.16	0.10	0.05	0.08	0.07	0.28	0.06	0.22
Sr-89	0.05	0.01	0.04	0.50	0.04	0.04	0.11	0.06	0.07	0.03	0.72	0.67	0.16	2.66
Sr-90	0.02	0.02	0.02	0.03	0.03	0.02	0.05	0.01	0.06	0.07	0.07	0.13	0.10	0.14
Zr-95/Nb	0.01	0.24	0.64	1.14	1.44	1.54	0.24	0.02	0.10	0.72	1.40	1.04	0.56	0.17
Ru-106/-103	0.50	0.18	0.61	0.67	0.42	0.17	0.05	0.25	0.36	1.45	0.95	0.52	0.18	0.13
Sb-124	-	-	-	0.05	0.03	0.05	0.03	-	-	0.02	0.12	0.14	0.05	0.07
Sb-125	0.15	0.03	-	0.01	0.01	0.06	0.07	0.31	0.03	0.01	0.03	0.01	0.02	0.12
I-131	0.10	0.06	0.42	0.28	0.22	0.14	0.03	0.16	0.22	0.08	0.41	0.13	0.14	0.45
Cs-137/-134	0.78	0.83	0.79	0.63	0.94	1.96	0.59	0.30	1.49	0.53	0.92	1.04	0.68	3.42
Ba-140/La	-	0.06	0.14	0.03	0.01	0.04	0.13	-	0.24	0.08	0.11	0.07	0.07	0.35
Ce-144/-141	0.23	0.62	0.75	1.10	0.96	0.82	0.22	0.19	0.39	1.16	1.18	1.05	0.64	0.19
Ra-226	0.18	0.19	0.13	0.11	0.10	0.11	0.23	0.12	0.19	0.20	0.17	0.18	0.19	0.19

TABLE V. CONCENTRATION FACTORS OF RADIONUCLIDES IN SEDIMENTS, SUSPENDED SOLIDS, MUSCLE AND BONE OF DANUBE FISH AND WATER PLANTS IN 1973
Average (minima and maxima in parentheses): - indicates not determined

Nuclide	Sediments	Suspended solids	Fish		Water plants
			Muscle	Bone	
K-40	1300 (650-2080)	1070 (520-2010))	700 (570-1580)	1560 (970-2450)	320 (220-490)
Mn-54	1520 (570-2430)	1030 (500-1890)	230 (60- 440)	- -	240 (65-430)
Fe-59	660 (660)	- -	540 (540)	- -	- -
Co-58	890 (310-2420)	2860 (290-15440)	130 (30- 310)	220 (100- 310)	530 (15-1800)
Co-60	2830 (540-6070)	5380 (470-12040)	130 (10- 380)	310 (140- 420)	220 (25-520)
Zn-65	480 (230- 740)	3030 (90-8820)	510 (190- 780)	- -	150 (150)
Sr-89	140 (20- 350)	1250 (1- 8340)	20 (10- 60)	40 (10- 60)	130 (2-510)
Sr-90	810 (60-1660)	2460 (10-10560)	20 (5- 40)	360 (19-1090)	20 (3- 50)
Zr-95	1430 (340-3570)	1250 (180- 3600)	160 (40- 440)	500 (180-1220)	150 (65-460)
Nb-95	1850 (710- 3970)	870 (80- 2700)	180 (40- 460)	- -	100 (20-250)
Ru-103	450 (40-1190)	870 (360- 2420)	170 (40- 270)	- -	100 (25-250)
Ru-106	1190 (260-2790)	4730 (140-10760)	70 (15- 170)	580 (300-1430)	30 (10- 90)
Sb-124	220 (180- 270)	780 (760- 800)	90 (70- 120)	810 (350-1100)	30 (30)
Sb-125	560 (230-1000)	4300 (480-28000)	100 (30- 170)	470 (470)	140 (60- 330)
I-131	6800 (510-13080)	- -	520 (250- 780)	- -	- -
Cs-134	3040 (840- 5360)	13950 (1000-58210)	200 (30- 380)	240 (60- 500)	320 (10-1530)
Cs-137	3160 (1340- 5950)	18780 (1210-93140)	170 (35- 350)	210 (20- 640)	180 (10-1000)
Ba-140/La	- -	530 (320- 780)	- -	- -	- -
Ce-141	750 (290-1080)	1020 (340- 1860)	150 (30- 560)	- -	270 (50- 810)
Ce-144	1010 (260-2180)	1330 (60- 8160)	60 (20- 200)	390 (140- 760)	310 (50- 630)
Ra-226	2070 (1390-2570)	2410 (480-10160)	70 (25- 110)	510 (170- 810)	190 (25- 690)
Th-232	3160 (810-5950)	1900 (280- 5000)	80 (40- 110)	190 (190)	12L (90- 180)

TABLE VI. CALCULATED NUCLIDE CONCENTRATION BASED ON 3000 pCi^{3}H/l AND 30 pCi/l OF OTHER RADIONUCLIDES WITH THE DISTRIBUTION STATED FOR GUNDREMMINGEN

Nuclide	(pCi/l)
H-3	3000
Co-58	2.4
Co-60	0.9
Sr-89	13.8
Sr-90	1.2
I-131	7.5
Cs-134	0.5
Cs-137	1.9
Ba-La-140	1.8
Total (without H-3)	30.0

The most important critical pathway to man for radioactive waste products is here the consumption of Danube fish. The average consumption of fish flesh in the Federal Republic of Germany is 1 g per person per day. But some people eat more than 150 g/day. Other Danube countries have other eating habits, so there are no consistent conditions for evaluating radiation doses.

Another problem is that, in the event that new nuclear power stations are built on the Danube, the composition of the radionuclides from these future reactors is not known.

The evaluation of the radiation risk for human beings is based on maximum values. Because of the very different physical and chemical properties of the various river or water systems, the following statements only hold for boiling water reactors and the Danube river system.

In the Federal Republic of Germany the conditions for radioactive waste release are based on the recommendations of the national Atomic Energy Commission (13.10.1969). These Recommendations have set the maximum permissible genetic dose to the population as 5 rem/30 years. However, reactor plants should only supply up to 2 rem out of the permitted 5 rem/30 years, and furthermore one-half should be from liquid and one-half from gaseous radioactive components. These recommendations are in line with those of the ICRP.

TABLE VII. RADIATION DOSE (approx. mrem/a) TO THE POPULATION DUE TO RADIOACTIVITY IN DANUBE WATER AND SEDIMENT

(a) Drinking water		1.3
(b) Fish consumption		
100 g/head • day		18.0
10 g/head • day		1.8
(c) Cattle watering (milk)		0.1
(d) Irrigation:		
Milk		(1.5)
Total nutrition		3.0
(e) External dose [a]		0.1 - 10
	Total	6.3 - 32.4

[a] Depending on exposure.

The share of the radioactive dose received via drinking water, the food chain or other radiation sources (external gamma radiation) all together shall not be higher than 30 mrem/a (excluding radioactivity from fallout and medical treatment).

In view of the possibility that the Danube may be used for providing drinking water, for irrigating the fields and meadows, for fishing, swimming and other recreational purposes, our thinking has been based on the MNQ values (arithmetic mean of the low water flow over several years) of the Danube.

If a number of future nuclear power stations additionally discharge radioactive waste products into one water system, it will first be necessary to work from a definite agreed radionuclide composition. To calculate the radiation risk, we started from the radionuclide composition as it is known to us from the Gundremmingen power station (see Table VI). For drinking water (obtained from surface water) we took a smaller decontamination factor, about the same as that of the filtration on the river bank. As a basis for evaluating contamination of food we took concentration factors from fish (Table V) and for evaluating the external gamma radiation load we also used concentration factors as shown in Table V.

From Table VII, in which the results of our calculations are given, it can be seen that — if one is a great fish-eater — it is possible to exceed the concentration limit of 30 mrem/a for the whole-body or gonad dose, but this would be an extreme case. These values in Table VII are only valid for the Bavarian part of the Danube, since in the middle and lower Danube little radioecological information is available.

The MNQ values were made a basis of our estimates of the radiological capacity of the Danube, because the river for several months is within the range of the low water flow mark. The so-called flow standard of 100 pCi/l was fixed at 30 pCi/l (^3H : 3000 pCi/l) to reduce the radiation dose to individual members of the critical population groups. Thus for the German part of the Danube there is an annual waste removal capacity of 560 Ci in the form of fission and activation products (without ^3H) and of 56 000 Ci of ^3H.

With the future planned larger nuclear power stations there will be an annual release rate of 10 Ci in the form of fission and activation products (without ^3H) and 1000 Ci of ^3H per 1000 MW(e) installed. Thus nuclear power plants with a total output of about 50 000 MW(e) could be built on the German part of the Danube. However, the need for such an amount of installed nuclear capacity for Bavaria does not appear to exist.

BIBLIOGRAPHY

Erste Verordnung über den Schutz vor Schäden durch Strahlen radioaktiver Stoffe. Erste Strahlenschutzverordnung, Bundesgesetzblatt I, Fed. Rep. Germany (1965) 61.

HERBST, W., Radioaktive Kontaminierung von Ernährungsketten, Ges. Strahlen-Umweltforschung Rep. GSF K-52 (1970).

HÜBEL, K., et al., "Radioecological investigations on the upper reaches of the River Danube", Comparative Studies of Food and Environmental Contamination (Proc. Symp. Otaniemi, 1973) IAEA, Vienna (1974) 105.

ICRP, Report of Committee II on Permissible Dose for Internal Radiation (1959), Recommendations of the Internal Commission on Radiological Protection, ICRP Publication 2, Pergamon Press, New York, London (1959) pp. XXV-XXVI.

RAJEWSKY, B., Strahlendosis und Strahlenwirkung, Thieme, Stuttgart (1956).

STRUXNESS, E.G., et al., Comprehensive Report of the Clinch River Study, USAEC Rep. ORNL-4035-VC-70 Waste Disposal and Processing (1967).

DISCUSSION

G. STOTT: I think this paper is important, not only for what it says but also for its implications. Where there are a number of nuclear plants discharging waste into a large river system flowing through several countries, some international body, such as IAEA, UNEP or the United Nations itself, should be entrusted with the task of controlling the discharges to ensure that other countries further downstream can – if they wish – operate nuclear power plants and safely discharge waste themselves. I suggest that the time for such guidance or control in a continent such as Europe is now ripe. In Paper IAEA-SM-198/57 Mr. Preston suggested a register of discharges.

On a more specific matter I should like to point out that this paper considers the effect of radioactive waste discharges at rates similar to that for Gundremmingen. The capacity of the latter plant, however, is only 237 MW(e), and nowadays 660 MW(e) plants are commonly being built. Perhaps your findings would not apply to these larger plants.

H. HERRMANN: By 'waste discharges at rates similar to those at Gundremmingen' we were thinking less of the total release activity than of the nuclide composition of the release mixture. Here I would like to point out, as we did at the end of the paper, that we work on the basis of 10 Ci of fission and activation products (without ^{3}H) and 1000 Ci of ^{3}H per 1000 MW(e) installed.

Y. NISHIWAKI: In this same connection I should like to add that in order to assess the radiological impacts of releases into inland water systems, it is very important to have some basic hydrographical data, e.g. the mean, maximum and minimum flow-rates at the point of discharge and the seasonal variation. In this particular case it would be interesting to have such data for the River Danube at the point of discharge.

P. BEAU: Can the author explain the high contents of ^{89}Sr and even ^{90}Sr at Gundremmingen? Have any determinations of ^{32}P been made in the effluents from this power station?

H. HERRMANN: We believe that the relatively high ^{89}Sr and also ^{90}Sr values are due to the good solubility of these isotopes in water. Possibly the reason has something to do with the very old fuel elements still in use in this reactor. We have not so far tried to identify ^{32}P.

Y. NISHIWAKI: Since Mr. Stott referred to the establishment of an international register of marine radioactive waste disposals, as suggested in Mr. Preston's paper, I should like briefly to explain the situation regarding this matter.

In 1958-59 the IAEA convened a panel of experts which produced a report entitled "Radioactive Waste Disposal into the Sea". One of the panel's recommendations was that the IAEA should maintain an adequate register of radioactive waste disposals into the sea.

In August 1961, the Director General of the IAEA invited Member States to provide information on disposals made in 1959 and 1960. In December 1967 a letter was sent to all Member States requesting further information on disposals into the sea. Replies were received from 22 countries, of which eight stated that they had made such disposals and a further eight indicated that they would provide information about any disposals that they might make. The remaining replies were non-committal.

Some countries that were discharging only small quantities of radioactive waste into the sea gave very precise information down to the level of microcuries or millicuries, while no reply was received from other countries that seemed to be discharging substantial amounts. At that time, some countries seemed to have been afraid that their data might receive undue international publicity.

However, since the IAEA Symposium on Environmental Aspects of Nuclear Power Stations, held at UN Headquarters, New York, in August 1970, it has been suggested that the Agency should establish an international register of significant releases of radioactive wastes from the peaceful uses of nuclear energy not only into the sea, but also into the environment as a whole.

At the panel on Procedures for Establishing Limits for Radionuclides in the Sea, held by the Agency in Vienna in November 1970, the following recommendations were made:

> "Because of the international character of the seas and oceans, there has for some years been an interest in the possibility of a system of international surveillance of releases of radioactive waste to the sea. Such a system can conveniently be considered in two major stages. The first stage involves the central collection of all information on the amounts of radioactive material released to different parts of the

marine environment. It would be appropriate for the IAEA to undertake the collection and registration of these data. Because of its interests in such matters, the World Health Organization (WHO) has also expressed an interest in co-operating in this undertaking. Comparison of this information with the known order of magnitude of the capacity of the relevant sectors of the marine environment would indicate whether the second phase of surveillance would be necessary or meaningful. This second phase would involve some form of international registration or collection of environmental monitoring data. The present rates of introduction of radioactive wastes into the marine environment justify the initiation of a register of releases, but do not require the establishment of the second stage of surveillance at present."

The panel further indicated what factors should be considered in the register of radioactive wastes discharged to the sea:

1. Types of waste and their physical and chemical characteristics (solid and liquid).
2. Types of radioactivity (gross alpha, gross beta and specific isotopes, when available. Natural radioisotopes released from the mining and milling industry would be included, but potassium-40 should be excluded).
3. Sector of the marine environment (the location of major releases to the marine environment should be listed, whether they are to the open sea, to coastal waters or via rivers).
4. Quantities of waste reportable (e.g. minimum of 1000 Ci of tritium per year per point of discharge; or 100 Ci per year per point of discharge excluding tritium; or comparable quantities reaching the marine environment via a river).
5. Frequency of reporting (annually).

In the light of the various suggestions and proposals made, a consultants' meeting on a Register of Releases of Radioactive Wastes into the Environment was convened by the IAEA in co-operation with WHO; the meeting was held in Vienna from 28 to 30 July 1971. The consultants were of the opinion that the establishment within the United Nations system of an internationl register of releases of environmental pollutants, of which the register of radioactive releases might become a part, should be encouraged. They recommended that a register should be established by IAEA in co-operation with WHO for the collection, on an international basis, of information on radioactive waste releases and their locations. Based on these recommendations, possible procedures to be employed in establishing and maintaining such a register have been discussed. However, quite recently a new proposal was made in the Agency to prepare two international registers: (1) International Register of radionuclides committed to storage and disposal and (2) International Register of radionuclides released to the biosphere. Extensive discussions are now taking place at the IAEA on this new proposal for which financial support from UNEP is requested.

THE FISH/WATER ACCUMULATION FACTOR – AN IMPORTANT PARAMETER FOR DETERMINING THE ENVIRONMENTAL CAPACITY OF SURFACE WATERS

E. ETTENHUBER, W. RÖHNSCH
Board of Nuclear Safety and Radiation Protection,
Berlin, German Democratic Republic

Abstract

THE FISH/WATER ACCUMULATION FACTOR – AN IMPORTANT PARAMETER FOR DETERMINING THE ENVIRONMENTAL CAPACITY OF SURFACE WATERS.

The fish/water accumulation factors are important parameters in determining the receiving capacity of surface waters and the maximum permissible concentration of ^{137}Cs in waters used for fishing. In addition, through the pathway water-fish-man, it is possible to evaluate the body burden of man resulting from the ^{137}Cs concentration in waters. For the present work, the ^{137}Cs accumulation factors were determined in certain species of fish from five waters in the German Democratic Republic. Equations are given for determining the reception capacity of waters and the body burden from the ^{137}Cs contamination of water through the ecological chain water-fish-man. Using the experimentally determined accumulation factors, the authors evaluate the body burdens of the individual members of the public and the whole population resulting from the ^{137}Cs contamination of water through this critical pathway. It is concluded that fish production in the contaminated cooling water of nuclear power plants is possible without reservations.

INTRODUCTION

Under the ecological conditions obtaining in the German Democratic Republic ^{137}Cs released with waste waters from nuclear facilities into the surface waters of the environment is the critical nuclide and the ecological chain water-fish-man is the critical pathway, because ^{137}Cs, which is chemically similar to potassium, is accumulated in fish and fish serves increasingly as food for man. The higher the contamination of fish by ^{137}Cs and the higher the consumption of fish by man the more important is the contribution of this pathway to the resulting body burden. On the one hand, the knowledge of radioecological relations and especially the knowledge of fish/water accumulation factors permits one to evaluate the body burden of man resulting from the ^{137}Cs concentration of surface waters through the pathway water-fish-man. On the other hand, proceeding from the maximum permissible values of body burden it is possible to evaluate the maximum permissible concentration of ^{137}Cs in waters used for fishing, to determine maximum permissible values of radioactive waste water releases, and to assess the actual body burden by radioactive releases of nuclear facilities.

METHOD AND PROCEDURE OF INVESTIGATIONS

The ^{137}Cs concentration was experimentally determined in important kinds of fish in five waters of the GDR to compute the ^{137}Cs fish/water accumulation factors. The waters were chosen as follows: they had to have a potassium concentration within the typical concentration range for waters in the GDR and they had to be actual or potential suppliers of cooling water for nuclear power plants. Thus, the ^{137}Cs contamination was determined in certain kinds of fish from the Greifswalder Bodden (a bay on the Baltic Sea), from the Stechlinsee (an oligotrophic lake), from the Elbe and Oder (rivers largely or partly polluted by industrial waste waters) and also from the Müggelsee (a river lake).

Those species of fish were selected which occur in all waters so that comparisons can be made. The fish had also to be typical of GDR waters. The ^{137}Cs accumulation was only determined in the edible parts (fish flesh), because from the viewpoint of radiation protection such accumulation is only of interest in those parts and not in the whole fish or in particular organs. The fish samples came from commercial catches taken over several years in spring and autumn. We took samples that met normal requirements for human consumption, so there was no sorting by size or age. We just took care to see that the samples did not consist of only one specimen and that they were not too small.

RESULTS

The results of the ^{137}Cs determination in fish flesh are given in column 3 of Table I. From these values, which are averages of five annual single values, the ^{137}Cs fish/water accumulation factors given in column 4 were estimated.

DISCUSSION OF RESULTS

These factors are in line with the accumulation factors determined in surface waters with similar potassium concentrations in Finland, Sweden and the Federal Republic of Germany [1-3]. The results demonstrate that differences exist not only between the ^{137}Cs accumulation factors of several species of fish in the same surface water but also sometimes between the same species of fish in different surface waters. The differences existing between the accumulation factors for several species of fish in the same surface water confirm the ecological rule that the accumulation factors for predatory fish feeding on small fish (e.g. pike, perch, pike-perch) are greater than for fish feeding on all bottom sediments or on small organisms (e.g. roach, bream).

When comparing the ^{137}Cs figures for the same species of fish in different waters, considerable differences can be observed. The figures confirm the general ecological observation for surface waters, that the value of the ^{137}Cs fish/water accumulation factor is lower, the greater the potassium concentration. This is illustrated in Fig.1.

CONCLUSIONS

The investigations show that an average ^{137}Cs accumulation factor of 1000 should be applied when evaluating the body burden of man through the ecological chain water-fish-man and the receiving capacity of surface

TABLE I. ^{137}Cs CONTAMINATION OF FISH FLESH AND THE ^{137}Cs ACCUMULATION FACTORS

Surface water	Species of fish	^{137}Cs contamination of fish flesh (pCi/kg)	^{137}Cs accumulation factor
Greifswalder Bodden	Esox lucius	475	370
	Lucioperca lucioperca	333	260
	Perca fluviatilis	348	272
	Leuciscus rutilus	79	62
	Clupea harengus	65	51
	Pleuronectes flesus	109	85
	Anguilla anguilla	208	170
Stechlinsee	Esox lucius	5040	2520
	Perca fluviatilis	6170	3090
	Leuciscus rutilus	2068	1034
	Coregonus albula typica	3200	1600
	Anguilla anguilla	3580	1790
Oder	Esox lucius	308	1100
	Aspius rapax	280	1040
	Tinca vulgaris	180	575
	Leuciscus rutilus	140	480
	Anguilla anguilla	235	870
Elbe	Leuciscus rutilus	71	356
	Leuciscus idus	49	200
	Abramis brama	64	260
	Anguilla anguilla	126	587
Müggelsee	Lucioperca lucioperca	270	2070
	Perca fluviatilis	190	1480
	Abramis brama	93	715
	Anguilla anguilla	84	645

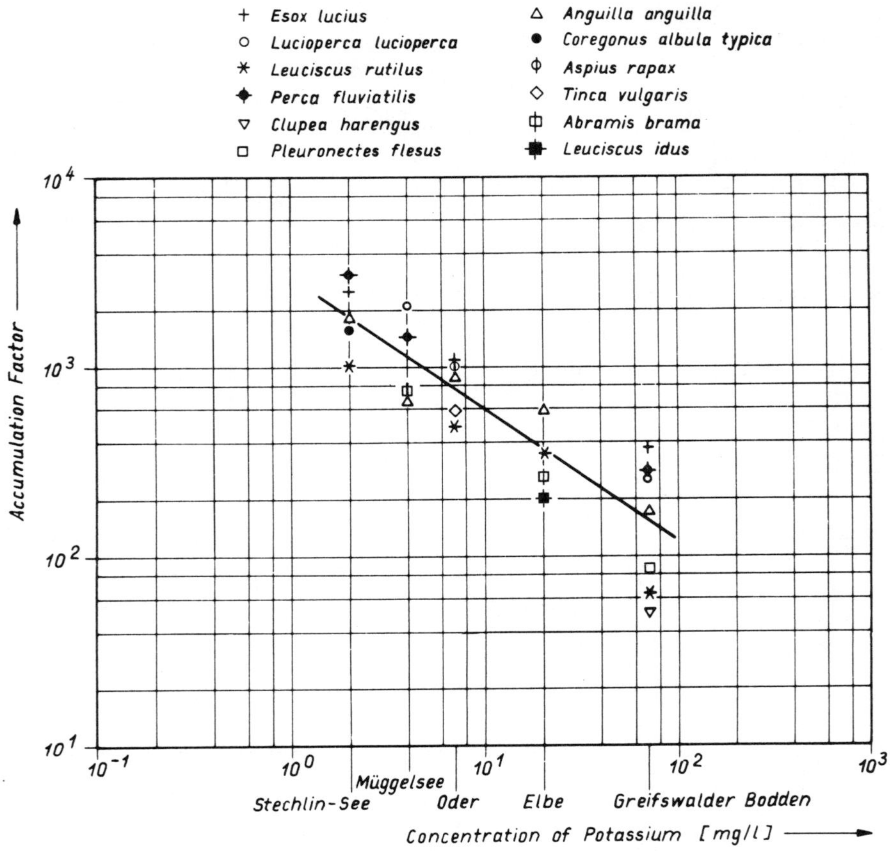

FIG. 1. Dependence of ^{137}Cs accumulation factor on potassium concentration of surface waters.

water for ^{137}Cs. The following equation can be used to calculate the body burden D through this ecological chain on the basis of the average accumulation factor of 1000:

$$D \text{ [in millirems per year]} = \frac{C \times K \times R \times MPD}{MPAI} \quad (1)$$

where C is the ^{137}Cs concentration in water (Ci/l)
K is the fish/water accumulation factor (l/kg)
R is the consumption rate (kg/a)
MPD is the maximum permissible dose (in this case for the whole population) (mrem/a)
MPAI is the maximum permissible annual intake by man (mCi/a) corresponding to the MPD.

TABLE II. BODY BURDEN BY CONSUMPTION OF FISH FROM FALLOUT

	Consumption rate (kg/a)	Accumulation factor (1/kg)	^{137}C concentration of waters (Ci/l)	Body burden (mrem/a)
Whole population	10	1000	1×10^{-12}	0.4
Individual members of public	50	2000	2×10^{-12}	8.0

In Table II the possible body burdens are given that result from ^{137}Cs contamination by fallout through the ecological chain water-fish-man under the specific conditions of the GDR. These body burden values are theoretical maximum values.

But the actual body burden of the whole population is much smaller, because the requirements of the GDR population with regard to fish are covered to a large extent by seafish with smaller ^{137}Cs contamination. The body burden of individual members of the public is likewise a maximum value conceivable only in exceptional cases.

On the basis of the average accumulation factor for conditions in the GDR it is possible to determine the receiving capacity MZQ for ^{137}Cs exclusively with respect to the ecological chain water-fish-man:

$$\text{MZQ [in curies]} = 10^3 \times \frac{V}{R \times K} \times \text{MPAI} \tag{2}$$

where V is the dilution capacity of the surface water (m^3).

In the GDR this general relation is valid only for rivers or lakes with inlet and outlet, because according to the principles of the Radiation Protection Ordinance [4] and the Water Protection Regulation [5] the discharge of radioactive waste waters into secluded surface waters is not permitted.

In the GDR the dilution capacity of rivers is defined as the average value of the lowest water levels over many years (MNQ value, in m^3/s). The following relation thus results from Eq.(2):

$$\text{MZQ [here in curies per second]} = 10^3 \times \frac{\text{MNQ}}{R \times K} \times \text{MPAI} \tag{3}$$

In Fig.2 the relation is shown between the dilution capacity of surface water and the receiving capacity on the basis of the average consumption rate in the GDR of 10 kg/a and the average accumulation factor of 1000 already mentioned. From this figure the ^{137}Cs reception capacity can be determined if the MNQ values of any water are known. On the basis of the values given for R of 10 kg/a and K of 1000 l/kg, a relation is shown in Fig.3 between the ^{137}Cs releases into surface waters and the ^{137}Cs intake by man for MNQ values of 10, 100 and 1000 m^3/s.

For example, Fig.3 shows that the ^{137}Cs releases of 380 Ci/a give the maximum permissible annual intake by man of 1.2×10^{-6} Ci/a, which

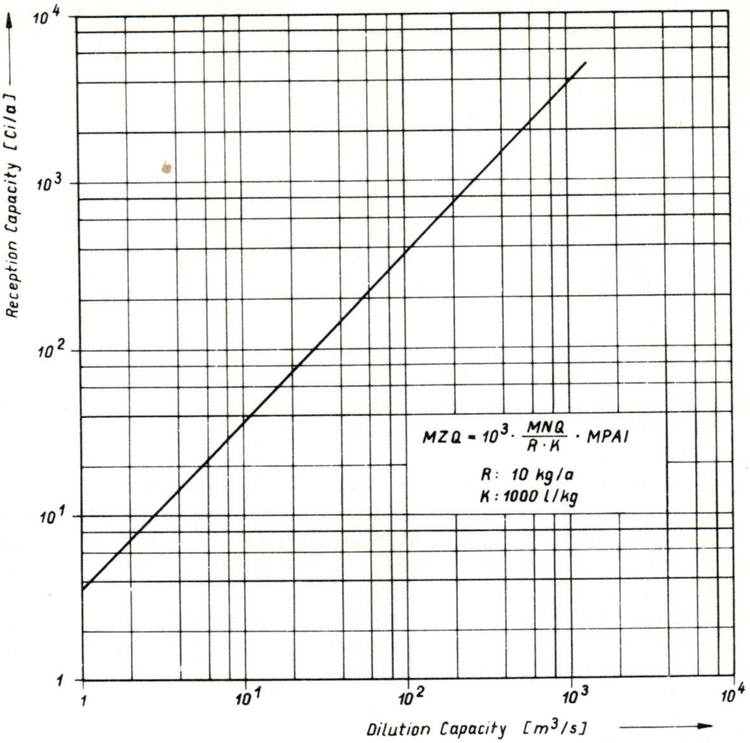

FIG. 2. Relation between the dilution capacity of surface waters and the reception capacity.

corresponds to a body burden of 50 mrem/a through the chain water-fish-man if the MNQ value is 100 m^3/s.

At present the radioactivity releases from the waste waters of nuclear facilities into surface waters amount to about 1 Ci/a.

By taking dilution capacity of 100 m^3/s, the accumulation factor of 1000, using a maximum, average and minimum consumption rate (50, 10, 1 kg/a respectively), and on the pessimistic assumption that the releases only consist of ^{137}Cs, the following ^{137}Cs intake values by man can be given, respectively, as: 3.2×10^{-10}Ci/a, 3.2×10^{-9} Ci/a and 1.5×10^{-8} Ci/a (see also Fig.4).

The relations and results shown in Figs 3 and 4 allow one to answer the question, where and to what extent the fish production on an industrial scale in surface waters contaminated by radioactive releases from nuclear facilities contributes to the body burden of the whole population. As an example the critical case is chosen where fish cultivation is carried on directly in contaminated cooling water with natural fish feeding. With once-through cooling the present requirement for the cooling water of nuclear power plants under construction or in operation is about 100 m^3/s. So Fig.4 can be used for determining the body burden.

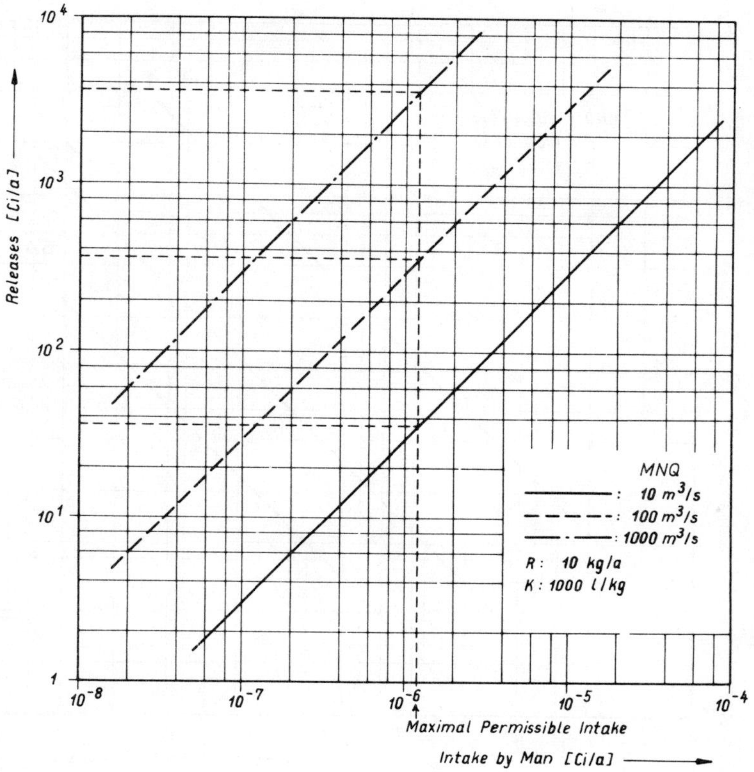

FIG. 3. Relation between the releases of ^{137}Cs from nuclear power plants and intake by man.

On the basis of the actual normal releases of 1 Ci/a and on the assumption that the releases contain only ^{137}Cs, body burdens, for the most pessimistic case, result which are similar to that burden which is possible from contamination of waters by global fallout through the ecological chain water-fish-man. The resulting value is about 2 mrem/a; this is only 4% of the maximum permissible value for the whole population in the GDR. But this resulting body burden is a theoretically possible maximum value, because ^{137}Cs amounts to less than 0.1% of the total radioactivity released with waste waters from nuclear power plants in the GDR.

Even smaller body burdens occur in all other possible cases: e.g. from fish cultivation in low-contaminated cooling water where artificial feeding is used, and from fishing in estuarine areas where cooling waters are released into rivers, seas or oceans.

It is therefore possible without reservation to use the cooling waters of nuclear facilities for fish production. Emergency situations causing short-time higher releases of radionuclides hardly affect the contamination of fish; Aten [6] shows that a higher concentration of radionuclides in water causes a higher concentration of radionuclides in fish only after

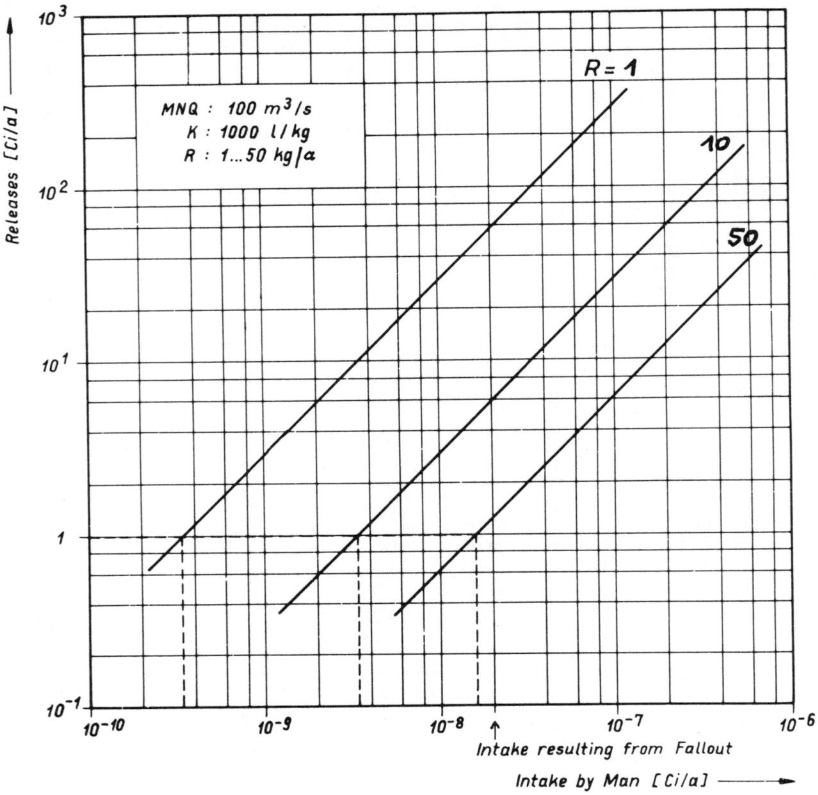

FIG. 4. Dependence of ^{137}Cs intake by man on consumption rate R.

some time, say after a fortnight. A higher contamination of fish decreases again to a lower value comparable to a normal radionuclide contamination of water in a comparable period. So emergency situations need not be taken into consideration.

SUMMARY

According to the principles and methods outlined and to the evaluated accumulation factors the receiving capacities of surface waters for the release of radioactive waste waters are determined for GDR nuclear power plants. In addition to these capacity values of surface waters, the permissible radioactive releases into the surface waters are limited by technological conditions and economic requirements. Thus it is ensured that the further use of waters for producing fish is possible without reservations.

REFERENCES

[1] KOLEHMAINEN, E., HAESONEN, E., MIETTINEN, J.K., Health Phys. 12 (1966) 912.

[2] AGNEDAL, P.O., Swedish Report AE-224 (1965).

[3] RUF, M., Zbl. Veterinärmedizin B 12 (1965) 605.

[4] Verordnung über den Schutz vor der schädigenden Einwirkung ionisierender Strahlung — Strahlenschutzverordnung — vom 26. Nov. 1969, GBl. II, Nr. 99 S. 627.

[5] Gesetz über den Schutz, die Nutzung und die Instandhaltung der Gewässer und den Schutz vor Hochwassergefahren — Wassergesetz — vom 17. April 1963 GBl. I, Nr. 5 S. 77.

[6] ATEN, A.H.W., in 2nd Int. Conf. Peaceful Uses Atom. Energy (Proc. Conf. Geneva, 1958) 18, UN, New York (1958) 414.

DISCUSSION

A.R. BERG: I do not understand your statement, in the Conclusions, that emergency situations need not be taken into consideration. If such a situation is of long duration, one must take into account not only the biological half-life of the radionuclide in fish but also the growth rate of the fish, which might result in a strongly accelerated uptake. In this context an accident occurring during a period of fast growth, such as the spring season, will be more dangerous than if it occurs in winter.

E. ETTENHUBER: We were referring to short-term situations in which the radioactivity release is increased for only a very short time.

Y. NISHIWAKI: Perhaps I can make a few comments in this connection. On 1 March 1954 the Japanese fishing boat "Lucky Dragon No.5" was showered by highly radioactive fall-out from H-bomb tests in the Bikini area. I was one of the Japanese scientists who first identified the strong radioactive contamination of the fish brought back to Japan by this fishing boat in the middle of March 1954. When examining the fish we noticed that strong radioactive contamination was present only on the skin, and little contamination was found in the internal edible parts at this stage. Upon examining more fish brought back to Japan later by other fishing boats we found that when the radioactive concentration of the local surface seawater had decreased, some time after the nuclear detonation, stronger contamination appeared in the edible parts of the fish.

I think it is extremely important to note this phenomenon in order not to waste precious food in the very initial stages of contamination of the environment in the case of a nuclear emergency. Under such transient conditions the concept of equilibrium-state concentration factors may not be applicable.

L. OTTENDORFER (Chairman): The biological half-life is involved here. If the fish are exposed initially to relatively high concentrations, after which they are surrounded by water of low activity, then they will not take up much activity.

F.O. HOFFMAN: What maximum permissible dose to individuals of the general public is adopted in the German Democratic Republic for regulating discharges from nuclear facilities into surface waters? In the Federal Republic of Germany a total whole-body dose of 30 mrem/a for all relevant pathways of exposure from all sources of release is in current use as a dose limit for liquid effluents.

E. ETTENHUBER: A maximum permissible dose of 50 mrem/a is adopted in establishing the maximum permissible releases to the various environments.

F.O. HOFFMAN: Do your concentration factors represent a state of equilibrium between water that has been filtered and fish flesh (net weight)?

E. ETTENHUBER: Yes, they do.

F.O. HOFFMAN: The determination of fish consumption by so-called "critical" individuals has been found to be fraught with a high degree of uncertainty. Could you give us some additional information on how your individual consumption rate of 50 kg/a was obtained?

E. ETTENHUBER: We interrogated the critical group, i.e. the fishermen.

Y. NISHIWAKI: Is there any co-ordination between the German Democratic Republic and neighbouring countries for assigning discharge rates into international river systems? For instance, in determining the discharge rate into the River Elbe, do you liaise with the German Federal Republic or with any other countries upstream or downstream? I think that good co-ordination with neighbouring countries is extremely important in matters of environmental safety.

E. ETTENHUBER: We have co-ordination arrangements with many countries. There is no need for co-ordination as regards discharges to the River Elbe, because no nuclear power plants use the waters of that river for cooling and no radioactive discharges are made into it.

Y. NISHIWAKI: As regards the River Danube, I should like to mention that the IAEA is planning to hold an advisory group meeting at Belgrade in November 1975, in cooperation with the Government of Yugoslavia, to study questions of mutual cooperation between countries in the Danube catchment area.

The task of the Advisory Group will be to prepare a programme giving particular attention to topics identified at the IAEA/WHO European Study Group Meeting on Radiological and Environmental Protection held at Budapest, Hungary, in September 1973; such topics include evaluation and definition of the radiological capacity of the river, summaries of its uses, data on discharges of radioactivity, mutually acceptable dose limits and release limits, a monitoring system, training and research.

It is considered important to start with a joint programme of the Danube riparian states to study, evaluate and define the environmental capacity of the Danube. Each country taking part in this programme is requested to provide summaries of the uses of the Danube waters (e.g. fishing, drinking water, irrigation, etc.), of quantities used and of the range of population groups involved, including future trends and developments.

RELEASE PATHWAYS FOR DEEP SEABED DISPOSAL OF RADIOACTIVE WASTES*

D.R. ANDERSON, W.P. BISHOP, V.T. BOWEN, J.P. BRANNEN, W.N. CAUDLE, R.J. DETRY, T.E. EWART, D.E. HAYES, G.R. HEATH, R.R. HESSLER, C.D. HOLLISTER, K. KEIL, J.A. McGOWAN, R.W. ROHDE, W.P. SCHIMMEL, C.L. SCHUSTER, A.J. SILVA, W.H. SMYRL, B.A. TAFT, D.M. TALBERT
Sandia Laboratories,
Albuquerque, New Mexico,
United States of America

Abstract

RELEASE PATHWAYS FOR DEEP SEABED DISPOSAL OF RADIOACTIVE WASTES.

The concept of disposal of high-level solidified and encapsulated radioactive wastes into the deep sea floor has recently been discussed. Such a scheme has conceptual advantages in that the areas of the mid-plate/mid-gyre regions of the oceans are relatively unproductive biologically and relatively devoid of cataclysmic events, and natural processes there are generally quite slow. Given a lack of singular events, a set of barriers against the dispersion of the radioisotopes may be defined. The inverse of these barriers is the set of mechanisms by which the isotopes are transported from the burial site through the barriers to parts of the ocean of immediate significance to mankind. These include: corrosion of the cask; leaching of the waste material; upward transport of the isotopes with the upward moving pore water (mediated by ion-exchange processes); biological transport through bioturbition in the upper sediment layers and lowest water layer; the slow throughput currents of the deep basins; advection and diffusion through the water column; thermally driven transport through the sediments or the water column; biological transport of incorporated isotopes across the seabed or upward through the water column. In principle, the rates of all these processes are measurable or capable of being estimated. Such estimates are given on the basis of present knowledge of the processes in the deep basins. A methodology is discussed for the analytical treatment of the set of processes to give the amount of the isotopes reaching some part of the environment (e.g., an oceanic regime of immediate significance to man) as a function of time. The authors conclude that disposal in the deep seabed is conceptually attractive because of the stability and predictability of the environment, but that it is not possible to give a firm estimate of the safety of such a scheme from the current knowledge of the mid-plate/mid-gyre regions.

INTRODUCTION

One of the critical problems to be solved in the development of nuclear power is the disposal of the potentially dangerous wastes from the nuclear fuels. While these wastes arise from several sectors of the nuclear fuel cycle, and take a large number of forms, we have limited our discussions to the disposal of the high-level wastes from the reprocessing of the fuel. These may, in fact, be the worst wastes because of the combination of intense radiation and heat from the relatively short-lived fission product nuclides plus the length of time required for the transuranic nuclides to decay.

---

* This work was supported by the U.S. Energy Research and Development Administration. The authors have been listed in alphabetical order. Messrs Anderson, Bishop, Brannen, Caudle, Detry, Rohde, Schimmel, Schuster, Smyrl and Talbert are at the Sandia Laboratories, Albuquerque, New Mexico; Messrs Bowen and Hollister are at the Woods Hole Oceanographic Institution, Woods Hole, Massachusetts; Messrs Ewart and Taft are at the University of Washington, Seattle, Washington; Mr Hayes is at Columbia University, Lamont Doherty Geological Observatory, Palisades, New York; Mr Heath is at the University of Rhode Island, Kingston, Rhode Island; Messrs Hessler and McGowan are at the Scripps Institution of Oceanography, La Jolla, California; Mr Keil is at the University of New Mexico, Albuquerque, New Mexico; and Mr Silva is at the Worcester Polytechnic Institute, Worcester, Massachusetts.

In our planning for the program and in the thinking about the several options for disposal, we have assumed that the high-level wastes will be those arising from the conventional existing fuel reprocessing technology and from the present generation of light water reactors. This means, in practice, that some of the transuranic nuclides will remain in the wastes, and that the times for the decay will be the order of 1000 years for the thermal output and severe radiation and of the order of a million years for the alpha-active transuranic nuclides.

In addition, we have assumed a growth of the nuclear industry in the U.S. consistent with projections of the U.S. ERDA [Ref. 3], giving in the year 2000 a total of about 10^4 cubic meters of solidified high-level wastes with an activity of about 10^5MCi and a thermal output of about 10^3 megawatts.

These are boundary conditions for any disposal program, and they quickly lead to one implication - that dispersal of the wastes into the environment is not acceptable.

Alternatives for the handling of the high-level wastes include both storage (usually temporary) and disposal options [Refs. 1, 4]. There may be uses for all or part of the wastes for heat or radiation sources; engineered storage is certainly possible though perhaps socially unacceptable in the longer term, and transmutation or space disposal of parts of the wastes may prove feasible given the development of a partitioning capability. Even with their development, however, some or all of the wastes will likely require disposal on the planet earth.

A number of geological media have been considered for disposal. The best known of these and the only one in present use is salt; however, clay, shale, crystalline rocks of several kinds, and similar media beneath the seabed are under consideration either as prime disposal candidates or alternates. The assessment of the capability of sub-seabed disposal is the subject of this discussion.

SEABED DISPOSAL

Since seventy percent (70%) of the surface of the earth is covered by the oceans, this vast area should in our opinion be considered as a part of the set of alternatives for disposal of the high-level wastes. Recent understanding of global geological and climatological patterns, including the description of plate tectonics which is due in part to recent advances in oceanographic research techniques (e.g., Deep Sea Drilling Project), gives some confidence that seabed disposal may be both feasible and safe.

Some parts of the sea floor have special properties which are certainly desirable characteristics for a disposal regime [Ref. 2]. The most seismicly stable areas of the earth are under the oceans in the centers of the great oceanic plates; the floors of the major basins are the most remote places from man's normal activities; parts of the ocean floor are severely resource-limited; and, in general, processes in parts of the oceans are slow, gentle, predictable, and depositional (processes on continents tend to be erosional). Along with this apparent simplicity, it must be understood that our knowledge of the details of the conditions of the ocean floors has not reached the same level as that of the continents.

Our knowledge of the ocean floors is sufficient, however, to catalogue and evaluate the several major ocean regimes with regard to their capabilities as waste repositories [Ref. 2]. The criteria which we have used in the selection of the areas for further study are:

-- Geological Stability: This may be estimated by observing the record of the past geological events held in its entirety in the sediment column and through a thorough understanding of the processes responsible for the record as we see it.
-- Lack of Resources: Waste disposal should not seriously interfere with the acquisition of resources, organic or inorganic.
-- Adequate Area: An adequate area should be available to handle all of the wastes for the world for the forseeable future.
-- Inaccessible: The area(s) should be as far removed as possible from the normal and expected activities of mankind.
-- Permanent: Recovery of the material at a later date need not be a requirement (but should not be precluded, all other factors being similar).
-- International: If feasible, seabed disposal can provide an international solution to an international problem. So, areas outside direct national jurisdiction are preferable.

Comparing the major ocean provinces using the above criteria (see Fig. 1a and 1b and Table I), we find that two of the provinces are best-suited for disposal. They are, in order of suitability: the abyssal hills and the flanks of the Mid-Oceanic Ridges. Those which also occur in the middles of the great oceanic gyres are especially attractive because of their low biological productivity. Thus, the areas in the middle of the tectonic plates and the middle of the gyres (mid-plate/mid-gyre) are the targets for our further studies.

While it is beyond the scope of this paper to present the possible engineering schemes by which emplacement of the waste could be made in the ocean floors, there are several which might prove both feasible and effective. They include:

-- Drilling and backfill.
-- Trenching and backfill.
-- Gravity penetration (with a large mass as driving force).
-- Free-fall penetration (with a high velocity as driving force).

The last two techniques require a plastic sediment which will collapse to fill the hole in a reasonable time. Such conditions may, in fact, exist in most surficial sediments.

With this background, the bases for a program to investigate the feasibility and safety of seabed disposal can be delineated, and they are:

-- Emplacement in the seabed (not in the water column).
-- Mid-plate/mid-gyre and flanks of mid-oceanic ridge regions (the most stable, resource-limited, biologically inactive, international and extensive regions).

TABLE I. COMPARISON OF THE MAJOR OCEANIC REGIMES

		Stability	Resources	Area $km^2 x 10^6$	International	Accessibility	Sediment Thickness km	Productivity	Ranking
Mid-Oceanic Ridge	Fracture Zone	low-high	low	small	yes	low-very low	0-2	high to low	2
	Flanks	high	very low	large (120)	yes	low-very low	0.5-1		
	Crest	low	low	small	yes	medium-high	0-0.1		
	Rift Valley	very low	low	small	yes	medium	0		
Continental Margin	Shelf	very low	very high	15-25	no	very high	2-20	high	
	Slope	very low	medium	15-25	no	very high	2-20	medium	
	Rise	medium	medium	20	both	low	5-10	low	
	Delta and Fans	very low	high	5-10	both	medium	5-10	high	
Ocean Basin Floors	Seamounts	very low	low	10	yes	very high	0-0.5		
	Trenches	very low	very low	6	both	very low	1-5	medium	
	Abyssal Hills (Mid-Plate)	high	very low	130	yes	very low	0.1-0.3	very low	1
	Abyssal Plains	low	medium	10-20	both	very low	0.5-2	very low	

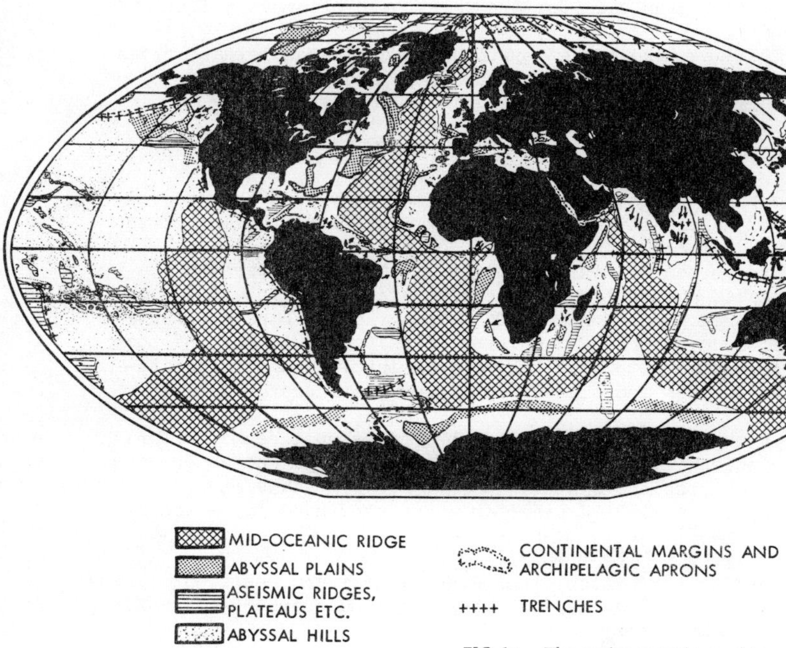

FIG. 1a. The major oceanic provinces (from Ref. [2]).

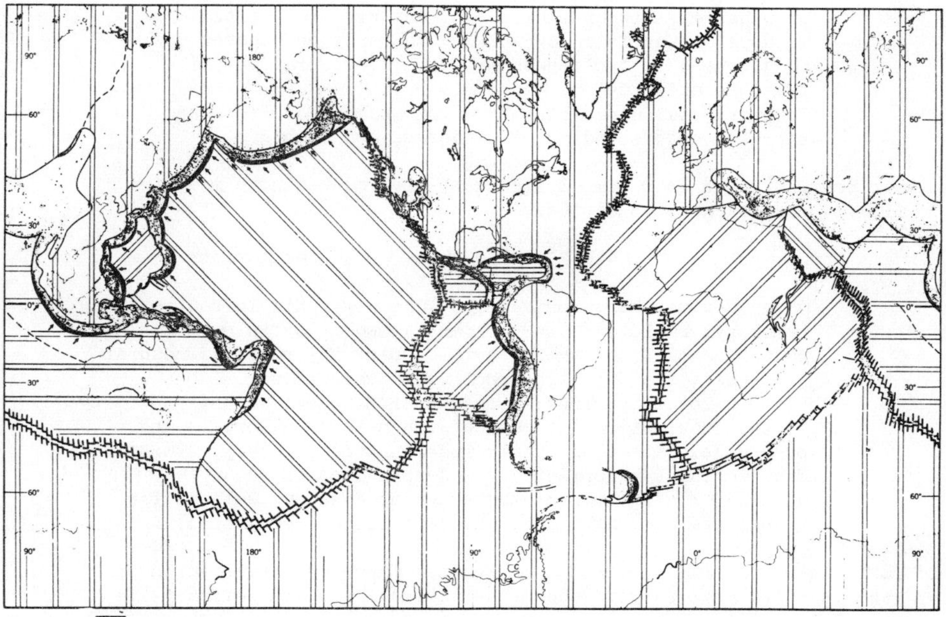

FIG. 1b. The tectonic plates (from Ref. [10]).

- -- Permanent disposal (for the worst case of about a million years).
- -- No release (the target system should provide release rates such that release occurs only after nuclide decay is virtually complete).

SEABED DISPOSAL PROGRAM

On the basis of the reasoning outlined above and the present knowledge of the mid-plate/mid-gyre and ridge flank regions of the oceans, a program has been under way for several years in the U.S. to provide the essential information necessary to assess the viability of seabed disposal and to develop any new engineering capabilities required. The content of the program has been documented in detail elsewhere [Refs. 2, 8], but it is useful to outline that content here and to note that the program involves major institutions in the nuclear and oceanographic communities.

While a number of intermediate objectives must be met before we have the scientific information needed to design and assess the ultimate seabed disposal system, overall program goals may be stated in approximately the order in which they need to be accomplished:

- -- To define the system, including natural processes, processes imposed by emplacement, engineering and operational requirements, and costs.
- -- To determine feasibility of seabed disposal from firm technical information.
- -- To determine the safety (i.e., the absence of releases or other effects on mankind) of any proposed system.
- -- To develop the total capability to carry out seabed disposal based on the findings of the above program (if feasible and safe).
- -- To prove that capability by a combination of tests and thorough understanding of processes and events.

Our program is at this time designed to meet or begin to meet the first three of these objectives.

The program may be described in three broad categories:

- -- Ocean Regime Characterization: Oceanographic investigations (geology, biology, water column) aimed at defining the natural conditions and materials, plus the rates of natural processes, that will interact with the wastes.
- -- Effects Investigations: Studies directed at an understanding of the interactions of wastes with natural materials and processes, including system modeling and efficacy assessment.
- -- Engineering Development: The development of technology needed for carrying out the above studies, as well as for transportation, handling, and emplacement operations.

These three program elements comprise the important tasks which provide the data necessary to judge the advisability and efficacy of seabed disposal.

Two classes of effort occupied the first years of the program: the collection of relevant historical data about a first study area (the central North Pacific mid-plate/mid-gyre region) and the development of important new data. Historical data were collected in the following disciplines and have been analyzed: physical oceanography, water-column biology, bathymetry, geology/geophysics, geotechnical or engineering properties of the sediments, chemical properties of manganese nodules, physical and chemical properties of the basalt, and corrosion processes.

Among new information derived were the following: benthic biological survey, bottom-current measurement (5 days to 18 months), radioisotope concentrations in water and sediments, engineering properties of the upper few meters of sea-floor material, detailed bathymetry and sub-bottom structure, photographic survey of nodules and large organisms, pelagic near-surface biology, physical oceanography, manganese nodule geochemistry, and thermal processes in porous media. Part of our effort has been directed toward the development of needed engineering or instrumentation capabilities in four areas: measurement of water-column dynamics (meso-scale and micro-scale), measurement of sediment structures (large-scale), measurement of sediment properties (laboratory and in-situ), and manipulation of equipment and experiments on the bottom in deep water.

Results of these studies provide the information required to compare the rates of the processes leading toward release of the nuclides from the disposal site against the rate of decay of the nuclides. The methodology of this comparison is the subject of the remainder of this discussion.

THE BARRIER MODEL

To compare the rates of decay of the radionuclides against the rates of migration of the nuclides toward mankind, we have adopted a formalism based on a set of sequential barriers to that release. Each of these barriers will inexorably be breached either by singular events or by the slow and steady action of physical, chemical and biological processes to dissolve the waste and transport the nuclides through the medium.

Each of these events or processes may be assigned a probability or a rate, and the overall rate of impact upon mankind will be the sum of the rates of the parallel paths of sequential events or processes less the rate of decay of the nuclides.

Conceptually, and to a great extent in the real case, the sequence of barriers can be grouped into regimes. These are shown in the flow diagram in Fig. 2, and include the basement rock, the layers of unconsolidated to lithified sediments, the "benthic boundary layer" (that section of the water and sediment columns in which major chemical and physical gradients occur -- i.e., from about 10 m above to about 1 m below the sediment/water interface), the water column and, of course, the atmosphere and the surficial waters of the continents.

Contact with man of his immediate environment will be far less for the rock and the sediments than for any of the other regimes and, as we shall see, the rates of passage through barriers in the solid formations are likely to be slower than in the mobile media.

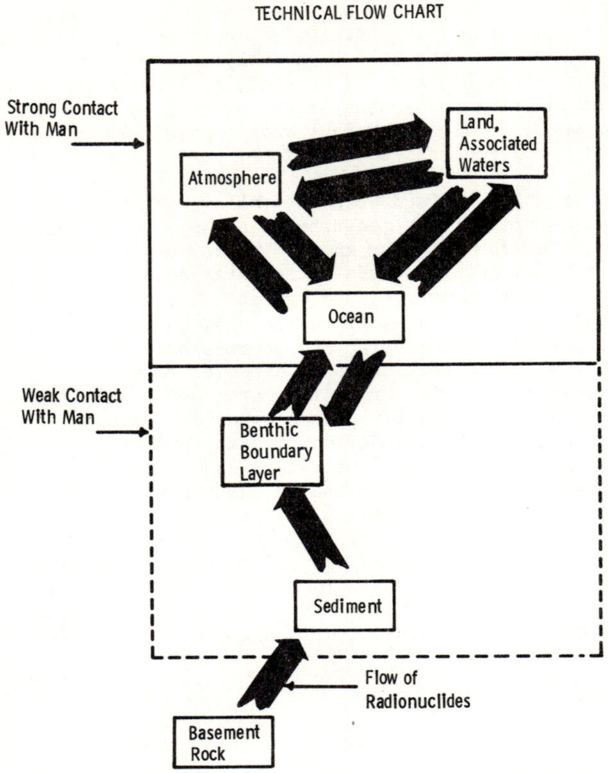

FIG. 2. Technical flow chart.

Because the process of emplacement of the wastes into any one of the regimes will require passage through the upper regimes, radionuclides can enter each regime either by passage up through the barriers below, or from a mishap during the emplacement. Lacking such a mishap, however, the input of nuclides will occur from the emplaced, solidified, encapsulated waste as shown in Fig. 3. Such release from the cask into the rock can occur through the action of slow corrosion and leaching processes or through the melting of the waste and the surrounding rock if the thermal output of the waste is sufficient. Alternatively, a singular event such as a massive geological dislocation or intrusion by man can release the nuclides into the rock.

Movement of the nuclides through the basement rock can occur through the action of a number of mechanisms outlined in Fig. 3. These are mainly transport of material by geological events or by the natural, chemical, or thermally induced motion of water. Of course, some processes, such as the decay of the nuclides and the absorption of the species on the rock surfaces (ion exchange), lessen the effect of transport of the nuclides. All of these processes and events will ultimately be included in the model used to evaluate the projected performance of a candidate seabed disposal scheme which involves emplacement in basement rock.

Nuclides passing out of the basement rock will enter the sediment column. In the model, a subroutine will handle the transport of the nuclides through the sediments and will take as an input the output of the deep-rock regime subroutine. In addition, input of nuclides may be derived from emplacement of the wastes in the sediments, in which case the input to the subroutine will be the corrosion and leaching processes, and events which will lead to the breaching of the sediment barrier are indicated in Fig. 4.

The major transport mechanisms will involve the movement of pore water either through the natural advection caused by compaction of the sediments or through movement induced by disturbance of the sediments during emplacement or induced by the radiogenic heat. Some processes which will tend to reduce the rate of migration of the nuclides are ion-exchange with the sediments and possible accretion of manganese or silicon oxides which might either coat a cask or incorporate the nuclides in less mobile media. Of course, decay of the nuclides will occur during the time of transport through the sediments.

From available data it now seems likely that for the slow and steady process of transport through the sediments, escape times may be considerably in excess of that required for the decay of the nuclides. However, some of the rates for processes shown in Fig. 4 are not as yet available, and the results noted below are for a crude model of the water flow and ion-exchange properties alone.

Within the benthic boundary layer gradients in chemical composition of the water occur both in the water column and in the sediment, and the biological processes of interest to this model are concentrated in this regime.

As shown in Fig. 5, radionuclides may enter the benthic boundary layer (and thus the subroutine in the model) from both directions and from singular events. Recovery of natural resources by mining and fishing must be included in the possible events leading to breaching of the sequence of barriers in this regime. Mitigating processes include absorption on the constantly falling debris, biological fixation in less mobile creatures or in feces, and perhaps in a simple precipitation of the dissolved nuclides from the water column.

We are at present limiting our modeling efforts to the three regimes discussed above and the water column itself. Nuclides put into the water column (see Fig. 6), either from transport through the benthic boundary layer or by some singular event, will be transported by a large variety of mechanisms ranging from bulk diffusion to storm-like eddies in the water column and transport by the biological community. Some models exist using simplified and perhaps

To Sediment

↑

BASEMENT ROCK

Transport Through the Basement Rock Via --

- Motion of Rock Mass (Seismic, Volcanic)
- Motion of Molten Waste
  - Movement of Entire Molten Region
  - "Gas" Generation
  - Penetration into Fissures
- Convective Motion of Water Through Fractured Rock (2D Flow Through Porous Medium)
- Hole Closure
- Radioactive Decay

NOTE: No Biological Transport

↑

Radioactivity is Released Via --
- Corrosion of Cask, Leaching from Ceramic or Glass Waste Form
- Melting of Waste, Surrounding Rock with Subsequent Mixing and Movement of Molten Mass
- Failure of Container During Emplacement
- Natural Events (Seismic, Volcanic)
- Intrusion by Man (Accidental, Overt)

FIG. 3. Basement rock regime.

To Benthic Boundary Layer

↑

SEDIMENT (Below One Meter)

Transport Through the Sediment Via --
- Convective Motion of Pore Water
  Nonionic Forms Move with Water
  Ionic Forms are Retarded by Ion Exchange
- Chemical Potential
- Accretion, MnO_2, SiO_2
- Layering of Sediments
- Hole Closure
- Radioactive Decay
- Heat-Changing Flow Mechanism
- Authigenesis
- Sediment Column Length
- Mining

↑

Radioactivity is Released Via --
- Transport Through Basement Rock to Interface
- Failure of Container During Emplacement in Sediment or Basement Rock
- Corrosion of Cask, Leaching of Waste
- Natural Events (Seismic, Volcanic)
- Intrusion by Man (Accidental, Overt)

FIG. 4. Sediment regime.

To Ocean
- Resuspension
- Upwelling
- Mixing
- Biological Transport
- Diffusion

From Ocean
- Precipitation
- Biological Transport into BBL
- Adsorption

BENTHIC BOUNDARY LAYER

Transport Through the Benthic Boundary Layer Via --
- Water Motion
- Accretion, MNO_2, SiO_2
- Diffusion
- Thermal Plume
- Biological Movement
- Sediment Motion
- Radioactive Decay
- Fishing
- Mining

NOTE: Possibility of Strong Fixation of Radionuclides Within Benthic Boundary Layer Exists

Radioactivity Enters Via --
- Transport Through Sediment to Interface
- Failure During Emplacement
- Loss of Container in Water Column with Subsequent Sinking into Benthic Boundary Layer
- Corrosion of a Cask in the Benthic Boundary Layer with Leaching of Waste Form
- Natural Events
- Intrusion by Man

FIG. 5. Benthic boundary layer regime.

To Man Via Atmosphere and Land

From Atmosphere and Land

OCEAN

Transport Through the Ocean Via --
- Deep Horizontal Advection
- Deep Vertical Mixing
- Surface Currents
- Biological Concentration
- Biological Movement Horizontal and Vertical
- Heated Plume
- Adsorption onto Falling Debris
- Fishing
- Dilution
- Internal Structure (eddies, internal waves, etc.)
- Enhanced Biological Productivity

Radionuclide Enters Via --
- Transport Through Benthic Boundary Layer
- Failure During Transport or Emplacement
- Corrosion and Leaching of Cask and Waste Form

FIG. 6. Ocean regime.

TABLE II. UNSUBSTANTIATED ESTIMATES OF BARRIER EFFICACY

Barrier	Estimated Years To Breakthrough*	Notes
Cannister		
- metal (Zr, Mo, Ti)	$> 10^3$ **	- for complete uniform corrosion
- ceramic	?	- manufacture difficult
Waste Form	$\geq 10^4$ **	- for complete dissolution at $\sim 10^{-5}$ $gm/cm^2/day$
Rock		
- locally melted	?	- requires properties be measured and samples collected
- monolithic crystalline	?	
- pillow structure	$\gg 10^1$	
Sediment Column		
- at rate of pore water	$> 10^6$	- neglects thermally driven transport
- with ion-exchange	?	- varies with ion; highest for transuranics
Benthic Boundary Layer	$\geq 10^2$	- varies with ion—both biological and chemical mechanisms either fix or release
Water Column		
- mid-water	$\leq 10^3$	- many questions regarding internal structure and biological mechanisms
- mixed layer (top 800m)	10^0	- physical and biological movements rapid
Short Circuits/Added Protection		
- plug	?	- little experience in deep ocean
- accretion (MnO_2, SiO_2)	?	- little evidence available
- thermal effects	?	- generally accelerate all processes
- biology	?	- little evidence available

*Definition of these times is underway and new estimates are continually being made available.
**Depends on water circulation in medium and temperature.

unrealistic assumptions about the movement of the water [see, for example, Ref. 9]. We shall be attempting, although not as a first priority, to construct or adapt a model with inclusion of all or most of the possible mechanisms.

The methodology represented by the sequence of barriers (or processes breaching these barriers) seems to us amenable to mathematical interpretation, and such a model is under development. Interpretation of some of the barriers individually is possible in a crude manner, and some of the mathematical representation is now capable of giving early information based on rather restrictive assumptions. These pieces of information and some of their implications are cited below.

RESULTS TO DATE

In assessing the efficacy of the barriers, we have taken two approaches to the estimation of the times for release of the nuclides. These are 1) estimation from existing data of the transport times for water through the barriers and 2) the mathematical representation of the barriers including all of the relevant processes.

From the historical data summarized elsewhere [Ref. 8] and new data collected from our North Pacific study sites, estimates have been made of the times to breakthrough of water which could bear nuclides for several classes of barriers (see Table II). These are based on insufficient data and, at best, can be considered only as limits within which the correct answer may lie. In many cases, they are no more than intelligent guesses. They do serve two functions, however, by pointing out the need for further data in several crucial areas and by demonstrating that a combination of a cask which can survive the time of strong thermal output plus the sediments could provide a sufficient barrier to the release of the nuclides.

Regardless of the media in which the radioactive waste will come to rest, an insoluble waste form and a cask which will remain intact during emplacement are necessary. Glassy or ceramic materials have been produced with simulated high-level wastes which exhibit leach rates (or dissolution rates) less than 10^{-5} g/cm^2/day at ambient conditions, and laboratory electrochemical experiments provide measurement of corrosion rates for metallic casks which extrapolate to lifetimes of about 1000 years. Both of these rates are likely to be less in a medium in which rapid movement of water is precluded, but both will likely be accelerated in the presence of radiogenic heat. Neither the problem of the construction of a ceramic cask nor of its decomposition has been addressed, but we know that ceramic artifacts have survived millennia in ocean environments.

BASEMENT ROCK REGIME

The basaltic rock underlying the deep ocean sediments is the most poorly understood media because of the paucity of data and of measurements of its bulk properties. Recent developments both in the field (Deep Sea Drilling Project) and in the laboratory do, however, shed some light on the nature of basaltic crust and on the relevant processes there.

Initial information from the Deep Sea Drilling Project, which has obtained cores from a few kilometers below the surface of the sediment to several hundred meters below the basement rock/sediment interface, has revealed an increasing compaction of the sediments toward limestone or shale, then an interbedding of pillow lava with fine-grained pelagic sediment to a depth of hundreds of meters below the sediment/basement rock interface. This is underlain by layers of more monolithic basalt. The deeper sections of basalt showed more differentiation of the minerals and larger crystals resulting from slower cooling. Although the bit penetrated some hundreds of meters into the basalt, the total length of the core was much less, indicating that in some areas the rock may have been weak and possibly fractured. As the bit passed through these regions, the material was pulverized and lost.

The rate of water movement through monolithic basalt beneath the deep ocean is unknown but almost certainly decreased in the deep crystalline (slowly cooled) formations. In the interbedded structure of the top few hundred meters of basalt, water probably moves more rapidly and may travel horizontally as well as vertically.

If the thermal output of wastes emplaced in the basement rock is sufficient, melting of the waste and the surrounding rock is possible [Ref. 5]. In this case, the cannister need only survive the emplacement procedure.

Laboratory studies [Ref. 5] of melting of a crystalline rock chemically similar to basalt (dolerite) have provided some important insight into understanding of this melting phenomenon. The laboratory specimen (about 1 m^3) cracked due to thermal stresses and the molten waste/rock material flowed for a few centimeters into the cracks and solidified through resealing the openings. Calculations of thermal processes [Ref. 6] predicted the measured thermal profiles and the presence of a convection cell. Tests of the leaching properties of the resultant mixed, solidified rock-waste matrix showed rates not dissimilar to those for borosilicate glass waste forms. In a program parallel to ours, the extension of these tests to other rocks and to actual wastes is planned.

The mathematical representation of the motion of nuclides through basement rock must await a clearer understanding of the processes and the structure of the basaltic crust of the ocean floors. We shall continue to follow the progress of the Deep Sea Drilling Program's Interstitial Phase of Ocean Drilling (IPOD) and we hope to derive some understanding of thermal processes from laboratory tests.

THE SEDIMENT REGIME

If emplacement in the sediment is the chosen disposal method, then the cannister barrier becomes important until the thermal output has been reduced to levels which produce minimal water transport (approximately 200 years to 1% of fresh waste). Initial considerations indicate that metal cannisters, when kept below 100°C, can be built which will last for the time period of interest. Laboratory and in-situ corrosion studies are under way to verify our estimates. After the thermally driven mechanism has been eliminated, the movement of the nuclides will be controlled by interstitial water movement, chemical potential, and the ion-exchange properties of the sediments.

Interstitial water moves upward through the sediments during compaction. The rates of movement are very slow and may approach the average rate of sediment accumulation. The data we have on permeabilities of sediment as measured in the laboratory suggest that as water moves, by compaction processes, toward the sediment, the sediment water interfaces at rates of tens of meters in millions of years.

A model for radionuclide transport through soil by ground water [Ref. 7] has been modified to simulate vertical transport in the sediments. Important parameters for this model are the velocity of the water moving through the porous media, the dispersion coefficients of the flow as controlled by particle size and water velocity, and the distribution and absorption coefficients of the sediment for each ion. Some of the present problems inherent in this model are that it assumes trace concentrations, a homogeneous media through which the water flows, linear absorption isotherms, and does not take into account the presence of competing ions.

As an example, for a mixture containing quantities of each of the radionuclides in the following decay chain passing through a 6 Km column of sandy terrestial soil with a water velocity of 0.11 Km/y, and a dispersion coefficient of 4.2×10^{-7} Km^2/y, and absorption coefficient of ^{246}Cm = 3000, ^{242}Pu = 10 000, ^{238}U = 14 000, ^{234}U = 14 000, and ^{230}Th = 50 000, the first radioactivity exits at 7.5×10^4 years and the peak output occurs at 2.0×10^6 years.

$$^{246}Cm \rightarrow {}^{242}Pu \rightarrow {}^{238}U \rightarrow {}^{234}U \rightarrow {}^{230}Th \rightarrow {}^{226}Ra$$

Since the most important retention parameter in relation to ocean sediments are the ion exchange rates of the sediments and the water movement, parametric studies as well as laboratory measurements are under way to quantify their influence. Our initial calculation using one radionuclide ^{238}U and the same inputs as for the above sandy terrestrial soil calculation except for the column length and water velocity produced some interesting results. For a 100 m column and a flow rate of 0.1 m/y, the radionuclide will be contained long enough to be reduced to below background levels. Since the flow rate in the calculation is 10^5 faster than anticipated and the assumed absorption coefficient approximately ten times too small, a much shorter column should contain the ^{238}U. Improvements to the model are under way and new estimates of the absorption coefficients and flow rates are being made.

The heated cask may alter the rate of radionuclide movement through the sediment column. Using preliminary geotechnical and thermal properties of the red clay sediments and assuming either total conduction or total convection, cask temperature will reach 160°C or 3°C, respectively. Even though the sediment restricts the flow, some material transport may be possible if a temperature of near 160°C is attained. Laboratory experiments are under way to determine the relative amount of conduction and convection in homogeneous red clay heated at laboratory pressures. A representative flow model will then be constructed. Some of the mechanisms involving thermally driven instabilities and the problems of sediment disturbance during emplacement and hole closure are also being addressed.

Field experiments are under way to characterize the topography and structure of the sediments at a study site. Sediment layers which contain less pore water could act as a lid to help contain the waste.

Finally, there are some factors still to be assessed. These are the intriguing possibility that the cask can be cleverly designed to be the nucleus of some accretionary process (chert and manganese oxides naturally accrete in layers or nodules within or at the surface of the sediments). Thermal effects on all the processes are likely to alter rates, and these driving forces are just beginning to be assessed. All of these additional factors will be included in the final model.

BENTHIC BOUNDARY LAYER REGIME

There are indications that the sediment surface has special chemical properties. This is not surprising since chemical gradients exist in the top few centimeters of the sediments, most biological activity takes place in this zone, and manganese nodules are found mainly on or near the sediment surface. Preliminary conclusions from our field experiments indicate that the plutonium once collected in or on the sediment surface passes back into the water only slowly, but it does move through the sediment more rapidly. The mechanism of movement at this time is unknown but could result from chemical absorption and desorption on the sediment particles by transport of the ion-laden sediment particles, by water, or by transport of either the ion or the particles on which it is absorbed by the biological mechanisms. Instruments to measure both particle movement and radionuclide movement are being developed.

Initial experiments to characterize and estimate rates of biological mobility using baited cameras have shown that in very short times many creatures come to the bait, quickly devour the bait, and then move away. How they locate the food remains a mystery. At different depths, different species are encountered. In the mid-plate/mid-gyre regions we find rattail, fish, and amphipods; however, a recent baited camera placed in a trench has photographed large numbers of amphipods. The special chemical, biological, and physical gradients of the benthic boundary layer will be used in our model of nuclide transport.

WATER COLUMN

If one defines "release" as "transport to mankind," the water column itself provides a limited barrier to release. Material balance arguments and the age of bottom water in the great basins indicate that movement of the dense water from the regions of formation in high latitudes to areas where this water is returned to the surface layers takes from 1000 to 2000 years. Measurements of long-term (9-month) deep currents (100 m off the bottom) at a site in the North Pacific indicate an average current velocity of 1 cm/s (see Figs. 7, 8, and 9). Measurements at 2 stations 1200 m off the bottom showed lower speeds than those at 100 m (approximately 95% of the currents were below threshold which is about 2.5 $cm \cdot s^{-1}$). Overall, the internal structure and movement of the abyssal water was complex, and the indicated horizontal transport was more rapid than originally expected.

In addition, while still in part a mystery, the biological community either in the surface waters or on the bottom may provide a path for transport either laterally or vertically. In short, the

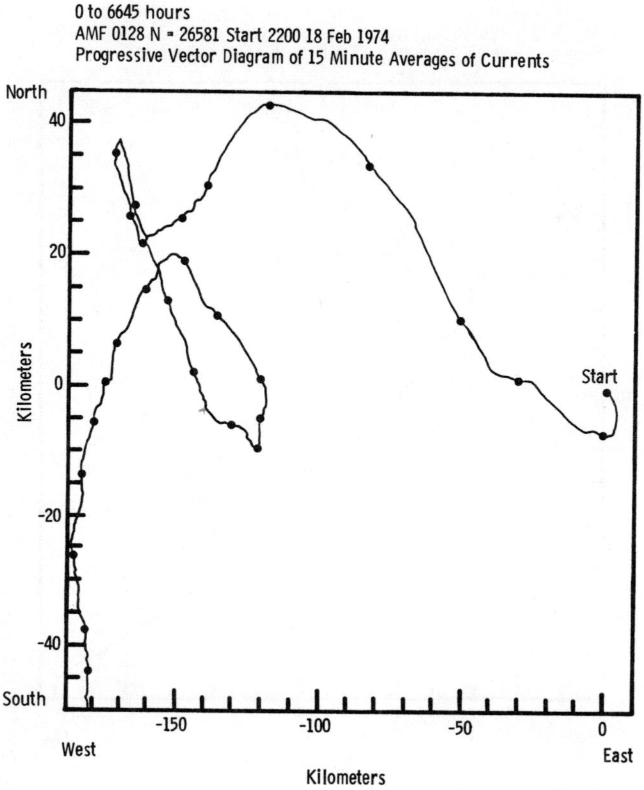

FIG. 7. Long-term (9 month) current meter record at 31N-158W separated from Figs 8 and 9 meters by 30 km.

water column is likely to be a poor barrier for large quantities of nuclides but provides some protection against inadvertent release of smaller amounts, such as from a single cask.

The model for the transport of the nuclides through the water column is far from defined. We merely feel certain now that the model will need to include not only simple bulk diffusion and advection coefficients but the effects of internal structures of the water column (e.g., eddies), major periodic motions, (e.g., tides and internal waves), biological transport either through the food chain or through diurnal and seasonal migrations, and even mechanisms driven by the activities of mankind.

CONCLUSIONS

In assessing the areas of our planet for the purpose of disposal of high-level nuclear wastes, there are compelling reasons to look closely at the floors of the sea for a possible solution. Of that 70% of the earth's surface, the areas in the middle of the oceanic

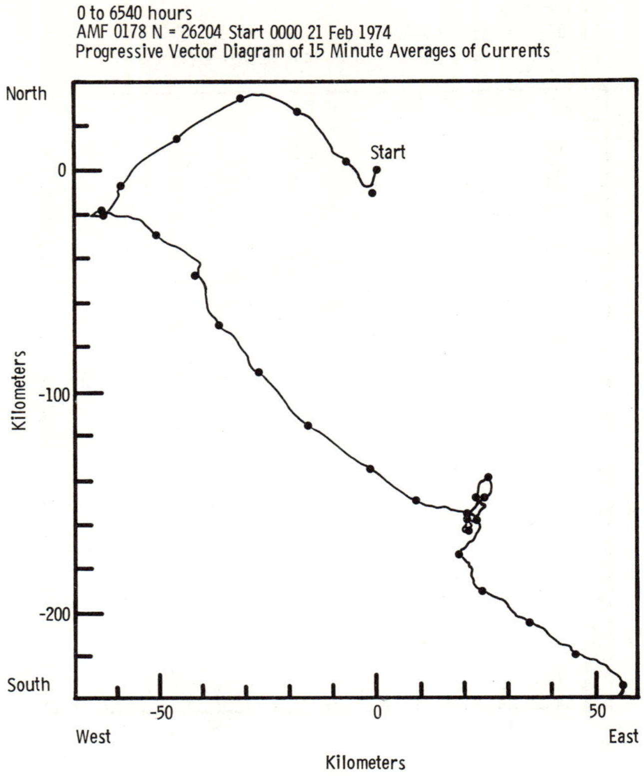

FIG. 8. Long-term (9 month) current meter record at 31N-158W separated from Figs 7 and 9 meters by 30 km.

plates and the great water gyres (mid-plate/mid-gyre regions) seem to offer the most promise and the technology for a disposal system in these areas is nearly at hand.

Assuming emplacement of the wastes below the surface of the sea floor sediments, either within the sediments or in the basement rock below, it is possible to construct a methodology for the assessment of the performance of the disposal system--a set of barriers to the release of the nuclides from the disposal site. We have established this methodology schematically and have prepared a set of preliminary estimates of the performance of the barriers. These estimates are encouraging in that they indicate containment times in excess of those required for decay of the radionuclides.

We are preparing a mathematical model of the barriers, and one part of that model has been tried in a crude manner. This roughly confirms the earlier estimate of the length of time for movement of the nuclides through the sediment column (more than a million years).

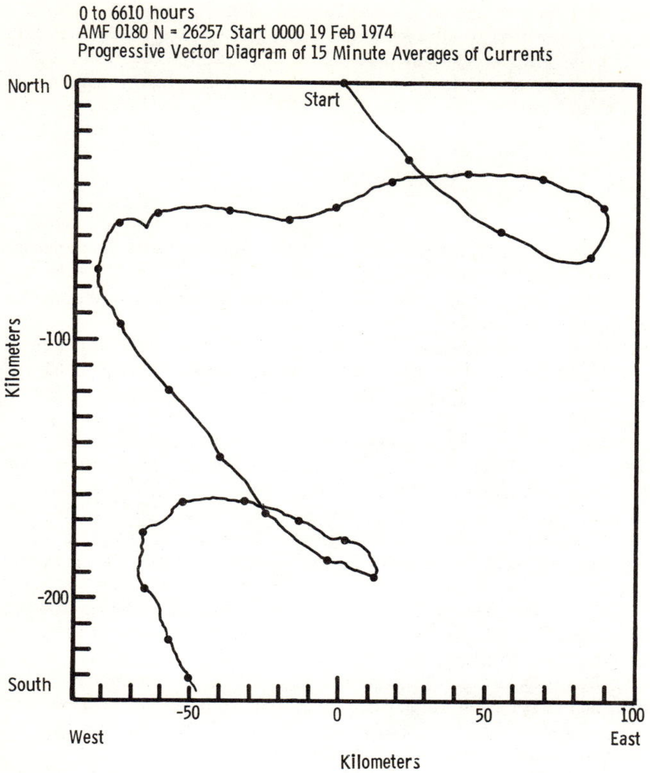

FIG. 9. Long-term (9 month) current meter record at 31N-158W separated from Figs 7 and 8 meters by 30 km.

Further refinements of this sediment subroutine and of the other parts of the model will be accomplished in the next phase of our on-going program as new data become available.

Data regarding the several aspects of the seabed disposal system are being provided by ongoing studies of oceanographic parameters, effects, and engineering requirements. The first estimates from the model indicate a need for refined information regarding the physical, biological, thermal, and chemical properties of the sediments. The acquisition of this data will be the focus of the field and laboratory work for the coming period.

Though we remain cautious, to date we have found no serious technical reason to doubt the possibility of successful seabed disposal of high-level nuclear wastes. We are now convinced that a means of assessing such disposal in a clear and rational manner can be developed and that the engineering requirements are not insurmountable. While we are not yet ready to seriously advocate the initiation of disposal of such wastes into the seabed, we are firm in our earlier conclusion that there are stable and otherwise "useless" areas of the sea floor which deserve serious attention as possible disposal sites, and that a system for disposal of nuclear wastes into the seabed can be developed and assessed in a timely manner.

REFERENCES

[1] BNWL 1900, High-Level Radioactive Waste Management Alternatives, Richland, Washington 99352.

[2] Bishop, W. P., Hollister, C. D., "Seabed Disposal - Where to Look," Nuclear Technology 24 (1974).

[3] Blomeke, J. O., "Management of Wastes From the Nuclear Fuel Cycle," presented at the American Nuclear Society Optical Meeting, "Commercial Nuclear Fuel Technology Today," Toronto, Ontario, April 27-30, 1975.

[4] See for example Proc. Symp. Management of Radioactive Wastes from Fuel Processing, Nov. 27-Dec. 1, 1972; Organization for Economic Cooperation and Development, Paris (1973); Kubo, A. S., Rose, D. J., "Disposal of Nuclear Waste," Sci. 183 (4188) (1973) 1205; Zeller, E. J., Saunders, D. F., Argino, E. E., "Putting Radioactive Wastes on Ice," Bull. Atomic Scientists (1973) 4.

[5] Klett, R. D., "Deep Rock Nuclear Waste Disposal Test: Design and Operation," SAND 74-0042, UC-70, Sandia Laboratories, Albuquerque, New Mexico, September 1974.

[6] Gaski, J. D., Lewis, D. R., Thompson, L. R., "Chrysler Improved Numerical Differencing Analyzer for 3rd Generation Computers (CINDA-3G) TN-AP-67-287," Chrysler Corporation, Space Division, New Orleans, Louisiana, October 1967.

[7] Lester, D. H., Jonsen, G., Burkholder, H. C., "Migration of Radionuclide Chains Through an Absorbing Medium," presented at the American Institute of Chemical Engineers, March 16-20, 1975, Houston, Texas, BNWL-SA-5079, Battelle Pacific Northest Laboratories, Richland, Washington.

[8] Bishop, W. P., Ed., "Seabed Disposal Program--A First-Year Report December 1974," SAND 74-0410, Sandia Laboratories, Albuquerque, New Mexico, March 1975.

[9] Webb, G. A. M., Morley, F., "A Model for the Evaluation of the Deep Ocean Dispoaal of Radioactive Waste," National Radiological Protection Board, Harwell, Didcot, Berks, NRPB-R14, July 1973.

[10] Heezen, B. C., Hollister, C. D., The Face of the Deep, Oxford University Press (1971).

DISCUSSION

R. BITTEL: It might be of interest to study what use can be made, in connection with the disposal of radioactive wastes, of the results of the mining research being performed in respect of the ocean bed.

B.R. HOOKWAY: I think this paper may well be the most important one presented at this symposium. The future of the nuclear industry stands or falls by our ability to dispose safely of our high-level wastes, and a well-thought-out method such as the one described offers a gleam of hope. What we have to bear in mind, however, is that we are already under pressure to reduce the limits of even the low-level wastes at present being dumped in the ocean. This shows that we have some way to go before we can get public and political acceptance of disposals of high-level waste, even if these prove to be scientifically acceptable. I suggest that we should all go away from this symposium determined to do everything possible to secure acceptance of ideas such as those contained in this paper.

H. KAUTSKY: I should just like to point out that current meters used in the deep seas give information about current strength and direction, and their time-dependence, at the measuring point only. They say almost nothing about the real water mass transport. It could thus be possible for released radioactive material to enter areas with upwelling properties and hence return to regions of importance to man.

D.R. ANDERSON: I agree with you that the current meters only tell what is happening at a point source and do not indicate where the water goes. This is shown by a comparison of the readings of the three meters. However, we do not envisage releases of nuclides into the water column.

H. KAUTSKY: What do you think about the possibility of dumping long-lived radioelements such as the transuranics and radium? I think that the sea has to be viewed as a whole, with a limited capacity for the dumping of such long-lived elements. This may be contrasted with the possibility of dumping relatively high amounts of short-lived isotopes.

D.H. ANDERSON: We do not propose to dump, but rather to study the feasibility of using the ocean floors as possible disposal sites.

R. BITTEL: We cannot decide whether disposals can be made in the ocean floor without a knowledge of the biophysico-chemical characteristics of the deep-sea sediments. One should never forget the irreversible nature of disposal in the ocean deeps.

Yu.V. SIVINTSEV: In this same connection I should like to say that there is an important difference between the disposal of long-lived radioactive wastes deep beneath the earth's surface on the territory of a particular country, and disposal into the floors of international seas and oceans. If a risk is incurred even by the disposal of wastes into the River Danube, then very great caution indeed will be necessary in considering the possibility of using the sea and ocean beds for the final burial of long-lived radioactive isotopes.

V.A. KNIZHNIKOV: Your interesting paper advances various arguments for the burial of high-level radioactive wastes into the deep ocean bed, as opposed to ground disposal. In particular, you refer to the great geological stability of the ocean floor and the fact that it is little affected by human activities such as might result in the dispersal of radioactivity into the environment. However, do you not think that, in evaluating the prospects for waste-burial in the ocean floors, one must consider the opposite argument, namely that it is easier to exercise control in the case of ground disposal and easier to take appropriate measures against any dispersal which might result from errors and unforeseen circumstances? In the ocean floor it is practically impossible to localize dispersion.

J.K. MIETTINEN: You said that canisters can be embedded in ocean sediments for the time of interest. Can you give us any information about the material, size and construction of the canisters, the depth at which they would be embedded and the time for which they would remain intact?

D.R. ANDERSON: We have not completed a parameter study on the size and construction of canisters. We have studied samples of Ti, Mo, Zr and ceramics (Pyrex glass). The problems identified so far relate to the closure mechanisms of the cans. By the end of the year we hope to have an estimate of the depth in the sediment necessary to contain the nuclides.

K.C. PILLAI: I feel that the possible consequences of complexing of transuranic nuclides in the bottom layers should also be investigated, especially since you say that non-ionic species may be transported into the water.

D.R. ANDERSON: I agree. We have not yet considered this transport mechanism in detail.

GENERAL DISCUSSION

HELD AFTER SESSION VI AND AFTER SESSION X
Chairman: Yu.V. SIVINTSEV (Soviet Union)

Yu.V. SIVINTSEV (Chairman): We can see that there are many unsolved problems. If, at first sight, the difficulties before us seem to be incapable of solution, this is largely owing to the great variability of even simple hydrobiological systems and the intricate ways in which these are interconnected. I think that model experiments will help us to understand our difficulties but, unfortunately, not very much has been done to standardize experimental tanks and aquaria and to match them to real hydrosystems; much remains to be done in the near future in this connection.

Although it is clear that in model experiments it is necessary — as a minimum — to control the physical, chemical and physico-chemical conditions and to ensure that the results are reproducible, none of the papers presented at this symposium dealt with the important question of reproducibility of data. In my opinion, the vital question of what numerical values are to be given to the concentration coefficients will not be solved until we have established the conditions under which results obtained in one laboratory can be reproduced in another. In regard to these model experiments I should hope that, in future, more attention will be paid to studies of the stable and radioactive isotopes of the same elements in the hydrosystem, i.e. to the problem of isotopic dilution. No less important is the study of competition between different elements with similar chemical properties. Interesting papers were, in fact, presented on this important problem, but the experimental data gathered in this field are on the whole not sufficient.

Important papers were presented on the behaviour of plutonium in aquatic systems. Plutonium and other transuranic elements, which do not have stable isotopes in nature, must remain the centre of attention for years to come, since the development of nuclear power using fission reactors will necessarily involve the use of fast-breeder reactors and the wider utilization of plutonium. The leakage of plutonium to the external environment and its accumulation in various parts of the ecosystem may then become a critical consideration as regards the further evolution of nuclear power. The collection of objective information in this area is therefore a very important activity.

Nuclear power stations are being constructed in many countries throughout the world and this means that we shall have to continue and widen our radioecological research on fission fragments and activation products in particular localities. Such studies can be safely carried out with the low concentrations of radionuclides present in the vicinity of nuclear power stations; in particular we need to shed some light on the "radiation tightness", if one may use the expression, of various compartments of actual ecosystems.

I think that in our approach to the problem we are rather weak on theory. There is an obvious need for appropriate theories in radiobiology and radioecology and, in this connection, wider use must be made of mathematical statistics and programming for processing experimental data and obtaining semi-empirical and empirical formulae. Without this, it will not be possible to bridge the gap between theory and experiment and to set up what in electronics would be called a "feedback" system.

I feel that one of the main features of our symposium has been that none of the papers presented gave any clear evidence of radiation damage to ecosystems through present discharges of radioactive wastes from nuclear installations into aquatic environments. On the contrary, the data presented here emphasized the minimal risk connected with the operation of nuclear power plants. Unfortunately, there is still considerable opposition in many countries to the operation of such plants and it is not realized that conventional power production must inevitably make way for new methods, which are less dangerous for mankind and the environment. In this connection I think it would be useful for a forthcoming IAEA symposium to be devoted to comparative studies on the risks involved in nuclear power engineering and in other forms of human activity; at the same time the question of how the public can be objectively informed could be discussed.

I should now like to open the general discussion and, in particular, to hear the views of participants regarding the more important questions calling for study in the near future.

N. SELLERS: May I endorse the Chairman's remarks in support of work on risk assessment and the relation of nuclear risks to other risks to mankind. This could be a major step towards obtaining public acceptance for the development and exploitation of nuclear power throughout the world.

One aspect of this is related to long-term waste disposal problems as referred to in paper IAEA-SM-198/34 by Anderson et al. It was stated that disposal in international waters should be by international agreement and under international control, but it was implied that land disposals within national boundaries would remain under national control.

Looking at the changes in national boundaries over the last 2000 years and considering the time for which highly-active wastes must be securely contained, I firmly believe that long-term land disposals should also be by international agreement. International acceptance would be a major contribution assuring public opinion throughout the world that nuclear development is being safely and responsibly handled.

V.A. KNIZHNIKOV: In this same connection I noticed that the words "danger", "danger from radioactive disposals", etc. have kept recurring throughout the symposium, and I should like to emphasize that the unnecessarily frequent use of such terms is liable to give the public a false picture of the state of affairs. Every human activity involves certain dangers and risks, and even inactivity itself is dangerous. An objective appraisal of the situation requires comparative studies, but I would estimate that the risk of cancer being produced by an irradiation dose of 5 rem/a is equivalent to the risk incurred by smoking several cigarettes a day. Smoking one cigarette a day involves the same risk as irradiation by the limiting dose for individual members of the population. An hour spent in a smoky room, as for instance yesterday at the Government reception, involves a risk equivalent to that resulting from irradiation by the limiting dose for the whole population, but I think I can say that we are by no means opposed to attendance at such receptions. The dose — and risk — from a nuclear power station is at present factors of ten lower than this, but the public is nevertheless alarmed. I think there is a need for greater circumspection in referring to the danger from nuclear power station disposals.

Y. NISHIWAKI: Perhaps I can also add a few comments on the subject of risks. There are many different types of risk: natural risks and artificial risks, voluntary risks and involuntary risks, enjoyable risks and unenjoyable risks, individual risks and social or public risks, etc. I think it is important to try to establish a philosophy of risks and to give the general public a proper idea of the hazards associated with nuclear radiations as compared with other types of risk encountered in daily life. The rapid acceleration of technological progress has brought broad benefits to society, but the negative side-effects of technology — environmental problems and societal problems — have also become more important in recent years. This has established a need to assess both the positive and negative impacts of new technologies, including the nuclear power industry.

The International Institute for Applied Systems Analysis (IIASA), which was recently established at Laxenburg, near Vienna, is playing an important role in this connection. The IIASA is a non-governmental scientific institution whose membership consists of professional institutes from 13 countries, including the USA and USSR. The IIASA officially began its programme in 1973. Its basic mission is to apply the methods of applied systems analysis to problems of current concern in modern societies. Specific problem areas include: ecology, water resources, urban planning, biomedical systems, energy systems, etc. Systems analysis includes such methods as cost-benefit analysis or risk-benefit analysis. One of the early IIASA projects was the study of energy systems; the principal problems here have been identified as the interactions of energy production with the atmosphere, the hydrosphere, the ecosphere and the sociosphere. The interactions with the sociosphere, in particular the objective evaluation of risks and their public perception and acceptability, are of special interest now in the rapidly growing nuclear industries. This is a subject in which the Agency has some expertise and it therefore recently organized a joint IAEA/IIASA Research Project on Risk Assessment (Division of Nuclear Safety and Environmental Protection, IAEA, Vienna). This research is of direct interest to those concerned with the radiological impacts of releases from nuclear facilities, especially so in areas such as siting, public acceptance and environmental assessment. Anyone interested in these subjects may make inquiries at the Agency, or at the IIASA directly.

S. BONOTTO: Turning to a different matter, I should like to point out that a lot of valuable data and information never becomes generally available because of the language barrier. This applies in particular to the work of Chinese authors, which may possibly be available to Japanese scientists but never becomes known to European workers. It would be useful if details of such work could be published and made generally available. Full references to such work should also be given in reports and papers. I hope that the time is not far distant when the scientists from countries such as China will be able to participate in these symposia.

Y. NISHIWAKI: In response to Mr. Bonotto's comments I should like to say that the IAEA invites all its Member States and other related international organizations to attend its conferences and symposia. Some of the Chinese literature may indeed be available through various Member States such as Japan. However, I do not think that all the works of Chinese authors are easily available in Japan, again because of the language barrier. Chinese and Japanese are two quite distinct languages, in spite of the use of classical Chinese characters in both.

If Chinese papers on nuclear energy were to appear in official publications in some of the Agency's Member States, they could be introduced into the

International Nuclear Information System (INIS) and would thus be made more universally available.

Employing computer techniques for storage, correlation and retrieval of information, INIS — which came into operation in 1970 — is a world catalogue of technical information relating to the peaceful uses of atomic energy. Its coverage includes the following subjects: nuclear physics and chemistry; reactors and reactor materials; uranium production and fuel cycles; waste management; nuclear techniques in food and agriculture; health, safety and the environment; isotope production; industrial applications; effects of radiation and dosimetry; peaceful nuclear explosions; safeguards; legal and economic aspects. At present, 46 countries and 13 international organizations are members of this information system and currently 37 countries and 6 international organizations — which together account for more than 90% of world nuclear literature — are providing input. Its scope is continually being extended, and it is estimated that when in full operation about 85 000 items of new literature will be catalogued each year. Though the Agency serves as a focal point, responsible for the editing and collating of INIS material, more than 90% of the input comes from Member States. Regional and national organizations regularly send in data on newly-published literature in their areas. A bibliographic description of each item, and keywords to identify subject matter, are recorded on magnetic tape, and the abstract of each item is entered in a microfiche file. A microfiche is a small piece of film on which 60 to 72 pages of text can be recorded. The full texts of less readily available material — such as patent specifications, technical reports and theses — are also recorded on microfiche. About 5% of INIS output is supplied by the IAEA itself, describing its own publications and those of the other international organizations with which it works.

The output of the system consists of magnetic tapes giving bibliographic descriptions and keywords, and a printed bulletin — INIS Atomindex — containing the same information, both distributed regularly. Cumulative indexes are issued twice a year, and at all times Member States have access to complete sets of the abstracts and full texts of the less readily available information on microfiche.

This computerized information centre benefits Member States in speeding up the use of the results of scientific research, and in avoiding the duplication inevitable in less centralized information systems. To developing countries it makes quickly available information on developments in nuclear science to which they might not otherwise have access.

C.L. OSTERBERG: In the United States a major problem is what to do with nuclear wastes. I should like to ask the Chairman how this problem is being handled in the Soviet Union.

Yu.V. SIVINTSEV (Chairman): Studies are performed in the USSR on the burial of radioactive wastes and on methods of waste treatment, and medium and low-level radioactive wastes are being disposed of. One facility with which I am well acquainted is a centre for processing radioactive wastes from Moscow and the Moscow area. This centre is situated at a distance of approximately 150 km from Moscow, in a locality which was specially chosen for its hydrological and atmospheric characteristics. Wastes of relatively low activity from laboratories, institutes and various installations are brought to this centre, which has facilities for such processes as purifying and concentrating liquid wastes with the aid of ion-exchange resins (when the salt content is not too high), flocculation (for wastes with a high content of mineral impurities), and evaporation to produce concentrates containing the radioactivity. The concentrates are then incorporated in concrete blocks, which are buried in special disposal areas. In certain cases, where the activity involved is higher, the concentrate is vitrified, and the resulting vitreous or ceramic blocks are also stored on the grounds of this centre. The centre likewise collects and stores, for future processing, substances with which we do not yet know how to deal.

B.R. HOOKWAY: One wonders whether this symposium was the proper forum for some of the papers presented. Is an international symposium of this nature the place for purely descriptive papers reporting on work carried out in the laboratory, i.e. work which does not reflect events in the real world and does not show that the processes being studied could represent a hazard to man? If the answer is that insufficient papers of the required importance are offered, then perhaps the conclusion is that there are too many symposia.

S. BONOTTO: In my opinion, studies on the environment are useful in order to assess what really happens in nature, but they should be performed in conjunction with laboratory investigations. With reference to Mr. Hookway's comment, then, I consider that laboratory experiments are indeed necessary in order to understand complex environmental phenomena.

A.R. BERG: The comments I want to make are of a general nature. I hope the experience that has been gained in the radiological protection of man and his environment will not lead the authorities to relax their efforts, or to become unwary, in looking at the future development of nuclear energy.

The fact that the present radiological impact on the environment is satisfactory does not mean that it is going to remain so in the future when the proliferation of nuclear facilities will greatly increase the contamination of the environment in normal and, particularly, in emergency conditions. Let us not forget also that the impact on the aquatic environment represents only one aspect of the complicated problem of waste disposal and that, even in the aquatic environment, the modifications induced by man might bring some unpredictable changes in the levels of accumulation of radionuclides. As an example let us consider the 15-fold increase in radiostrontium uptake by fish in the presence of a complexing agent (EDTA) in water.

A good line of future research in radioecology, apart from the plutonium problem, which requires priority, would be to study and determine the interaction of radionuclides with substances introduced by man into the environment (including heat), with particular attention to ecological chains and to the physico-chemical forms of the elements.

H.P. JAMMET: There are a few points which I feel require stressing. First of all, experimental laboratory work and in situ observational studies are to be regarded as complementary. Secondly, in regard to radiological protection, I think we can say that this is founded on respect for dose limits, which should be optimized at the lowest level commensurate with economic and social considerations. Finally, comparison between the risks connected with atomic energy and those connected with other sources of energy should yield a generalized optimization.

K.C. PILLAI: This symposium has dealt with the behaviour of radionuclides discharged from nuclear facilities. I feel that transuranic elements, too, are of environmental significance in view of the accelerated development of fast-breeder reactors and fuel reprocessing plants. We have information on the distribution of some of these nuclides in the environment, but much more information is required on their nature in the environment: their bioaccumulation, biokinetics, possible complexing reactions in the environment and their significance, sorption and desorption mechanisms in sediments, etc. I feel that this information will be of great importance.

J.K. MIETTINEN: In addition to in situ studies, studies of global fall-out are also important. In situ studies give information on the special conditions prevailing at that particular site. Global fall-out is rather similar everywhere and its study gives information on environmental factors and makes generalizations possible. Laboratory studies are necessary for interpreting the results of environmental studies. There is a particular shortage of laboratory work on transuranic elements, particularly in aquatic animals. To my knowledge, this is the first international symposium where a fairly coherent picture on transuranics in hydrospheric food chains has been given. In my opinion the Agency made the right decision in accepting some papers which were outside the strict field indicated by the title of the symposium.

Yu.V. SIVINTSEV: Before concluding our discussion, I should like to stress that international collaboration between mathematicians, physicists, chemists, oceanologists, radiologists, etc., will be indispensable in trying to solve the problems facing us in ecology, beginning in fact with river systems.

CHAIRMEN OF SESSIONS

Session I	J.K. MIETTINEN	Finland
Session II	C.L. OSTERBERG	USA
Session III	M. SAIKI	Japan
Session IV	K.C. PILLAI	India
Session V	A. PRESTON	United Kingdom
Session VI	Yu.V. SIVINTSEV	USSR
Session VII	W. FELDT	Germany, Fed. Rep.
Session VIII	J. C. ANCELLIN	France
Session IX	P.O. AGNEDAL	Sweden
Session X	L.J. OTTENDORFER	Austria

SECRETARIAT OF THE SYMPOSIUM

Scientific Secretaries	Y. NISHIWAKI	Division of Nuclear Safety and Environmental Protection, IAEA
	S. FOWLER	International Laboratory of Marine Radioactivity, Monaco
Administrative Secretary	R. NAJAR	Division of External Relations, IAEA
Editor	L.A. SELF	Division of Publications, IAEA
Records Officer	D.J. MITCHELL	Division of Languages and Policy-making Organs, IAEA
Liaison Officer, Finnish Government	T. ASANTI	Technical Research Centre of Finland

LIST OF PARTICIPANTS

ARGENTINA

Cancio, D.	Comisión Nacional de Energía Atómica, Av. Libertador 8250, Buenos Aires

AUSTRALIA

Giles, M.S.	Australian Atomic Energy Commission, Private Mail Bag, Sutherland NSW 2232

AUSTRIA

Ottendorfer, L.J.	Federal Institute for Water Quality, Schiffmühlenstrasse 120, 1223 Vienna
Rank, D.O.	Bundesversuchs- und Forschungsanstalt, Arsenal, Objekt 210, 1030 Vienna

BELGIUM

Bonotto, S.	C.E.N.-S.C.K. Nuclear Centre Mol, 2400 Mol

CANADA

Niemela, V.E.	Environment Canada, Ottawa KIA OH3, Ontario

CZECHOSLOVAKIA

Kříž, Z.	Czechoslovak Atomic Energy Commission, Slezská 9, Prague 2

DENMARK

Hannibal, L.G.	National Health Service of Denmark, State Institute of Radiation, Frederikssundsvej 378, 2700 Brønshøj, Copenhagen

EGYPT

El Kashef, H.S.	Atomic Energy Establishment, Cairo

FINLAND

Åkesson, T.E.	Institute of Radiation Protection, P.O. Box 268, 00101 Helsinki 10
Aro, M.I.	Technical Research Centre of Finland, Lönnrotinkatu 37, 02150 Otaniemi
Blomqvist, L.G.	Ministry of Trade and Industry, Energy Division, Aleksanterink. 10, Helsinki 17
Castrén, O.I.	Institute of Radiation Protection, Box 268, 00101 Helsinki
Cumlander, L.-R.T.	Imatran Voima Power Company, PL 138, 00101 Helsinki 10
Hakanen, M.E.	University of Helsinki, Department of Radiochemistry, Unioninkatu 35, 00170 Helsinki 17
Heinonen, J.U.	Technical Research Centre of Finland, Lönnrotinkatu 37, 02150 Otaniemi
Heinonen, O.J.	University of Helsinki, Department of Radiochemistry, Unioninkatu 35, 00170 Helsinki 17
Jaakkola, T.V.A.	University of Helsinki, Department of Radiochemistry, Unioninkatu 35, 00170 Helsinki 17
Järvinen, Marja-Hannele	University of Helsinki, Department of Radiochemistry, Unioninkatu 35, 00170 Helsinki 17
Jauho, P.A.	Technical Research Centre of Finland, Lönnrotinkatu 37, 02150 Helsinki
Jussila, K.K.	Institute of Radiation Protection, Lönnrotinkatu 37, 02150 Helsinki
Karjala, J.J.	TVO Power Company, Kutojantie 8, 02610 Espoo 61
Kortelainen, P.I.	Helsinki Electricity Works, Kampinkuja 2, Helsinki
Kuusi, J.	Finnatom, Kasarminkatu 44, 00130 Helsinki 13
Launiainen, J.O.	Institute of Marine Research, Tähtitorninkatu 2, PL 166, Helsinki 14
Miettinen, J.K.	University of Helsinki, Department of Radiochemistry, Unioninkatu 35, 00170 Helsinki 17
Ojala, J.A.	Institute of Radiation Protection, PL 268, 00101 Helsinki 10

Paakkola, O.	Institute of Radiation Protection, PL 268, 00101 Helsinki 10
Palmén, B.	Ministry of Trade and Industry, Bulevardi 28, 00120 Helsinki 12
Patoharju, O.	Helsinki City Office, Pohjoisesplanadi 15-17, 00170 Helsinki 17
Pehkonen, Pirkko T.	Helsinki Electricity Works, Kampinkuja 2, 00100 Helsinki 10
Pesonen, Taina Orvokki	Helsinki City Water Works, Pasilankatu 41, 00410 Helsinki 41
Rahola, Tua Marianne	University of Helsinki, Department of Radiochemistry, Unioninkatu 35, 00170 Helsinki 17
Rissanen, Eeva Kristiina	Institute of Radiation Protection, PL 268, 00101 Helsinki 10
Rosenberg, R.J.	Technical Research Centre of Finland, Reactor Laboratory, 02150 Espoo 15
Ruohtula, M.	Institute of Radiation Protection, PL 268, 00101 Helsinki 10
Salo, Anneli Leila	Institute of Radiation Protection, PL 268, 00101 Helsinki 10
Salonen, Laina	University of Helsinki, Department of Radiochemistry, Unioninkatu 35, 00170 Helsinki 17
Savolainen, I.T.	Technical Research Centre of Finland, 02150 Espoo 15
Saxén, Ritva Liisa	Institute of Radiation Protection, PL 268, 00101 Helsinki 10
Sjöblom, Kirsti-Liisa	Institute of Radiation Protection, PL 268, 00101 Helsinki 10
Tamminen, N.A.	Imatra Power Company, P.O. Box 138, 00101 Helsinki 10
Tarjanne, R.A.	Lappeenranta University of Technology, Box 57, SF-53101 Lappeenranta
Tiainen, O.J.A.	Helsinki Electricity Works, Kampinkuja 2, 00100 Helsinki 10
Tillander, M.	University of Helsinki, Department of Radiochemistry, Unioninkatu 35, 00170 Helsinki 17

Väisänen, S.T.J.	Institute of Radiation Protection, PL 268, 00101 Helsinki 10
Vuori, S.J.	Technical Research Centre of Finland, 02150 Espoo 15
Vuorinen, A.P.U.	Institute of Radiation Protection, PL 268, 00101 Helsinki 10

FRANCE

Ancellin, J.C.	Laboratoire de Radioécologie Marine, CEA, Centre de La Hague, B.P. 209, 50100 Cherbourg
Beau, P.	Electricité de France, B.P. 120, 93203 Saint Denis
Bittel, R.	Commissariat à l'Energie Atomique, Dépt. de Protection (SPS), B.P. 6, 92260 Fontenay-aux-Roses
Bovard, P.L.	Centre d'Etudes Nucléaires, B.P. 6, 92260 Fontenay-aux-Roses
Coulon, R.B.	Centre d'Etudes Nucléaires, B.P. 6, 92260 Fontenay-aux-Roses
Douëllou, D.	CEN Valduc, B.P. 14, 21120 Is-sur-Tille
Farges, L.F.	CEA, 26 rue de la Fédération, 75015 Paris
Fraizier, A.	CEA, Dépt. de protection, Laboratoire de radioécologie marine, B.P. 209, 50107 Cherbourg
Grauby, A.	CEN Cadarache, B.P. 1, 13100 Saint-Paul-lez-Durance
Jammet, H.P.	CEN-FAR, B.P. 6, 92260 Fontenay-aux-Roses
Khalanski, M.N.	Electricité de France, Etudes et Recherches, 6, quai Watier, 78400 Chatou
Logre, P.L.	CEN G, B.P. 85, Centre de Tri, 38041 Grenoble Cedex
Matutano, Josette M.	CEN Saclay, B.P. 2, 91190 Gif-sur-Yvette
Schaeffer, R.	Ministère de l'Industrie et de la Recherche, Serv. Central de Sûreté des Installations Nucléaires, 13, rue de Bourgogne, 75007 Paris

Toureau, P.	Electricité de France, 3, rue de Messine, Paris 8
Vignes, Suzanne	Comité de Radioprotection, Electricité de France, 3, rue de Messine, 75008 Paris

GERMAN DEMOCRATIC REPUBLIC

Ettenhuber, E.	Board of Nuclear Safety and Radiation Protection of the GDR, Waldowallee 117, 1157 Berlin
Moldenhawer, F.H.	Board of Nuclear Safety and Radiation Protection of the GDR, Waldowallee 117, 1157 Berlin

GERMANY, FEDERAL REPUBLIC

Bruessermann, K.W.	Lehrstuhl für Reaktortechnik der RWTH Aachen, Postfach, 5100 Aachen
Feldt, W.	Isotopenlaboratorium der Bundesforschungsanstalt für Fischerei, Wüstland 2, 2000 Hamburg 55
Fendler, H.G.	TÜV-Baden, Richard Wagnerstrasse 2, 6800 Mannheim
Herrmann, H.	Bayerische Biologische Versuchsanstalt, Kaulbachstr. 37, 8000 München 22
Hoffman, F.O.	Institut für Reaktorsicherheit, Glockengasse 2, 5000 Köln 1
Huber, O.	Bundesgesundheitsamt, Ingolstädter Landstr. 1, 8042 Neuherberg
Kautsky, H.	Deutsches Hydrographisches Institut, Wüstland 2, 2000 Hamburg 55
Mundschenk, H.	Bundesanstalt für Gewässerkunde, Kaiserin-Augusta-Anlage 15, 5400 Koblenz
Murray, C.N.	Deutsches Hydrographisches Institut, Wüstland 2, 2000 Hamburg 55
Wutschig, R.W.	TÜV Stuttgart, Bebelstr. 48, 7000 Stuttgart

INDIA

Pillai, K.C.	Bhabha Atomic Research Centre, Trombay, Bombay 85

IRAN

Parnianpour, H.	Atomic Energy Organization of Iran, P.O. Box 1247, Teheran

ISRAEL

Alter, J.	Israel AEC – Licensing Division, P.O. Box 17120, Tel Aviv
Shafrir, N.H.	Department Nuclear Engineering, Technion – Israel Institute of Technology, Haifa

ITALY

Boeri, G.	CNEN, Viale Regina Margherita 125, 00198 Rome
Brofferio, Carla	CNEN, Viale Regina Margherita 125, 00198 Rome
Brondi, A.	CNEN, Lab. Geominerario, S. Maria di Galeria, 00060 Rome
Clementel, S.	Agip Nucleare, Corso di P. ta Romana 68, Milan
Rossiello Ancarani, Laura	CNEN, Viale Regina Margherita 125, 00198 Rome

JAPAN

Honda, Y.	Department of Nuclear Reactor Engineering, Faculty of Science and Technology, Kinki University, 3-4-1 Kowakae, Higashi - Osaka City
Ishiyama, T.	Radiation Center of Osaka Prefecture, Shinke-cho, Sakai-shi, Osaka
Kinoshita, M.	Power Reactor and Nuclear Fuel Development Corporation, Tokai Works, Tokai, Ibaraki
Nagaya, Y.	National Institute of Radiological Sciences, 9-1, Anagawa 4-chome, Chiba-shi
Saiki, M.	National Institute of Radiological Sciences, 9-1, Anagawa 4-chome, Chiba-shi
Segawa, A.	Power Reactor and Nuclear Fuel Development Corporation, 9-13, 1-chome, Akasaka, Minato-ku, Tokyo

NETHERLANDS

Baas, J.L.	Ministry of Health and Environmental Protection, Dr. Reyersstraat 10, Leidschendam/Den Haag
de Regt, Hendrika J.	Postbus 17, Lelystad
Strackee, L.	National Institute of Public Health, P.O. Box 1, Bilthoven

van Erkelens, P.C.	Health Council, J.C. van Markenlaan 5, Rijswijk
van Weers, A.W.	Reactor Centrum Nederland, Petten

NORWAY

Garder, Karen	Institutt for Atomenergi, Box 40, 2007 Kjeller

SWEDEN

Agnedal, P.-O.	AB Atomenergi, Fack, 61101 Nyköping
Boge, R.	National Institute of Radiation Protection, Fack, 10401 Stockholm 60
Edgren, M.	Swedish Environment Protection Board, 17011 Drottningholm
Grimås, U.G.A.	Swedish Environment Protection Board, 17120 Solna
Holmberg, K.E.O.	AB Atomenergi, Studsvik, 61101 Nyköping
Mandahl, B.	Oskarshamnsverkets Kraftgrupp AB, Birg. Jarsgatan 41 A, Stockholm
Norrby, S.	National Institute of Radiation Protection, Fack, 10401 Stockholm 60
Papp, T.	Swedish State Power Board, Fack, Vällingby
Pettersson, Ulla, Anna Maria	AB Atomenergi, Fack, 61101 Nyköping
Schelin, P.E.	Swedish Environment Protection Board, 17011 Drottningholm
Snihs, J.O.H.	National Institute of Radiation Protection, Fack, 10401 Stockholm 60

TURKEY

Sarici, E.L.	T.E.K. Nuclear Energy Division, No. 26/7, Sihhiye-Ankara

UNION OF SOVIET SOCIALIST REPUBLICS

Knizhnikov, V.A.	Institute of Biophysics, Ministry of Public Health, Moscow 98, Zhivopisnaya 46
Pisarev, V.V.	Institute of Applied Geophysics, Moscow
Sivintsev, Yu.V.	Institute of Atomic Energy, Moscow

UNITED KINGDOM

Beattie, J.R.	Safety and Reliability Directorate, UKAEA, Wigshaw Lane, Culcheth, Warrington, WA3 4HJ
Broadley, J.S.	Dounreay Experimental Reactor Establishment, Caithness, Scotland
Hamilton, E.I.	Natural Environment Research Council, Institute of Marine Environmental Research, 67 Citadel Rd., Plymouth, Devon
Harris, N.M.	Nuclear Installations Inspectorate, Health and Safety Executive, Thames House South, Millbank, London SW1
Hetherington, J.A.	Fisheries Radiobiological Laboratory, Hamilton Dock, Lowestoft, Suffolk
Hinchliff, D.	Department of Environment, Northern Ireland, Stormont, Belfast
Hookway, B.R.	Department of Environment, Queen Anne's Chambers, Broadway, London SW1
Howells, Gwyneth D.	Central Electricity Research Laboratories, Leatherhead, Surrey
Howells, H.	British Nuclear Fuels Ltd, Windscale Works, Sellafield, Seascale, Cumbria
Lakey, J.R.A.	Royal Naval College, Greenwich, London SE 10
Marshall, J.D.	N.R.P.S., Institute of Naval Medicine, Alverstoke, Gosport, Hants
Owers, M.J.	Central Electricity Generating Board (SE Region Scientific Services Dept.), Suffolk Rd., Gravesend, Kent
Preston, A.	Ministry of Agriculture, Fisheries and Food, Fisheries Laboratory, Lowestoft, Suffolk NR33 OHT
Rawlins, L.V.	Rolls-Royce and Associates Ltd, P.O. Box 31, Derby
Sellers, N.	Hartlepool Power Station, Tees Road, Hartlepool, Cleveland
Stott, G.	H.M. Industrial Pollution Inspectorate for Scotland, Scottish Development Department, 21 Hill St., Edinburgh EH2 3JY
White, K.E.	Water Research Centre, Stevenage Laboratory, Elder Way, Stevenage, Herts SG1 1TH

UNITED STATES OF AMERICA

Blaylock, B.G.	Environmental Sciences Division, Bldg. 2001, Oak Ridge National Laboratory, Oak Ridge, Tenn. 37830
Edgington, D.N.	Radiological and Environmental Research Division, Argonne National Laboratory, Argonne, Illinois 60515
Nyquist, D.	Environics Division, Civil Engineering Research Facility, University of New Mexico, P.P. Box 188, University Station, Albuquerque, New Mexico 87131
Osterberg, C.L.	Division of Biomedical and Environmental Research, US Energy Research and Development Administration, Washington, D.C. 20545
Pinder, J.E. III	Savannah River Ecology Laboratory, P.O. Drawer E, Aiken, South Carolina 29801
Thompson, M.A.	Dow Chemical Co., Rocky Flats Plant, P.P. Box 888, Golden, Colorado 80401
Wahlgren, M.A.	Argonne National Laboratory, D-203 E 161, Argonne, Illinois 60439

YUGOSLAVIA

Terček, V.	Nuclear Power Plant, Krško

ORGANIZATIONS

COMMISSION OF THE EUROPEAN COMMUNITIES

Berg, A.R.	Groupe Biologie, C.C.R. Euratom, I-21020 Ispra, Italy
Hampe, E.A.	CEC Directorate for Sanitary Protection, Centre Louvigny, Luxembourg

OECD NUCLEAR ENERGY AGENCY

Olivier, J.-P.	38, Bd. Suchet, F-75016 Paris, France

INTERNATIONAL ATOMIC ENERGY AGENCY

Ohkuma, K.	At present: Division of Nuclear Safety and Environmental Protection, IAEA 1011 Vienna, Austria

AUTHOR INDEX

(including Participants in Discussions)

The numbers underlined indicate the first page of a paper by an author. Other numbers denote the page numbers of discussion comments.

TRANSLITERATION INDEX

The following conversion table is provided for the convenience of readers and to encourage the use of SI units.

FACTORS FOR CONVERTING UNITS TO SI SYSTEM EQUIVALENTS *

SI base units are the metre (m), kilogram (kg), second (s), ampere (A), kelvin (K), candela (cd) and mole (mol). [For further information, see International Standards ISO 1000 (1973), and ISO 31/0 (1974) and its several parts]

Multiply		*by*	*to obtain*
Mass			
pound mass (avoirdupois)	1 lbm	= 4.536×10^{-1}	kg
ounce mass (avoirdupois)	1 ozm	= 2.835×10^{1}	g
ton (long) (= 2240 lbm)	1 ton	= 1.016×10^{3}	kg
ton (short) (= 2000 lbm)	1 short ton	= 9.072×10^{2}	kg
tonne (= metric ton)	1 t	= 1.00×10^{3}	kg
Length			
statute mile	1 mile	= 1.609×10^{0}	km
yard	1 yd	= 9.144×10^{-1}	m
foot	1 ft	= 3.048×10^{-1}	m
inch	1 in	= 2.54×10^{-2}	m
mil (= 10^{-3} in)	1 mil	= 2.54×10^{-2}	mm
Area			
hectare	1 ha	= 1.00×10^{4}	m^2
(statute mile)^2	1 mile^2	= 2.590×10^{0}	km^2
acre	1 acre	= 4.047×10^{3}	m^2
yard^2	1 yd^2	= 8.361×10^{-1}	m^2
foot^2	1 ft^2	= 9.290×10^{-2}	m^2
inch^2	1 in^2	= 6.452×10^{2}	mm^2
Volume			
yard^3	1 yd^3	= 7.646×10^{-1}	m^3
foot^3	1 ft^3	= 2.832×10^{-2}	m^3
inch^3	1 in^3	= 1.639×10^{4}	mm^3
gallon (Brit. or Imp.)	1 gal (Brit)	= 4.546×10^{-3}	m^3
gallon (US liquid)	1 gal (US)	= 3.785×10^{-3}	m^3
litre	1 l	= 1.00×10^{-3}	m^3
Force			
dyne	1 dyn	= 1.00×10^{-5}	N
kilogram force	1 kgf	= 9.807×10^{0}	N
poundal	1 pdl	= 1.383×10^{-1}	N
pound force (avoirdupois)	1 lbf	= 4.448×10^{0}	N
ounce force (avoirdupois)	1 ozf	= 2.780×10^{-1}	N
Power			
British thermal unit/second	1 Btu/s	= 1.054×10^{3}	W
calorie/second	1 cal/s	= 4.184×10^{0}	W
foot-pound force/second	1 ft·lbf/s	= 1.356×10^{0}	W
horsepower (electric)	1 hp	= 7.46×10^{2}	W
horsepower (metric) (= ps)	1 ps	= 7.355×10^{2}	W
horsepower (550 ft·lbf/s)	1 hp	= 7.457×10^{2}	W

* Factors are given exactly or to a maximum of 4 significant figures

Multiply		by	to obtain
Density			
pound mass/inch^3	1 lbm/in^3	= 2.768×10^4	kg/m^3
pound mass/foot^3	1 lbm/ft^3	= 1.602×10^1	kg/m^3
Energy			
British thermal unit	1 Btu	= 1.054×10^3	J
calorie	1 cal	= 4.184×10^0	J
electron-volt	1 eV	$\simeq 1.602 \times 10^{-19}$	J
erg	1 erg	= 1.00×10^{-7}	J
foot-pound force	1 ft·lbf	= 1.356×10^0	J
kilowatt-hour	1 kW·h	= 3.60×10^6	J
Pressure			
newtons/metre^2	1 N/m^2	= 1.00	Pa
atmosphere[a]	1 atm	= 1.013×10^5	Pa
bar	1 bar	= 1.00×10^5	Pa
centimetres of mercury (0°C)	1 cmHg	= 1.333×10^3	Pa
dyne/centimetre^2	1 dyn/cm^2	= 1.00×10^{-1}	Pa
feet of water (4°C)	1 ftH_2O	= 2.989×10^3	Pa
inches of mercury (0°C)	1 inHg	= 3.386×10^3	Pa
inches of water (4°C)	1 inH_2O	= 2.491×10^2	Pa
kilogram force/centimetre^2	1 kgf/cm^2	= 9.807×10^4	Pa
pound force/foot^2	1 lbf/ft^2	= 4.788×10^1	Pa
pound force/inch^2 (= psi)[b]	1 lbf/in^2	= 6.895×10^3	Pa
torr (0°C) (= mmHg)	1 torr	= 1.333×10^2	Pa
Velocity, acceleration			
inch/second	1 in/s	= 2.54×10^1	mm/s
foot/second (= fps)	1 ft/s	= 3.048×10^{-1}	m/s
foot/minute	1 ft/min	= 5.08×10^{-3}	m/s
mile/hour (= mph)	1 mile/h	= 4.470×10^{-1}	m/s
		1.609×10^0	km/h
knot	1 knot	= 1.852×10^0	km/h
free fall, standard (= g)		= 9.807×10^0	m/s^2
foot/second^2	1 ft/s^2	= 3.048×10^{-1}	m/s^2
Temperature, thermal conductivity, energy/area·time			
Fahrenheit, degrees −32	°F − 32	$\frac{5}{9}$	°C
Rankine	°R	$\frac{5}{9}$	K
1 Btu·in/ft^2·s·°F		= 5.189×10^2	W/m·K
1 Btu/ft·s·°F		= 6.226×10^1	W/m·K
1 cal/cm·s·°C		= 4.184×10^2	W/m·K
1 Btu/ft^2·s		= 1.135×10^4	W/m^2
1 cal/cm^2·min		= 6.973×10^2	W/m^2
Miscellaneous			
foot^3/second	1 ft^3/s	= 2.832×10^{-2}	m^3/s
foot^3/minute	1 ft^3/min	= 4.719×10^{-4}	m^3/s
rad	rad	= 1.00×10^{-2}	J/kg
roentgen	R	= 2.580×10^{-4}	C/kg
curie	Ci	= 3.70×10^{10}	disintegration/s

[a] atm abs: atmospheres absolute;
atm (g): atmospheres gauge.

[b] lbf/in^2 (g) (= psig): gauge pressure;
lbf/in^2 abs (= psia): absolute pressure.

HOW TO ORDER IAEA PUBLICATIONS

Exclusive sales agents for IAEA publications, to whom all orders and inquiries should be addressed, have been appointed in the following countries:

UNITED KINGDOM	Her Majesty's Stationery Office, P.O. Box 569, London SE 1 9NH
UNITED STATES OF AMERICA	UNIPUB, P.O. Box 433, Murray Hill Station, New York, N.Y. 10016

In the following countries IAEA publications may be purchased from the sales agents or booksellers listed or through your major local booksellers. Payment can be made in local currency or with UNESCO coupons.

ARGENTINA	Comisión Nacional de Energía Atómica, Avenida del Libertador 8250, Buenos Aires
AUSTRALIA	Hunter Publications, 58 A Gipps Street, Collingwood, Victoria 3066
BELGIUM	Service du Courrier de l'UNESCO, 112, Rue du Trône, B-1050 Brussels
CANADA	Information Canada, 171 Slater Street, Ottawa, Ont. K 1 A OS 9
C.S.S.R.	S.N.T.L., Spálená 51, CS-110 00 Prague
	Alfa, Publishers, Hurbanovo námestie 6, CS-800 00 Bratislava
FRANCE	Office International de Documentation et Librairie, 48, rue Gay-Lussac, F-75005 Paris
HUNGARY	Kultura, Hungarian Trading Company for Books and Newspapers, P.O. Box 149, H-1011 Budapest 62
INDIA	Oxford Book and Stationery Comp., 17, Park Street, Calcutta 16
ISRAEL	Heiliger and Co., 3, Nathan Strauss Str., Jerusalem
ITALY	Libreria Scientifica, Dott. de Biasio Lucio "aeiou", Via Meravigli 16, I-20123 Milan
JAPAN	Maruzen Company, Ltd., P.O.Box 5050, 100-31 Tokyo International
NETHERLANDS	Marinus Nijhoff N.V., Lange Voorhout 9-11, P.O. Box 269, The Hague
PAKISTAN	Mirza Book Agency, 65, The Mall, P.O.Box 729, Lahore-3
POLAND	Ars Polona, Centrala Handlu Zagranicznego, Krakowskie Przedmiescie 7, PL-00-068 Warsaw
ROMANIA	Cartimex, 3-5 13 Decembrie Street, P.O.Box 134-135, Bucarest
SOUTH AFRICA	Van Schaik's Bookstore, P.O.Box 724, Pretoria
	Universitas Books (Pty) Ltd., P.O.Box 1557, Pretoria
SPAIN	Nautrónica, S.A., Pérez Ayuso 16, Madrid-2
SWEDEN	C.E. Fritzes Kungl. Hovbokhandel, Fredsgatan 2, S-103 07 Stockholm
U.S.S.R.	Mezhdunarodnaya Kniga, Smolenskaya-Sennaya 32-34, Moscow G-200
YUGOSLAVIA	Jugoslovenska Knjiga, Terazije 27, YU-11000 Belgrade

---

Orders from countries where sales agents have not yet been appointed and requests for information should be addressed directly to:

Publishing Section,
International Atomic Energy Agency,
Kärntner Ring 11, P.O.Box 590, A-1011 Vienna, Austria

QH543.5.A1 I57 1975 ULGS
Impacts of nuclear release

U005 001847510

75- 09342